华为 HCIE-Datacom 认证实验指导

（视频讲解+在线刷题）

刘伟　王鹏　周航　阳惠娇◎编著

清華大学出版社
北　京

内容简介

本书以新版华为网络技术职业认证 HCIE–Datacom（考试代码为 H12–891、H12–892）为基础，以 eNSP 模拟器为仿真平台，从行业实际应用出发组织全部内容。全书共分为 17 章，主要内容包括 OSPF 的高级特性、IS–IS 的高级特性、BGP 的高级特性、网络安全技术、MPLS 技术 、MPLS VPN 技术、MPLS VPN 跨域、EVPN 技术、IPv6 路由、IPv6 过渡技术、QoS 技术、VXLAN 技术、准入控制技术、广域网 VPN 技术、Segment Routing、SRv6 和网络自动化等。

本书可以作为华为 ICT 学院的配套实验教材，用来提升学生的实际动手能力，也可以作为计算机网络相关专业的实验指导书，还可以作为相关企业的培训教材，对于从事网络管理和运维的技术人员来说，本书也是一本很实用的技术参考书。

图书在版编目（CIP）数据

华为 HCIE–Datacom 认证实验指导：视频讲解 + 在线刷题 / 刘伟等编著 .
北京：清华大学出版社，2024.11. -- ISBN 978–7–302–67604–1

Ⅰ. TP393

中国国家版本馆 CIP 数据核字第 202443RY01 号

责任编辑：袁金敏
封面设计：刘　超
责任校对：徐俊伟
责任印制：杨　艳
出版发行：清华大学出版社
　　网　　址：https://www.tup.com.cn，https://www.wqxuetang.com
　　地　　址：北京清华大学学研大厦 A 座　　**邮　　编**：100084
　　社 总 机：010–83470000　　**邮　　购**：010–62786544
　　投稿与读者服务：010–62776969，c–service@tup.tsinghua.edu.cn
　　质 量 反 馈：010–62772015，zhiliang@tup.tsinghua.edu.cn
印 装 者：定州启航印刷有限公司
经　　销：全国新华书店
开　　本：190mm × 235mm　　**印　　张**：29.5　　**字　　数**：797 千字
版　　次：2024 年 11 月第 1 版　　**印　　次**：2024 年 11 月第 1 次印刷
定　　价：128.00 元

产品编号：107046–01

前　言

华为作为全球领先的通信设备供应商，产品涉及路由、交换、安全、无线、存储、云计算等诸多方面。而华为推出的系列职业认证 HCIA、HCIP、HCIE 无疑是 IT 领域最成功的职业认证之一。本书作者从事教育工作多年，曾在很多大学授课，也从事过人力资源工作，对学生的技能水平和企业的用人需求都很了解，很多计算机网络专业的学生在大学毕业后找工作，有两点比较欠缺：一是对理论知识的把握程度，二是操作具体设备的能力。基于此，本书作者结合实践及教学经验编写了此书，主要以 HCIE-Datacom 职业认证为依托，从实际应用的角度出发，以华为官方考试大纲为背景设计拓扑，详细介绍了新版 HCIE 中的技术内容。

本书特色

（1）内容完善，系统全面。本书以新版华为网络技术职业认证 HCIE-Datacom 为基础，以 eNSP 模拟器为仿真平台，从行业实际应用出发系统全面地组织本书内容。

（2）目标导向，实践为王。本书以实际应用为目标，采用案例驱动的方式，真实模拟企业环境。这不仅培养了读者的网络设计、配置、分析和排错能力，而且可以为他们未来的职业生涯打下坚实基础。

（3）与时俱进，紧跟前沿。本书内容与最新版的华为 HCIE-Datacom 认证大纲紧密结合，确保读者在学习过程中既能掌握前沿知识，又能顺利通过认证考试。对于重点和难点内容，我们进行了深入的剖析和解读，确保读者能够真正理解和掌握。

（4）学练一体，完美融合。本书不仅提供了详尽的理论知识梳理，更通过大量的实验案例让读者在实践中学习和成长。每个步骤都有详细的操作指导和分析，真正做到了学练一体，确保学习效果的最大化。

（5）视频教学，直击核心。除了文字内容，我们还额外提供了实操教学视频。这些视频不仅可以指导读者如何进行实际操作，还结合网络工程师的职业规划、技术难点和工作项目等内容，为读者提供全方位的教学指导。

主要内容

本书共17章，知识结构如下：

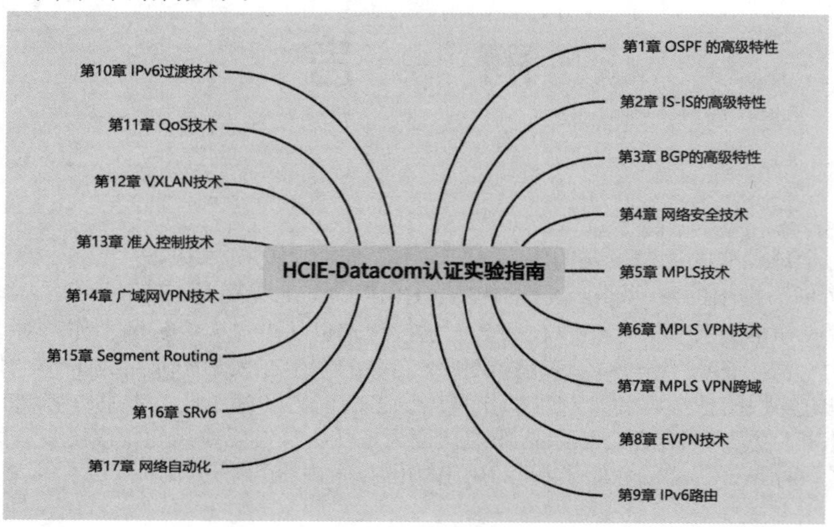

读者对象

本书面向多层次读者，满足多样化需求。

（1）华为ICT学院学员的最佳拍档。作为学院的配套教材，本书为学员提供全面、深入的ICT知识体系，助力学员掌握前沿技术。

（2）计算机网络专业学生的进阶指南。无论你是初学者还是希望提升技能的学子，本书都是你学习路上的得力助手，助你深入理解晦涩难懂的知识，提升技能。

（3）企业培训的必备教材。针对企业培训需求，本书提供了系统化的培训内容，帮助企业快速提升员工或学员的ICT技能。

（4）网络技术人员的实用手册。对于正在从事或希望深入此领域的技术人员，本书提供了实用的技术参考和解决方案，帮助你解决实际问题。

作者寄语

“读书之法，在循序而渐进，熟读而深思”，建议读者在学习本书时，参考以下学习方法。

1. 理论知识要先学会总结，然后去理解和记忆

华为相关技术的知识点特别多，有的读者学完以后，去找相关的工作，面试官问的问题，他都觉得学过，但就是答不上来，所以读者在学习的过程中，一定要对所学的知识点进行提炼和总结，然后去记忆，这样才能在面试时做到从容面对。

2. 多做实验，提高动手能力和排错能力

华为的职业认证比较注重学员的动手能力，但是企业觉得很多新入职人员的动手能力、分析和解决问题的能力太差，所以大家在平时的学习中要加强实践操作能力和排错能力。俗话说：“熟读唐诗三百首，不会作诗也会吟。”本书大部分的篇幅在讲解实验，就是希望读者通过实践提高动手能力和排错能力。

3. 多问为什么，每一个知识点的问题都要及时解决

许多学生刚开始学一门技术时，很有激情，会全身心地投入。但是一旦遇到问题，他觉得问同学和老师是一件很可耻的事情，等问题积累得越来越多，慢慢就听不懂老师所讲的内容了，也做不出来实验了，最后对这门课就失去了信心。所以大家一定要多请教，有问题就马上解决，这样才能时刻保持追求技术的激情，才能把一门技术学好、学透。

4. 不理解的内容多看几遍，反复学，肯定可以学会

HCIE-Datacom 属于华为数通认证系列的最高级，若前期基础没有打好，就会感觉难度比较大。建议大家一定要打好 HCIA 和 HCIP 的基础，并且坚持学习，多看多学，肯定可以学会。

5. 感谢支持，不吝赐教

自我们的第一本书《华为 HCIA-Datacom 认证实验指南》出版以来，得到了众多读者的喜爱和支持。后陆续出版的《HCIP-Datacom 认证实验指南》《HCIA-Datacom 认证学习指南》等，均获得了大家一如既往的肯定。在此由衷地感谢大家的喜爱。

本书资源

本书提供关键知识点的教学视频，请使用手机扫描书中的二维码观看相关教学视频。

本书提供在线刷题，请扫描以下本书服务二维码，按照说明进入在线刷题平台。

若您在学习本书的过程中发现疑问或错漏之处，也请您通过扫描以下二维码与我们取得联系。您可以进入读者交流群，与更多读者在线交流学习，也可以通过技术支持或者售后服务与我们取得联系，感谢您的支持。

本书服务二维码

致谢

本书由长沙卓应教育咨询有限公司的刘伟编写并统稿，参加编写工作的还有王鹏、周航、阳惠娇等。针对庞大的华为网络及其复杂技术编写一本适合学生的实验教材确实不是一件容易的事情，衷心感谢长沙卓应教育咨询有限公司各位领导的支持和指导。本书的顺利出版也离不开清华大学出版社编辑的支持与指导，在此一并表示衷心的感谢。

尽管本书经过了作者与出版社编辑的精心审读与校对，但限于时间、篇幅，难免存在疏漏之处，敬请各位读者不吝赐教。

编者

2024 年 10 月

目　　录

第1章

OSPF 的高级特性

1.1 OSPF 的高级特性概述

OSPF（Open Shortest Path First，开放式最短路径优先）是基于链路状态的内部网关路由协议，当网络拓扑出现问题时，OSPF 提供多种快速收敛和保护机制，在大型网络中，OSPF 还可以使用路由选路及路由信息控制来减小特定的路由表的大小。

1.2 OSPF 快速收敛

1.2.1 PRC

（1）PRC 的工作原理：当网络上路由发生变化时，PRC（Partial Route Calculation，部分路由计算）只对发生变化的路由进行重新计算。

（2）PRC 不计算节点路径，而是根据 SPF（Shortest Path First，最短路径优先）算法算出来的最短路径树来更新路由。

（3）在路由计算过程中，节点代表路由器，叶子代表路由，PRC 只处理变化的叶子信息。

备注：树没有发生变化，只是加了一片叶子。典型的例子就是加一个环回口。

1.2.2 智能定时器

（1）控制 LSA（Link–State Advertisement，链路状态通告）的生成和接收。

①网络动荡时：LSA 的更新时间间隔为 5 s，接收时间间隔为 1 s。

备注：控制 LSA 频繁地发和收时占用的设备资源（可以举例）。

②网络稳定时：LSA 的更新时间间隔为 0 s，加快收敛速度。

（2）控制路由计算。

①问题：网络频繁变化→ LSDB（Link State DataBase，链路状态数据库）变化→重新计算 SPF →占用资源。

②办法：设置合理的 SPF 算法的时间间隔。

（3）配置命令。

①更新时间间隔。

[Huawei–ospf–1]lsa–originate–interval { 0 |{ intelligent–timer max–interval start–interval hold–interval | other–type interval } }。参数含义见表 1–1。

表 1–1　ospf LSA 更新时间间隔

参　数	描　述	默认值
max–interval	最长时间间隔	5 s
start–interval	初始时间间隔	0.5 s
hold–interval	基数时间间隔	1 s

【技术要点1】

（1）初次更新LSA的时间间隔由start-interval参数指定。

（2）第n（$n \geq 2$）次更新LSA的时间间隔为hold-interval × $2(n-2)$。

（3）当hold-interval × $2(n-2)$达到指定的最长时间间隔max-interval时，OSPF连续三次更新LSA的时间间隔都是最长时间间隔，再次返回步骤（1），按照初始时间间隔start-interval更新LSA。

②接收时间间隔。

[Huawei-ospf-1]lsa-arrival-interval { interval | intelligent-timer max-interval start-interval hold-interval }。参数含义见表 1-2。

表 1-2 OSPF LSA 接收时间间隔

参 数	描 述	默认值
max-interval	最长时间间隔	1 s
start-interval	初始时间间隔	0.5 s
hold-interval	基数时间间隔	5 s

【技术要点2】

（1）初次接收LSA的时间间隔由start-interval参数指定。

（2）第n（$n\geqslant 2$）次接收LSA的时间间隔为hold-interval × 2(n–2)。

（3）当hold-interval × 2(n–2)达到指定的最长时间间隔max-interval时，OSPF连续三次接收LSA的时间间隔都是最长时间间隔，再次返回步骤（1），按照初始时间间隔start-interval接收LSA。

③路由计算时间间隔。

[Huawei-ospf-1]spf-schedule-interval { interval1 | intelligent-timer max-interval start-interval hold-interval | millisecond interval2 }。参数含义见表 1-3。

表 1-3 OSPF 路由计算时间间隔

参 数	描 述	默认值
max-interval	SPF 最长时间间隔	10 s
start-interval	SPF 初始时间间隔	0.5 s
hold-interval	SPF 基数时间间隔	1 s

【技术要点3】

（1）初次计算SPF的时间间隔由start-interval参数指定。

（2）第n（$n\geqslant 2$）次计算SPF的时间间隔为hold-interval × 2(n–2)。

（3）当hold-interval × 2(n–2)达到指定的最长时间间隔max-interval时，OSPF连续三次计算SPF的时间间隔都是最长时间间隔，再次返回步骤（1），按照初始时间间隔start-interval计算SPF。

1.2.3 OSPF 故障快速收敛

1.OSPF IP FRR（Fast ReRoute，快速重路由）

OSPF IP FRR 是动态 IP FRR，利用 LFA（Loop-Free Alternates，无环路交替）算法预先计算出备份路径并保存在转发表中，以备在出现故障时将流量快速切换到备份链路上，保证流量不中断，从而达到流量保护的目的，该功能可将故障恢复时间降低到 50 ms 以内。

LFA 计算备份链路的基本思路：以可提供备份链路的邻居为根节点，首先利用 SPF 算法计算出到目的节点的最短距离，然后按照不等式计算出开销最小且无环的备份链路。

（1）链路保护。如图 1-1 所示，流量从设备 S 到 D 进行转发，网络开销值满足链路保护公式，保证当主链路出现故障后，设备S将流量切换到备份链路S到N后可以继续向下游转发，确保流量中断小于50 ms。

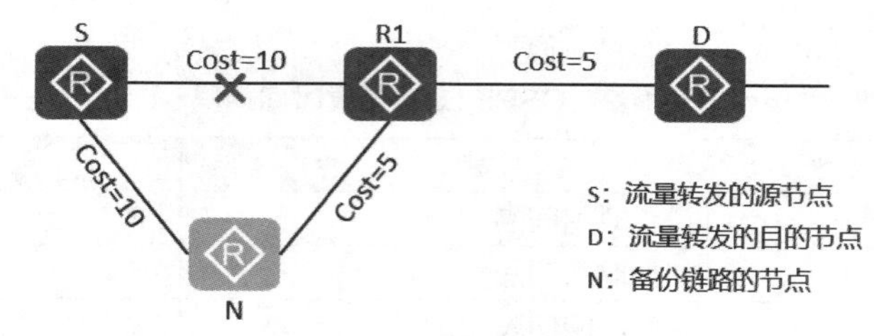

图 1-1　链路保护

链路保护公式：Distance_opt(N,D)<Distance_opt(N,S)+Distance_opt(S,D)。

保证从节点N到节点D的流量不会再经过节点S，即保证没有环路。

（2）节点链路双保护。如图1-2所示，流量从设备S到D进行转发，网络开销值满足链路保护公式和节点保护公式，当节点E出现故障后，可以切换到节点N。

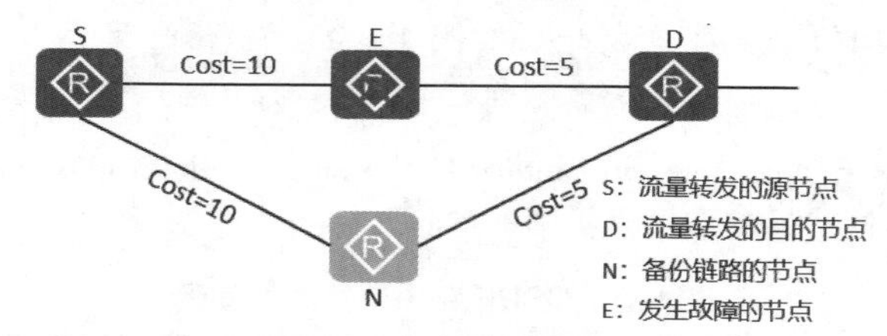

图 1-2　节点链路双保护

链路保护公式：Distance_opt(N,D)<Distance_opt(N,S)+Distance_opt(S,D)。

节点保护公式：Distance_opt(N,D)<Distance_opt(N,E)+Distance_opt(E,D)。

保证从节点N到节点D的流量不会经过节点S和节点E，即保证没有环路。

2. OSPF 与 BFD 联动

OSPF与BFD（Bidirectional Forwarding Detection，用于检测两个转发点之间故障的网络协议）联动就是将BFD和OSPF关联起来，一旦与邻居之间的链路出现故障，BFD对链路故障的快速感应能够加快OSPF对于网络拓扑变化的响应。

1.3　OSPF 路由控制

1.3.1　等价路由

扫一扫，看视频

（1）当路由表中存在到达同一目的地址且同一路由协议发现的多条路由时，若这几条路由的开销值相同，那么这些路由是等价路由，可以实现负载分担。

（2）设置进行负载分担的等价路由的最大数量：[Huawei-ospf-1] maximum load-balancing number。

1.3.2　默认路由

OSPF的默认路由的相关信息见表1-4。

表 1-4　OSPF 的默认路由

区域类型	产生条件	发布方式	LSA 类型	泛洪范围
普通区域	default-route-advertise 命令配置	ASBR 发布	5	普通区域
Stub 区域和 Totally Stub 区域	自动产生	ABR 发布	3	Stub 区域
NSSA 区域	nssa [default-route-advertise] 命令配置	ASBR 发布	7	NSSA 区域
Totally NSSA 区域	自动产生	ABR 发布	3	NSSA 区域

1.3.3　对发送的 LSA 进行过滤

（1）当两台路由器之间存在多条链路时，可以在某些链路上通过对发送的 LSA 进行过滤，减少不必要的重传，节省带宽资源。

（2）通过对 OSPF 接口出方向的 LSA 进行过滤以不向邻居发送无用的 LSA，从而减少邻居 LSDB 的大小，提高网络收敛速度。

（3）配置对 OSPF 接口出方向的 LSA 进行过滤。

[Huawei-GigabitEthernet0/0/1] ospf filter-lsa-out { all | { summary [acl { acl-number | acl-name }] | ase [acl { acl-number | acl-name }] | nssa [acl { acl-number | acl-name }] } }。参数详情见表 1-5。

表 1-5　OSPF LSA 过滤参数

参　数	说　明
all	对除 Grace LSA 外的所有 LSA 进行过滤
summary	对 Network Summary LSA（Type3 LSA）进行过滤
ase	对 AS External LSA（Type5 LSA）进行过滤
nssa	对 NSSA LSA（Type7 LSA）进行过滤
acl acl-number	指定基本访问控制列表编号。整数形式，取值范围为 2000 ~ 2999
acl acl-name	指定访问控制列表名称

1.3.4　对 ABR Type3 LSA 进行过滤

（1）对区域内出 / 入方向的 ABR Type3 LSA（Summary LSA）设置过滤条件，只有通过过滤的 LSA，才能被发布和接收。

（2）通过对区域内的 LSA 进行过滤可以不向邻居发送无用的 LSA，从而减少 LSDB 的大小，提高网络收敛速度。

（3）配置。

① 出方向：[Huawei-ospf-1-area-0.0.0.1] filter { acl-number | acl-name acl-name | ip-prefix ip-prefix-name | route-policy route-policy-name } export。

② 入方向：[Huawei-ospf-1-area-0.0.0.1] filter { acl-number | acl-name acl-name | ip-prefix ip-prefix-name | route-policy route-policy-name } import。

1.3.5　OSPF Database Overflow

（1）OSPF 要求同一个区域中的路由器保存相同的 LSDB。随着网络上路由数量的不断增加，一些路由器由于系统资源有限，不能再承载如此多的路由信息，这种状态就称为数据库超限（OSPF Database Overflow）。

（2）对于路由信息不断增加导致路由器系统资源耗尽而失效的问题，可以通过配置Stub或NSSA区域来解决，但Stub或NSSA区域的方案不能解决动态路由增长导致的数据库超限问题。为了解决数据库超限引发的问题，通过设置LSDB中External LSA的最大条目数，可以动态限制数据库的规模。

（3）OSPF Database Overflow工作原理流程图如图1-3所示。

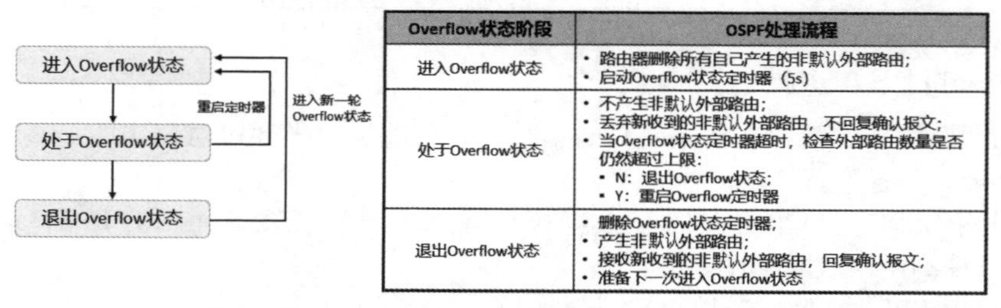

Overflow状态阶段	OSPF处理流程
进入Overflow状态	• 路由器删除所有自己产生的非默认外部路由； • 启动Overflow状态定时器（5s）
处于Overflow状态	• 不产生非默认外部路由； • 丢弃新收到的非默认外部路由，不回复确认报文； • 当Overflow状态定时器超时，检查外部路由数量是否仍然超过上限： 　• N：退出Overflow状态； 　• Y：重启Overflow定时器
退出Overflow状态	• 删除Overflow状态定时器； • 产生非默认外部路由； • 接收新收到的非默认外部路由，回复确认报文； • 准备下一次进入Overflow状态

图1-3　OSPF Database Overflow工作原理流程图

（4）设置OSPF的LSDB中External LSA的最大条目数：[Huawei-ospf-1] lsdb-overflow-limit number。

1.4　OSPF的其他特性

1.4.1　OSPF多进程

（1）OSPF支持多进程，在同一台路由器上可以运行多个不同的OSPF进程，它们之间互不影响，彼此独立。不同OSPF进程之间的路由交互相当于不同路由协议之间的路由交互。

（2）路由器的一个接口只能属于某一个OSPF进程。

1.4.2　OSPF与BGP的联动

（1）故障：当有新的设备加入网络中，或者设备重启时，可能会出现在BGP（Border Gateway Protocol，边界网关协议）收敛期间内网络流量丢失的现象。这是IGP收敛速度比BGP快造成的。

（2）故障举例，如图1-4所示。

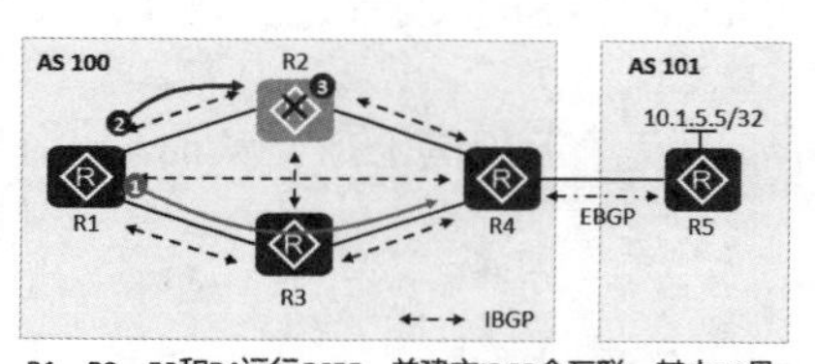

图1-4　OSPF与BGP联动图（1）

（3）故障解决：OSPF与BGP联动，如图1-5所示。

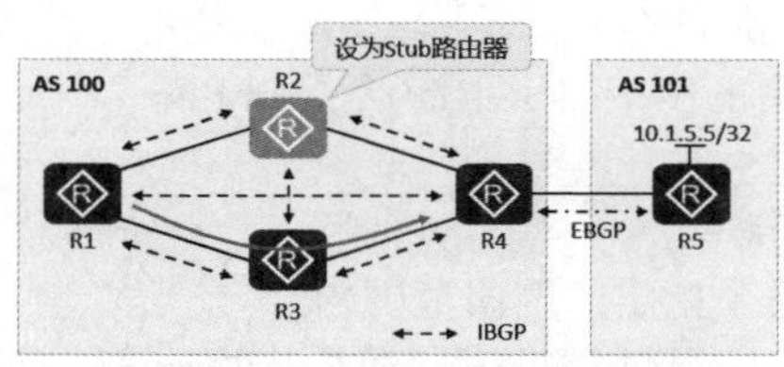

在R2上能使BGP联动，这样，在BGP收敛完成前，R1不把流量转发到R2上，而是继续使用备份设备R3转发流量，直到R2上的BGP路由完成收敛。

配置Stub路由器：

[Huawei-ospf-1] **stub-router** [**on-startup** [*interval*]]

- 配置Stub路由器是一种特殊的路由选路，配置了Stub路由器的路径不被优选。
- 实现方法是将度量值设为最大（65535），尽量避免数据从此路由器转发。用于保护此路由器链路，通常使用在升级等维护操作的场景。

图 1–5　OSPF 与 BGP 联动图（2）

1.4.3　FA

在 5 类 LSA 中，FA（Forwarding Address，转发地址）的引入使 OSPF 在某些特殊的场景下可以避免次优路径问题。

（1）次优路径举例，如图 1–6 所示。

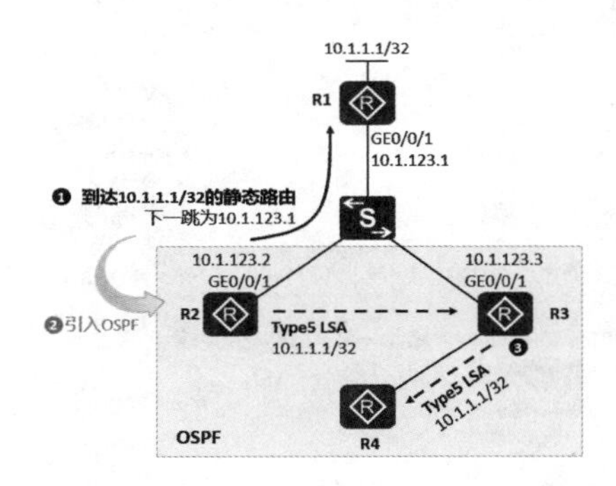

- R2、R3和R4运行OSPF，均部署在Area0中。其中R2和R3的GE0/0/1接口都激活OSPF并建立邻接关系，但是两者与外部路由器R1并不建立OSPF邻接关系。
  1. R2配置到达10.1.1.1/32的静态路由，下一跳为10.1.123.1。
  2. R2将静态路由引入OSPF，产生Type5 LSA在区域内泛洪。
  3. R3接收到R2产生的5类LSA，计算出到达10.1.1.1/32的外部路由，并且将路由的下一跳指定为R2（10.1.123.2）。
- OSPF域内的路由器如R4到达10.1.1.1/32的路径是R4→R3→R2→R1，该路径是次优路径。

图 1–6　OSPF 次优路径

（2）利用 FA 解决次优路径问题，如图 1–7 所示。

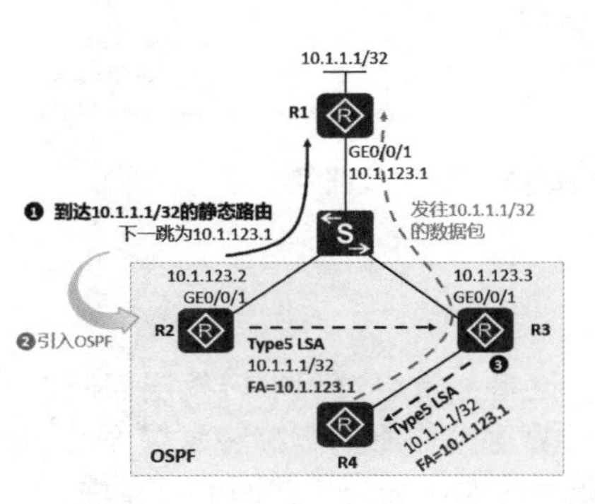

- R2向OSPF区域内通告到达10.1.1.1/32的外部路由时，为对应的Type5 LSA设置FA，值为其自己到达该外部路由的下一跳：10.1.123.1。
- 当R3收到该LSA后，计算到达10.1.1.1/32的路由时，发现FA为非0，因此它认为到达目标地址10.1.1.1/32的下一跳为FA所指定的地址，即10.1.123.1。

图 1–7　OSPF FA 地址解决次优路径

（3）当 ASBR 引入外部路由时，若 Type5 LSA 中的 FA 字段为 0，表示路由器认为到达目的网段的数据包应该发往该 ASBR；若 Type5 LSA 中的 FA 字段不为 0，表示路由器认为到达目的网段的数据包应该发往这个 FA 所标识的设备。当以下条件全部满足时，FA 字段才可以被设置为非 0。

① ASBR 在其连接外部网络的接口（外部路由的出接口）上激活了 OSPF。

②该接口没有被配置为 Silent-Interface。

③该接口的 OSPF 网络类型为 Broadcast 或 NBMA。

④该接口的 IP 地址在 OSPF 配置的 network 命令指定的网段范围内。

⑤到达 FA 地址的路由必须是 OSPF 区域内部路由或区域间路由，这样接收到该外部 LSA 的路由器才能够加载该 LSA 进入路由表。加载的外部 LSA 生成的路由条目下一跳与到达 FA 地址的下一跳相同。

在 7 类 LSA 中，FA 的引入使 OSPF 在某些特殊的场景下可以避免次优路径问题。

（1）次优路径举例和解决办法，NSSA 区域 OSPF 次优路径如图 1-8 所示。

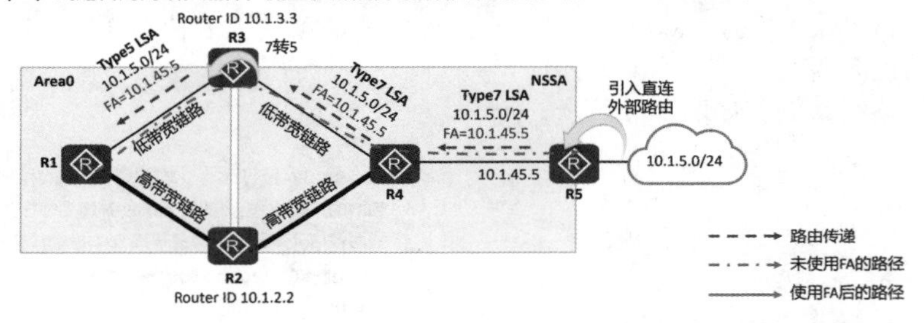

图 1-8　NSSA 区域 OSPF 次优路径

（2）解决办法：FA 地址解决 NSSA 区域次优路径，如图 1-9 所示。

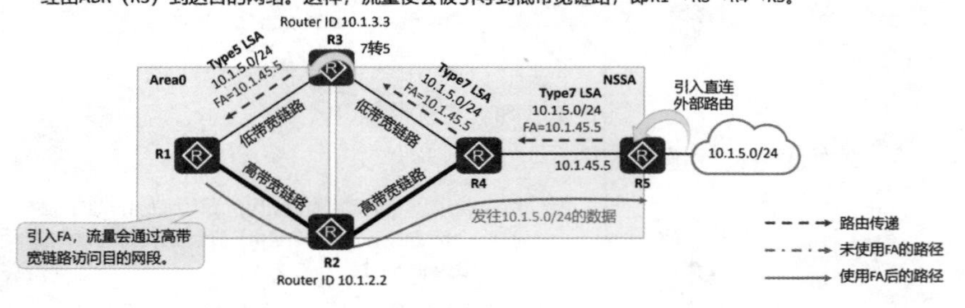

图 1-9　FA 地址解决 NSSA 区域次优路径

1.4.4　GR

1. 技术介绍

GR（Graceful Restart，平滑重启）技术保证了设备在重启过程中转发层面能够继续指导数据的转发，同时控制层面邻居关系的重建以及路由计算等动作不会影响转发层面的功能，从而避免了路由振荡引发的业务中断，保证了关键业务的数据转发，提高了整网的可靠性。

2. OSPF 的报文格式

OSPF 通过新增 Grace-LSA 来支持 GR 功能，这种 LSA 用于在开始 GR 和退出 GR 时向邻居通告 GR 的时间、原因以及接口地址等内容。OSPF 的 9 类 LSA 报文格式见表 1-6。

表 1-6　OSPF 的 9 类 LSA 报文格式

LS Age	Options	LS Type=9
Opaque Type=3	0	
Advertising Router		
LS Sequence Number		
LS Checksum	Length	
TLVs（Type-Length-Value）		

表 1-6 中的各项说明如下。

（1）LS Age：LSA 产生后所经过的时间，以秒为单位。

（2）Options：可选项。

（3）LS Type：类型为 9，代表不透明 LSA。

（4）Opaque Type：Opaque 类型。

（5）Opaque ID：为 0。

（6）Advertising Router：产生此 LSA 的路由器的 Router ID。

（7）LS Sequence Number：LSA 的序列号。根据这个值可以判断哪个 LSA 是最新的。

（8）LS Checksum：校验和。

（9）Length：LSA 的总长度。

（10）TLVs：重启时间、原因、接口地址。

3. OSPF GR 的过程

OSPF GR 的过程如图 1-10 所示。

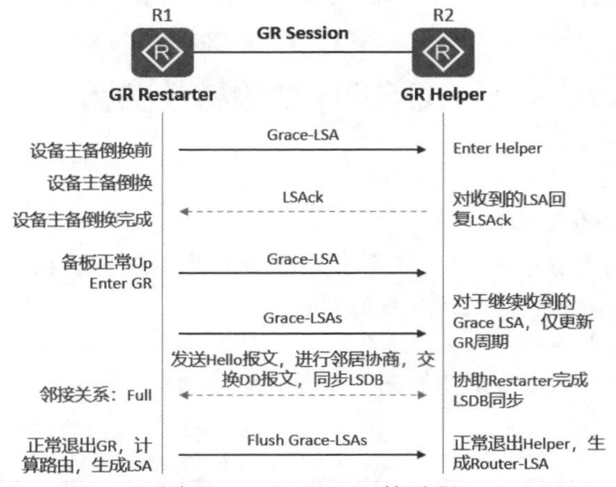

图 1-10　OSPF GR 的过程

4. 配置

（1）使能 opaque-LSA 特性。

```
[Huawei-ospf-1]opaque-capability enable
```

（2）Restarter 端使能 OSPF GR 特性，并配置 Restarter 端 GR 的会话参数。

```
[Huawei-ospf-1]graceful-restart[period period | planned-only | partial]
```

Period 默认为 120 s，取值范围为 1~1800 s。

1.4.5 NSR

1. NSR 的专业术语

（1）HA（High Availability）：高可靠性 / 高实用性的简称，这里指主备板间的备份通道。

（2）NSF（Non-Stop Forwarding）：不间断转发。

（3）NSR（Non-Stop Routing）：不间断路由。一种在系统控制平面发生故障且存在备用控制平面的场景下，邻居控制平面不感知的技术。

（4）AMB（Active Main Board）和 SMB（Slave Main Board）：主用主控板和备用主控板，单板上承载控制平面进程。

（5）LPU（Line Interface Process Unit）：接口板，单板上承载转发控制进程。

2. NSR 的原理

NSR 的原理主要包括以下三个过程。

（1）批量备份：当备用主控板启动后，NSR 功能将使能。此时，主用主控板将路由信息和转发信息批量备份到备用主控板上。批量备份过程在实时备份过程之前进行，此时 NSR 无法实施主备倒换过程。

（2）实时备份：当批量备份过程结束后，系统进入实时备份阶段。任何在控制平面和转发平面的改变都将实时从主用主控板备份到备用主控板上。在该阶段，备用主控板能够随时代替主用主控板工作。

（3）主备倒换：在已经完成备份的 NSR 系统主用主控板发生故障时，备用主控板会通过硬件状态感知到主用主控板故障，并成为新的主用主控板。备用主控板升为主用主控板后，该单板会切换接口板的报文上送通道。由于倒换时间足够短，路由协议在主备切换的过程中不会和邻居节点断连。

1.5 OSPF 高级特性实验

1.5.1 OSPF IP FRR

扫一扫，看视频

1. 实验目的

全网运行 OSPF 协议，并且在 AR1 上开启 OSPF IP FRR，当 AR1 → AR3 → AR4 的路径出现故障时，能实现备份路径的快速切换。

2. 实验拓扑

OSPF IP FRR 实验拓扑如图 1-11 所示。

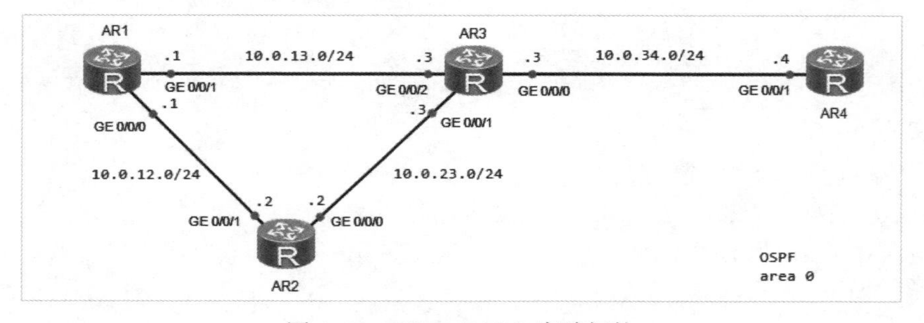

图 1-11　OSPF IP FRR 实验拓扑

3. 实验步骤

步骤 1： 配置 IP 地址。

AR1 的配置：

```
<Huawei>system-view
Enter system view, return user view with Ctrl+Z.
[Huawei]undo info-center enable
[Huawei]sysname AR1
[AR1]interface g0/0/0
[AR1-GigabitEthernet0/0/0]ip address 10.0.12.1 24
[AR1-GigabitEthernet0/0/0]quit
[AR1]interface g0/0/1
[AR1-GigabitEthernet0/0/1]ip address 10.0.13.1 24
[AR1-GigabitEthernet0/0/1]quit
[AR1]interface LoopBack 0
[AR1-LoopBack0]ip address 1.1.1.1 32
[AR1-LoopBack0]quit
```

AR2 的配置：

```
<Huawei>system-view
Enter system view, return user view with Ctrl+Z.
[Huawei]undo info-center enable
Info: Information center is disabled.
[Huawei]sysname AR2
[AR2]interface g0/0/0
[AR2-GigabitEthernet0/0/0]ip address 10.0.23.2 24
[AR2-GigabitEthernet0/0/0]quit
[AR2]interface g0/0/1
[AR2-GigabitEthernet0/0/1]ip address 10.0.12.2 24
[AR2-GigabitEthernet0/0/1]quit
[AR2]interface LoopBack 0
[AR2-LoopBack0]ip address 2.2.2.2 32
[AR2-LoopBack0]quit
```

AR3 的配置：

```
<Huawei>system-view
Enter system view, return user view with Ctrl+Z.
[Huawei]undo info-center enable
Info: Information center is disabled.
[Huawei]sysname AR3
[AR3]interface g0/0/0
[AR3-GigabitEthernet0/0/0]ip address 10.0.34.3 24
[AR3-GigabitEthernet0/0/0]quit
[AR3]interface g0/0/1
[AR3-GigabitEthernet0/0/1]ip address 10.0.23.3 24
[AR3-GigabitEthernet0/0/1]quit
[AR3]interface g0/0/2
[AR3-GigabitEthernet0/0/2]ip address 10.0.13.3 24
[AR3-GigabitEthernet0/0/2]quit
```

AR4 的配置：

```
<Huawei>system-view
Enter system view, return user view with Ctrl+Z.
[Huawei]undo info-center enable
Info: Information center is disabled.
[Huawei]sysname AR4
[AR4]interface g0/0/1
[AR4-GigabitEthernet0/0/1]ip address 10.0.34.4 24
[AR4-GigabitEthernet0/0/1]quit
[AR4]interface LoopBack 0
[AR4-LoopBack0]ip address 4.4.4.4 32
[AR4-LoopBack0]quit
```

步骤 2： 运行 OSPF 路由协议。

AR1 的配置：

```
[AR1]ospf router-id 1.1.1.1
[AR1-ospf-1]area 0
[AR1-ospf-1-area-0.0.0.0]network 10.0.12.0 0.0.0.255
[AR1-ospf-1-area-0.0.0.0]network 10.0.13.0 0.0.0.255
[AR1-ospf-1-area-0.0.0.0]network 1.1.1.1 0.0.0.0
[AR1-ospf-1-area-0.0.0.0]quit
```

AR2 的配置：

```
[AR2]ospf router-id 2.2.2.2
[AR2-ospf-1]area 0
[AR2-ospf-1-area-0.0.0.0]network 10.0.12.0 0.0.0.255
[AR2-ospf-1-area-0.0.0.0]network 10.0.23.0 0.0.0.255
[AR2-ospf-1-area-0.0.0.0]network 2.2.2.2 0.0.0.0
[AR2-ospf-1-area-0.0.0.0]quit
```

AR3 的配置：

```
[AR3]ospf router-id 3.3.3.3
[AR3-ospf-1]area 0
[AR3-ospf-1-area-0.0.0.0]network 10.0.13.0 0.0.0.255
[AR3-ospf-1-area-0.0.0.0]network 10.0.23.0 0.0.0.255
[AR3-ospf-1-area-0.0.0.0]network 10.0.34.0 0.0.0.255
[AR3-ospf-1-area-0.0.0.0]network 3.3.3.3 0.0.0.0
[AR3-ospf-1-area-0.0.0.0]quit
```

AR4 的配置：

```
[AR4]ospf router-id 4.4.4.4
[AR4-ospf-1]area 0
[AR4-ospf-1-area-0.0.0.0]network 10.0.34.0 0.0.0.255
[AR4-ospf-1-area-0.0.0.0]network 4.4.4.4 0.0.0.0
[AR4-ospf-1-area-0.0.0.0]quit
```

步骤 3： 实验调试。

在 AR1 上查看关于 4.4.4.4 的 OSPF 的路由。

```
<AR1>display ospf routing 4.4.4.4

         OSPF Process 1 with Router ID 1.1.1.1

 Destination : 4.4.4.4/32
 AdverRouter : 4.4.4.4                  Area      : 0.0.0.0
 Cost        : 2                        Type      : Stub
 NextHop     : 10.0.13.3                Interface : GigabitEthernet0/0/1
 Priority    : Medium                   Age       : 00h01m16s
```

通过以上输出可以看出，AR1 访问 AR4 只有一条路径，即 AR1 → AR3 → AR4，当 AR1 与 AR3 之间的链路出现故障以后，OSPF 会运行 SPF 算法，然后切换到 AR1 → AR2 → AR3 → AR4。这个故障的切换时间比较长。

在 AR1 中配置 FRR：

```
[AR1-ospf-1]frr                         //开启 FRR 功能
[AR1-ospf-1-frr]loop-free-alternate     //采用 LFA 算法
[AR1-ospf-1-frr]quit
```

【技术说明】

链路保护公式：

Distance_opt(AR2,AR4)<Distance_opt(AR2,AR1)+Distance_opt(AR1, AR4)

在此实验中，2<1+2，满足条件。

再次在 AR1 上查看关于 4.4.4.4 的 OSPF 路由：

```
[AR1]display ospf routing 4.4.4.4

         OSPF Process 1 with Router ID 1.1.1.1

 Destination    : 4.4.4.4/32
 AdverRouter    : 4.4.4.4               Area      : 0.0.0.0
 Cost           : 2                     Type      : Stub
 NextHop        : 10.0.13.3             Interface : GigabitEthernet0/0/1
 Priority       : Medium                Age       : 00h00m09s
 Backup Nexthop : 10.0.12.2             Backup Interface: GigabitEthernet0/0/0
 Backup Type    : LFA LINK
```

通过以上输出可以看出，OSPF 生成了一条备份链路。当 AR1 与 AR3 之间的链路出现故障以后，马上切换到 AR1–AR2–AR3–AR4，时间在 50 ms 以内。

1.5.2　OSPF 与 BFD 联动配置

1. 实验目的

全网运行 OSPF，并且开启 OSPF 与 BFD 的联动，实现 OSPF 邻居故障时能够快速发现故障。

2. 实验拓扑

OSPF 与 BFD 联动配置实验拓扑如图 1–12 所示。

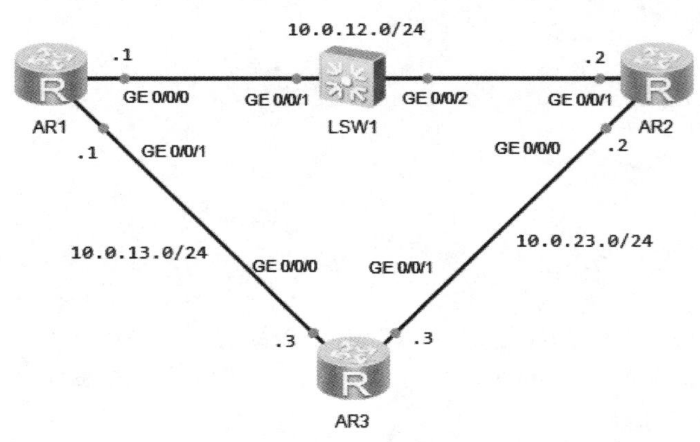

图 1-12　OSPF 与 BFD 联动配置实验拓扑

3. 实验步骤

步骤 1：IP 地址的配置。

AR1 的配置：

```
<Huawei>system-view
Enter system view, return user view with Ctrl+Z.
[Huawei]undo info-center enable
Info: Information center is disabled.
[Huawei]sysname AR1
[AR1]interface G0/0/0
[AR1-GigabitEthernet0/0/0]ip address 10.0.12.1 24
[AR1-GigabitEthernet0/0/0]quit
[AR1]interface g0/0/1
[AR1-GigabitEthernet0/0/1]ip address 10.0.13.1 24
[AR1-GigabitEthernet0/0/1]quit
[AR1]interface LoopBack 0
[AR1-LoopBack0]ip address 1.1.1.1 32
[AR1-LoopBack0]quit
```

AR2 的配置：

```
<Huawei>system-view
Enter system view, return user view with Ctrl+Z.
[Huawei]undo info-center enable
Info: Information center is disabled.
[Huawei]sysname AR2
[AR2]interface g0/0/0
[AR2-GigabitEthernet0/0/0]ip address 10.0.23.2 24
[AR2-GigabitEthernet0/0/0]quit
[AR2]interface g0/0/1
[AR2-GigabitEthernet0/0/1]ip address 10.0.12.2 24
[AR2-GigabitEthernet0/0/1]quit
[AR2]interface LoopBack 0
[AR2-LoopBack0]ip address 2.2.2.2 32
[AR2-LoopBack0]quit
```

AR3 的配置：

```
<Huwei>system-view
Enter system view, return user view with Ctrl+Z.
[Huwei]undo info-center enable
Info: Information center is disabled.
[Huwei]sysname AR3
[AR3]interface g0/0/0
[AR3-GigabitEthernet0/0/0]ip address 10.0.13.3 24
[AR3-GigabitEthernet0/0/0]quit
[AR3]interface g0/0/1
[AR3-GigabitEthernet0/0/1]ip address 10.0.23.3 24
[AR3-GigabitEthernet0/0/1]quit
[AR3]interface LoopBack 0
[AR3-LoopBack0]ip address 3.3.3.3 32
[AR3-LoopBack0]quit
```

步骤 2: 运行 OSPF。

AR1 的配置：

```
[AR1]ospf router-id 1.1.1.1
[AR1-ospf-1]area 0
[AR1-ospf-1-area-0.0.0.0]network 10.0.12.0 0.0.0.255
[AR1-ospf-1-area-0.0.0.0]network 10.0.13.0 0.0.0.255
[AR1-ospf-1-area-0.0.0.0]network 1.1.1.1 0.0.0.0
[AR1-ospf-1-area-0.0.0.0]quit
```

AR2 的配置：

```
[AR2]ospf router-id 2.2.2.2
[AR2-ospf-1]area 0
[AR2-ospf-1-area-0.0.0.0]network 10.0.12.0 0.0.0.255
[AR2-ospf-1-area-0.0.0.0]network 10.0.23.0 0.0.0.255
[AR2-ospf-1-area-0.0.0.0]network 2.2.2.2 0.0.0.0
[AR2-ospf-1-area-0.0.0.0]quit
```

AR3 的配置：

```
[AR3]ospf router-id 3.3.3.3
[AR3-ospf-1]area 0
[AR3-ospf-1-area-0.0.0.0]network 10.0.13.0 0.0.0.255
[AR3-ospf-1-area-0.0.0.0]network 10.0.23.0 0.0.0.255
[AR3-ospf-1-area-0.0.0.0]network 3.3.3.3 0.0.0.0
[AR3-ospf-1-area-0.0.0.0]quit
```

步骤 3: 配置 BFD。

AR1 的配置：

```
[AR1]bfd                                  //全局启用 BFD
[AR1-bfd]quit
[AR1]ospf
[AR1-ospf-1]bfd all-interfaces enable   //所有接口启用 BFD
[AR1-ospf-1]bfd all-interfaces  min-rx-interval 100 min-tx-interval 100 detect-
```

```
multiplier 3   //本地最小接收时间间隔为 100 ms，本地最小发送时间间隔为 100 ms，BFD 检测倍数为 3
[AR1-ospf-1]quit
```

AR2 的配置：

```
[AR2]bfd
[AR2-bfd]quit
[AR2]ospf
[AR2-ospf-1]bfd all-interfaces enable
[AR2-ospf-1]bfd all-interfaces min-rx-interval 100 min-tx-interval 100 detect-
multiplier 3
[AR2-ospf-1]quit
```

【技术要点】BFD会话检测时间

BFD会话检测时间由TX（Desired Min TX Interval）、RX（Required Min RX Interval）和DM（Detect Multi）三个参数决定。BFD报文的实际发送时间间隔和实际接收时间间隔由BFD会话协商决定。

（1）本地BFD报文的实际发送时间间隔 = MAX {本地配置的发送时间间隔，对端配置的接收时间间隔}。

（2）本地BFD报文的实际接收时间间隔 = MAX {对端配置的发送时间间隔，本地配置的接收时间间隔}。

（3）本地BFD报文的实际检测时间。

①异步模式：本地BFD报文的实际检测时间 = 本地BFD报文的实际接收时间间隔 × 对端配置的BFD检测倍数。

②查询模式：本地BFD报文的实际检测时间 = 本地BFD报文的实际接收时间间隔 × 本端配置的BFD检测倍数。

步骤 4: 实验调试。

查看 BFD 的详细信息：

```
[AR1]display bfd session all verbose
--------------------------------------------------------------------------------
Session MIndex : 512        (One Hop) State : Up        Name : dyn_8192
--------------------------------------------------------------------------------
  Local Discriminator       : 8192             Remote Discriminator    : 8193
  Session Detect Mode       : Asynchronous Mode Without Echo Function
  BFD Bind Type             : Interface(GigabitEthernet0/0/0)
  Bind Session Type         : Dynamic
  Bind Peer IP Address      : 10.0.12.2
  NextHop Ip Address        : 10.0.12.2
  Bind Interface            : GigabitEthernet0/0/0
  FSM Board Id              : 0                TOS-EXP                 : 7
  Min Tx Interval (ms)      : 100              Min Rx Interval (ms)    : 100
  Actual Tx Interval (ms): 100                 Actual Rx Interval (ms): 100
  Local Detect Multi        : 3                Detect Interval (ms)    : 300
  Echo Passive              : Disable          Acl Number              : -
  Destination Port          : 3784             TTL                     : 255
```

```
  Proc Interface Status   : Disable          Process PST            : Disable
  WTR Interval (ms)       : -
  Active Multi            : 3
  Last Local Diagnostic   : No Diagnostic
  Bind Application        : OSPF
  Session TX TmrID        : -                Session Detect TmrID   : -
  Session Init TmrID      : -                Session WTR TmrID      : -
  Session Echo Tx TmrID   : -
  PDT Index               : FSM-0 | RCV-0 | IF-0 | TOKEN-0
  Session Description     : -
--------------------------------------------------------------------------------

--------------------------------------------------------------------------------
Session MIndex : 513       (One Hop) State : Down      Name : dyn_8193
--------------------------------------------------------------------------------
  Local Discriminator     : 8193             Remote Discriminator   : 0
  Session Detect Mode     : Asynchronous Mode Without Echo Function
  BFD Bind Type           : Interface(GigabitEthernet0/0/1)
  Bind Session Type       : Dynamic
  Bind Peer IP Address    : 10.0.13.3
  NextHop Ip Address      : 10.0.13.3
  Bind Interface          : GigabitEthernet0/0/1
  FSM Board Id            : 0                TOS-EXP                : 7
  Min Tx Interval (ms)    : 100              Min Rx Interval (ms)   : 100
  Actual Tx Interval (ms): 13500             Actual Rx Interval (ms): 13500
  Local Detect Multi      : 3                Detect Interval (ms)   : -
  Echo Passive            : Disable          Acl Number             : -
  Destination Port        : 3784             TTL                    : 255
  Proc Interface Status   : Disable          Process PST            : Disable
  WTR Interval (ms)       : -
  Active Multi            : -
  Last Local Diagnostic   : No Diagnostic
  Bind Application        : OSPF
  Session TX TmrID        : 2059             Session Detect TmrID   : -
  Session Init TmrID      : -                Session WTR TmrID      : -
  Session Echo Tx TmrID   : -
  PDT Index               : FSM-1 | RCV-0 | IF-0 | TOKEN-0
  Session Description     : -
--------------------------------------------------------------------------------

     Total UP/DOWN Session Number : 1/1
```

查看 OSPF 的 BFD 会话：

```
[AR1]display ospf bfd session all
         OSPF Process 1 with Router ID 1.1.1.1
 Area 0.0.0.0 interface 10.0.12.1(GigabitEthernet0/0/0)'s BFD Sessions
 NeighborId:2.2.2.2          AreaId:0.0.0.0          Interface:GigabitEthernet0/0/0
 BFDState:up                  rx    :100              tx       :100
```

```
 Multiplier:3                   BFD Local Dis:8192        LocalIpAdd:10.0.12.1
 RemoteIpAdd:10.0.12.2          Diagnostic Info:No diagnostic information

 Area 0.0.0.0 interface 10.0.13.1(GigabitEthernet0/0/1)'s BFD Sessions
 NeighborId:3.3.3.3             AreaId:0.0.0.0            Interface:GigabitEthernet0/0/1
 BFDState:down                  rx    :13500              tx       :13500
 Multiplier:0                   BFD Local Dis:8193        LocalIpAdd:10.0.13.1
 RemoteIpAdd:10.0.13.3          Diagnostic Info:No diagnostic information
```

【技术要点】

BFD与OSPF联动就是将BFD和OSPF协议关联起来，通过BFD对链路故障的快速感应进而通知OSPF协议，从而加快OSPF协议对于网络拓扑变化的响应。

关闭交换机的 G0/0/1：

```
<Huawei>system-view
Enter system view, return user view with Ctrl+Z.
[Huawei]undo info-center enable
Info: Information center is disabled.
[Huawei]sysname LSW1
[LSW1]interface g0/0/1
[LSW1-GigabitEthernet0/0/1]shutdown
[LSW1-GigabitEthernet0/0/1]quit
```

在 AR1 上查看 BFD 的信息：

```
[AR1]display bfd session all verbose
--------------------------------------------------------------------------------
Session MIndex : 513        (One Hop) State : Down        Name : dyn_8193
--------------------------------------------------------------------------------
  Local Discriminator       : 8193            Remote Discriminator     : 0
  Session Detect Mode       : Asynchronous Mode Without Echo Function
  BFD Bind Type             : Interface(GigabitEthernet0/0/1)
  Bind Session Type         : Dynamic
  Bind Peer IP Address      : 10.0.13.3
  NextHop Ip Address        : 10.0.13.3
  Bind Interface            : GigabitEthernet0/0/1
  FSM Board Id              : 0               TOS-EXP                  : 7
  Min Tx Interval (ms)      : 100             Min Rx Interval (ms)     : 100
  Actual Tx Interval (ms): 13500              Actual Rx Interval (ms): 13500
  Local Detect Multi        : 3               Detect Interval (ms)     : -
  Echo Passive              : Disable         Acl Number               : -
  Destination Port          : 3784            TTL                      : 255
  Proc Interface Status     : Disable         Process PST              : Disable
  WTR Interval (ms)         : -
  Active Multi              : -
  Last Local Diagnostic     : No Diagnostic
  Bind Application          : OSPF
  Session TX TmrID          : 2059            Session Detect TmrID     : -
  Session Init TmrID        : -               Session WTR TmrID        : -
```

```
  Session Echo Tx TmrID   : -
  PDT Index               : FSM-1 | RCV-0 | IF-0 | TOKEN-0
  Session Description     : -
------------------------------------------------------------------------
```

如果没有 BFD，OSPF 的邻居关系检测要 4 个 Hello 包的时间，配置了 BFD，检测是毫秒级的。

1.5.3　OSPF 与 BGP 联动配置

1. 实验目的

AS 100 内部的 IGP 协议配置为 OSPF，并且 AS 100 内部的设备配置 OSPF 全互联，AR4 和 AR5 之间建立 EBGP 邻居关系。通过修改 OSPF 开销的方式将 AR1 访问 AR5 的流量路径改为 AR1 → AR2 → AR4 → AR5。在 AR2 上配置 OSPF 与 BGP 的联动，实现当 AR2 故障恢复后不会出现数据丢包现象。

2. 实验拓扑

OSPF 与 BGP 联动配置的实验拓扑如图 1-13 所示。

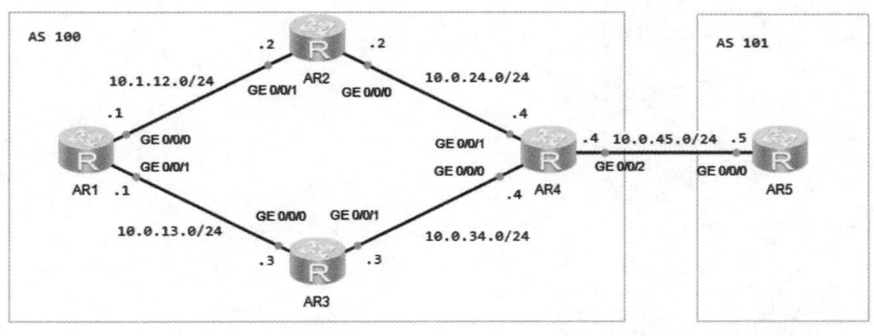

图 1-13　OSPF 与 BGP 联动配置的实验拓扑

3. 实验步骤

步骤 1：配置 IP 地址。

AR1 的配置：

```
<Huawei>system-view
Enter system view, return user view with Ctrl+Z.
[Huawei]undo info-center enable
Info: Information center is disabled.
[Huawei]sysname AR1
[AR1]interface g0/0/0
[AR1-GigabitEthernet0/0/0]ip address 10.0.12.1 24
[AR1-GigabitEthernet0/0/0]quit
[AR1]interface g0/0/1
[AR1-GigabitEthernet0/0/1]ip address 10.0.13.1 24
[AR1-GigabitEthernet0/0/1]quit
[AR1]interface LoopBack 0
[AR1-LoopBack0]ip address 1.1.1.1 32
[AR1-LoopBack0]quit
```

AR2 的配置：

```
<Huawei>system-view
Enter system view, return user view with Ctrl+Z.
[Huawei]undo info-center enable
Info: Information center is disabled.
[Huawei]sysname AR2
[AR2]interface g0/0/0
[AR2-GigabitEthernet0/0/0]ip address 10.0.24.2 24
[AR2-GigabitEthernet0/0/0]quit
[AR2]interface g0/0/1
[AR2-GigabitEthernet0/0/1]ip address 10.0.12.2 24
[AR2-GigabitEthernet0/0/1]quit
[AR2]interface LoopBack 0
[AR2-LoopBack0]ip address 2.2.2.2 32
[AR2-LoopBack0]quit
```

AR3 的配置：

```
<Huawei>system-view
Enter system view, return user view with Ctrl+Z.
[Huawei]undo info-center enable
Info: Information center is disabled.
[Huawei]sysname AR3
[AR3]interface g0/0/0
[AR3-GigabitEthernet0/0/0]ip address 10.0.13.3 24
[AR3-GigabitEthernet0/0/0]quit
[AR3]interface g0/0/1
[AR3-GigabitEthernet0/0/1]ip address 10.0.34.3 24
[AR3-GigabitEthernet0/0/1]quit
[AR3]interface LoopBack 0
[AR3-LoopBack0]ip address 3.3.3.3 32
[AR3-LoopBack0]quit
```

AR4 的配置：

```
<Huawei>system-view
Enter system view, return user view with Ctrl+Z.
[Huawei]undo info-center enable
Info: Information center is disabled.
[Huawei]sysname AR4
[AR4]interface g0/0/0
[AR4-GigabitEthernet0/0/0]ip address 10.0.34.4 24
[AR4-GigabitEthernet0/0/0]quit
[AR4]interface g0/0/1
[AR4-GigabitEthernet0/0/1]ip address 10.0.24.4 24
[AR4-GigabitEthernet0/0/1]quit
[AR4]interface g0/0/2
[AR4-GigabitEthernet0/0/2]ip address 10.0.45.4 24
[AR4-GigabitEthernet0/0/2]quit
```

```
[AR4]interface LoopBack 0
[AR4-LoopBack0]ip address 4.4.4.4 32
[AR4-LoopBack0]quit
```

AR5 的配置：

```
<Huawei>system-view
Enter system view, return user view with Ctrl+Z.
[Huawei]undo info-center enable
Info: Information center is disabled.
[Huawei]sysname AR5
[AR5]interface g0/0/0
[AR5-GigabitEthernet0/0/0]ip address 10.0.45.5 24
[AR5-GigabitEthernet0/0/0]quit
[AR5]interface LoopBack 0
[AR5-LoopBack0]ip address 5.5.5.5 32
[AR5-LoopBack0]quit
```

步骤 2：运行 IGP。

AR1 的配置：

```
[AR1]ospf router-id 1.1.1.1
[AR1-ospf-1]area 0
[AR1-ospf-1-area-0.0.0.0]network 10.0.12.0 0.0.0.255
[AR1-ospf-1-area-0.0.0.0]network 10.0.13.0 0.0.0.255
[AR1-ospf-1-area-0.0.0.0]network 1.1.1.1 0.0.0.0
[AR1-ospf-1-area-0.0.0.0]quit
```

AR2 的配置：

```
[AR2]ospf router-id 2.2.2.2
[AR2-ospf-1]area 0
[AR2-ospf-1-area-0.0.0.0]network 10.0.12.0 0.0.0.255
[AR2-ospf-1-area-0.0.0.0]network 10.0.24.0 0.0.0.255
[AR2-ospf-1-area-0.0.0.0]network 2.2.2.2 0.0.0.0
[AR2-ospf-1-area-0.0.0.0]quit
```

AR3 的配置：

```
[AR3]ospf router-id 3.3.3.3
[AR3-ospf-1]area 0
[AR3-ospf-1-area-0.0.0.0]network 10.0.13.0 0.0.0.255
[AR3-ospf-1-area-0.0.0.0]network 10.0.34.0 0.0.0.255
[AR3-ospf-1-area-0.0.0.0]network 3.3.3.3 0.0.0.0
[AR3-ospf-1-area-0.0.0.0]quit
```

AR4 的配置：

```
[AR4]ospf router-id 4.4.4.4
[AR4-ospf-1]area 0
[AR4-ospf-1-area-0.0.0.0]network 10.0.24.0 0.0.0.255
[AR4-ospf-1-area-0.0.0.0]network 10.0.34.0 0.0.0.255
[AR4-ospf-1-area-0.0.0.0]network 4.4.4.4 0.0.0.0
[AR4-ospf-1-area-0.0.0.0]quit
```

步骤 3: 运行 BGP。

AR1 的配置：

```
[AR1]bgp 100
[AR1-bgp]undo synchronization
[AR1-bgp]peer 2.2.2.2 as-number 100
[AR1-bgp]peer 2.2.2.2 connect-interface LoopBack 0
[AR1-bgp]peer 3.3.3.3 as-number 100
[AR1-bgp]peer 3.3.3.3 connect-interface LoopBack 0
[AR1-bgp]peer 4.4.4.4 as-number 100
[AR1-bgp]peer 4.4.4.4 connect-interface LoopBack 0
[AR1-bgp]network 1.1.1.1 32
[AR1-bgp]quit
```

AR2 的配置：

```
[AR2]bgp 100
[AR2-bgp]undo synchronization
[AR2-bgp]peer 1.1.1.1 as-number 100
[AR2-bgp]peer 1.1.1.1 connect-interface loo0
[AR2-bgp]peer 3.3.3.3 as-number 100
[AR2-bgp]peer 3.3.3.3 connect-interface LoopBack 0
[AR2-bgp]peer 4.4.4.4 as-number 100
[AR2-bgp]peer 4.4.4.4 connect-interface LoopBack 0
[AR2-bgp]quit
```

AR3 的配置：

```
[AR3]bgp 100
[AR3-bgp]undo synchronization
[AR3-bgp]peer 1.1.1.1 as-number 100
[AR3-bgp]peer 1.1.1.1 connect-interface LoopBack 0
[AR3-bgp]peer 2.2.2.2 as-number 100
[AR3-bgp]peer 2.2.2.2 connect-interface LoopBack 0
[AR3-bgp]peer 4.4.4.4 as-number 100
[AR3-bgp]peer 4.4.4.4 connect-interface LoopBack 0
[AR3-bgp]quit
```

AR4 的配置：

```
[AR4]bgp 100
[AR4-bgp]undo synchronization
[AR4-bgp]peer 1.1.1.1 as-number 100
[AR4-bgp]peer 1.1.1.1 connect-interface LoopBack 0
[AR4-bgp]peer 2.2.2.2 as-number 100
[AR4-bgp]peer 2.2.2.2 connect-interface LoopBack 0
[AR4-bgp]peer 3.3.3.3 as-number 100
[AR4-bgp]peer 3.3.3.3 connect-interface LoopBack 0
[AR4-bgp]peer 10.0.45.5 as-number 101
[AR4-bgp]peer 1.1.1.1 next-hop-local
[AR4-bgp]peer 2.2.2.2 next-hop-local
[AR4-bgp]peer 3.3.3.3 next-hop-local
```

```
[AR4-bgp]quit
```

AR5 的配置：

```
[AR5]bgp 101
[AR5-bgp]router-id 5.5.5.5
[AR5-bgp]undo synchronization
[AR5-bgp]peer 10.0.45.4 as-number 100
[AR5-bgp]network 5.5.5.5 32
[AR5-bgp]quit
```

步骤 4：在 AR1 上创建环回口 lo1，并宣告进 BGP。

```
[AR1]interface LoopBack 1
[AR1-LoopBack1]ip address 11.11.11.11 32
[AR1-LoopBack1]quit
[AR1]bgp 100
[AR1-bgp]network 11.11.11.11 32
[AR1-bgp]quit
```

步骤 5：实验调试。

在 AR3 上改开销，让 11.11.11.11 访问 5.5.5.5 走 AR1 → AR2 → AR4 → AR5。

```
[AR3]interface g0/0/1
[AR3-GigabitEthernet0/0/1]ospf cost 10
[AR3-GigabitEthernet0/0/1]quit
```

关闭 AR2 查看现象。

OSPF 的收敛速度快，BGP 的收敛速度慢，因此会造成数据丢失。

在 AR2 上配置：

```
[AR2]ospf
[AR2-ospf-1]stub-router
[AR2-ospf-1]quit
```

在 AR2 上查看 OSPF 的路由表：

```
[AR2]display ospf routing
         OSPF Process 1 with Router ID 2.2.2.2
                  Routing Tables
 Routing for Network
 Destination         Cost        Type          NextHop          AdvRouter       Area
 2.2.2.2/32          0           Stub          2.2.2.2          2.2.2.2         0.0.0.0
 10.0.12.0/24        65535       Transit       10.0.12.2        2.2.2.2         0.0.0.0
 10.0.24.0/24        65535       Transit       10.0.24.2        2.2.2.2         0.0.0.0
 1.1.1.1/32          65535       Stub          10.0.12.1        1.1.1.1         0.0.0.0
 3.3.3.3/32          65536       Stub          10.0.24.4        3.3.3.3         0.0.0.0
 3.3.3.3/32          65536       Stub          10.0.12.1        3.3.3.3         0.0.0.0
 4.4.4.4/32          65535       Stub          10.0.24.4        4.4.4.4         0.0.0.0
 10.0.13.0/24        65536       Transit       10.0.12.1        1.1.1.1         0.0.0.0
 10.0.34.0/24        65536       Transit       10.0.24.4        3.3.3.3         0.0.0.0
 Total Nets: 9
 Intra Area: 9  Inter Area: 0  ASE: 0  NSSA: 0
```

如果把路由的开销设置为 65535，就不会选这条路，等故障收敛完成后，再改回来。

1.5.4 OSPF LSA 过滤

1. 实验目的

AR1、AR2、AR3 之间运行 OSPF，在 AR2 的接口以及 OSPF 区域视图中配置 OSPF 的 LSA 过滤，在 AR2 的接口中实现外部路由的过滤，在 AR2 的 OPSF 区域视图中实现区域间的路由过滤。

2. 实验拓扑

OSPF LSA 过滤的实验拓扑如图 1-14 所示。

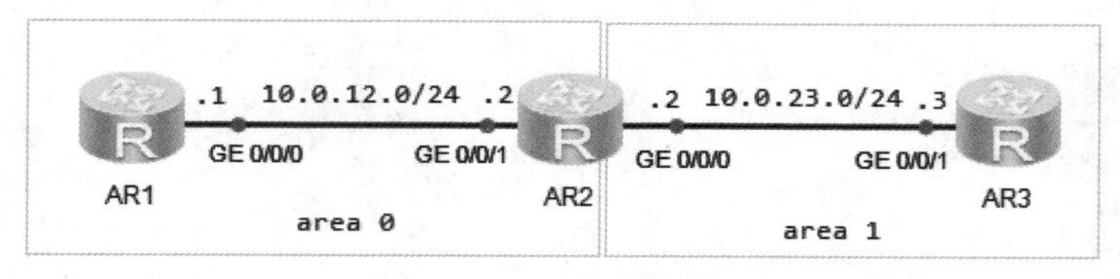

图 1-14　OSPF LSA 过滤的实验拓扑

3. 实验步骤

步骤 1： 配置 IP 地址。

AR1 的配置：

```
<Huawei>system-view
Enter system view, return user view with Ctrl+Z.
[Huawei]undo info-center enable
Info: Information center is disabled.
[Huawei]sysname AR1
[AR1]interface g0/0/0
[AR1-GigabitEthernet0/0/0]ip address 10.0.12.1 24
[AR1-GigabitEthernet0/0/0]quit
[AR1]interface LoopBack 0
[AR1-LoopBack0]ip address 1.1.1.1 32
[AR1-LoopBack0]quit
```

AR2 的配置：

```
<Huawei>system-view
Enter system view, return user view with Ctrl+Z.
[Huawei]undo info-center enable
Info: Information center is disabled.
[Huawei]sysname AR2
[AR2]interface g0/0/0
[AR2-GigabitEthernet0/0/0]ip address 10.0.23.2 24
[AR2-GigabitEthernet0/0/0]quit
[AR2]interface g0/0/1
[AR2-GigabitEthernet0/0/1]ip address 10.0.12.2 24
[AR2-GigabitEthernet0/0/1]quit
[AR2]interface LoopBack 0
[AR2-LoopBack0]ip address 2.2.2.2 32
[AR2-LoopBack0]quit
```

AR3 的配置：

```
<Huawei>system-view
Enter system view, return user view with Ctrl+Z.
[Huawei]undo info-center enable
Info: Information center is disabled.
[Huawei]sysname AR3
[AR3]interface g0/0/1
[AR3-GigabitEthernet0/0/1]ip address 10.0.23.3 24
[AR3-GigabitEthernet0/0/1]quit
[AR3]interface LoopBack 0
[AR3-LoopBack0]ip address 3.3.3.3 32
[AR3-LoopBack0]quit
```

步骤 2： 配置 OSPF。

AR1 的配置：

```
[AR1]ospf router-id 1.1.1.1
[AR1-ospf-1]area 0
[AR1-ospf-1-area-0.0.0.0]network 10.0.12.0 0.0.0.255
[AR1-ospf-1-area-0.0.0.0]network 1.1.1.1 0.0.0.0
[AR1-ospf-1-area-0.0.0.0]quit
```

AR2 的配置：

```
[AR2]ospf router-id 2.2.2.2
[AR2-ospf-1]area 0
[AR2-ospf-1-area-0.0.0.0]network 10.0.12.0 0.0.0.255
[AR2-ospf-1-area-0.0.0.0]network 2.2.2.2 0.0.0.0
[AR2-ospf-1]area 1
[AR2-ospf-1-area-0.0.0.1]network 10.0.23.0 0.0.0.255
[AR2-ospf-1-area-0.0.0.0]quit
```

AR3 的配置：

```
[AR3]ospf router-id 3.3.3.3
[AR3-ospf-1]area 1
[AR3-ospf-1-area-0.0.0.1]network 10.0.23.0 0.0.0.255
[AR3-ospf-1-area-0.0.0.1]network 3.3.3.3 0.0.0.0
[AR3-ospf-1-area-0.0.0.1]quit
```

步骤 3： 实验调试。

在 AR1 上创建两个环回口，并把它通过 5 类 LSA 引入 OSPF。

```
[AR1]interface LoopBack 8
[AR1-LoopBack8]ip address 8.8.8.8 32
[AR1-LoopBack8]quit
[AR1]interface LoopBack 9
[AR1-LoopBack9]ip address 9.9.9.9 32
[AR1-LoopBack9]quit
[AR1]ospf
[AR1-ospf-1]import-route direct
[AR1-ospf-1]quit
```

在 AR2 上查看 OSPF 的 LSDB：

```
[AR2]display ospf lsdb
         OSPF Process 1 with Router ID 2.2.2.2
                 Link State Database
                         Area: 0.0.0.0
 Type       LinkState ID    AdvRouter      Age    Len    Sequence    Metric
 Router     2.2.2.2         2.2.2.2        211    60     80000013    1
 Router     1.1.1.1         1.1.1.1        146    48     8000000F    1
 Router     3.3.3.3         3.3.3.3        721    48     80000004    1
 Network    10.0.23.2       2.2.2.2        718    32     80000002    0
 Network    10.0.12.2       2.2.2.2        211    32     80000004    0

                 AS External Database
 Type       LinkState ID    AdvRouter      Age    Len    Sequence    Metric
 External   10.0.12.0       1.1.1.1        146    36     80000001    1
 External   9.9.9.9         1.1.1.1        146    36     80000001    1
 External   8.8.8.8         1.1.1.1        146    36     80000001    1
 External   1.1.1.1         1.1.1.1        146    36     80000001    1
```

通过以上输出可以看出，在 AR2 上收到了 4 条外部路由。

在 AR1 的接口下，对发送的 LSA 进行过滤，让 AR2 只能收到 8.8.8.8 和 10.0.12.0 这两条外部路由。

```
[AR1]acl 2000
[AR1-acl-basic-2000]rule 10 deny source 1.1.1.1 0
[AR1-acl-basic-2000]rule 20 deny source 9.9.9.9 0
[AR1-acl-basic-2000]rule 30 permit source any
[AR1-acl-basic-2000]quit
[AR1]interface g0/0/0
[AR1-GigabitEthernet0/0/0]ospf filter-lsa-out ase acl 2000
[AR1-GigabitEthernet0/0/0]quit
```

保存配置，重新启动设备，查看 AR2 的 OSPF 的 LSDB。

```
<AR2>display ospf lsdb
         OSPF Process 1 with Router ID 2.2.2.2
                 Link State Database
                         Area: 0.0.0.0
 Type       LinkState ID    AdvRouter      Age    Len    Sequence    Metric
 Router     2.2.2.2         2.2.2.2        47     60     000000A     1
 Router     1.1.1.1         1.1.1.1        59     48     0000006     1
 Router     3.3.3.3         3.3.3.3        48     48     0000006     1
 Network    10.0.23.3       3.3.3.3        48     32     0000002     0
 Network    10.0.12.2       2.2.2.2        50     32     0000002     0

                 AS External Database
 Type       LinkState ID    AdvRouter      Age    Len    Sequence    Metric
 External   10.0.12.0       1.1.1.1        98     36     0000001     1
 External   8.8.8.8         1.1.1.1        123    36     0000001     1
```

通过以上输出可以看出，已完成实验要求。

【技术要点】为什么要重启设备？

对于已经发送的LSA，要到3600 s才能达到老化时间。

在 AR2 上查看 OSPF 的 LSDB：

```
<AR2>display ospf lsdb
        OSPF Process 1 with Router ID 2.2.2.2
                Link State Database
                        Area: 0.0.0.0
 Type      LinkState ID    AdvRouter      Age   Len    Sequence   Metric
 Router    2.2.2.2         2.2.2.2        271   60     000000A    1
 Router    1.1.1.1         1.1.1.1        283   48     0000006    1
 Router    3.3.3.3         3.3.3.3        272   48     0000006    1
 Network   10.0.23.3       3.3.3.3        272   32     0000002    0
 Network   10.0.12.2       2.2.2.2        274   32     0000002    0

                AS External Database
 Type      LinkState ID    AdvRouter      Age   Len    Sequence   Metric
 External  10.0.12.0       1.1.1.1        322   36     0000001    1
 External  8.8.8.8         1.1.1.1        347   36     0000001    1
```

在 AR2 上通过对 ABR Type3 LSA 进行过滤，让 AR3 收不到 1.1.1.1 的路由。

```
[AR2]acl 2000
[AR2-acl-basic-2000]rule deny source 1.1.1.1 0
[AR2-acl-basic-2000]rule permit source any
[AR2-acl-basic-2000]quit
[AR2]ospf
[AR2-ospf-1]area 1
[AR2-ospf-1-area-0.0.0.1]filter 2000 import // 拒绝 1.1.1.1 进入区域 1
[AR2-ospf-1-area-0.0.0.1]quit
```

【思考】什么时候用import？什么时候用export？

在本实验中，如果在AR2的区域1中配置filter 2000 import，则是在进入区域1时去掉1.1.1.1；如果在AR2的区域0中配置filter 2000 export，则是在从区域0出去时去掉1.1.1.1。

再次查看 AR3 的 LSDB：

```
<AR3>display ospf lsdb
        OSPF Process 1 with Router ID 10.0.23.3
                Link State Database
                        Area: 0.0.0.1
 Type      LinkState ID    AdvRouter      Age   Len    Sequence   Metric
 Router    10.0.23.3       10.0.23.3      620   48     0000006    1
 Router    2.2.2.2         2.2.2.2        621   36     0000003    1
 Network   10.0.23.3       10.0.23.3      620   32     0000001    0
 Sum-Net   10.0.12.0       2.2.2.2        664   28     0000001    1
 Sum-Net   2.2.2.2         2.2.2.2        664   28     0000001    0
 Sum-Asbr  1.1.1.1         2.2.2.2        625   28     0000001    1
```

```
                  AS External Database
 Type       LinkState ID    AdvRouter       Age     Len     Sequence     Metric
 External   10.0.12.0       1.1.1.1         669     36      0000001      1
 External   8.8.8.8         1.1.1.1         697     36      0000001      1
```

通过以上输出可以看出，没有关于 1.1.1.1 这条路由的 3 类 LSA。

1.5.5 在 5 类 LSA 中利用 FA 解决次优路径问题

1. 实验目的

了解 OSPF 的 5 类 LSA 的 FA 地址的作用。

2. 实验拓扑

5 类 LSA FA 解决次优路径问题的实验拓扑如图 1-15 所示。

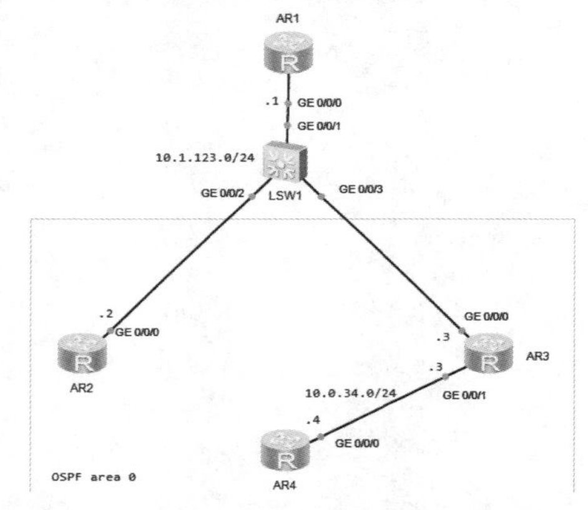

图 1-15　5 类 LSA FA 解决次优路径问题的实验拓扑

3. 实验步骤

步骤 1：配置 IP 地址。

AR1 的配置：

```
<Huawei>system-view
Enter system view, return user view with Ctrl+Z.
[Huawei]undo info-center enable
Info: Information center is disabled.
[Huawei]sysname AR1
[AR1]interface g0/0/0
[AR1-GigabitEthernet0/0/0]ip address 10.1.123.1 24
[AR1-GigabitEthernet0/0/0]quit
[AR1]interface LoopBack 0
[AR1-LoopBack0]ip address 1.1.1.1 32
[AR1-LoopBack0]quit
```

AR2 的配置：

```
<Huawei>system-view
Enter system view, return user view with Ctrl+Z.
[Huawei]undo info-center enable
```

```
Info: Information center is disabled.
[Huawei]sysname AR2
[AR2]interface g0/0/0
[AR2-GigabitEthernet0/0/0]ip address 10.1.123.2 24
[AR2-GigabitEthernet0/0/0]quit
[AR2]interface LoopBack 0
[AR2-LoopBack0]ip address 2.2.2.2 32
[AR2-LoopBack0]quit
```

AR3 的配置：

```
<Huawei>system-view
Enter system view, return user view with Ctrl+Z.
[Huawei]undo info-center enable
Info: Information center is disabled.
[Huawei]sysname AR3
[AR3]interface g0/0/0
[AR3-GigabitEthernet0/0/0]ip address 10.1.123.3 24
[AR3-GigabitEthernet0/0/0]quit
[AR3]interface g0/0/1
[AR3-GigabitEthernet0/0/1]ip address 10.0.34.3 24
[AR3-GigabitEthernet0/0/1]quit
[AR3]interface LoopBack 0
[AR3-LoopBack0]ip address 3.3.3.3 32
[AR3-LoopBack0]quit
```

AR4 的配置：

```
<Huawei>system-view
Enter system view, return user view with Ctrl+Z.
[Huawei]undo info-center enable
Info: Information center is disabled.
[Huawei]sysname AR4
[AR4]interface g0/0/0
[AR4-GigabitEthernet0/0/0]ip address 10.0.34.4 24
[AR4-GigabitEthernet0/0/0]quit
[AR4]interface LoopBack 0
[AR4-LoopBack0]ip address 4.4.4.4 32
[AR4-LoopBack0]quit
```

步骤 2：运行 OSPF。

AR2 的配置：

```
[AR2]ospf router-id 2.2.2.2
[AR2-ospf-1]area 0
[AR2-ospf-1-area-0.0.0.0]network 10.1.123.0 0.0.0.255
[AR2-ospf-1-area-0.0.0.0]network 2.2.2.2 0.0.0.0
[AR2-ospf-1-area-0.0.0.0]quit
```

AR3 的配置：

```
[AR3]ospf router-id 3.3.3.3
[AR3-ospf-1]area 0
```

```
[AR3-ospf-1-area-0.0.0.0]network 10.1.123.0 0.0.0.255
[AR3-ospf-1-area-0.0.0.0]network 10.0.34.0 0.0.0.255
[AR3-ospf-1-area-0.0.0.0]network 3.3.3.3 0.0.0.0
[AR3-ospf-1-area-0.0.0.0]quit
```

AR4 的配置：

```
[AR4]ospf router-id 4.4.4.4
[AR4-ospf-1]area 0
[AR4-ospf-1-area-0.0.0.0]network 10.0.34.0 0.0.0.255
[AR4-ospf-1-area-0.0.0.0]network 4.4.4.4 0.0.0.0
[AR4-ospf-1-area-0.0.0.0]quit
```

步骤 3：实验调试。

在 AR2 上配置到达 1.1.1.1/32 的静态路由时，下一跳为 10.1.123.1。

```
[AR2]ip route-static 1.1.1.1 32 10.1.123.1
[AR2]ospf
[AR2-ospf-1]import-route static
[AR2-ospf-1]quit
```

在 AR2 上查看关于 1.1.1.1 这条 5 类 LSA 的详细信息：

```
[AR2]display ospf lsdb ase 1.1.1.1
         OSPF Process 1 with Router ID 2.2.2.2
                 Link State Database
  Type       : External
  Ls id      : 1.1.1.1
  Adv rtr    : 2.2.2.2
  Ls age     : 110
  Len        : 36
  Options    : E
  seq#       : 80000001
  chksum     : 0x62da
  Net mask   : 255.255.255.255
  TOS 0  Metric: 1
  E type     : 2
  Forwarding Address : 10.1.123.1 //FA 地址不为 0
  Tag        : 1
  Priority   : Low
```

【技术要点】

当ASBR引入外部路由时，若Type5 LSA中的FA字段为0，则表示路由器认为到达目的网段的数据包应该发往该ASBR；若Type5 LSA中的FA字段不为0，则表示路由器认为到达目的网段的数据包应该发往这个FA所标识的设备。当以下条件全部满足时，FA字段才可以被设置为非0。

（1）ASBR在其连接外部网络的接口（外部路由的出接口）上激活了OSPF。

（2）该接口没有被配置为Silent-Interface。

（3）该接口的OSPF网络类型为Broadcast或NBMA。

（4）该接口的IP地址在OSPF配置的network命令指定的网段范围内。

（5）到达FA地址的路由必须是OSPF区域内部路由或区域间路由，这样接收到该外部LSA

的路由器才能够加载该LSA进入路由表。加载的外部LSA生成的路由条目的下一跳与到达FA地址的下一跳相同。

在 AR4 上查看关于 1.1.1.1 这条 5 类 LSA 的详细信息：

```
<AR4>display ospf lsdb ase 1.1.1.1
         OSPF Process 1 with Router ID 4.4.4.4
                 Link State Database
  Type       : External
  Ls id      : 1.1.1.1
  Adv rtr    : 2.2.2.2
  Ls age     : 490
  Len        : 36
  Options    : E
  seq#       : 80000001
  chksum     : 0x62da
  Net mask   : 255.255.255.255
  TOS 0  Metric: 1
  E type     : 2
  Forwarding Address : 10.1.123.1
  Tag        : 1
  Priority   : Medium
```

通过以上输出可以看出，AR4 访问 1.1.1.1 的路径为 AR4 → AR3 → AR1。

【思考】如果FA为0.0.0.0，会产生什么样的问题呢?

如果FA为0.0.0.0，那么AR4访问1.1.1.1会走AR4→AR3→AR2→AR1，将产生次优路径。

1.5.6　在 7 类 LSA 中利用 FA 解决次优路径问题

1. 实验目的

了解 OSPF 的 7 类 LSA 的 FA 地址的作用。

2. 实验拓扑

7 类 LSA FA 解决次优路径问题的实验拓扑如图 1–16 所示。

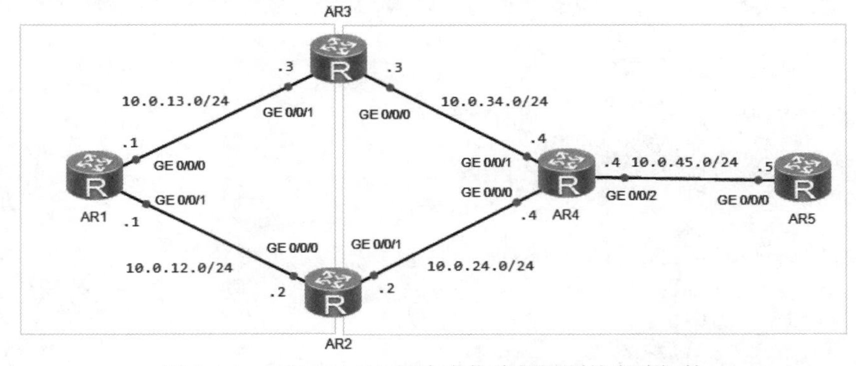

图 1–16　7 类 LSA FA 解决次优路径问题的实验拓扑

3. 实验步骤

步骤 1：配置 IP 地址。

AR1 的配置：

```
<Huawei>system-view
Enter system view, return user view with Ctrl+Z.
[Huawei]undo info-center enable
Info: Information center is disabled.
[Huawei]sysname AR1
[AR1]interface g0/0/0
[AR1-GigabitEthernet0/0/0]ip address 10.0.13.1 24
[AR1-GigabitEthernet0/0/0]quit
[AR1]interface g0/0/1
[AR1-GigabitEthernet0/0/1]ip address 10.0.12.1 24
[AR1-GigabitEthernet0/0/1]quit
[AR1]interface LoopBack 0
[AR1-LoopBack0]ip address 1.1.1.1 32
[AR1-LoopBack0]quit
```

AR2 的配置：

```
<Huawei>system-view
Enter system view, return user view with Ctrl+Z.
[Huawei]undo info-center enable
Info: Information center is disabled.
[Huawei]sysname AR2
[AR2]interface g0/0/0
[AR2-GigabitEthernet0/0/0]ip address 10.0.12.2 24
[AR2-GigabitEthernet0/0/0]quit
[AR2]interface g0/0/1
[AR2-GigabitEthernet0/0/1]ip address 10.0.24.2 24
[AR2-GigabitEthernet0/0/1]quit
[AR2]interface LoopBack 0
[AR2-LoopBack0]ip address 2.2.2.2 32
[AR2-LoopBack0]quit
```

AR3 的配置：

```
<Huawei>system-view
Enter system view, return user view with Ctrl+Z.
[Huawei]undo info-center enable
Info: Information center is disabled.
[Huawei]sysname AR3
[AR3]interface g0/0/1
[AR3-GigabitEthernet0/0/1]ip address 10.0.13.3 24
[AR3-GigabitEthernet0/0/1]quit
[AR3]interface g0/0/0
[AR3-GigabitEthernet0/0/0]ip address 10.0.34.3 24
[AR3-GigabitEthernet0/0/0]quit
[AR3]interface LoopBack 0
```

```
[AR3-LoopBack0]ip address 3.3.3.3 32
[AR3-LoopBack0]quit
```

AR4 的配置：

```
<Huawei>system-view
Enter system view, return user view with Ctrl+Z.
[Huawei]undo info-center enable
Info: Information center is disabled.
[Huawei]sysname AR4
[AR4]interface g0/0/0
[AR4-GigabitEthernet0/0/0]ip address 10.0.24.4 24
[AR4-GigabitEthernet0/0/0]quit
[AR4]interface g0/0/1
[AR4-GigabitEthernet0/0/1]ip address 10.0.34.4 24
[AR4-GigabitEthernet0/0/1]quit
[AR4]interface g0/0/2
[AR4-GigabitEthernet0/0/2]ip address 10.0.45.4 24
[AR4-GigabitEthernet0/0/2]quit
[AR4]interface LoopBack 0
[AR4-LoopBack0]ip address 4.4.4.4 32
[AR4-LoopBack0]quit
```

AR5 的配置：

```
<Huawei>system-view
Enter system view, return user view with Ctrl+Z.
[Huawei]undo info-center enable
Info: Information center is disabled.
[Huawei]sysname AR5
[AR5]interface g0/0/0
[AR5-GigabitEthernet0/0/0]ip address 10.0.45.5 24
[AR5-GigabitEthernet0/0/0]quit
[AR5]interface LoopBack 0
[AR5-LoopBack0]ip address 5.5.5.5 32
[AR5-LoopBack0]quit
```

步骤 2：运行 OSPF。

AR1 的配置：

```
[AR1]ospf router-id 1.1.1.1
[AR1-ospf-1]area 0
[AR1-ospf-1-area-0.0.0.0]network 10.0.13.0 0.0.0.255
[AR1-ospf-1-area-0.0.0.0]network 10.0.12.0 0.0.0.255
[AR1-ospf-1-area-0.0.0.0]network 1.1.1.1 0.0.0.0
[AR1-ospf-1-area-0.0.0.0]quit
```

AR2 的配置：

```
[AR2]ospf router-id 2.2.2.2
[AR2-ospf-1]area 0
[AR2-ospf-1-area-0.0.0.0]net
[AR2-ospf-1-area-0.0.0.0]network 10.0.12.0 0.0.0.255
```

```
[AR2-ospf-1-area-0.0.0.0]network 2.2.2.2 0.0.0.0
[AR2-ospf-1-area-0.0.0.0]quit
[AR2-ospf-1]area 1
[AR2-ospf-1-area-0.0.0.1]network 10.0.24.0 0.0.0.255
[AR2-ospf-1-area-0.0.0.1]nssa
[AR2-ospf-1-area-0.0.0.1]quit
```

AR3 的配置：

```
[AR3]ospf router-id 3.3.3.3
[AR3-ospf-1]area 0
[AR3-ospf-1-area-0.0.0.0]network 10.0.13.0 0.0.0.255
[AR3-ospf-1-area-0.0.0.0]network 3.3.3.3 0.0.0.0 a
[AR3-ospf-1-area-0.0.0.0]network 3.3.3.3 0.0.0.0
[AR3-ospf-1-area-0.0.0.0]quit
[AR3-ospf-1]area 1
[AR3-ospf-1-area-0.0.0.1]network 10.0.34.0 0.0.0.255
[AR3-ospf-1-area-0.0.0.1]nssa
[AR3-ospf-1-area-0.0.0.1]quit
```

AR4 的配置：

```
[AR4]ospf router-id 4.4.4.4
[AR4-ospf-1]area 1
[AR4-ospf-1-area-0.0.0.1]network 10.0.24.0 0.0.0.255
[AR4-ospf-1-area-0.0.0.1]network 10.0.34.0 0.0.0.255
[AR4-ospf-1-area-0.0.0.1]network 10.0.45.0 0.0.0.255
[AR4-ospf-1-area-0.0.0.1]network 4.4.4.4 0.0.0.0
[AR4-ospf-1-area-0.0.0.1]nssa
[AR4-ospf-1-area-0.0.0.1]quit
```

AR5 的配置：

```
[AR5]ospf router-id 5.5.5.5
[AR5-ospf-1]area 1
[AR5-ospf-1-area-0.0.0.1]network 10.0.45.0 0.0.0.255
[AR5-ospf-1-area-0.0.0.1]network 5.5.5.5 0.0.0.0
[AR5-ospf-1-area-0.0.0.1]nssa
[AR5-ospf-1-area-0.0.0.1]quit
```

步骤 3: 实验调试。

在 AR1 上查看路由表：

```
[AR1]display ip routing-table
Route Flags: R - relay, D - download to fib
------------------------------------------------------------------------------
Routing Tables: Public
         Destinations : 18       Routes : 21
Destination/Mask    Proto   Pre  Cost     Flags  NextHop        Interface
        1.1.1.1/32  Direct  0    0        D      127.0.0.1      LoopBack0
        2.2.2.2/32  OSPF    10   1        D      10.0.12.2      GigabitEthernet0/0/1
        3.3.3.3/32  OSPF    10   1        D      10.0.13.3      GigabitEthernet0/0/0
        4.4.4.4/32  OSPF    10   2        D      10.0.12.2      GigabitEthernet0/0/1
```

```
                   OSPF    10   2      D     10.0.13.3    GigabitEthernet0/0/0
        5.5.5.5/32 OSPF    10   3      D     10.0.12.2    GigabitEthernet0/0/1
                   OSPF    10   3      D     10.0.13.3    GigabitEthernet0/0/0
      10.0.12.0/24 Direct  0    0      D     10.0.12.1    GigabitEthernet0/0/1
      10.0.12.1/32 Direct  0    0      D     127.0.0.1    GigabitEthernet0/0/1
    10.0.12.255/32 Direct  0    0      D     127.0.0.1    GigabitEthernet0/0/1
      10.0.13.0/24 Direct  0    0      D     10.0.13.1    GigabitEthernet0/0/0
      10.0.13.1/32 Direct  0    0      D     127.0.0.1    GigabitEthernet0/0/0
    10.0.13.255/32 Direct  0    0      D     127.0.0.1    GigabitEthernet0/0/0
      10.0.24.0/24 OSPF    10   2      D     10.0.12.2    GigabitEthernet0/0/1
      10.0.34.0/24 OSPF    10   2      D     10.0.13.3    GigabitEthernet0/0/0
      10.0.45.0/24 OSPF    10   3      D     10.0.12.2    GigabitEthernet0/0/1
                   OSPF    10   3      D     10.0.13.3    GigabitEthernet0/0/0
       127.0.0.0/8 Direct  0    0      D     127.0.0.1    InLoopBack0
      127.0.0.1/32 Direct  0    0      D     127.0.0.1    InLoopBack0
127.255.255.255/32 Direct  0    0      D     127.0.0.1    InLoopBack0
255.255.255.255/32 Direct  0    0      D     127.0.0.1    InLoopBack0
```

通过以上输出可以看出，到达 5.5.5.5 有两条路由，负载均衡。

修改 AR1 的开销：

```
[AR1]interface g0/0/0
[AR1-GigabitEthernet0/0/0]ospf cost 2
[AR1-GigabitEthernet0/0/0]quit
```

在 AR1 上再次查看路由表：

```
[AR1]display ip routing-table
Route Flags: R - relay, D - download to fib
------------------------------------------------------------------------------
Routing Tables: Public
        Destinations : 18       Routes : 19
Destination/Mask    Proto   Pre  Cost     Flags   NextHop        Interface
        1.1.1.1/32  Direct  0    0         D      127.0.0.1      LoopBack0
        2.2.2.2/32  OSPF    10   1         D      10.0.12.2      GigabitEthernet0/0/1
        3.3.3.3/32  OSPF    10   2         D      10.0.13.3      GigabitEthernet0/0/0
        4.4.4.4/32  OSPF    10   2         D      10.0.12.2      GigabitEthernet0/0/1
        5.5.5.5/32  OSPF    10   3         D      10.0.12.2      GigabitEthernet0/0/1
      10.0.12.0/24  Direct  0    0         D      10.0.12.1      GigabitEthernet0/0/1
      10.0.12.1/32  Direct  0    0         D      127.0.0.1      GigabitEthernet0/0/1
    10.0.12.255/32  Direct  0    0         D      127.0.0.1      GigabitEthernet0/0/1
      10.0.13.0/24  Direct  0    0         D      10.0.13.1      GigabitEthernet0/0/0
      10.0.13.1/32  Direct  0    0         D      127.0.0.1      GigabitEthernet0/0/0
    10.0.13.255/32  Direct  0    0         D      127.0.0.1      GigabitEthernet0/0/0
      10.0.24.0/24  OSPF    10   2         D      10.0.12.2      GigabitEthernet0/0/1
      10.0.34.0/24  OSPF    10   3         D      10.0.13.3      GigabitEthernet0/0/0
                    OSPF    10   3         D      10.0.12.2      GigabitEthernet0/0/1
      10.0.45.0/24  OSPF    10   3         D      10.0.12.2      GigabitEthernet0/0/1
       127.0.0.0/8  Direct  0    0         D      127.0.0.1      InLoopBack0
      127.0.0.1/32  Direct  0    0         D      127.0.0.1      InLoopBack0
```

```
127.255.255.255/32  Direct  0    0          D   127.0.0.1       InLoopBack0
255.255.255.255/32  Direct  0    0          D   127.0.0.1       InLoopBack0
```

通过以上输出可以看出，AR1 访问 AR5 的路径为 AR1 → AR2 → AR4 → AR5。

在 AR5 上创建一个环回口，引入 OSPF：

```
[AR5]interface LoopBack 100
[AR5-LoopBack100]ip address 100.100.100.100 32
[AR5-LoopBack100]quit
[AR5]ospf
[AR5-ospf-1]import-route direct
[AR5-ospf-1]quit
```

在 AR5 上查看 7 类 LSA：

```
[AR5]display ospf lsdb nssa 100.100.100.100
         OSPF Process 1 with Router ID 5.5.5.5
                         Area: 0.0.0.1
                 Link State Database
  Type        : NSSA
  Ls id       : 100.100.100.100
  Adv rtr     : 5.5.5.5
  Ls age      : 96
  Len         : 36
  Options     : NP
  seq#        : 80000001
  chksum      : 0xe22c
  Net mask    : 255.255.255.255
  TOS 0  Metric: 1
  E type      : 2
  Forwarding Address : 5.5.5.5   //FA 地址为 5.5.5.5
  Tag         : 1
  Priority    : Low
```

在 AR1 上查看 5 类 LSA：

```
[AR1]display ospf lsdb ase 100.100.100.100
         OSPF Process 1 with Router ID 1.1.1.1
                 Link State Database
  Type        : External
  Ls id       : 100.100.100.100
  Adv rtr     : 5.5.5.5
  Ls age      : 1050
  Len         : 36
  Options     : E
  seq#        : 80000001
  chksum      : 0x5ecc
  Net mask    : 255.255.255.255
  TOS 0  Metric: 1
  E type      : 2
  Forwarding Address : 0.0.0.0
  Tag         : 1
```

```
    Priority   : Medium

    Type       : External
    Ls id      : 100.100.100.100
    Adv rtr    : 3.3.3.3
    Ls age     : 223
    Len        : 36
    Options    : E
    seq#       : 80000001
    chksum     : 0x9589
    Net mask   : 255.255.255.255
    TOS 0  Metric: 1
    E type     : 2
    Forwarding Address : 5.5.5.5
    Tag        : 1
    Priority   : Medium
```

通过以上输出可以看出，由于AR3的Router ID比AR2大，因此它将执行7转5的动作，但是它的Forwarding Address为5.5.5.5。

在AR1上查看OSPF的路由表：

```
[AR1]display ip routing-table
Route Flags: R - relay, D - download to fib
------------------------------------------------------------------------------
Routing Tables: Public
         Destinations : 19       Routes : 20

Destination/Mask    Proto   Pre  Cost   Flags NextHop         Interface

        1.1.1.1/32  Direct  0    0        D   127.0.0.1       LoopBack0
        2.2.2.2/32  OSPF    10   1        D   10.0.12.2       GigabitEthernet0/0/1
        3.3.3.3/32  OSPF    10   2        D   10.0.13.3       GigabitEthernet0/0/0
        4.4.4.4/32  OSPF    10   2        D   10.0.12.2       GigabitEthernet0/0/1
        5.5.5.5/32  OSPF    10   3        D   10.0.12.2       GigabitEthernet0/0/1
      10.0.12.0/24  Direct  0    0        D   10.0.12.1       GigabitEthernet0/0/1
      10.0.12.1/32  Direct  0    0        D   127.0.0.1       GigabitEthernet0/0/1
    10.0.12.255/32  Direct  0    0        D   127.0.0.1       GigabitEthernet0/0/1
      10.0.13.0/24  Direct  0    0        D   10.0.13.1       GigabitEthernet0/0/0
      10.0.13.1/32  Direct  0    0        D   127.0.0.1       GigabitEthernet0/0/0
    10.0.13.255/32  Direct  0    0        D   127.0.0.1       GigabitEthernet0/0/0
      10.0.24.0/24  OSPF    10   2        D   10.0.12.2       GigabitEthernet0/0/1
      10.0.34.0/24  OSPF    10   3        D   10.0.13.3       GigabitEthernet0/0/0
                    OSPF    10   3        D   10.0.12.2       GigabitEthernet0/0/1
      10.0.45.0/24  OSPF    10   3        D   10.0.12.2       GigabitEthernet0/0/1
100.100.100.100/32  O_ASE   150  1        D   10.0.12.2       GigabitEthernet0/0/1
      127.0.0.0/8   Direct  0    0        D   127.0.0.1       InLoopBack0
      127.0.0.1/32  Direct  0    0        D   127.0.0.1       InLoopBack0
127.255.255.255/32  Direct  0    0        D   127.0.0.1       InLoopBack0
255.255.255.255/32  Direct  0    0        D   127.0.0.1       InLoopBack0
```

通过以上输出可以看出，AR1 访问 100.100.100.100 走的是 AR2，没有走 AR3，虽然 AR3 生成了 5 类 LSA，这是因为存在 FA 地址。

【思考】如果FA的地址为0.0.0.0，那么数据该怎么走?

如果FA的地址为0.0.0.0，那么AR1访问100.100.100.100的转发路径为AR1→AR3→AR4→AR5，此路径为次优路径。

1.5.7 OSPF GR

1. 实验目的

在设备之间运行 OSPF 协议，并且在 AR1 上配置 OSPF GR，观察实验现象。

2. 实验拓扑

OSPF GR 的实验拓扑如图 1-17 所示。

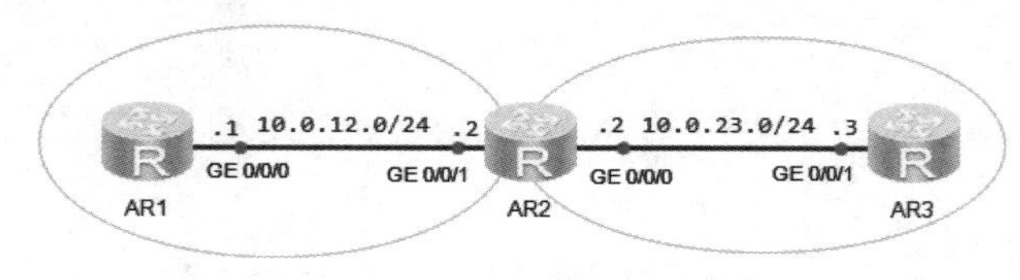

图 1-17　OSPF GR 的实验拓扑

3. 实验步骤

步骤 1：配置 IP 地址。

AR1 的配置：

```
<Huawei>system-view
Enter system view, return user view with Ctrl+Z.
[Huawei]undo info-center enable
Info: Information center is disabled.
[Huawei]sysname AR1
[AR1]interface g0/0/0
[AR1-GigabitEthernet0/0/0]ip address 10.0.12.1 24
[AR1-GigabitEthernet0/0/0]quit
[AR1]interface LoopBack 0
[AR1-LoopBack0]ip address 1.1.1.1 32
[AR1-LoopBack0]quit
```

AR2 的配置：

```
<Huawei>system-view
Enter system view, return user view with Ctrl+Z.
[Huawei]undo info-center enable
Info: Information center is disabled.
[Huawei]sysname AR2
[AR2]interface g0/0/1
[AR2-GigabitEthernet0/0/1]ip address 10.0.12.2 24
[AR2-GigabitEthernet0/0/1]quit
[AR2]interface g0/0/0
[AR2-GigabitEthernet0/0/0]ip address 10.0.23.2 24
```

```
[AR2-GigabitEthernet0/0/0]quit
[AR2]interface LoopBack 0
[AR2-LoopBack0]ip address 2.2.2.2 32
[AR2-LoopBack0]quit
```

AR3 的配置：

```
<Huawei>system-view
Enter system view, return user view with Ctrl+Z.
[Huawei]undo info-center enable
Info: Information center is disabled.
[Huawei]sysname AR3
[AR3]interface g0/0/1
[AR3-GigabitEthernet0/0/1]ip address 10.0.23.3 24
[AR3-GigabitEthernet0/0/1]quit
[AR3]interface LoopBack 0
[AR3-LoopBack0]ip address 3.3.3.3 32
[AR3-LoopBack0]quit
```

步骤 2：运行 OSPF。

AR1 的配置：

```
[AR1]ospf router-id 1.1.1.1
[AR1-ospf-1]area 0
[AR1-ospf-1-area-0.0.0.0]network 10.0.12.0 0.0.0.255
[AR1-ospf-1-area-0.0.0.0]network 1.1.1.1 0.0.0.0
[AR1-ospf-1-area-0.0.0.0]quit
```

AR2 的配置：

```
[AR2]ospf router-id 2.2.2.2
[AR2-ospf-1]area 0
[AR2-ospf-1-area-0.0.0.0]network 10.0.12.0 0.0.0.255
[AR2-ospf-1-area-0.0.0.0]network 2.2.2.2 0.0.0.0
[AR2-ospf-1-area-0.0.0.0]quit
[AR2-ospf-1]area 1
[AR2-ospf-1-area-0.0.0.1]network 10.0.23.0 0.0.0.255
[AR2-ospf-1-area-0.0.0.1]quit
```

AR3 的配置：

```
[AR3]ospf router-id 3.3.3.3
[AR3-ospf-1]area 1
[AR3-ospf-1-area-0.0.0.1]network 10.0.23.0 0.0.0.255
[AR3-ospf-1-area-0.0.0.1]network 3.3.3.3 0.0.0.0
[AR3-ospf-1-area-0.0.0.1]quit
```

步骤 3：开启 GR。

AR1 的配置：

```
[AR1]ospf 1
[AR1-ospf-1]opaque-capability enable   //使能 opaque-LSA 特性
[AR1-ospf-1]graceful-restart           //使能 OSPF GR 特性
[AR1-ospf-1]quit
```

AR2 的配置：

```
[AR2]ospf
[AR2-ospf-1]opaque-capability enable
[AR2-ospf-1]graceful-restart
[AR2-ospf-1]quit
```

AR3 的配置：

```
[AR3]ospf
[AR3-ospf-1]opaque-capability enable
[AR3-ospf-1]graceful-restart
[AR3-ospf-1]quit
```

步骤 4: 实验调试。

在 AR1 上查看路由器上 OSPF GR 的状态：

```
[AR1]display ospf 1 graceful-restart
         OSPF Process 1 with Router ID 1.1.1.1
 Graceful-restart capability      : enabled
 Graceful-restart support         : planned and un-planned, totally
 Helper-policy support            : planned and un-planned, strict lsa check
 Current GR state                 : normal
 Graceful-restart period          : 120 seconds
 Number of neighbors under helper:
  Normal neighbors     : 0
  Virtual neighbors    : 0
  Sham-link neighbors : 0
  Total neighbors      : 0
 Number of restarting neighbors : 0
 Last exit reason:
  On graceful restart : none
  On Helper           : none
```

通过以上输出可以看出，OSPF 的 GR 状态为 enabled。

在 AR1 用户视图下执行 reset ospf 1 process graceful-restart，重启 OSPF 进程 1：

```
<AR1>reset ospf 1 process graceful-restart
```

同时在 AR2 上执行 display ospf 1 peer，查看与 AR1 的 OSPF 邻接关系：

```
[AR2]display ospf 1 peer
         OSPF Process 1 with Router ID 2.2.2.2
                 Neighbors
 Area 0.0.0.0 interface 10.0.12.2(GigabitEthernet0/0/1)'s neighbors
 Router ID: 1.1.1.1          Address: 10.0.12.1        GR State: Normal
   State: Full  Mode:Nbr is  Slave  Priority: 1
   DR: 10.0.12.1  BDR: 10.0.12.2  MTU: 0
   Dead timer due in 37  sec
   Retrans timer interval: 5
   Neighbor is up for 00:00:19
   Authentication Sequence: [ 0 ]
                 Neighbors
```

```
 Area 0.0.0.1 interface 10.0.23.2(GigabitEthernet0/0/0)'s neighbors
 Router ID: 10.0.23.3        Address: 10.0.23.3        GR State: Normal
   State: Full  Mode:Nbr is  Master  Priority: 1
   DR: 10.0.23.2  BDR: 10.0.23.3  MTU: 0
   Dead timer due in 35  sec
   Retrans timer interval: 5
   Neighbor is up for 00:06:30
   Authentication Sequence: [ 0 ]
```

通过以上输出可以看出，OSPF 的邻居状态为 Full，GR State 为 Normal。

查看路由器上 OSPF GR 的状态：AR1 正常退出 GR；AR2 正常退出 Helper。

```
<AR1>display ospf 1 graceful-restart
        OSPF Process 1 with Router ID 1.1.1.1
 Graceful-restart capability     : enabled
 Graceful-restart support        : planned and un-planned, totally
 Helper-policy support           : planned and un-planned, strict lsa check
 Current GR state                : normal
 Graceful-restart period         : 120 seconds
 Number of neighbors under helper:
  Normal neighbors     : 0
  Virtual neighbors    : 0
  Sham-link neighbors  : 0
  Total neighbors      : 0
 Number of restarting neighbors : 0
 Last exit reason:
  On graceful restart : successful exit
  On Helper           : none
[AR2]display ospf 1 graceful-restart
        OSPF Process 1 with Router ID 2.2.2.2
 Graceful-restart capability     : enabled
 Graceful-restart support        : planned and un-planned, totally
 Helper-policy support           : planned and un-planned, strict lsa check
 Current GR state                : normal
 Graceful-restart period         : 120 seconds
 Number of neighbors under helper:
  Normal neighbors     : 0
  Virtual neighbors    : 0
  Sham-link neighbors  : 0
  Total neighbors      : 0
 Number of restarting neighbors : 0
 Last exit reason:
  On graceful restart : none
  On Helper           : successful exit
```

在 AR1 的接口抓包。

AR1 发送给 224.0.0.5 的 9 类 LSA 如图 1-18 所示。

```
 7 26.656000    10.0.12.1          224.0.0.5     …   106 LS Update

> Frame 7: 106 bytes on wire (848 bits), 106 bytes captured (848 bits) on interface 0
> Ethernet II, Src: HuaweiTe_95:2a:48 (00:e0:fc:95:2a:48), Dst: IPv4mcast_05 (01:00:5e:00:00:05)
> Internet Protocol Version 4, Src: 10.0.12.1, Dst: 224.0.0.5
v Open Shortest Path First
  > OSPF Header
  v LS Update Packet
      Number of LSAs: 1
    v LSA-type 9 (Opaque LSA, Link-local scope), len 44
        .000 0000 0000 0001 = LS Age (seconds): 1
        0... .... .... .... = Do Not Age Flag: 0
      > Options: 0x02, (E) External Routing
        LS Type: Opaque LSA, Link-local scope (9) LS的类型为不透明LSA
        Link State ID Opaque Type: grace-LSA (3)
        Link State ID Opaque ID: 0
        Advertising Router: 1.1.1.1
        Sequence Number: 0x80000001
        Checksum: 0xbae5
        Length: 44
      > Grace Period: 120 seconds 保持时间为120s
      > Restart Reason: Software Restart (1) 重启原因：软件重启
      > Restart IP: 10.0.12.1 重启的IP地址为10.0.12.1
```

图 1–18　AR1 发送给 224.0.0.5 的 9 类 LSA

AR2 回应的 LSACK 如图 1–19 所示。

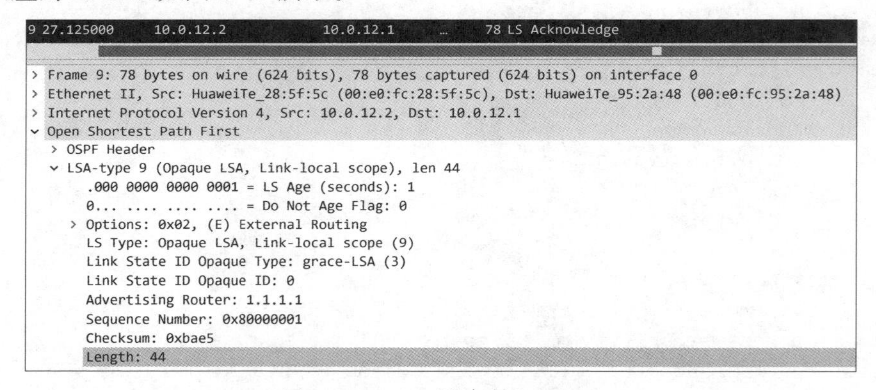

图 1–19　AR2 回应的 LSACK

第 2 章

IS-IS 的高级特性

2.1 IS-IS 的高级特性概述

IS-IS 的高级特性包括 IS-IS 的快速收敛、路由控制以及其他特性。IS-IS 的快速收敛是为了提高路由的收敛速度而产生的扩展特性，包括 I-SPF（Incremental SPF，增量最短路径优先算法）、PRC、智能定时器、LSP（Label Switched Path，标签交换路径）快速扩散。为了实现优化 IS-IS 网络和便于流量管理，需要对网络中的路由进行更加精确的控制。

2.2 IS-IS 的快速收敛

2.2.1 I-SPF

（1）I-SPF 的工作原理。当网络拓扑改变时，SPF 只对受影响的节点进行路由计算，而不是对全部节点重新进行路由计算，从而加快了路由的计算。

（2）I-SPF 算法的过程如图 2-1 所示。

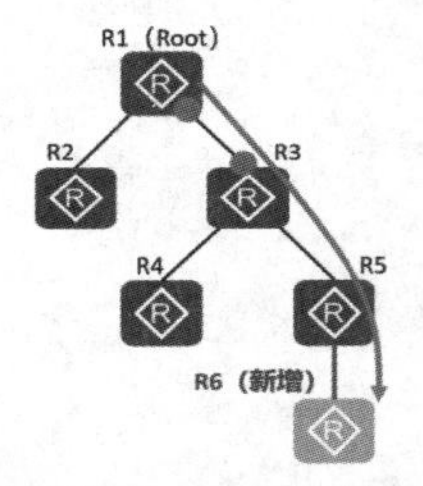

图 2-1 I-SPF 算法的过程

2.2.2 LSP 快速扩散

（1）正常情况下，当 IS-IS 路由器收到其他路由器发来的 LSP 时，如果此 LSP 比本地 LSDB 中相应的 LSP 要新，则更新 LSDB 中的 LSP，并用一个定时器定期将 LSDB 内已更新的 LSP 扩散出去。

（2）LSP 快速扩散特性改进了这种方式，使能了此特性的设备收到一个或多个较新的 LSP 时，在路由计算之前，先将小于指定数目的 LSP 扩散出去，加快 LSDB 的同步过程。这种方式在很大程度上可以提高整个网络的收敛速度。

（3）配置 LSP 快速扩散。

[Huawei-isis-1] flash-flood [lsp-count | max-timer-interval interval | [level-1 | level-2]]。参数详情见表 2-1。

表 2-1 配置 LSP 快速扩散

参 数	说 明
lsp-count	指定每个接口一次扩散 LSP 的最大数量，取值范围为 1 ~ 15，默认为 5
max-timer-interval interval	指定 LSP 扩散的最大时间间隔。整数形式，取值范围为 10 ~ 50000，单位为 ms，默认值为 10 ms

2.3　IS-IS 的路由控制

2.3.1　等价路由

（1）当 IS-IS 网络中有多条冗余链路时，可能会出现多条等价路由，此时可以采取两种方式。

①配置负载分担。流量会被均匀分配到每条链路上。

②配置等价路由优先级。针对等价路由中的每一条路由，明确指定其优先级，优先级高的路由将被优选，优先级低的路由可以作为备用链路。

（2）配置 IS-IS 对等价路由的处理方式。

①配置 IS-IS 路由负载分担：

```
[Huawei-isis-1]maximum load-balancing number    //最大负载均衡的链路数量
```

②配置 IS-IS 等价路由的优先级：

```
[Huawei-isis-1]nexthop ip-address weight value  //下一跳地址的权重范围为1~254
```

（3）当组网中存在的等价路由数量大于 maximum load-balancing 命令配置的等价路由数量时，按照以下原则选取有效路由进行负载分担。

①路由优先级：选取优先级低（优先级高）的路由进行负载分担。

②下一跳设备的 System ID：如果路由的优先级相同，则比较下一跳设备的 System ID，选取 System ID 小的路由进行负载分担。

③本地设备出接口索引：如果路由优先级和下一跳设备的 System ID 都相同，则比较出接口的接口索引，选取接口索引较小的路由进行负载分担。

2.3.2　默认路由

（1）在 IS-IS 中，主要通过以下 3 种方式控制默认路由的生成和发布。

①在 Level-1-2 设备上，控制其产生的 Level-1 LSP 中 ATT 位的置位情况。

②在 Level-1 设备上，通过配置使其即使收到 ATT 位置位的 Level-1 LSP，也不会自动产生默认路由。

③在 IS-IS 中发布默认路由。

（2）默认路由生成的配置方法。

①在 Level-1-2 设备上，设置 IS-IS LSP 报文的 ATT 比特位置位规则：

```
[Huawei-isis-1]attached-bit advertise {always | never} //只需在L1-2上配置
```

②在 Level-1 设备上，控制 Level-1 设备不因为 ATT 位下发默认路由到路由表：

```
[Huawei-isis-1]attached-bit avoid-learning            //要在L1-2和L-1上都配置
```

③配置运行 IS-IS 的设备生成默认路由：

```
[Huawei-isis-1]default-route-advertise [always|match default|route-policy route-
policy-name][cost cost|tag tag|[level-1|level-1-2|level-2]][avoid-learning]
```

2.4　IS-IS 的其他特性

2.4.1　IS-IS 多实例和多进程

（1）IS-IS 多实例是指在同一台路由器上可以配置多个 VPN 实例与多个 IS-IS 进程相关联。

（2）IS-IS 多进程是指在同一个 VPN 实例下（或者同在公网下）可以创建多个 IS-IS 进程，每个进程之间互不影响，彼此独立。不同进程之间的路由交互相当于不同路由协议之间的路由交互。

2.4.2 LSP 分片

（1）LSP 分片的原因。当 IS-IS 要发布的 PDU 中的信息量太大时，IS-IS 路由器将会生成多个 LSP 分片，用于携带更多的 IS-IS 信息。

（2）LSP 的报文格式如图 2-2 所示。

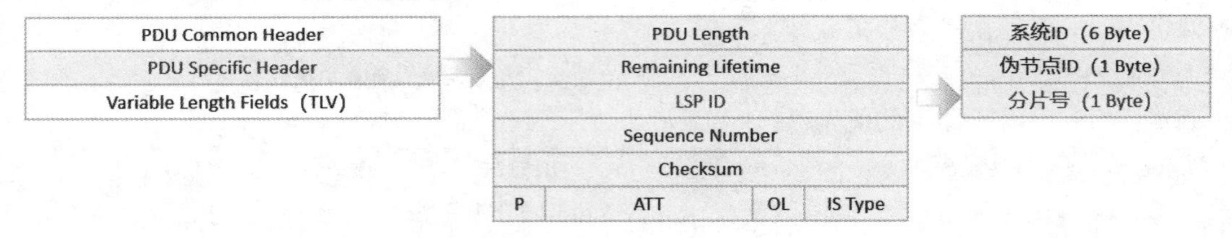

图 2-2　LSP 的报文格式

（3）LSP 分片扩展的基本术语。

①初始系统（Originating System）：初始系统是实际运行 IS-IS 的路由器。允许一个单独的 IS-IS 进程像多个虚拟路由器一样发布 LSP，而初始系统是指那个“真正”的 IS-IS 进程。

②系统 ID（Normal System-ID）：初始系统的系统 ID。

③虚拟系统（Virtual System）：由附加系统 ID（Additional System-ID）标识的系统，生成扩展 LSP 分片。这些分片在其 LSP ID 中携带附加系统 ID。

④附加系统 ID：虚拟系统的系统 ID，由网络管理器统一分配。每个附加系统 ID 都允许生成 256 个扩展的 LSP 分片。

⑤ 24 号 TLV（IS Alias ID TLV）：LSP 分片携带该 TLV 信息，用于表示初始系统与虚拟系统的关系。

（4）LSP 分片扩展的工作原理。

①在 IS-IS 中，每个系统 ID 都标识一个系统，每个系统都最多可生成 256 个 LSP 分片。通过增加附加系统 ID，可以最多配置 50 个虚拟系统，从而使 IS-IS 进程最多可生成 13 056 个 LSP 分片。

② IS-IS 路由器可以在两种模式下运行 LSP 分片扩展特性。

- mode-1：用于网络中部分路由器不支持 LSP 分片扩展特性的情况。LSP 分片扩展 mode-1 如图 2-3 所示。

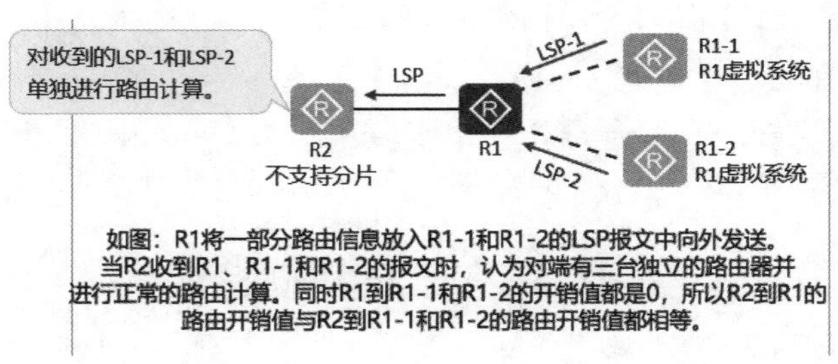

图 2-3　LSP 分片扩展 mode-1

● mode-2：用于网络中所有路由器都支持 LSP 分片扩展特性的情况。LSP 分片扩展 mode-2 如图 2-4 所示。

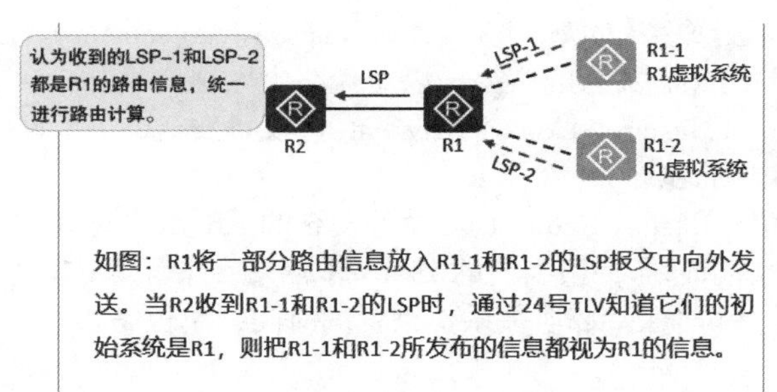

图 2-4　LSP 分片扩展 mode-2

（5）LSP 分片扩展的基本配置命令。

①使能 IS-IS 进程的 LSP 分片扩展。

```
[Huawei-isis-1]lsp-fragments-extend [ [ level-1 | level-2 | level-1-2 ]| [ mode-1 |
mode-2 ]]// 默认为 mode-1 和 level-1-2
```

②配置一个虚拟系统。

```
[Huawei-isis-1]virtual-system virtual-system-id  // 指定 IS-IS 进程的虚拟系统 ID
```

需要注意的是，以上两条命令是配合使用的，只有使能了 LSP 分片扩展，并用 reset isis all 命令重启了 IS-IS 进程后，配置的虚拟系统 ID 才会生效。

2.4.3　IS-IS GR

1.IS-IS GR 简介

IS-IS GR 是一种支持 GR 能力的高可靠性技术，可以实现数据的不间断转发。设备发生主备倒换后，由于没有保存主备倒换前的邻居信息，因此一开始发送的 Hello 报文中不包含邻居列表。此时邻居设备收到后检查邻居关系，发现在重启设备的 Hello 报文的邻居列表中没有自己，这样邻居关系将会断掉。同时，邻居设备通过生成新的 LSP 报文，将拓扑变化的信息泛洪给区域内的其他设备。区域内的其他设备会基于新的链路状态数据库进行路由计算，从而造成路由中断或者路由震荡。IETF（The Internet Engineering Task Force，国际互联网工程任务组）针对这种情况为 IS-IS 制定了 GR 规范，避免协议重启带来的路由震荡和流量转发中断的现象。

2. 基本概念

（1）GR-Restarter：具备 GR 能力且要进行 GR 的设备。

（2）GR-Helper：具备 GR 能力，辅助 GR 设备完成 GR 功能的设备。GR-Restarter 一定具有 GR-Helper 的能力。

（3）IS-IS GR 过程由 GR-Restarter 和 GR-Helper 配合完成。

3. Restart TLV

Restart TLV 是包含在 IIH（IS-to-IS Hello PDUs）报文中的扩展部分。支持 IS-IS GR 能力的设备的所有 IIH 报文都包含 Restart TLV。Restart TLV 中携带了协议重启的一些参数。其报文格式如图 2-5 所示。

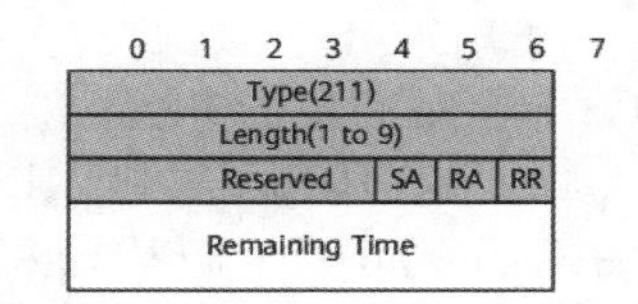

图 2-5　Restart TLV 的报文格式

（1）Type：TLV 的类型，值为 211，表示 Restart TLV。

（2）Length：TLV 值的长度。

（3）SA：抑制发布邻居关系位（Suppress Adjacency Advertisement）。用于发生 Starting 的设备请求邻居抑制与自己相关的邻居关系的广播，以避免路由黑洞。

（4）RA：重启应答位（Restart Acknowledgement）。设备发送的 RA 置位的 Hello 报文用于通告邻居确认收到了 RR 置位的报文。

（5）RR：重启请求位（Restart Request）。设备发送的 RR 置位的 Hello 报文用于通告邻居自己发生 Restarting/Starting，请求邻居保留当前的 IS–IS 邻居关系并返回 CSNP 报文。

（6）Remaining Time：邻居保持邻居关系不重置的时间，单位是秒。

4. 定时器

在 IS–IS 的 GR 能力扩展中引入了三个定时器，分别是 T1 定时器、T2 定时器和 T3 定时器。

（1）T1 定时器：如果 GR Restarter 已发送 RR 置位的 IIH 报文，但直到 T1 定时器超时还没有收到 GR Helper 的包含 Restart TLV 且 RA 置位的 IIH 报文的确认消息时，会重置 T1 定时器并继续发送包含 Restart TLV 的 IIH 报文。当收到确认报文或者 T1 定时器已超时 3 次时，取消 T1 定时器。使能了 IS–IS GR 特性的进程，在每个接口都会维护一个 T1 定时器。在 Level–1–2 路由设备上，广播网接口为每个 Level 维护一个 T1 定时器，默认为 3 s。

（2）T2 定时器：GR Restarter 从重启开始到本 Level 所有设备 LSDB 完成同步的时间。T2 定时器是系统等待各层 LSDB 同步的最长时间。Level–1 和 Level–2 的 LSDB 各维护一个 T2 定时器，默认为 60 s。

（3）T3 定时器：GR Restarter 成功完成 GR 所允许的最大时间。T3 定时器的初始值为 65535 s，但在收到邻居回应的 RA 置位的 IIH 报文后，取值会变为各个 IIH 报文的 Remaining Time 字段值中的最小者。T3 定时器超时表示 GR 失败，整个系统维护一个 T3 定时器，默认为 300 s。

5. IS–IS Restarting 过程

主备倒换或重启 IS–IS 进程触发的 GR 过程称为 Restarting，转发表保持不变。IS–IS Restarting 过程如图 2–6 所示。

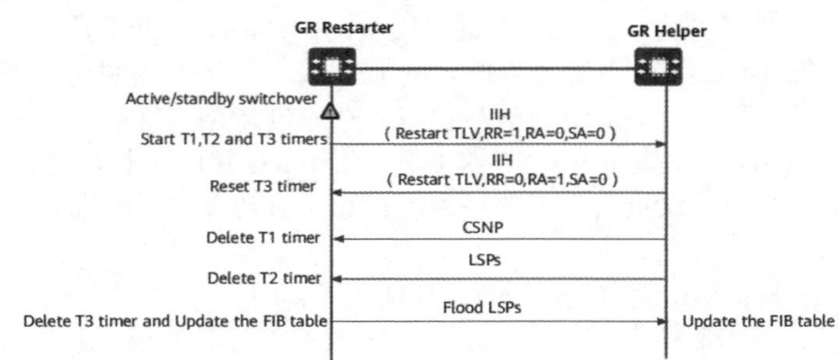

图 2–6　IS–IS Restarting 过程

（1）GR Restarter 进行主备倒换后，GR Restarter 启动 T1、T2 和 T3 定时器，从所有接口发送包含 Restart TLV 的 IIH 报文，其中 RR 置位，RA 和 SA 位清除。

（2）GR Helper 收到 IIH 报文以后，GR Helper 维持邻居关系，刷新当前的 Holdtime，回送一个包含 Restart TLV 的 IIH 报文（RR 清除，RA 置位，Remaining Time 是从现在到 Holdtime 超时的时间间隔），发送 CSNP 报文和所有 LSP 报文给 GR Restarter。

（3）GR Restarter 接收到邻居的 IIH 回应报文后（RR 清除、RA 置位），进行如下处理：

①把 T3 的当前值和报文中的 Remaining Time 比较，取其中较小者作为 T3 的值。

②如果在接口收到确认报文和 CSNP 报文之后，则取消该接口的 T1 定时器。

③如果该接口没有收到确认报文和 CSNP 报文，T1 会不停地重置，重发含 Restart TLV 的 IIH 报文。如果 T1 超时次数超过阈值，则 GR Restarter 强制取消 T1 定时器，启动正常的 IS-IS 处理流程。

（4）当 GR Restarter 所有接口上的 T1 定时器都取消，CSNP 列表清空且收集全所有的 LSP 报文后，可以认为和所有的邻居都完成了同步，取消 T2 定时器。

（5）T2 定时器被取消，表示本 Level 的 LSDB 已经同步。如果是单 Level 系统，则直接触发 SPF 计算。如果是 Level-1-2 系统，此时判断另一个 Level 的 T2 定时器是否也被取消。如果两个 Level 的 T2 定时器都被取消，则触发 SPF 计算；否则等待另一个 Level 的 T2 定时器超时。

（6）各层的 T2 定时器都被取消后，GR Restarter 取消 T3 定时器，更新 FIB 表。GR Restarter 可以重新生成各层的 LSP 并泛洪，在同步过程中收到的自己重启前生成的 LSP 此时也可以被删除。

6. IS-IS Starting 过程

对于 Starting 设备，因为没有保留 FIB 表项，所以一方面希望在 Starting 之前重置与邻居关系为 Up 的邻居的关系，同时希望邻居能在一段时间内抑制和自己的邻居关系的发布。其处理过程与 Restarting 不同，如图 2-7 所示。

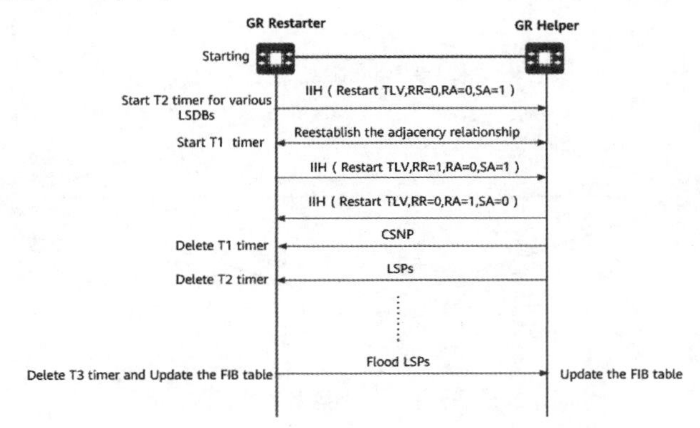

图 2-7　IS-IS Starting 过程

（1）GR Restarter Starting 后，为每层 LSDB 的同步启动 T2 定时器，从各个接口发送携带 Restart TLV 的 IIH 报文，其中 RR 位清除，SA 位置位。

（2）邻居收到携带 Restart TLV 的 IIH 报文后，根据设备是否支持 GR，进行如下处理。

①支持 GR：重新初始化邻居关系。在发送的 LSP 中取消和 GR Restarter 邻居关系的描述，进行 SPF 计算时也不考虑和 GR Restarter 相连的链路，直到收到 SA 位清除的 IIH 为止。

②不支持 GR：邻居忽略 Restart TLV，重置和 GR Restarter 之间的邻居关系。回应一个不含 Restart TLV 的 IIH 报文，转入正常的 IS-IS 处理流程。这时不会抑制和 GR Restarter 的邻居关系的发布。在点到点链路上，还会发送一个 CSNP 报文。

（3）邻居关系重新初始化之后，在每个接口上 GR Restarter 都和邻居重建邻居关系。当有一个邻居关系到达 Up 状态后，GR Restarter 为该接口启动 T1 定时器。

（4）在 T1 定时器超时后，GR Restarter 发送 RR 置位、SA 置位的 IIH 报文。

（5）邻居收到 RR 置位和 SA 置位的 IIH 报文后，发送一个 RR 清除、RA 置位的 IIH 报文作为确认报文，并发送 CSNP 报文。

（6）GR Restarter 收到邻居的 IIH 确认报文和 CSNP 报文以后，取消 T1 定时器。如果没有收到 IIH 报文或者 CSNP 报文，就不停重置 T1 定时器，重发 RR 置位、SA 置位的 IIH 报文。如果 T1 超时次数超过阈值，GR Restarter 就强制取消 T1 定时器，进入正常的 IS-IS 处理流程完成 LSDB 同步。

（7）GR Restarter 收到 Helper 端的 CSNP 以后，开始同步 LSDB。

（8）本 Level 的 LSDB 同步完成后，GR Restarter 取消 T2 定时器。

（9）所有的 T2 定时器都被取消以后，启动 SPF 计算，重新生成 LSP 并泛洪。

2.5 IS-IS 实验

2.5.1 IS-IS 等价路由

1. 实验目的

所有设备配置 IS-IS 协议，并且建立 IS-IS 邻居关系，在 AR1 上配置等价路由的最大数量为 2，并且通过配置负载分担优先级，实现 AR1 访问 AR5，使用 AR2 作为下一跳。

2. 实验拓扑

IS-IS 等价路由的实验拓扑如图 2-8 所示。

扫一扫，看视频

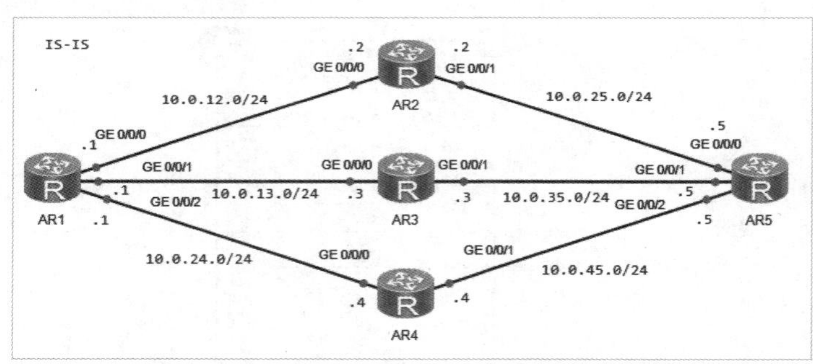

图 2-8 IS-IS 等价路由的实验拓扑

3. 实验步骤

步骤 1： 配置 IP 地址。

AR1 的配置：

```
<Huawei>system-view
Enter system view, return user view with Ctrl+Z.
[Huawei]undo info-center enable
Info: Information center is disabled.
[Huawei]sysname AR1
[AR1]interface g0/0/0
[AR1-GigabitEthernet0/0/0]ip address 10.0.12.1 24
[AR1-GigabitEthernet0/0/0]quit
```

```
[AR1]interface g0/0/1
[AR1-GigabitEthernet0/0/1]ip address 10.0.13.1 24
[AR1-GigabitEthernet0/0/1]quit
[AR1]interface g0/0/2
[AR1-GigabitEthernet0/0/2]ip address 10.0.24.1 24
[AR1-GigabitEthernet0/0/2]quit
[AR1]interface LoopBack 0
[AR1-LoopBack0]ip address 1.1.1.1 32
[AR1-LoopBack0]quit
```

AR2 的配置：

```
<Huawei>system-view
Enter system view, return user view with Ctrl+Z.
[Huawei]undo info-center enable
Info: Information center is disabled.
[Huawei]sysname AR2
[AR2]interface g0/0/0
[AR2-GigabitEthernet0/0/0]ip address 10.0.12.2 24
[AR2-GigabitEthernet0/0/0]quit
[AR2]interface g0/0/1
[AR2-GigabitEthernet0/0/1]ip address 10.0.25.2 24
[AR2-GigabitEthernet0/0/1]quit
[AR2]interface LoopBack 0
[AR2-LoopBack0]ip address 2.2.2.2 32
[AR2-LoopBack0]quit
```

AR3 的配置：

```
<Huawei>system-view
Enter system view, return user view with Ctrl+Z.
[Huawei]undo info-center enable
Info: Information center is disabled.
[Huawei]sysname AR3
[AR3]interface g0/0/0
[AR3-GigabitEthernet0/0/0]ip address 10.0.13.3 24
[AR3-GigabitEthernet0/0/0]quit
[AR3]interface g0/0/1
[AR3-GigabitEthernet0/0/1]ip address 10.0.35.3 24
[AR3-GigabitEthernet0/0/1]quit
[AR3]interface LoopBack 0
[AR3-LoopBack0]ip address 3.3.3.3 32
[AR3-LoopBack0]quit
```

AR4 的配置：

```
<Huawei>system-view
Enter system view, return user view with Ctrl+Z.
[Huawei]undo info-center enable
Info: Information center is disabled.
[Huawei]sysname AR4
```

```
[AR4]interface g0/0/0
[AR4-GigabitEthernet0/0/0]ip address 10.0.24.4 24
[AR4-GigabitEthernet0/0/0]quit
[AR4]interface g0/0/1
[AR4-GigabitEthernet0/0/1]ip address 10.0.45.4 24
[AR4-GigabitEthernet0/0/1]quit
[AR4]interface LoopBack 0
[AR4-LoopBack0]ip address 4.4.4.4 32
[AR4-LoopBack0]quit
```

AR5 的配置：

```
<Huawei>system-view
Enter system view, return user view with Ctrl+Z.
[Huawei]undo info-center enable
Info: Information center is disabled.
[Huawei]sysname AR5
[AR5]interface g0/0/0
[AR5-GigabitEthernet0/0/0]ip address 10.0.25.5 24
[AR5-GigabitEthernet0/0/0]quit
[AR5]interface g0/0/1
[AR5-GigabitEthernet0/0/1]ip address 10.0.35.5 24
[AR5-GigabitEthernet0/0/1]quit
[AR5]interface g0/0/2
[AR5-GigabitEthernet0/0/2]ip address 10.0.45.5 24
[AR5-GigabitEthernet0/0/2]quit
[AR5]interface LoopBack 0
[AR5-LoopBack0]ip address 5.5.5.5 32
[AR5-LoopBack0]quit
```

步骤 2：运行 IS-IS。

AR1 的配置：

```
[AR1]isis
[AR1-isis-1]network-entity 49.0001.0000.0000.0001.00
[AR1-isis-1]is-level level-2
[AR1-isis-1]quit
[AR1]interface g0/0/0
[AR1-GigabitEthernet0/0/0]isi enable
[AR1-GigabitEthernet0/0/0]quit
[AR1]interface g0/0/1
[AR1-GigabitEthernet0/0/1]isis enable
[AR1-GigabitEthernet0/0/1]quit
[AR1]interface g0/0/2
[AR1-GigabitEthernet0/0/2]isis enable
[AR1-GigabitEthernet0/0/2]quit
[AR1]interface LoopBack 0
[AR1-LoopBack0]isis enable
[AR1-LoopBack0]quit
```

AR2 的配置：

```
[AR2]isis
[AR2-isis-1]network-entity 49.0001.0000.0000.0002.00
[AR2-isis-1]is-level level-2
[AR2-isis-1]quit
[AR2]interface g0/0/0
[AR2-GigabitEthernet0/0/0]isis enable
[AR2-GigabitEthernet0/0/0]quit
[AR2]interface g0/0/1
[AR2-GigabitEthernet0/0/1]isis enable
[AR2-GigabitEthernet0/0/1]quit
[AR2]interface LoopBack 0
[AR2-LoopBack0]isis enable
[AR2-LoopBack0]quit
```

AR3 的配置：

```
[AR3]isis
[AR3-isis-1]network-entity 49.0001.0000.0000.0003.00
[AR3-isis-1]is-level level-2
[AR3-isis-1]quit
[AR3]interface g0/0/0
[AR3-GigabitEthernet0/0/0]isis enable
[AR3-GigabitEthernet0/0/0]quit
[AR3]interface g0/0/1
[AR3-GigabitEthernet0/0/1]isis enable
[AR3-GigabitEthernet0/0/1]quit
[AR3]interface LoopBack 0
[AR3-LoopBack0]isis enable
[AR3-LoopBack0]quit
```

AR4 的配置：

```
[AR4]isis
[AR4-isis-1]network-entity 49.0001.0000.0000.0004.00
[AR4-isis-1]is-level level-2
[AR4-isis-1]quit
[AR4]interface g0/0/0
[AR4-GigabitEthernet0/0/0]isis enable
[AR4-GigabitEthernet0/0/0]quit
[AR4]interface g0/0/1
[AR4-GigabitEthernet0/0/1]isis enable
[AR4-GigabitEthernet0/0/1]quit
[AR4]interface LoopBack 0
[AR4-LoopBack0]isis enable
[AR4-LoopBack0]quit
```

AR5 的配置：

```
[AR5]isis
[AR5-isis-1]network-entity 49.0001.0000.0000.0005.00
```

```
[AR5-isis-1]is-level level-2
Info: IS Level Changed, Resetting ISIS...
[AR5-isis-1]quit
[AR5]interface g0/0/0
[AR5-GigabitEthernet0/0/0]isis enable
[AR5-GigabitEthernet0/0/0]quit
[AR5]interface g0/0/1
[AR5-GigabitEthernet0/0/1]isis enable
[AR5-GigabitEthernet0/0/1]quit
[AR5]interface g0/0/2
[AR5-GigabitEthernet0/0/2]isis enable
[AR5-GigabitEthernet0/0/2]quit
[AR5]interface LoopBack 0
[AR5-LoopBack0]isis enable
[AR5-LoopBack0]quit
```

步骤 3: 实验调试。

在 AR1 上查看路由：

```
<AR1>display ip routing-table
Route Flags: R - relay, D - download to fib
------------------------------------------------------------------------------
Routing Tables: Public
         Destinations : 21        Routes : 23
Destination/Mask    Proto     Pre   Cost   Flags  NextHop       Interface
       1.1.1.1/32   Direct    0     0      D      127.0.0.1     LoopBack0
       2.2.2.2/32   ISIS-L2   15    10     D      10.0.12.2     GigabitEthernet0/0/0
       3.3.3.3/32   ISIS-L2   15    10     D      10.0.13.3     GigabitEthernet0/0/1
       4.4.4.4/32   ISIS-L2   15    10     D      10.0.24.4     GigabitEthernet0/0/2
       5.5.5.5/32   ISIS-L2   15    20     D      10.0.12.2     GigabitEthernet0/0/0
                    ISIS-L2   15    20     D      10.0.13.3     GigabitEthernet0/0/1
                    ISIS-L2   15    20     D      10.0.24.4     GigabitEthernet0/0/2
     10.0.12.0/24   Direct    0     0      D      10.0.12.1     GigabitEthernet0/0/0
```

通过以上输出可以看出，AR1 访问 5.5.5.5 有 3 条路径。

通过命令让 AR1 访问 5.5.5.5 只有两条路径：

```
[AR1]isis
[AR1-isis-1]maximum load-balancing 2  //最大负载均衡为两条
[AR1-isis-1]quit
```

在 AR1 上查看关于 5.5.5.5 的路由表：

```
[AR1]display ip routing-table 5.5.5.5
Route Flags: R - relay, D - download to fib
------------------------------------------------------------------------------
Routing Table : Public
Summary Count : 2
Destination/Mask    Proto     Pre   Cost   Flags  NextHop       Interface
       5.5.5.5/32   ISIS-L2   15    20     D      10.0.12.2     GigabitEthernet0/0/0
                    ISIS-L2   15    20     D      10.0.13.3     GigabitEthernet0/0/1
```

【思考】为什么选择AR2和AR3，没有选择AR4?

当组网中存在的等价路由数量大于maximum load-balancing命令配置的等价路由数量时，按照以下原则选取有效路由进行负载分担。

（1）路由优先级：选取优先级低（优先级高）的路由进行负载分担。

（2）下一跳设备的System ID：如果路由的优先级相同，则比较下一跳设备的System ID，选取System ID小的路由进行负载分担。

（3）本地设备出接口索引：如果路由优先级和下一跳设备的System ID都相同，则比较出接口的接口索引，选取接口索引较小的路由进行负载分担。

因为三条路径的路由优先级都相同（取值范围为1~254），所以比较第2条：下一跳的System ID，因为AR2的System ID为0000.0000.0002，AR3的System ID为0000.0000.0003，AR4的System ID为0000.0000.0004，所以选择了AR2和AR3，没有选择AR4。

在 AR1 上配置路由负载分担优先级，让数据走 AR2：

```
[AR1]ISIS
[AR1-isis-1]nexthop 10.0.12.2 weight 10
[AR1-isis-1]nexthop 10.0.13.3 weight 20
[AR1-isis-1]quit
```

【技术要点】weight

（1）值越小越优，取值范围为1~254。

（2）谁最小就优选谁，不能负载均衡。

（3）该方式可以在不修改原有配置的基础上指定某条路由被优选，便于业务的管理，同时提高网络的可靠性。

再次在 AR1 上查看关于 5.5.5.5 路由表：

```
[AR1]display ip routing-table 5.5.5.5
Route Flags: R - relay, D - download to fib
------------------------------------------------------------------------------
Routing Table : Public
Summary Count : 1
Destination/Mask    Proto    Pre  Cost  Flags  NextHop     Interface
       5.5.5.5/32   ISIS-L2  15   20    D      10.0.12.2   GigabitEthernet0/0/0
```

通过以上输出可以看出实验成功。

2.5.2　IS-IS 默认路由

1. 实验目的

全网运行 IS-IS，AR1 为 Level2 设备，属于区域 49.0001，AR2 为 Level1-2 设备，AR3 为 Level 设备，属于区域 49.0023。通过配置了解 IS-IS 默认路由的特性。

扫一扫，看视频

2. 实验拓扑

IS-IS 默认路由的实验拓扑如图 2-9 所示。

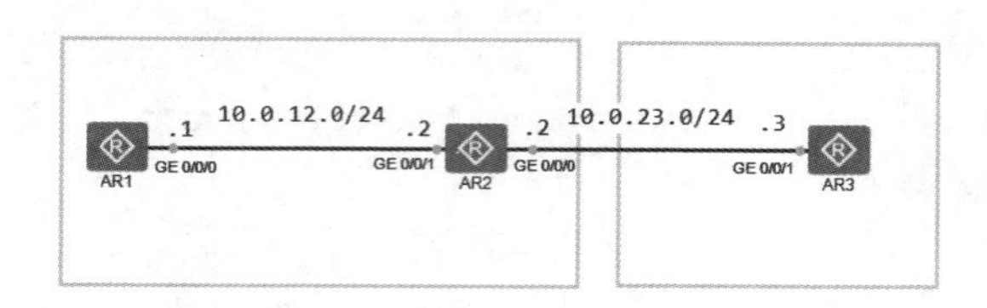

图 2-9　IS-IS 默认路由的实验拓扑

3. 实验步骤

步骤 1： 配置 IP 地址。

AR1 的配置：

```
<Huawei>system-view
Enter system view, return user view with Ctrl+Z.
[Huawei]undo info-center enable
Info: Information center is disabled.
[Huawei]sysname AR1
[AR1]interface g0/0/0
[AR1-GigabitEthernet0/0/0]ip address 10.0.12.1 24
[AR1-GigabitEthernet0/0/0]quit
[AR1]interface LoopBack 0
[AR1-LoopBack0]ip address 1.1.1.1 32
[AR1-LoopBack0]quit
```

AR2 的配置：

```
<Huawei>system-view
Enter system view, return user view with Ctrl+Z.
[Huawei]undo info-center enable
Info: Information center is disabled.
[Huawei]sysname AR2
[AR2]interface g0/0/1
[AR2-GigabitEthernet0/0/1]ip address 10.0.12.2 24
[AR2-GigabitEthernet0/0/1]quit
[AR2]interface g0/0/0
[AR2-GigabitEthernet0/0/0]ip address 10.0.23.2 24
[AR2-GigabitEthernet0/0/0]quit
[AR2]interface LoopBack 0
[AR2-LoopBack0]ip address 2.2.2.2 32
[AR2-LoopBack0]quit
```

AR3 的配置：

```
<Huawei>system-view
Enter system view, return user view with Ctrl+Z.
[Huawei]undo info-center enable
Info: Information center is disabled.
[Huawei]sysname AR3
[AR3]interface g0/0/1
[AR3-GigabitEthernet0/0/1]ip address 10.0.23.3 24
```

```
[AR3-GigabitEthernet0/0/1]quit
[AR3]interface LoopBack 0
[AR3-LoopBack0]ip address 3.3.3.3 32
[AR3-LoopBack0]quit
```

步骤 2: 配置 IS-IS。

AR1 的配置:

```
[AR1]isis
[AR1-isis-1]network-entity 49.0001.0000.0000.0001.00
[AR1-isis-1]is-level level-2
[AR1-isis-1]quit
[AR1]interface g0/0/0
[AR1-GigabitEthernet0/0/0]isis enable
[AR1-GigabitEthernet0/0/0]quit
[AR1]interface LoopBack 0
[AR1-LoopBack0]isis enable
[AR1-LoopBack0]quit
```

AR2 的配置:

```
[AR2]isis
[AR2-isis-1]network-entity 49.0023.0000.0000.0002.00
[AR2-isis-1]quit
[AR2]interface g0/0/1
[AR2-GigabitEthernet0/0/1]isis enable
[AR2-GigabitEthernet0/0/1]quit
[AR2]interface g0/0/0
[AR2-GigabitEthernet0/0/0]isis enable
[AR2-GigabitEthernet0/0/0]quit
[AR2]interface LoopBack 0
[AR2-LoopBack0]isis enable
[AR2-LoopBack0]quit
```

AR3 的配置:

```
[AR3]isis
[AR3-isis-1]network-entity 49.0023.0000.0000.0003.00
[AR3-isis-1]is-level level-1
[AR3-isis-1]quit
[AR3]interface g0/0/1
[AR3-GigabitEthernet0/0/1]isis enable
[AR3-GigabitEthernet0/0/1]quit
[AR3]interface LoopBack 0
[AR3-LoopBack0]isis enable
[AR3-LoopBack0]quit
```

4. 实验调试

查看 AR3 上的路由表:

```
[AR3]display ip routing-table
Route Flags: R - relay, D - download to fib
```

```
-------------------------------------------------------------------------------
Routing Tables: Public
         Destinations : 11        Routes : 11
Destination/Mask    Proto    Pre   Cost   Flags NextHop         Interface
       0.0.0.0/0    ISIS-L1  15    10     D     10.0.23.2       GigabitEthernet0/0/1
       2.2.2.2/32   ISIS-L1  15    10     D     10.0.23.2       GigabitEthernet0/0/1
       3.3.3.3/32   Direct   0     0      D     127.0.0.1       LoopBack0
     10.0.12.0/24   ISIS-L1  15    20     D     10.0.23.2       GigabitEthernet0/0/1
     10.0.23.0/24   Direct   0     0      D     10.0.23.3       GigabitEthernet0/0/1
     10.0.23.3/32   Direct   0     0      D     127.0.0.1       GigabitEthernet0/0/1
   10.0.23.255/32   Direct   0     0      D     127.0.0.1       GigabitEthernet0/0/1
     127.0.0.0/8    Direct   0     0      D     127.0.0.1       InLoopBack0
     127.0.0.1/32   Direct   0     0      D     127.0.0.1       InLoopBack0
127.255.255.255/32  Direct   0     0      D     127.0.0.1       InLoopBack0
255.255.255.255/32  Direct   0     0      D     127.0.0.1       InLoopBack0
```

通过以上输出可以看出，AR3 上有一条默认路由，下一跳为 10.0.23.2。

关闭 AR3 的 g0/0/1，然后打开，开启抓包，结果如图 2-10 所示。

```
[AR3]interface g0/0/1
[AR3-GigabitEthernet0/0/1]shutdown
[AR3-GigabitEthernet0/0/1]quit
[AR3]interface g0/0/1
[AR3-GigabitEthernet0/0/1]undo shutdown
[AR3-GigabitEthernet0/0/1]quit
```

```
> Frame 91: 105 bytes on wire (840 bits), 105 bytes captured (840 bits) on interface 0
> IEEE 802.3 Ethernet
> Logical-Link Control
> ISO 10589 ISIS InTRA Domain Routeing Information Exchange Protocol
v ISO 10589 ISIS Link State Protocol Data Unit
    PDU length: 88
    Remaining lifetime: 1199
    LSP-ID: 0000.0000.0002.00-00
    Sequence number: 0x0000000b
    Checksum: 0x359e [correct]
    [Checksum Status: Good]
  > Type block(0x0b): Partition Repair:0, Attached bits:1, Overload bit:0, IS type:3
  > Protocols supported (t=129, l=1)
  > Area address(es) (t=1, l=4)
  > IP Interface address(es) (t=132, l=12)
  > IP Internal reachability (t=128, l=36)
```

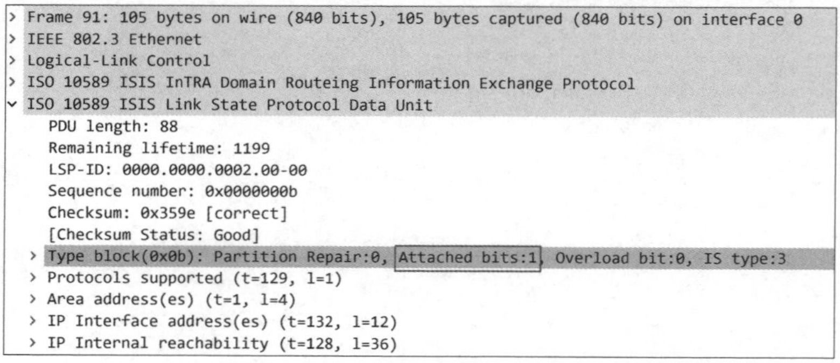

图 2-10　AR3 的 g0/0/1 接口抓包结果

通过以上输出可以看出，ATT 为 1，是 Level1-2 的路由 AR2 发送给 AR3 的，由此产生默认路由。

在 AR3 上设置命令，让其拒绝接收默认路由：

```
[AR3]isis
[AR3-isis-1]attached-bit avoid-learning    //控制 Level-1 设备不因为 ATT 位下发默认路由
                                           //到路由表
[AR3-isis-1]quit
```

再次查看 AR3 的路由表：

```
[AR3]display ip routing-table
Route Flags: R - relay, D - download to fib
------------------------------------------------------------------------------
Routing Tables: Public
         Destinations : 10       Routes : 10
  Destination/Mask    Proto   Pre  Cost  Flags  NextHop         Interface
        2.2.2.2/32   ISIS-L1  15   10     D     10.0.23.2       GigabitEthernet0/0/1
        3.3.3.3/32   Direct   0    0      D     127.0.0.1       LoopBack0
      10.0.12.0/24   ISIS-L1  15   20     D     10.0.23.2       GigabitEthernet0/0/1
      10.0.23.0/24   Direct   0    0      D     10.0.23.3       GigabitEthernet0/0/1
      10.0.23.3/32   Direct   0    0      D     127.0.0.1       GigabitEthernet0/0/1
    1 0.0.23.255/32  Direct   0    0      D     127.0.0.1       GigabitEthernet0/0/1
      127.0.0.0/8    Direct   0    0      D     127.0.0.1       InLoopBack0
     127.0.0.1/32    Direct   0    0      D     127.0.0.1       InLoopBack0
127.255.255.255/32   Direct   0    0      D     127.0.0.1       InLoopBack0
255.255.255.255/32   Direct   0    0      D     127.0.0.1       InLoopBack0
```

通过以上输出可以看出，默认路由消失了。

删除 ATT 拒绝默认路由功能：

```
[AR3]isis
[AR3-isis-1]undo attached-bit avoid-learning
[AR3-isis-1]quit
```

在 AR2 上设置拒绝发送默认路由：

```
[AR2]isis
[AR2-isis-1]attached-bit advertise never
[AR2-isis-1]quit
```

在 AR3 上查看路由表：

```
[AR3]display ip routing-table
Route Flags: R - relay, D - download to fib
------------------------------------------------------------------------------
Routing Tables: Public
         Destinations : 10       Routes : 10
  Destination/Mask    Proto   Pre  Cost  Flags  NextHop         Interface
        2.2.2.2/32   ISIS-L1  15   10     D     10.0.23.2       GigabitEthernet0/0/1
        3.3.3.3/32   Direct   0    0      D     127.0.0.1       LoopBack0
      10.0.12.0/24   ISIS-L1  15   20     D     10.0.23.2       GigabitEthernet0/0/1
      10.0.23.0/24   Direct   0    0      D     10.0.23.3       GigabitEthernet0/0/1
    10.0.23.3/32      Direct   0    0      D     127.0.0.1       GigabitEthernet0/0/1
   10.0.23.255/32     Direct   0    0      D     127.0.0.1       GigabitEthernet0/0/1
     127.0.0.0/8      Direct   0    0      D     127.0.0.1       InLoopBack0
    127.0.0.1/32      Direct   0    0      D     127.0.0.1       InLoopBack0
127.255.255.255/32    Direct   0    0      D     127.0.0.1       InLoopBack0
255.255.255.255/32    Direct   0    0      D     127.0.0.1       InLoopBack0
```

通过以上输出可以看出，默认路由消失了。

配置 AR2 下发 ATT 置位：

```
[AR2]isis
[AR2-isis-1]attached-bit advertise always
[AR2-isis-1]quit
```

【技术要点】控制默认路由

（1）在Level1-2路由器上关闭发送默认路由的功能：attached-bit advertise never。
（2）在Level1路由器上关闭接收默认路由的功能：attached-bit avoid-learning。

在 AR1 上设置一条默认路由指向 NULL 0，在 AR1 上下发默认路由：

```
[AR1]ip route-static 0.0.0.0 0.0.0.0 NULL 0
[AR1]isis
[AR1-isis-1]default-route-advertise
[AR1-isis-1]quit
```

在 AR2 上查看路由表：

```
[AR2]display ip routing-table
Route Flags: R - relay, D - download to fib
------------------------------------------------------------------------------
Routing Tables: Public
         Destinations : 14             Routes : 14
 Destination/Mask     Proto    Pre  Cost  Flags  NextHop        Interface
        0.0.0.0/0     ISIS-L2  15   10    D      10.0.12.1      GigabitEthernet0/0/1
       1.1.1.1/32     ISIS-L2  15   10    D      10.0.12.1      GigabitEthernet0/0/1
       2.2.2.2/32     Direct   0    0     D      127.0.0.1      LoopBack0
       3.3.3.3/32     ISIS-L1  15   10    D      10.0.23.3      GigabitEthernet0/0/0
     10.0.12.0/24     Direct   0    0     D      10.0.12.2      GigabitEthernet0/0/1
     10.0.12.2/32     Direct   0    0     D      127.0.0.1      GigabitEthernet0/0/1
   10.0.12.255/32     Direct   0    0     D      127.0.0.1      GigabitEthernet0/0/1
     10.0.23.0/24     Direct   0    0     D      10.0.23.2      GigabitEthernet0/0/0
     10.0.23.2/32     Direct   0    0     D      127.0.0.1      GigabitEthernet0/0/0
   10.0.23.255/32     Direct   0    0     D      127.0.0.1      GigabitEthernet0/0/0
      127.0.0.0/8     Direct   0    0     D      127.0.0.1      InLoopBack0
     127.0.0.1/32     Direct   0    0     D      127.0.0.1      InLoopBack0
127.255.255.255/32    Direct   0    0     D      127.0.0.1      InLoopBack0
255.255.255.255/32    Direct   0    0     D      127.0.0.1      InLoopBack0
```

通过以上输出可以看出，AR2 上面有了一条默认路由。

第 3 章

BGP 的高级特性

3.1 BGP 的高级特性概述

在大型网络中通常会部署 BGP，相比于 IGP，BGP 拥有更加灵活的路由控制能力。每一条 BGP 路由都可以携带多个路径属性，针对其属性也有特有的路由匹配工具，包括 AS_Path Filter 和 Community Filter。根据实际组网需求，BGP 可以实施路由策略，控制路由的接收和发布。

同时，为了提升网络性能，BGP 提供了各种高级特性以及多种组网部署方案。

3.2 BGP 路由控制

3.2.1 正则表达式

正则表达式是按照一定的模板来匹配字符串的公式，由普通字符（如字符 a ~ z）和特殊字符组成。

（1）普通字符：匹配的对象是普通字符本身，包括所有的大写和小写字母、数字、标点符号以及一些特殊符号。例如，a 匹配 abc 中的 a，10 匹配 10.113.25.155 中的 10，@ 匹配 ×××@×××.com 中的 @。

（2）特殊字符：配合普通字符匹配复杂或特殊的字符串组合，即位于普通字符之前或之后，用于限制或扩充普通字符的独立控制字符或占位符。用于描述它前面的字符的重复使用方式，并且限定一个完整的范围。

①特殊字符类型 1 见表 3-1。

表 3-1　特殊字符类型 1

特殊字符	说　明	举　例
.	匹配任意单个字符，包括空格	0.0 匹配 0x0、020……
^	匹配行首的位置，即一个字符串的开始	^10 匹配 10.1.1.1，不匹配 20.1.1.1
$	匹配行尾的位置，即一个字符串的结束	1 $ 匹配 10.1.1.1，不匹配 10.1.1.2
_	下画线，匹配任意一个分隔符。 匹配一个逗号（，）、左花括号（{）、右花括号（}）、左圆括号（(）、右圆括号（)）。 匹配输入字符串的开始位置（同 ^）。 匹配输入字符串的结束位置（同 $）。 匹配一个空格	_10 匹配（10、{10、空格 10 等。 10_ 匹配 10）、10}、10 空格等
|	管道字符，逻辑或。例如，x | y 匹配 x 或 y	100 | 200 匹配 100 或者 200
\	转义字符，用于将下一个字符（特殊字符或普通字符）标记为普通字符	* 匹配 *

特殊字符类型 1 举例如下。

^a.$：匹配一个以字符 a 开始，以任意单一字符结束的字符串，如 a0、a!、ax 等。

^100_：匹配以 100 为起始的字符串，如 100、100200、100300400 等。

^100$：只匹配 100。

100$|400$：匹配以 100 或 400 结束的字符串，如 100、1400、300400 等。

^\(65000\)$：只匹配 (65000)。

②特殊字符类型 2 见表 3–2。

表 3–2　特殊字符类型 2

特殊字符	说　明	举　例	
*	匹配前面的子正则表达式 0 次或多次	10* 匹配 1、10、100、1000……	(10)* 匹配空、10、1010、101010……
+	匹配前面的子正则表达式 1 次或多次	10+ 匹配 10、100、1000……	(10)+ 匹配 10、1010、101010……
?	匹配前面的子正则表达式 0 次或 1 次	10? 匹配 1 或 10	(10)? 匹配空或 10

特殊字符类型 2 举例如下。

abc*d：匹配 c 字符 0 次或多次，如 abd、abcd、abccd、abcccd、abccccd 等。

abc+d：匹配 c 字符 1 次或多次，如 abcd、abccd、abcccd、abccccd 等。

abc?d：匹配 c 字符 0 次或 1 次，如 abd、abcd。

a(bc)?d：匹配 bc 字符串 0 次或 1 次，如 ad、abcd。

③特殊字符类型 3 见表 3–3。

表 3–3　特殊字符类型 3

特殊字符	说　明	举　例
[xyz]	匹配正则表达式中包含的任意一个字符	[123] 匹配 255 中的 2
[^xyz]	匹配正则表达式中未包含的字符	[^123] 匹配 123 之外的任意字符
[a–z]	匹配正则表达式指定范围内的任意字符	[0–9] 匹配 0~9 之间的所有数字
[^a–z]	匹配正则表达式指定范围外的任意字符	[^0–9] 匹配所有非数字字符（匹配 0~9 之外的任意字符）

特殊字符类型 3 举例如下。

[abcd]：匹配 abcd 中任意一个字符，即只要出现了 a、b、c、d 中的任意字符即可，如 ax、b!、abc、d0 等。

[a–c 1–2]$：匹配以字符 a、b、c、1、2 结束的字符串，如 a、a1、62、xb、7ac 等。

[^act]$：匹配不以字符 a、c、t 结束的字符串，如 ax、b!、d 等。

[123].[7–9]：匹配如 1 7、2x9、348 等。

备注：读者可以通过本书所附的网址进行测试。

【思考】下列正则表达式代表的含义

^$	匹配不包含任何AS号的AS_Path，也就是本AS内的路由
.*	匹配所有，任何路由
^10[012349]$	匹配100、101、102、103、104、109
^10[^0-6]$	匹配除100~106以外的AS_Path
^10.	匹配100~109，以及10
^12(_34)?_56$	匹配12 56，以及12 34 56

3.2.2 AS_Path Filter

（1）AS_Path Filter 是将 BGP 中的 AS_Path 属性作为匹配条件的过滤器，利用 BGP 路由携带的 AS_Path 列表对路由进行过滤。

（2）当不希望接收某些 AS 的路由时，可以利用 AS_Path Filter 对携带这些 AS 号的路由进行过滤，从而实现拒绝某些路由。

3.2.3 Community Filter

1.Community（团体）属性

团体属性一是一组有相同特征的目的地址的集合。团体属性用一组以 4 字节为单位的列表来表示，其格式是 aa:nn 或团体属性号。

（1）aa:nn：aa 和 nn 的取值范围都是 0 ～ 65535，管理员可根据实际情况设置具体数值。通常 aa 表示 AS（Autonomous System，自治系统）编号，nn 是管理员定义的团体属性标识。例如，对于来自 AS 100 的一条路由，如果管理员定义的团体属性标识是 1，则该路由的团体属性格式是 100:1。

（2）团体属性号：团体属性号是0～4294967295的整数。标准协议中定义，0（0x00000000）～65535（0x0000FFFF）和4294901760（0xFFFF0000）～4294967295（0xFFFFFFFF）是预留的。

2. 公认团体属性

公认团体属性见表 3-4。

表 3-4　公认团体属性

团体属性名称	团体属性号	说　明
Internet	0（0x00000000）	设备在收到具有此属性的路由后，可以向任何 BGP 对等体发送该路由。默认情况下，所有的路由都属于 Internet 团体
No_Advertise	4294967042（0xFFFFFF02）	设备在收到具有此属性的路由后，将不向任何 BGP 对等体发送该路由
No_Export	4294967041（0xFFFFFF01）	设备在收到具有此属性的路由后，将不向 AS 外发送该路由
No_Export_Subconfed	4294967043（0xFFFFFF03）	设备在收到具有此属性的路由后，将不向 AS 外发送该路由。如果使用了联盟，也不向联盟内其他子 AS 发布该路由

3.3　BGP 特性简介

3.3.1 ORF

如果设备只接收自己需要的路由，但对端设备又无法针对每个与它连接的设备维护不同的出口策略，就可以通过配置 BGP 基于前缀的 ORF（Outbound Route Filters，出口路由过滤器）来满足两端设备的需求。

BGP 基于前缀的 ORF 能力，能将本端设备配置的基于前缀的入口策略通过路由刷新报文发送给 BGP 邻居。BGP 邻居根据这些策略（刷新报文中）构造出口策略，在路由发送时对路由进行过滤。

这样不仅可以避免因本端设备接收大量无用的路由而降低本端设备的 CPU 使用率，还可以有效减少 BGP 邻居的配置工作，从而降低链路带宽的占用率。

3.3.2　BGP 对等体组

对等体组（Peer Group）是一些具有某些相同策略的对等体的集合。当一个对等体加入对等体组中时，该对等体将获得与所在对等体组相同的配置。当对等体组的配置改变时，组内成员的配置也相应改变。

在大型 BGP 网络中，对等体的数量会很多，其中很多对等体具有相同的策略，在配置时会重复使用一些命令，利用对等体组可以简化配置。

3.3.3　BGP 认证

1. 常见 BGP 攻击

常见 BGP 攻击主要有以下两种。

（1）建立非法 BGP 邻居关系，通告非法路由条目，干扰正常路由表。

（2）发送大量非法 BGP 报文，路由器收到后报送 CPU，导致 CPU 利用率升高。

2. BGP 认证

BGP 认证分为 MD5 认证和 Keychain 认证，对 BGP 对等体关系进行认证可以预防非法 BGP 邻居建立。

（1）MD5 认证。BGP 使用 TCP 作为传输层协议，为提高 BGP 的安全性，可以在建立 TCP 连接时进行 MD5 认证。BGP 的 MD5 认证只是为 TCP 连接设置 MD5 认证密码，由 TCP 完成认证，如图 3–1 所示。

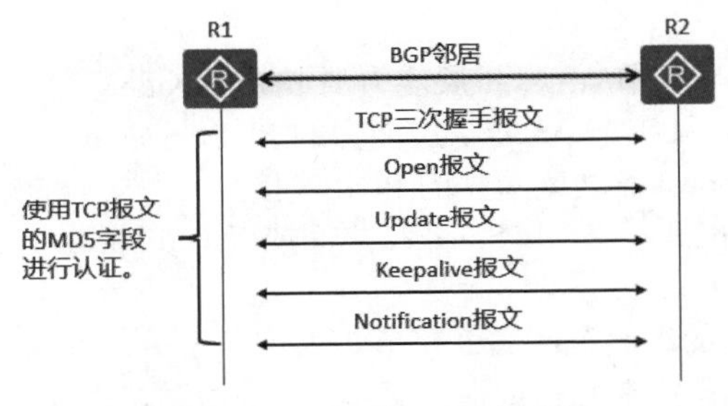

图 3–1　BGP 的 MD5 认证

（2）Keychain 认证。BGP 对等体两端必须都配置针对使用 TCP 连接的应用程序的 Keychain 认证，并且配置的 Keychain 必须使用相同的加密算法和密码，才能正常建立 TCP 连接，交互 BGP 消息，如图 3–2 所示。

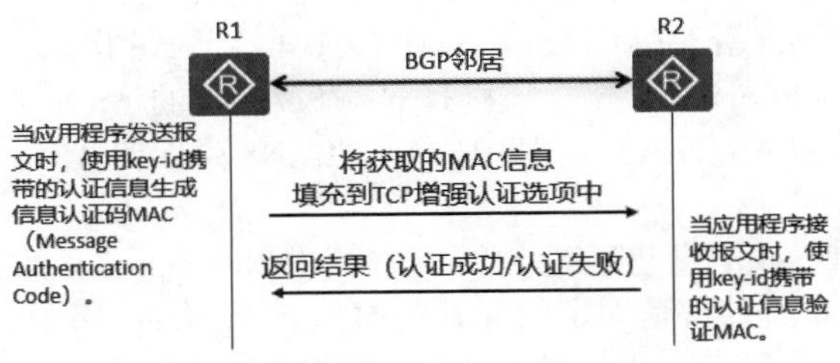

图 3–2　BGP 的 Keychain 认证

【技术要点】MD5与Keychain的区别

（1）MD5算法配置简单，配置后生成单一密码，需要人为干预才可以更换密码。

（2）Keychain具有一组密码，可以根据配置自动切换，但是配置过程较为复杂，适用于对安全性能要求比较高的网络。

（3）BGP的MD5认证与Keychain认证互斥。

3. GTSM

BGP 的 GTSM（Generalized TTL Security Mechanism，通用 TTL 安全保护机制）功能用于检测 IP 报文头中的 TTL（Time To Live）值是否在一个预先设置好的特定范围内，并对不符合 TTL 值范围的报文进行丢弃，这样就避免了网络攻击者模拟“合法”BGP 报文攻击设备。

当攻击者模拟合法的 BGP 报文，对 R2 不断地发送非法报文进行攻击时，TTL 值必然小于 255。如果 R2 使能 BGP 的 GTSM 功能，将 IBGP（Internal/Interior BGP，域内 BGP）对等体报文的 TTL 的有效范围设为 [255，255]，系统会对所有 BGP 报文的 TTL 值进行检查，丢弃 TTL 值小于 255 的攻击报文，从而避免了因网络攻击报文导致 CPU 占用率高的问题。

3.3.4 4 字节 AS 号

1. 目的

目前网络上使用的 AS 号范围为 1 ~ 65 535（2 字节），随着网络规模的扩大，可分配的 AS 号已经濒临枯竭，需要将 AS 号范围扩展为 1 ~ 4 294 967 295（4 字节），并且能够与仅支持 2 字节 AS 号的 Old Speaker 兼容。

2. 相关概念

（1）New Speaker：支持 4 字节 AS 号扩展能力的 BGP Speaker。

（2）Old Speaker：不支持 4 字节 AS 号扩展能力的 BGP Speaker。

（3）New Session：New Speaker 之间建立的 BGP 连接。

（4）Old Session：New Speaker 和 Old Speaker 之间或者 Old Speaker 之间建立的 BGP 连接。

3. 定义

4 字节 AS 号是将 AS 号的编码范围由 2 字节扩大为 4 字节，并通过定义新的能力码和新的可选过渡属性来协商 4 字节 AS 号能力和传递 4 字节 AS 号信息，使支持 4 字节能力的 New Speaker 之间、New Speaker 和只支持 2 字节 AS 号的 Old Speaker 之间能够进行通信。

为了支持 4 字节 AS 号，标准协议定义了一种新的 Open 能力码（0x41，代表本端支持 4 字节能力扩展）用于进行 BGP 连接的能力协商。

另外，标准协议还定义了两种新的可选过渡属性 AS4_Path（属性码为 0x11）和 AS4_Aggregator（属性码为 0x12），用于在 Old Session 上传递 4 字节 AS 信息。

如果 New Speaker 和 Old Speaker 建立连接且 New Speaker 的 AS 号大于 65 535，则需要在 Old Speaker 端指定对端 AS 号为 AS_TRANS。其中，AS_TRANS 是保留 AS 号，值为 23 456。

4. 基本原理

BGP 通过相互通告 Open 消息进行能力协商，其中 New Speaker 的 Open 消息格式示意图如图 3-3 所示。BGP 的 Open 消息头是固定的，其中 My AS Number 字段填写的是本地 AS 号，但是 My AS Number 字段只占有 2 字节，无法填充 4 字节的 AS 号。因此 New Speaker 在发送 Open 消息时，将 AS_TRANS 号 23456 填充到 My AS Number 字段，而将自己实际的 4AS Number 填写在可选能力字段，这样邻居间就能通过 Open 消息的可选能力字段获知对方是否支持 4 字节 AS 能力。

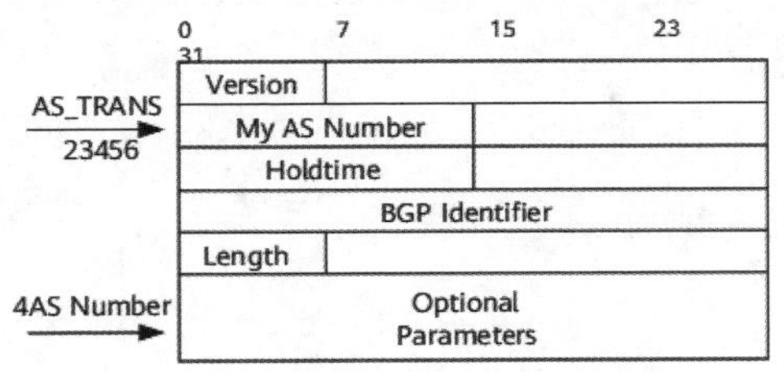

图 3-3　New Speaker 的 Open 消息格式示意图

New Speaker 与 New Speaker 之间、New Speaker 与 Old Speaker 之间的邻居建立过程示意图如图 3-4 所示。不同的 BGP Speaker 之间通过 Open 消息向对端通告是否支持 4AS 能力，能力协商完成后，New Speaker 之间建立 New Session，New Speaker 和 Old Speaker 之间建立 Old Session。

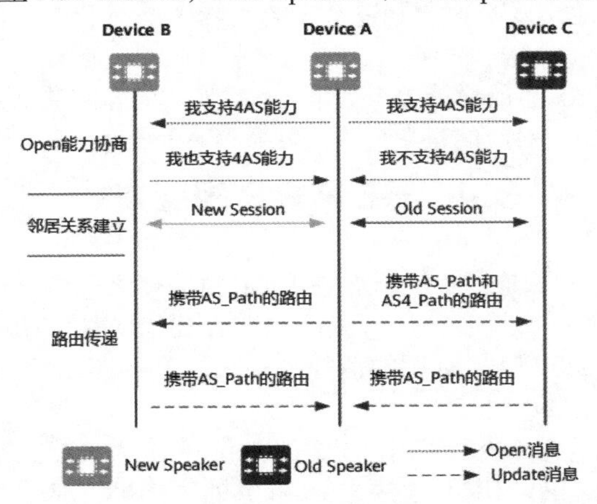

图 3-4　BGP 邻居建立过程示意图

在 New Speaker 之间的 Update 消息中，AS_Path 属性和 Aggregator 属性中的 AS 号按照 4 字节进行编码。另外，由于 Old Speaker 的 Update 消息中 AS_Path 属性和 Aggregator 属性的 AS 号都是按照 2 字节编码，因此：

（1）当 New Speaker 向 Old Speaker 发送 Update 消息时，如果存在大于 65535 的 AS 号信息，会使用 AS4_Path 属性和 AS4_Aggregator 属性辅助 AS_Path 属性和 AS_Aggregator 属性传递 4 字节 AS 号信息，AS4_Path 属性和 AS4_Aggregator 属性对 Old Speaker 来说是完全透明的。如图 3-5 所示，当 AS 2.2 的 New Speaker 向 AS 65002 的 Old Speaker 发送 Update 消息时，Update 消息中的 AS_Path 为（23456, 23456, 65001），AS4_Path 为（2.2, 1.1, 65001），其中 23456 就是使用 AS_TRANS 替代 1.1 和 2.2 的结果。AS 65002 的 Old Speaker 在向外发送路由时，不会对 AS4_Path 做任何改动，而是将（2.2, 1.1, 65001）透明传输给其他 AS。

（2）当 New Speaker 从 Old Speaker 收到带有 AS_Path 属性、AS4_Path 属性、AS_Aggregator 属性、AS4_Aggregator 属性的 Update 消息时，会根据重构算法重构出真正的 AS_Path 属性和 AS_Aggregator 属性。在图 3-5 中，当 AS 65003 的 New Speaker 从 AS 65002 的 Old Speaker 收到带有 AS_Path 属性（65002, 23456, 23456, 65001）和 AS4_Path 属性（2.2, 1.1, 65001）的 Update 消息后，重构出了真正的 AS_Path 属性（65002, 2.2, 1.1, 65001）。

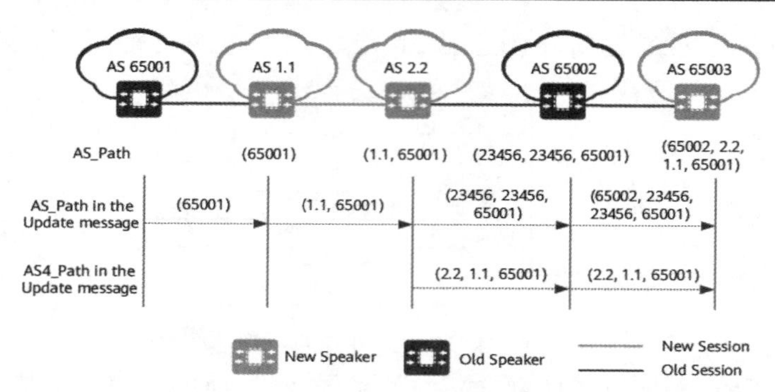

图 3-5　BGP Update 消息传递示意图

5. 4 字节 AS 的形式

4 字节 AS 分为两种形式：整数形式和点分形式。整数形式和点分形式的 4 字节 AS 号在系统内部都以无符号的整数形式存储。点分形式的 4 字节 AS 号一般为 $x.y$ 格式，整数形式的 4 字节 AS 号和点分形式的 4 字节 AS 号的换算关系如下：整数形式的 4 字节 AS 号 $= x \times 65536 + y$。例如，点分形式的 4 字节 AS 号 2.3，对应的整数形式的 4 字节 AS 号为 $2 \times 65536 + 3 = 131075$。

3.4　BGP RR 组网方式

1. BGP RR

为保证 IBGP 对等体之间的连通性，需要在 IBGP 对等体之间建立全连接（Full-mesh）关系。假设在一个 AS 内部有 n 台 BGP 设备，那么应该建立的 IBGP 连接数就为 $n(n-1)/2$。当 IBGP 对等体数目很多时，对网络资源和 CPU 资源的消耗都很大。利用路由反射器（Route Reflector，RR）可以解决这一问题。

如图 3-6 所示，在一个 AS 内，其中一台 BGP 设备作为 RR，其他 BGP 设备作为客户机（Client）。客户机与 RR 之间建立 IBGP 连接。RR 和它的客户机组成一个集群（Cluster）。RR 在客户机之间反射路由信息，客户机之间不需要建立 BGP 连接。既不是 RR 也不是客户机的 BGP 设备称为非客户机（Non-Client）。非客户机与 RR 之间，以及所有的非客户机之间仍然必须建立全连接关系。

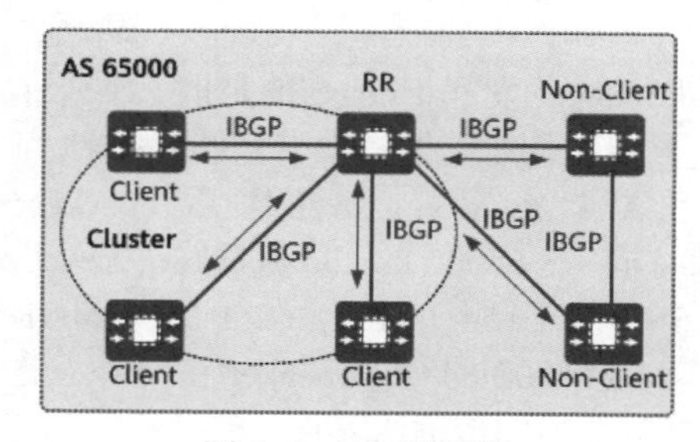

图 3-6　RR 示意图

2. 反射规则

当 RR 收到对等体发来的路由时，首先使用 BGP 选路策略来选择最佳路由。在向 IBGP 邻居发布学习到的路由信息时，RR 按照以下规则发布路由。

（1）将从非客户机 IBGP 对等体学到的路由发布给此 RR 的所有客户机。

（2）将从客户机学到的路由发布给此 RR 的所有非客户机和客户机。

（3）将从 EBGP（External BGP, 域间 BGP）对等体学到的路由发布给所有的非客户机和客户机。

RR 的配置方便，只需对作为反射器的 BGP 设备进行配置，客户机并不需要知道自己是客户机。在某些网络中，RR 的客户机之间已经建立了全连接，它们可以直接交换路由信息，此时客户机到客户机通过 RR 的路由反射是没有必要的，而且还占用带宽资源。可以通过配置来禁止客户机通过 RR 的路由反射，但客户机到非客户机之间的路由仍然可以被反射。默认情况下，允许客户机通过 RR 的路由反射。反射器可以修改 BGP 路由的各种属性，如 AS_Path 属性、MED 属性、本地优先级属性、团体属性等。

3. Originator_ID 属性

Originator_ID 属性长 4 字节，由 RR 产生，携带了本地 AS 内部路由发起者的 Router ID，用于防止集群内产生路由环路。

（1）当一条路由第 1 次被 RR 反射时，RR 将 Originator_ID 属性加入这条路由，标识这条路由的发起设备。如果一条路由中已经存在了 Originator_ID 属性，则 RR 将不会创建新的 Originator_ID。

（2）当其他 BGP Speaker 接收到这条路由时，将比较收到的 Originator_ID 和本地的 Router ID，如果两个 ID 相同，则 BGP Speaker 会忽略掉这条路由，不进行处理。

4. Cluster_List 属性

RR 和它的客户机组成一个集群。在一个 AS 内，每个 RR 使用唯一的 Cluster_ID 作为标识。

为防止产生路由环路，RR 使用 Cluster_List 记录反射路由经过的所有 Cluster_ID。

Cluster_List 由一系列 Cluster_ID 组成，描述了一条路由所经过的反射器路径，这和描述路由经过的 AS 路径的 AS_Path 属性有相似之处。Cluster_List 由路由反射器产生。

（1）当 RR 在它的客户机之间或客户机与非客户机之间反射路由时，RR 会把本地 Cluster_ID 添加到 Cluster_List 的前面。如果 Cluster_List 为空，RR 就创建一个。

（2）当 RR 接收到一条更新路由时，RR 会检查 Cluster_List。如果 Cluster_List 中已经有本地 Cluster_ID，就丢弃该路由；如果没有本地 Cluster_ID，就将其加入 Cluster_List，然后反射该更新路由。

5. 备份 RR

为增加网络的可靠性，防止单点故障，有时需要在一个集群中配置一个以上的 RR。这时，相同集群中的 RR 要共享相同的 Cluster_ID，以避免路由环路。在冗余的环境里，客户机会收到不同反射器发来的到达同一目的地的多条路由，这时客户机应用 BGP 选择路由的策略来选择最佳路由。

如图 3-7 所示，RR1 和 RR2 在同一个 Cluster 内。RR1 和 RR2 之间配置 IBGP 连接，即两个反射器互为非客户机。

（1）当客户机 Client1 从外部对等体接收到一条更新路由后，它通过 IBGP 向 RR1 和 RR2 通告这条路由。

（2）当 RR1 接收到该更新路由后，它向其他的客户机（Client1、Client2、Client3）和非客户机（RR2）反射，同时将本地 Cluster_ID 添加到 Cluster_List 前面。

（3）当 RR2 接收到该反射路由后，检查 Cluster_List，发现自己的 Cluster_ID 已经包含在 Cluster_List 中。因此，它丢弃该更新路由，不再向自己的客户机反射。

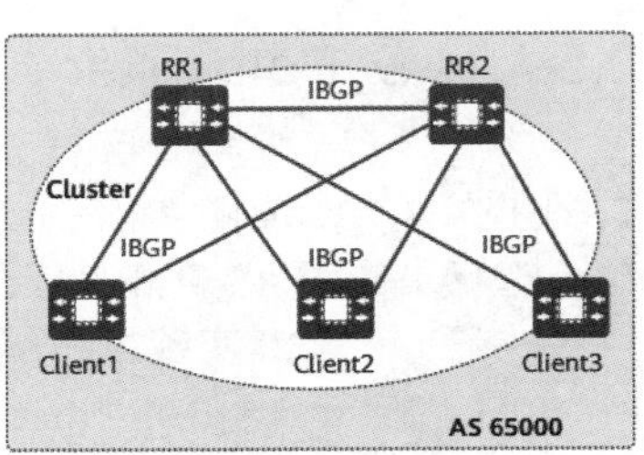

图 3-7　BGP RR 示例

如果RR1和RR2配置不同的Cluster_ID，各RR除了收到来自客户机的路由，还将接收到另外一个RR反射的更新路由。这时，通过给RR1和RR2配置相同的Cluster_ID，可以减少各RR接收的路由数量，从而节省内存开销。

6. AS中的多个集群

一个AS中可能存在多个集群。各个RR之间是IBGP对等体的关系，一个RR可以把另一个RR配置成自己的客户机或非客户机。因此，用户可以灵活地配置AS内部集群与集群之间的关系。

例如，一个骨干网被分成多个反射集群，每个RR将其他的RR配置成非客户机，各RR之间建立全连接。每个客户机只与所在集群的RR建立IBGP连接。这样该AS内的所有BGP设备都会收到反射路由信息，如图3–8所示。

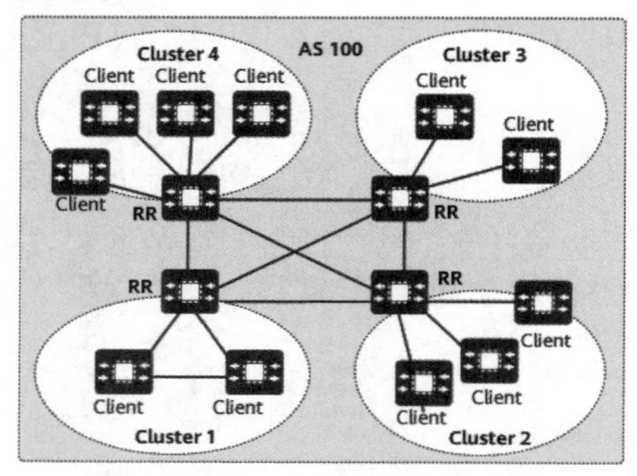

图3–8　BGP RR多集群场景

7. 分级反射器

在实际的反射器部署中，常用的是分级反射器的场景。如图3–9所示，ISP为AS 100提供Internet路由，ISP与AS 100内建立双出口EBGP连接。AS 100内部分为两个集群，其中Cluster1内的4台BGP设备是核心设备。

（1）Cluster1中部署了两个一级RR（RR–1），这种冗余结构保证了AS 100内部网络核心层的可靠性。核心层其余两台BGP设备作为RR–1的客户机。

（2）Cluster2中部署了一个二级RR（RR–2），这个RR–2同时也是RR–1的客户机。

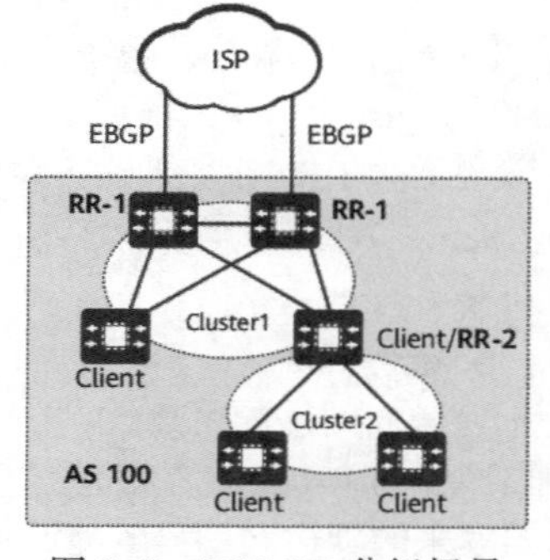

图3–9　BGP RR分级场景

3.5　BGP高级特性实验

3.5.1　AS_Path Filter

扫一扫，看视频

1. 实验目的

如图3–10所示，AR1~AR4的AS号分别属于AS 101~AS 104，要求在AR3上面使用AS_Path Filter将始发于AS 101的路由过滤。

2. 实验拓扑

AS_Path Filter的实验拓扑如图3–10所示。

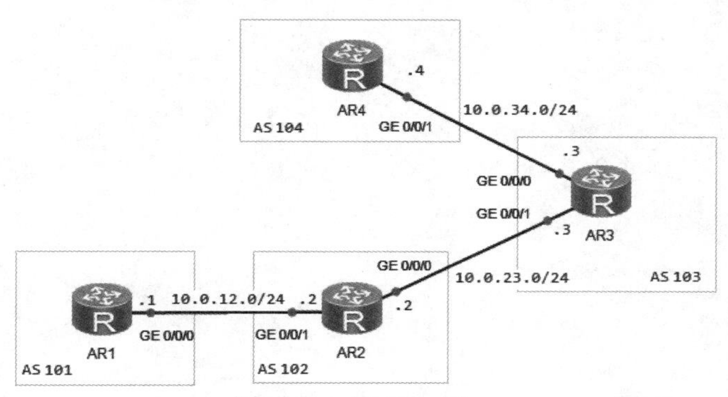

图 3-10　AS_Path Filter

3. 实验步骤

步骤 1：配置 IP 地址。

AR1 的配置：

```
<Huawei>system-view
Enter system view, return user view with Ctrl+Z.
[Huawei]undo info-center enable
[Huawei]sysname AR1
[AR1]interface g0/0/0
[AR1-GigabitEthernet0/0/0]ip address 10.0.12.1 24
[AR1-GigabitEthernet0/0/0]quit
[AR1]interface LoopBack 0
[AR1-LoopBack0]ip address 1.1.1.1 32
[AR1-LoopBack0]quit
```

AR2 的配置：

```
<Huawei>system-view
Enter system view, return user view with Ctrl+Z.
[Huawei]undo info-center enable
Info: Information center is disabled.
[Huawei]sysname AR2
[AR2]interface g0/0/1
[AR2-GigabitEthernet0/0/1]ip address 10.0.12.2 24
[AR2-GigabitEthernet0/0/1]quit
[AR2]interface g0/0/0
[AR2-GigabitEthernet0/0/0]ip address 10.0.23.2 24
[AR2-GigabitEthernet0/0/0]quit
[AR2]interface loo
[AR2]interface LoopBack 0
[AR2-LoopBack0]ip address 2.2.2.2 32
[AR2-LoopBack0]quit
```

AR3 的配置：

```
<Huawei>system-view
Enter system view, return user view with Ctrl+Z.
[Huawei]undo info-center enable
```

```
Info: Information center is disabled.
[Huawei]sysname AR3
[AR3]interface g0/0/1
[AR3-GigabitEthernet0/0/1]ip address 10.0.23.3 24
[AR3-GigabitEthernet0/0/1]quit
[AR3]interface g0/0/0
[AR3-GigabitEthernet0/0/0]ip address 10.0.34.3 24
[AR3-GigabitEthernet0/0/0]quit
[AR3]interface LoopBack 0
[AR3-LoopBack0]ip address 3.3.3.3 32
[AR3-LoopBack0]quit
```

AR4 的配置：

```
<Huawei>system-view
Enter system view, return user view with Ctrl+Z.
[Huawei]undo info-center enable
Info: Information center is disabled.
[Huawei]sysname AR4
[AR4]interface g0/0/1
[AR4-GigabitEthernet0/0/1]ip address 10.0.34.4 24
[AR4-GigabitEthernet0/0/1]quit
[AR4]interface LoopBack 0
[AR4-LoopBack0]ip address 4.4.4.4 32
[AR4-LoopBack0]quit
```

步骤 2：配置 BGP。

AR1 的配置：

```
[AR1]bgp 101
[AR1-bgp]router-id 1.1.1.1
[AR1-bgp]peer 10.0.12.2 as-number 102
[AR1-bgp]quit
```

AR2 的配置：

```
[AR2]bgp 102
[AR2-bgp]router-id 2.2.2.2
[AR2-bgp]peer 10.0.12.1 as-number 101
[AR2-bgp]peer 10.0.23.3 as-number 103
[AR2-bgp]quit
```

AR3 的配置：

```
[AR3]bgp 103
[AR3-bgp]router-id 3.3.3.3
[AR3-bgp]peer 10.0.23.2 as-number 102
[AR3-bgp]peer 10.0.34.4 as-number 104
[AR3-bgp]quit
```

AR4 的配置：

```
[AR4]bgp 104
[AR4-bgp]router-id 4.4.4.4
```

```
[AR4-bgp]peer 10.0.34.3 as-number 103
[AR4-bgp]quit
```

在 AR1 上宣告 1.1.1.1：

```
[AR1]bgp 101
[AR1-bgp]network 1.1.1.1 32
[AR1-bgp]quit
```

在 AR2 上宣告 2.2.2.2：

```
[AR2]bgp 102
[AR2-bgp]network 2.2.2.2 32
[AR2-bgp]quit
```

步骤 3：实验调试。

在 AR3 上查看 BGP 的路由表：

```
<AR3>display bgp routing-table
 BGP Local router ID is 3.3.3.3
 Status codes: * - valid, > - best, d - damped,
               h - history,  i - internal, s - suppressed, S - Stale
               Origin : i - IGP, e - EGP, ? - incomplete
 Total Number of Routes: 2
      Network          NextHop        MED       LocPrf       PrefVal      Path/Ogn
 *>   1.1.1.1/32       10.0.23.2                             0            102 101i
 *>   2.2.2.2/32       10.0.23.2      0                      0            102i
```

通过以上输出可以看出，AR3 上收到了两条 BGP 的路由，分别为 1.1.1.1 和 2.2.2.2。其中，1.1.1.1 的 AS_Path 为 102 101 。

通过 AS_Path Filter 拒绝始发于 AS 101 的路由，但其他路由可以通过。

①创建 AS_Path Filter：

```
[AR2]ip as-path-filter 1 deny _101$       //创建 as-path-filter 编号为 1，拒绝始发于 AS 101
                                          //的路由通过
[AR2]ip as-path-filter 1 permit .*        //创建 as-path-filter 编号为 1，允许所有路由通过
```

②应用 AS_Path Filter：

```
[AR2]bgp 102
[AR2-bgp]peer 10.0.23.3 as-path-filter 1 export
[AR2-bgp]quit
```

【技术要点】

创建了AS_Path Filter之后，可以直接调用，也可以用路由策略的方式，效果一样。具体操作方法如下：

```
[AR2]route-policy joinlabs permit node 10
[AR2-route-policy]if-match as-path-filter 1
[AR2-route-policy]quit
[AR2]bgp 102
[AR2-bgp]peer 10.0.23.3 route-policy joinlabs export
[AR2-bgp]quit
```

再次在 AR3 上查看 BGP 的路由表：

```
<AR3>display bgp routing-table
 BGP Local router ID is 3.3.3.3
 Status codes: * - valid, > - best, d - damped,
               h - history,  i - internal, s - suppressed, S - Stale
               Origin : i - IGP, e - EGP, ? - incomplete
 Total Number of Routes: 1
      Network          NextHop        MED        LocPrf     PrefVal     Path/Ogn
 *>   2.2.2.2/32       10.0.23.2      0                     0           102i
```

通过以上输出可以看出，始发于 AS 101 的 1.1.1.1 这条路由被过滤了。

查看 AS_Path Filter：

```
[AR2]display ip as-path-filter  1
As path filter number: 1
        deny            _101$
        permit          .*
```

显示 BGP 表中所有 AS_Path 被该正则表达式匹配的路由：

```
[AR2]display bgp routing-table regular-expression _101$
 Total Number of Routes: 1
 BGP Local router ID is 2.2.2.2
 Status codes: * - valid, > - best, d - damped,
               h - history,  i - internal, s - suppressed, S - Stale
               Origin : i - IGP, e - EGP, ? - incomplete
      Network          NextHop        MED        LocPrf     PrefVal     Path/Ogn
 *>   1.1.1.1/32       10.0.12.1      0                     0           101i
```

通过以上输出可以看出，1.1.1.1 这条路由被过滤了。

3.5.2　配置 Community 属性与 Community Filter

扫一扫，看视频

1. 实验目的

在设备之间建立 EBGP 邻居关系，通过在 AR1 上给路由设置自定义的 Community 属性值，并且在 AR2 上使用 Community Filter 实现路由过滤。

2. 实验拓扑

配置 Community 属性与 Community Filter 的实验拓扑如图 3-11 所示。

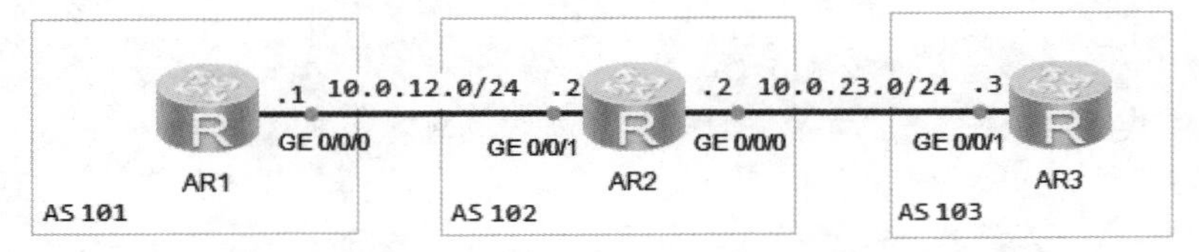

图 3-11　配置 Community 属性与 Community Filter 的实验拓扑

3. 实验步骤

步骤 1： 配置 IP 地址。

AR1 的配置：

```
<Huawei>system-view
```

```
Enter system view, return user view with Ctrl+Z.
[Huawei]undo info-center enable
Info: Information center is disabled.
[Huawei]sysname AR1
[AR1]interface g0/0/0
[AR1-GigabitEthernet0/0/0]ip address 10.0.12.1 24
[AR1-GigabitEthernet0/0/0]quit
[AR1]interface LoopBack 0
[AR1-LoopBack0]ip address 1.1.1.1 32
[AR1-LoopBack0]quit
```

AR2 的配置：

```
<Huawei>system-view
Enter system view, return user view with Ctrl+Z.
[Huawei]undo info-center enable
Info: Information center is disabled.
[Huawei]sysname AR2
[AR2]interface g0/0/1
[AR2-GigabitEthernet0/0/1]ip address 10.0.12.2 24
[AR2-GigabitEthernet0/0/1]quit
[AR2]interface g0/0/0
[AR2-GigabitEthernet0/0/0]ip address 10.0.23.2 24
[AR2-GigabitEthernet0/0/0]quit
[AR2]interface LoopBack 0
[AR2-LoopBack0]ip address 2.2.2.2 32
[AR2-LoopBack0]quit
```

AR3 的配置：

```
<Huawei>system-view
Enter system view, return user view with Ctrl+Z.
[Huawei]undo info-center enable
Info: Information center is disabled.
[Huawei]sysname AR3
[AR3]interface g0/0/1
[AR3-GigabitEthernet0/0/1]ip address 10.0.23.3 24
[AR3-GigabitEthernet0/0/1]quit
[AR3]interface LoopBack 0
[AR3-LoopBack0]ip address 3.3.3.3 32
[AR3-LoopBack0]quit
```

步骤 2：配置 BGP。

AR1 的配置：

```
[AR1]bgp 101
[AR1-bgp]peer 10.0.12.2 as-number 102
[AR1-bgp]quit
```

AR2 的配置：

```
[AR2]bgp 102
[AR2-bgp]peer 10.0.12.1 as-number 101
```

```
[AR2-bgp]peer 10.0.23.3 as-number 103
[AR2-bgp]quit
```

AR3 的配置：

```
[AR3]bgp 103
[AR3-bgp]peer 10.0.23.2 as-number 102
[AR3-bgp]quit
```

宣告 BGP 路由：

```
[AR1]interface LoopBack 1
[AR1-LoopBack1]ip address 10.1.1.1 32
[AR1-LoopBack1]quit
[AR1]interface LoopBack 2
[AR1-LoopBack2]ip address 10.2.2.2 32
[AR1-LoopBack2]quit
[AR1]bgp 101
[AR1-bgp]network 10.1.1.1 32
[AR1-bgp]network 10.2.2.2 32
[AR1-bgp]quit
```

步骤 3：实验调试。

在 AR1 上部署路由策略，使其在通告 BGP 路由 10.1.1.1/32 时携带 Community 属性值 100：1，使其在通告 BGP 路由 10.2.2.2 时携带 Community 属性值 101：1。

```
[AR1]ip ip-prefix 1 permit 10.1.1.1 32                    //前缀列表1匹配10.1.1.1
[AR1]ip ip-prefix 2 permit 10.2.2.2 32                    //前缀列表2匹配10.2.2.2
[AR1]route-policy joinlabs permit node 10
[AR1-route-policy]if-match ip-prefix 1
[AR1-route-policy]apply community 100:1
[AR1-route-policy]quit
[AR1]route-policy joinlabs permit node 20
[AR1-route-policy]if-match ip-prefix 2
[AR1-route-policy]apply community 101:1
[AR1-route-policy]quit
[AR1]route-policy joinlabs permit node 30
[AR1-route-policy]quit
[AR1]bgp 101
[AR1-bgp]peer 10.0.12.2 route-policy joinlabs export
[AR1-bgp]peer 10.0.12.2 advertise-community         //路由更新时发送Community属性
[AR1-bgp]quit
[AR2]bgp 102
[AR2-bgp]peer 10.0.23.3 advertise-community
[AR2-bgp]quit
```

在 AR3 上查看 10.1.1.1 和 10.2.2.2 的 BGP 详细路由：

```
<AR3>display bgp routing-table 10.1.1.1
BGP local router ID : 10.0.23.3
Local AS number : 103
Paths:   1 available, 1 best, 1 select
```

```
BGP routing table entry information of 10.1.1.1/32:
From: 10.0.23.2 (10.0.23.2)
Route Duration: 00h06m22s
Direct Out-interface: GigabitEthernet0/0/1
Original nexthop: 10.0.23.2
Qos information : 0x0
Community:<100:1>
AS-path 102 101, origin igp, pref-val 0, valid, external, best, select, active, pre 255
Not advertised to any peer yet
<AR3>display bgp routing-table 10.2.2.2

BGP local router ID : 10.0.23.3
Local AS number : 103
Paths:   1 available, 1 best, 1 select
BGP routing table entry information of 10.2.2.2/32:
From: 10.0.23.2 (10.0.23.2)
Route Duration: 00h07m07s
Direct Out-interface: GigabitEthernet0/0/1
Original nexthop: 10.0.23.2
Qos information : 0x0
Community:<101:1>
AS-path 102 101, origin igp, pref-val 0, valid, external, best, select, active, pre 255
Not advertised to any peer yet
```

通过以上输出可以看出，10.1.1.1 携带了 100 : 1 的 Community 属性，10.2.2.2 携带了 101 : 1 的 Community 属性。

① AR2 传递路由给 EBGP 对等体 AR3，在 AR2 上部署路由策略，过滤掉携带 101 : 1 的 Community 属性值的路由。

配置 Community Filter，匹配 Community 属性中包含 100 : 1 的路由：

```
[AR2]ip community-filter  1 permit 100:1
```

调用 Community Filter：

```
[AR2]route-policy joinlabs deny node 10
[AR2-route-policy]if-match community-filter 1
[AR2-route-policy]quit
[AR2]route-policy joinlabs permit node 20
[AR2-route-policy]quit
[AR2]bgp 102
[AR2-bgp]peer 10.0.23.3 route-policy joinlabs export
[AR2-bgp]quit
```

②在 AR3 上查看 BGP 的路由表。

```
<AR3>display bgp routing-table
BGP Local router ID is 10.0.23.3
Status codes: * - valid, > - best, d - damped,
              h - history,  i - internal, s - suppressed, S - Stale
              Origin : i - IGP, e - EGP, ? - incomplete
Total Number of Routes: 1
```

```
     Network          NextHop        MED        LocPrf      PrefVal       Path/Ogn
 *>  10.2.2.2/32      10.0.23.2                             0             102 101i
```

通过以上输出可以看出，10.1.1.1 的路由不见了，它的 Community 属性值为 100：1。

3.5.3 配置 ORF

1. 实验目的

AR1 和 AR2 建立 EBGP 邻居，在 AR2 上配置 ORF 功能，实现 AR1 按需发送部分 BGP 路由给 AR2。

2. 实验拓扑

配置 ORF 的实验拓扑如图 3-12 所示。

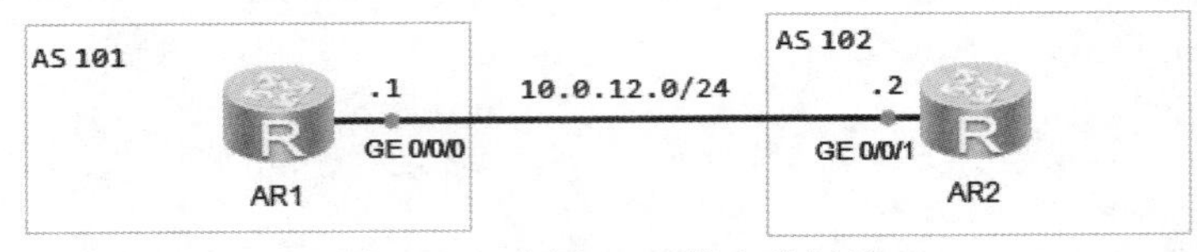

图 3-12　配置 ORF 的实验拓扑

3. 实验步骤

步骤 1： 配置 IP 地址。

AR1 的配置：

```
<Huawei>system-view
Enter system view, return user view with Ctrl+Z.
[Huawei]undo info-center enable
[Huawei]sysname AR1
[AR1]interface g0/0/0
[AR1-GigabitEthernet0/0/0]ip address 10.0.12.1 24
[AR1-GigabitEthernet0/0/0]quit
[AR1]interface LoopBack 0
[AR1-LoopBack0]ip address 1.1.1.1 32
[AR1-LoopBack0]quit
```

AR2 的配置：

```
<Huawei>system-view
Enter system view, return user view with Ctrl+Z.
[Huawei]undo info-center enable
[Huawei]sysname AR2
[AR2]interface g0/0/1
[AR2-GigabitEthernet0/0/1]ip address 10.0.12.2 24
[AR2-GigabitEthernet0/0/1]quit
[AR2]interface LoopBack 0
[AR2-LoopBack0]ip address 2.2.2.2 32
[AR2-LoopBack0]quit
```

步骤 2： 配置 BGP。

AR1 的配置：

```
[AR1]bgp 101
[AR1-bgp]peer 10.0.12.2 as-number 102
[AR1-bgp]quit
```

AR2 的配置：

```
[AR2]bgp 102
[AR2-bgp]peer 10.0.12.1 as-number 101
[AR2-bgp]quit
```

在 AR1 上创建 3 个环回口，分别为 10.1.1.1、10.2.2.2、10.3.3.3，并宣告进 BGP 中。

```
[AR1]interface LoopBack 1
[AR1-LoopBack1]ip address 10.1.1.1 32
[AR1-LoopBack1]quit
[AR1]interface LoopBack 2
[AR1-LoopBack2]ip address 10.2.2.2 32
[AR1-LoopBack2]quit
[AR1]interface LoopBack 3
[AR1-LoopBack3]ip address 10.3.3.3 32
[AR1-LoopBack3]quit
[AR1]bgp 101
[AR1-bgp]network 10.1.1.1 32
[AR1-bgp]network 10.2.2.2 32
[AR1-bgp]network 10.3.3.3 32
[AR1-bgp]quit
```

步骤 3：实验调试。

在 AR2 上查看 BGP 的路由表：

```
[AR2]display bgp routing-table
 BGP Local router ID is 10.0.12.2
 Status codes: * - valid, > - best, d - damped,
               h - history,  i - internal, s - suppressed, S - Stale
               Origin : i - IGP, e - EGP, ? - incomplete
 Total Number of Routes: 3
      Network          NextHop        MED        LocPrf       PrefVal      Path/Ogn
 *>   10.1.1.1/32      10.0.12.1      0                       0            101i
 *>   10.2.2.2/32      10.0.12.1      0                       0            101i
 *>   10.3.3.3/32      10.0.12.1      0                       0            101i
```

通过以上输出可以看出，3 条 BGP 的路由都是由 AR1 产生的。

在 AR2 上配置基于 10.1.1.1/32 的路由过滤策略，并使能 ORF 功能，发送 ORF 报文：

```
[AR2]ip ip-prefix 1 permit 10.1.1.1 32
[AR2]bgp 102
[AR2-bgp]peer 10.0.12.1 ip-prefix 1 import
[AR2-bgp]peer 10.0.12.1 capability-advertise orf ip-prefix send
                                  //使能基于地址前缀的 ORF 功能（只允许发送 ORF 报文）
[AR2-bgp]quit
```

在 AR1 上使能 ORF 功能，接收 ORF 报文：

```
[AR1]bgp 101
[AR1-bgp]peer 10.0.12.2 capability-advertise orf ip-prefix  receive
                                  //使能基于地址前缀的 ORF 功能（只允许接收 ORF 报文）
```

在 AR1 上查看从 10.0.12.2 收到的基于地址前缀的 ORF 信息。

```
<AR1>display bgp peer 10.0.12.2 orf ip-prefix
 Total number of ip-prefix received: 1
 Index        Action       Prefix          MaskLen         MinLen          MaxLen
 10           Permit       10.1.1.1        32
```

通过以上输出可以看出，前缀列表只允许 10.1.1.1。

在 AR3 上查看 BGP 的路由表：

```
<AR2>display bgp routing-table
 BGP Local router ID is 10.0.12.2
 Status codes: * - valid, > - best, d - damped,
               h - history,  i - internal, s - suppressed, S - Stale
               Origin : i - IGP, e - EGP, ? - incomplete
 Total Number of Routes: 1
      Network          NextHop        MED        LocPrf         PrefVal        Path/Ogn
 *>   10.1.1.1/32      10.0.12.1      0          0              101i
```

通过以上输出可以看出，AR2 只有 10.1.1.1 这一条路由，实现了实验要求。

3.5.4 配置 BGP 对等体组

1. 实验目的

AR4 需要和 AS 101 中的设备建立 EBGP 邻居关系，并且需要和 AS 104 中的设备建立 IBGP 邻居关系，要求通过以对等体组的方式建立邻居关系，简化 AR4 的 BGP 配置。

2. 实验拓扑

配置 BGP 对等体组的实验拓扑如图 3-13 所示。

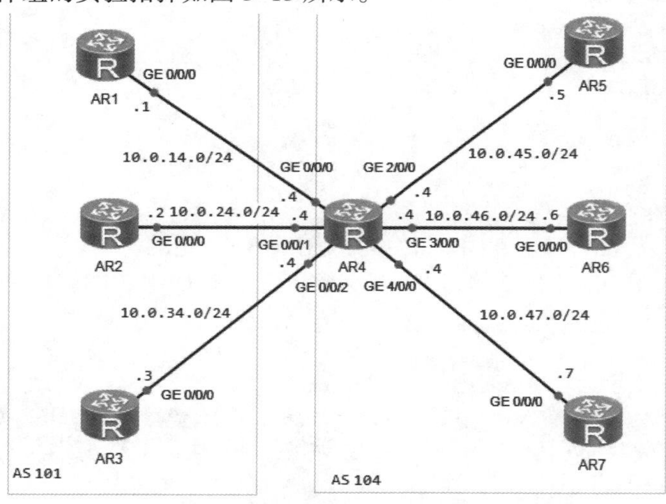

图 3-13　配置 BGP 对等体组的实验拓扑

3. 实验步骤

步骤 1：配置 IP 地址。

AR1 的配置：

```
<Huawei>system-view
```

```
[Huawei]undo info-center enable
[AR1]interface G0/0/0
[AR1-GigabitEthernet0/0/0]ip address 10.0.14.1 24
[AR1-GigabitEthernet0/0/0]quit
[AR1]interface LoopBack 0
[AR1-LoopBack0]ip address 1.1.1.1 32
[AR1-LoopBack0]quit
```

AR2 的配置：

```
<Huawei>system-view
Enter system view, return user view with Ctrl+Z.
[Huawei]undo info-center enable
Info: Information center is disabled.
[Huawei]sysname AR2
[AR2]interface g0/0/0
[AR2-GigabitEthernet0/0/0]ip address 10.0.24.2 24
[AR2-GigabitEthernet0/0/0]quit
[AR2]interface LoopBack 0
[AR2-LoopBack0]ip address 2.2.2.2 32
[AR2-LoopBack0]quit
```

AR3 的配置：

```
<Huawei>system-view
Enter system view, return user view with Ctrl+Z.
[Huawei]undo info-center enable
[Huawei]sysname AR3
[AR3]interface g0/0/0
[AR3-GigabitEthernet0/0/0]ip address 10.0.34.3 24
[AR3-GigabitEthernet0/0/0]quit
[AR3]interface LoopBack 0
[AR3-LoopBack0]ip address 3.3.3.3 32
[AR3-LoopBack0]quit
```

AR4 的配置：

```
<Huawei>system-view
Enter system view, return user view with Ctrl+Z.
[Huawei]undo info-center enable
Info: Information center is disabled.
[Huawei]sysname AR4
[AR4]interface g0/0/0
[AR4-GigabitEthernet0/0/0]ip address 10.0.14.4 24
[AR4-GigabitEthernet0/0/0]quit
[AR4]interface g0/0/1
[AR4-GigabitEthernet0/0/1]ip address 10.0.24.4 24
[AR4-GigabitEthernet0/0/1]quit
[AR4]interface g0/0/2
[AR4-GigabitEthernet0/0/2]ip address 10.0.34.4 24
[AR4-GigabitEthernet0/0/2]quit
```

```
[AR4]interface g2/0/0
[AR4-GigabitEthernet2/0/0]ip address 10.0.45.4 24
[AR4-GigabitEthernet2/0/0]quit
[AR4]interface g3/0/0
[AR4-GigabitEthernet3/0/0]ip address 10.0.46.4 24
[AR4-GigabitEthernet3/0/0]quit
[AR4]interface g4/0/0
[AR4-GigabitEthernet4/0/0]ip address 10.0.47.4 24
[AR4-GigabitEthernet4/0/0]quit
[AR4]interface LoopBack 0
[AR4-LoopBack0]ip address 4.4.4.4 32
[AR4-LoopBack0]quit
```

AR5 的配置：

```
<Huawei>system-view
Enter system view, return user view with Ctrl+Z.
[Huawei]undo info-center enable
Info: Information center is disabled.
[Huawei]sysname AR5
[AR5]interface g0/0/0
[AR5-GigabitEthernet0/0/0]ip address 10.0.45.5 24
[AR5-GigabitEthernet0/0/0]quit
[AR5]interface LoopBack 0
[AR5-LoopBack0]ip address 5.5.5.5 32
[AR5-LoopBack0]quit
```

AR6 的配置：

```
<Huawei>system-view
Enter system view, return user view with Ctrl+Z.
[Huawei]undo info-center enable
Info: Information center is disabled.
[Huawei]sysname AR6
[AR6]interface g0/0/0
[AR6-GigabitEthernet0/0/0]ip address 10.0.46.6 24
[AR6-GigabitEthernet0/0/0]quit
[AR6]interface LoopBack 0
[AR6-LoopBack0]ip address 6.6.6.6 32
[AR6-LoopBack0]quit
```

AR7 的配置：

```
<Huawei>system-view
Enter system view, return user view with Ctrl+Z.
[Huawei]undo info-center enable
Info: Information center is disabled.
[Huawei]sysname AR7
[AR7]interface g0/0/0
[AR7-GigabitEthernet0/0/0]ip address 10.0.47.7 24
[AR7-GigabitEthernet0/0/0]quit
```

```
[AR7]interface LoopBack 0
[AR7-LoopBack0]ip address 7.7.7.7 32
[AR7-LoopBack0]quit
```

步骤 2：配置 OSPF（在 AS 104 内部运行 OSPF）。

AR4 的配置：

```
[AR4]ospf router-id 4.4.4.4
[AR4-ospf-1]area 0
[AR4-ospf-1-area-0.0.0.0]network 10.0.45.0 0.0.0.255
[AR4-ospf-1-area-0.0.0.0]network 10.0.46.0 0.0.0.255
[AR4-ospf-1-area-0.0.0.0]network 10.0.47.0 0.0.0.255
[AR4-ospf-1-area-0.0.0.0]network 4.4.4.4 0.0.0.0
[AR4-ospf-1-area-0.0.0.0]quit
```

AR5 的配置：

```
[AR5]ospf router-id 5.5.5.5
[AR5-ospf-1]area 0
[AR5-ospf-1-area-0.0.0.0]network 10.0.45.0 0.0.0.255
[AR5-ospf-1-area-0.0.0.0]network 5.5.5.5 0.0.0.0
[AR5-ospf-1-area-0.0.0.0]quit
```

AR6 的配置：

```
[AR6]ospf router-id 6.6.6.6
[AR6-ospf-1]area 0
[AR6-ospf-1-area-0.0.0.0]network 10.0.46.0 0.0.0.255
[AR6-ospf-1-area-0.0.0.0]network 6.6.6.6 0.0.0.0
[AR6-ospf-1-area-0.0.0.0]quit
```

AR7 的配置：

```
[AR7]ospf router-id 7.7.7.7
[AR7-ospf-1]area 0
[AR7-ospf-1-area-0.0.0.0]network 10.0.47.0 0.0.0.255
[AR7-ospf-1-area-0.0.0.0]network 7.7.7.7 0.0.0.0
[AR7-ospf-1-area-0.0.0.0]quit
```

步骤 3：配置 IBGP。

AR4 的配置：

```
[AR4]bgp 104
[AR4-bgp]group 1 internal             //创建对等体组 1，为 IBGP 对等体组
[AR4-bgp]peer 5.5.5.5 group 1
[AR4-bgp]peer 6.6.6.6 group 1
[AR4-bgp]peer 7.7.7.7 group 1
[AR4-bgp]peer 1 connect-interface LoopBack 0
[AR4-bgp]peer 1 next-hop-local
[AR4-bgp]quit
```

AR5 的配置：

```
[AR5]bgp 104
[AR5-bgp]peer 4.4.4.4 as-number 104
```

```
[AR5-bgp]peer 4.4.4.4 connect-interface LoopBack 0
[AR5-bgp]quit
```

AR6 的配置：

```
[AR6]bgp 104
[AR6-bgp]peer 4.4.4.4 as-number 104
[AR6-bgp]peer 4.4.4.4 connect-interface LoopBack 0
[AR6-bgp]quit
```

AR7 的配置：

```
[AR7]bgp 104
[AR7-bgp]peer 4.4.4.4 as-number 104
[AR7-bgp]peer 4.4.4.4 connect-interface LoopBack 0
[AR7-bgp]quit
```

步骤 4：配置 EBGP。

AR4 的配置：

```
[AR4]bgp 104
[AR4-bgp]group 2 external            //创建对等体组 2，为 EBGP 对等体组
[AR4-bgp]peer 2 as-number 101
[AR4-bgp]peer 10.0.14.1 group 2
[AR4-bgp]peer 10.0.24.2 group 2
[AR4-bgp]peer 10.0.34.3 group 2
[AR4-bgp]quit
```

AR1 的配置：

```
[AR1]bgp 101
[AR1-bgp]pe
[AR1-bgp]peer 10.0.14.4 as
[AR1-bgp]peer 10.0.14.4 as-number 104
[AR1-bgp]quit
```

AR2 的配置：

```
[AR2]bgp 101
[AR2-bgp]peer 10.0.24.4 as-number 104
[AR2-bgp]quit
```

AR3 的配置：

```
[AR3]bgp 101
[AR3-bgp]peer 10.0.34.4 as-number 104
[AR3-bgp]quit
```

步骤 5：实验调试。

```
[AR4]display bgp peer
 BGP local router ID : 10.0.14.4
 Local AS number : 104
 Total number of peers : 6                Peers in established state : 6
       Peer     V    AS    MsgRcvd    MsgSent   OutQ    Up/Down     State         PrefRcv
    5.5.5.5     4    104   22         23        0       00:20:06    Established   0
    6.6.6.6     4    104   21         22        0       00:19:17    Established   0
```

```
     7.7.7.7    4    104    20        21          0          00:18:38    Established    0
   10.0.14.1    4    101    3         5           0          00:01:13    Established    0
   10.0.24.2    4    101    2         3           0          00:00:58    Established    0
   10.0.34.3    4    101    2         3           0          00:00:42    Established    0
```

通过以上输出可以看出，BGP 邻居关系都已建立成功。

3.5.5　配置 BGP 的安全性

1. 实验目的

AR1 和 AR2、AR3 设备需要建立 IBGP 邻居关系，要求在 AR1 和 AR2 之间配置 BGP 的 MD5 认证，在 AR1 和 AR3 之间配置 BGP 的 Keychain 认证保护邻居建立的安全性，并且在 AR2 和 AR3 之间配置 BGP 的 GTSM，以保证报文交互的安全性。

2. 实验拓扑

配置 BGP 的安全性的实验拓扑如图 3-14 所示。

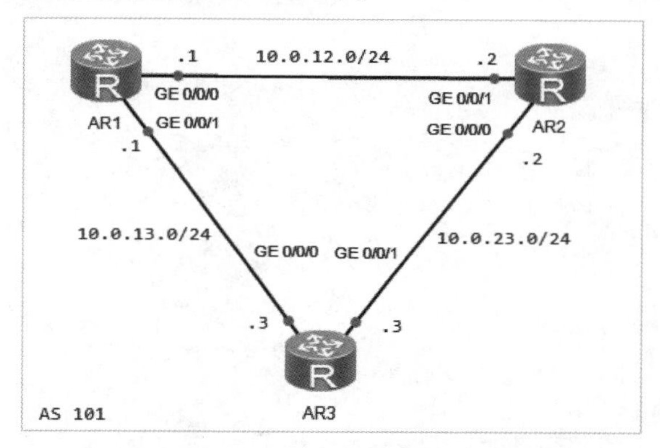

图 3-14　配置 BGP 的安全性的实验拓扑

3. 实验步骤

步骤 1： 配置 IP 地址。

AR1 的配置：

```
<Huawei>system-view
Enter system view, return user view with Ctrl+Z.
[Huawei]undo info-center enable
Info: Information center is disabled.
[Huawei]sysname AR1
[AR1]interface g0/0/0
[AR1-GigabitEthernet0/0/0]ip address 10.0.12.1 24
[AR1-GigabitEthernet0/0/0]quit
[AR1]interface g0/0/1
[AR1-GigabitEthernet0/0/1]ip address 10.0.13.1 24
[AR1-GigabitEthernet0/0/1]quit
[AR1]interface LoopBack 0
[AR1-LoopBack0]ip address 1.1.1.1 32
[AR1-LoopBack0]quit
```

AR2 的配置：

```
<Huawei>system-view
Enter system view, return user view with Ctrl+Z.
[Huawei]undo info-center enable
Info: Information center is disabled.
[Huawei]sysname AR2
[AR2]interface g0/0/1
[AR2-GigabitEthernet0/0/1]ip address 10.0.12.2 24
[AR2-GigabitEthernet0/0/1]quit
[AR2]interface g0/0/0
[AR2-GigabitEthernet0/0/0]ip address 10.0.23.2 24
[AR2-GigabitEthernet0/0/0]quit
[AR2]interface LoopBack 0
[AR2-LoopBack0]ip address 2.2.2.2 32
[AR2-LoopBack0]quit
```

AR3 的配置：

```
<Huawei>system-view
Enter system view, return user view with Ctrl+Z.
[Huawei]undo info-center enable
Info: Information center is disabled.
[Huawei]sysname AR3
[AR3]interface g0/0/0
[AR3-GigabitEthernet0/0/0]ip address 10.0.13.3 24
[AR3-GigabitEthernet0/0/0]quit
[AR3]interface g0/0/1
[AR3-GigabitEthernet0/0/1]ip address 10.0.23.3 24
[AR3-GigabitEthernet0/0/1]quit
[AR3]interface LoopBack 0
[AR3-LoopBack0]ip address 3.3.3.3 32
[AR3-LoopBack0]quit
```

步骤 2: 配置 OSPF。

AR1 的配置：

```
[AR1]ospf router-id 1.1.1.1
[AR1-ospf-1]area 0
[AR1-ospf-1-area-0.0.0.0]network 10.0.12.0 0.0.0.255
[AR1-ospf-1-area-0.0.0.0]network 10.0.13.0 0.0.0.255
[AR1-ospf-1-area-0.0.0.0]network 1.1.1.1 0.0.0.0
[AR1-ospf-1-area-0.0.0.0]quit
```

AR2 的配置：

```
[AR2]ospf router-id 2.2.2.2
[AR2-ospf-1]area 0
[AR2-ospf-1-area-0.0.0.0]network 10.0.12.0 0.0.0.255
[AR2-ospf-1-area-0.0.0.0]network 10.0.23.0 0.0.0.255
[AR2-ospf-1-area-0.0.0.0]network 2.2.2.2 0.0.0.0
[AR2-ospf-1-area-0.0.0.0]quit
```

AR3 的配置：

```
[AR3]ospf router-id 3.3.3.3
[AR3-ospf-1]area 0
[AR3-ospf-1-area-0.0.0.0]network 10.0.13.0 0.0.0.255
[AR3-ospf-1-area-0.0.0.0]network 10.0.23.0 0.0.0.255
[AR3-ospf-1-area-0.0.0.0]network 3.3.3.3 0.0.0.0
[AR3-ospf-1-area-0.0.0.0]quit
```

步骤 3: 配置 BGP。

AR1 的配置：

```
[AR1]bgp 101
[AR1-bgp]peer 2.2.2.2 as-number 101
[AR1-bgp]peer 2.2.2.2 connect-interface LoopBack 0
[AR1-bgp]peer 3.3.3.3 as-number 101
[AR1-bgp]peer 3.3.3.3 connect-interface LoopBack 0
[AR1-bgp]quit
```

AR2 的配置：

```
[AR2]bgp 101
[AR2-bgp]peer 1.1.1.1 as-number 101
[AR2-bgp]peer 1.1.1.1 connect-interface LoopBack 0
[AR2-bgp]peer 3.3.3.3 as-number 101
[AR2-bgp]peer 3.3.3.3 connect-interface LoopBack 0
[AR2-bgp]quit
```

AR3 的配置：

```
[AR3]bgp 101
[AR3-bgp]peer 1.1.1.1 as-number 101
[AR3-bgp]peer 2.2.2.2 as-number 101
[AR3-bgp]peer 1.1.1.1 connect-interface LoopBack 0
[AR3-bgp]peer 2.2.2.2 connect-interface LoopBack 0
[AR3-bgp]quit
```

步骤 4: 实验调试。

在 AR1 与 AR2 之间配置 MD5 认证：

AR1 的配置：

```
[AR1]bgp 101
[AR1-bgp]peer 2.2.2.2 password cipher 1234
[AR1-bgp]quit
```

AR2 的配置：

```
[AR2]bgp 101
[AR2-bgp]peer 1.1.1.1 password cipher 1234
[AR2-bgp]quit
```

【技术要点】MD5认证

在AR1上查看BGP对等体的认证信息：

```
[AR1]display bgp peer 2.2.2.2 verbose
        BGP Peer is 2.2.2.2,  remote AS 101
        Type: IBGP link
        BGP version 4, Remote router ID 10.0.23.2
        Update-group ID: 1
        BGP current state: Established, Up for 00h15m15s
        BGP current event: RecvKeepalive
        BGP last state: OpenConfirm
        BGP Peer Up count: 1
        Received total routes: 0
        Received active routes total: 0
        Advertised total routes: 0
        Port:  Local - 49215    Remote - 179
        Configured: Connect-retry Time: 32 sec
        Configured: Active Hold Time: 180 sec   Keepalive Time:60 sec
        Received  : Active Hold Time: 180 sec
        Negotiated: Active Hold Time: 180 sec   Keepalive Time:60 sec
        Peer optional capabilities:
        Peer supports bgp multi-protocol extension
        Peer supports bgp route refresh capability
        Peer supports bgp 4-byte-as capability
        Address family IPv4 Unicast: advertised and received
Received: Total 17 messages
                Update messages                0
                Open messages                  1
                KeepAlive messages             16
                Notification messages          0
                Refresh messages               0
Sent: Total 17 messages
                Update messages                0
                Open messages                  1
                KeepAlive messages             16
                Notification messages          0
                Refresh messages               0
Authentication type configured: MD5
Last keepalive received: 2023/01/30 15:30:24 UTC-08:00
Last keepalive sent    : 2023/01/30 15:30:23 UTC-08:00
Minimum route advertisement interval is 15 seconds
Optional capabilities:
Route refresh capability has been enabled
4-byte-as capability has been enabled
Connect-interface has been configured
Peer Preferred Value: 0
Routing policy configured:
No routing policy is configured
```

通过以上输出可以看出，认证的类型为MD5。

在 AR1 与 AR3 上配置 Keychain 认证。

（1）创建 Keychain。

```
[AR1]keychain lw mode absolute      //创建一个名为lw的Keychain,模式为绝对时间(周期性,如每天)
[AR1-keychain]receive-tolerance 10 //接收容错时间为10 min
[AR1-keychain]tcp-kind 182
[AR1-keychain]tcp-algorithm-id hmac-md5 50
[AR1-keychain]quit
```

【技术要点1】receive-tolerance

（1）由于网络环境或者两端时间未同步等因素影响，Key ID发送的报文可能存在延迟，即接收端在非活跃时间段收到发送端发送的报文，将导致因接收端处于非活跃状态而丢弃该报文，使协议报文传输中断。为了解决该问题，用户可以配置接收容错时间，使接收端在配置的接收时间到达后再持续生效指定的时间，保证使接收端能处理发送端发送的所有报文，保证协议报文的正常处理。

（2）配置接收容错时间将在接收时间的起始时间和结束时间同时增加相应的时间。

【技术要点2】kind-value

（1）对于由TCP连接建立的认证交互，TCP使用增强的TCP认证选项。目前，不同的厂商使用不同的kind-value值代表增强的TCP认证选项。为了实现不同厂商设备之间的交互，用户可以选择配置kind-value，使通信两端设备的TCP类型一致。

（2）配置kind-value值时，采用Keychain认证的通信双方必须采用TCP建立连接的应用程序；否则配置的TCP认证不生效。

（3）通信双方的TCP类型配置一致后，还需要指定通信双方相同的认证算法对应的ID。

（2）配置 Keychain 中的 Key。

AR1 的配置：

```
[AR1-keychain]key-id 1  //key1
[AR1-keychain-keyid-1]algorithm hmac-md5                  //算法为hmac-md5
[AR1-keychain-keyid-1]key-string cipher joinlabs1         //密码为joinlabs1
[AR1-keychain-keyid-1]send-time utc 12:00 2023-01-30 to 17:00 2023-01-30   //发送时间
[AR1-keychain-keyid-1]receive-time utc 12:00 2023-01-30 to 17:00 2023-01-30  //接收时间
[AR1-keychain-keyid-1]quit
[AR1-keychain]key-id 2
[AR1-keychain-keyid-2]algorithm hmac-md5
[AR1-keychain-keyid-2]key-string cipher joinlabs2
[AR1-keychain-keyid-2]send-time utc 17:01 2023-01-30 to 18:00 2023-01-30
[AR1-keychain-keyid-2]receive-time utc 17:01 2023-01-30 to 18:00 2023-01-30
[AR1-keychain-keyid-2]quit
```

AR3 的配置：

```
[AR3]keychain lw mode absolute
[AR3-keychain]receive-tolerance 10
[AR3-keychain]tcp-kind 182
[AR3-keychain]tcp-algorithm-id hmac-md5 50
```

```
[AR3-keychain]quit
[AR3]keychain lw
[AR3-keychain]key-id 1
[AR3-keychain-keyid-1]algorithm hmac-md5
[AR3-keychain-keyid-1]key-string cipher joinlabs1
[AR3-keychain-keyid-1]send-time utc 12:00 2023-01-30 to 17:00 2023-01-30
[AR3-keychain-keyid-1]receive-time utc 12:00 2023-01-30 to 17:00 2023-01-30
[AR3-keychain-keyid-1]quit
[AR3-keychain]key-id 2
[AR3-keychain-keyid-2]algorithm hmac-md5
[AR3-keychain-keyid-2]key-string cipher joinlabs2
[AR3-keychain-keyid-2]send-time utc 17:01 2023-01-30 to 18:00 2023-01-30
[AR3-keychain-keyid-2]receive-time utc 17:01 2023-01-30 to 18:00 2023-01-30
[AR3-keychain-keyid-2]quit
```

（3）配置 Keychain 认证。

AR1 的配置：

```
[AR1]bgp 101
[AR1-bgp]peer 3.3.3.3 keychain lw
[AR1-bgp]quit
```

AR3 的配置：

```
[AR3]bgp 101
[AR3-bgp]peer 1.1.1.1 keychain lw
[AR3-bgp]quit
```

查看当前处于 Active 状态的 Key ID：

```
<AR1>display keychain lw
 Keychain Information:
 ----------------------
 Keychain Name                : lw
   Timer Mode                 : Absolute
   Receive Tolerance(min)     : 10
   TCP Kind                   : 182
   TCP Algorithm IDs          :
     HMAC-MD5                 : 50
     HMAC-SHA1-12             : 2
     HMAC-SHA1-20             : 6
     MD5                      : 3
     SHA1                     : 4
 Number of Key IDs            : 2
 Active Send Key ID           : 2
 Active Receive Key IDs       : 01 02
 Default send Key ID          : Not configured
 Key ID Information:
 --------------------
 Key ID                       : 1
   Key string                 : %$%$Z4^:U0Tz^.Lup[UrVa`+$UBU%$%$ (cipher)
```

```
   Algorithm                    : HMAC-MD5
   SEND TIMER                   :
     Start time                 : 2023-01-30 12:00
     End time                   : 2023-01-30 17:00
     Status                     : Inactive
   RECEIVE TIMER                :
     Start time                 : 2023-01-30 12:00
     End time                   : 2023-01-30 17:00
     Status                     : Active

 Key ID                         : 2
   Key string                   : %$%$H>04M1/+dUKY`5:znJ+/$Zh9%$%$ (cipher)
   Algorithm                    : HMAC-MD5
   SEND TIMER                   :
     Start time                 : 2023-01-30 17:01
     End time                   : 2023-01-30 18:00
     Status                     : Active
   RECEIVE TIMER                :
     Start time                 : 2023-01-30 17:01
     End time                   : 2023-01-30 18:00
     Status                     : Active
```

通过以上输出可以看出，Key1 是 Inactive，Key2 是 Active。

查看 BGP 对等体已配置的认证类型是否是 Keychain：

```
<AR1>display bgp peer 3.3.3.3 verbose
        BGP Peer is 3.3.3.3,  remote AS 101
        Type: IBGP link
        BGP version 4, Remote router ID 10.0.13.3
        Update-group ID: 1
        BGP current state: Established, Up for 02h06m29s
        BGP current event: KATimerExpired
        BGP last state: OpenConfirm
        BGP Peer Up count: 1
        Received total routes: 0
        Received active routes total: 0
        Advertised total routes: 0
        Port:  Local - 49215    Remote - 179
        Configured: Connect-retry Time: 32 sec
        Configured: Active Hold Time: 180 sec   Keepalive Time:60 sec
        Received  : Active Hold Time: 180 sec
        Negotiated: Active Hold Time: 180 sec   Keepalive Time:60 sec
        Peer optional capabilities:
        Peer supports bgp multi-protocol extension
        Peer supports bgp route refresh capability
        Peer supports bgp 4-byte-as capability
        Address family IPv4 Unicast: advertised and received
 Received: Total 128 messages
                 Update messages                0
```

```
                        Open messages                  1
                        KeepAlive messages             127
                        Notification messages          0
                        Refresh messages               0
        Sent: Total 128 messages
                        Update messages                0
                        Open messages                  1
                        KeepAlive messages             127
                        Notification messages          0
                        Refresh messages               0
        Authentication type configured: Keychain(lw)
        Last keepalive received: 2023/01/30 17:21:22 UTC-08:00
        Last keepalive sent    : 2023/01/30 17:21:23 UTC-08:00
        Minimum route advertisement interval is 15 seconds
        Optional capabilities:
        Route refresh capability has been enabled
        4-byte-as capability has been enabled
        Connect-interface has been configured
        Peer Preferred Value: 0
        Routing policy configured:
        No routing policy is configured
```

【技术要点】MD5与Keychain认证

MD5认证只能为TCP连接设置认证密码。

Keychain认证除了可以为TCP连接设置认证密码外，还可以对BGP协议报文进行认证。

BGP MD5认证与BGP Keychain认证互斥。

思考：过了这段时间，BGP连接是不是会失效？

在AR2和AR3上配置GTSM

```
[AR2]bgp 101
[AR2-bgp]peer 3.3.3.3 valid-ttl-hops 1  //在AR1与AR2之间开启GTSM。由于两台路由器直连，
//因此TTL到达对方的有效范围是[255,255]，此处的valid-ttl-hops值取1
[AR2-bgp]quit

[AR3]bgp 101
[AR3-bgp]peer 2.2.2.2 valid-ttl-hops 1
[AR3-bgp]quit
```

【技术要点】peer 3.3.3.3 valid-ttl-hops 1

（1）在向邻居3.3.3.3发送BGP报文时，TTL为255。

（2）在接收邻居3.3.3.3发来的BGP报文时，要求报文的TTL大于等于255-1+1。

（3）被检测报文的TTL值的有效范围为[255–hops+1, 255]。

例如，对于EBGP直连路由，hops的取值为1，即有效的TTL值设为255。默认情况下，参数hops取值为255，即TTL的有效值范围为[1, 255]。

第4章

网络安全技术

4.1 网络安全技术概述

目前网络中以太网技术的应用非常广泛。然而，各种网络攻击的存在（如针对 ARP、DHCP 等协议的攻击），不仅造成了网络合法用户无法正常访问网络资源，而且对网络信息安全构成了严重威胁，因此以太网交换的安全性越来越重要。

在大中型企业中通常会部署防火墙双机热备，这样可以保证当网络中的主用设备出现故障时，备用设备能够平滑地接替主用设备的工作，从而实现业务的不间断运行；防火墙虚拟系统是指将一台防火墙设备划分为多个虚拟系统，每台虚拟系统相当于一台真实的设备。

4.2 以太网交换安全

4.2.1 端口隔离

1. 隔离类型

（1）双向隔离。同一端口隔离组的接口之间隔离，不同端口隔离组的接口之间不隔离。端口隔离只是针对同一设备上的端口隔离组成员，对于不同设备上的接口而言，无法实现该功能。

（2）单向隔离。实现不同端口隔离组的接口之间的隔离。默认情况下，未配置端口单向隔离。

2. 隔离模式

（1）L2（二层隔离、三层互通）。隔离同一 VLAN 内的广播报文，但是不同端口下的用户还可以进行三层通信。默认情况下，端口隔离模式为二层隔离、三层互通。

采用二层隔离、三层互通的隔离模式时，在 VLANIF 接口上使能 VLAN 内 Proxy ARP 功能，配置 arp-proxy inner-sub-vlan-proxy enable，可以实现同一 VLAN 内的主机通信。

（2）ALL（二层、三层都隔离）。同一 VLAN 内的不同端口下，用户二层、三层彻底隔离，无法通信。

4.2.2 MAC 地址表安全

1. MAC 地址表项

（1）动态 MAC 地址表项：由接口通过报文中的源 MAC 地址学习获得，表项可老化。在系统复位、接口板热插拔或接口板复位后，动态表项会丢失。

（2）静态 MAC 地址表项：由用户手工配置并下发到各接口板，表项不老化。在系统复位、接口板热插拔或接口板复位后，保存的表项不会丢失。接口和 MAC 地址静态绑定后，其他接口收到源 MAC 是该 MAC 地址的报文将会被丢弃。

（3）黑洞 MAC 地址表项：由用户手工配置并下发到各接口板，表项不老化。配置黑洞 MAC 地址后，源 MAC 地址或目的 MAC 地址是该 MAC 地址的报文将会被丢弃。

2. MAC 地址表安全功能

（1）静态 MAC 地址表项：将一些固定的上行设备或信任用户的 MAC 地址配置为静态 MAC 地址表项，可以保证其安全通信。

（2）黑洞 MAC 地址表项：防止黑客通过 MAC 地址攻击网络，交换机对来自黑洞 MAC 或者去往黑洞 MAC 的报文采取丢弃处理。

（3）动态 MAC 地址表项的老化时间：合理配置动态 MAC 地址表项的老化时间，可以防止

MAC 地址爆炸式增长。

（4）禁止 MAC 地址学习功能：对于网络环境固定的场景或者已经明确转发路径的场景，通过配置禁止 MAC 地址学习功能，可以限制非信任用户接入，防止 MAC 地址攻击，提高网络安全性。

（5）限制 MAC 地址学习数：在安全性较差的网络环境中，通过限制 MAC 地址学习数，可以防止攻击者通过变换 MAC 地址进行攻击。

4.2.3　端口安全

安全 MAC 地址类型见表 4–1。

表 4–1　安全 MAC 地址类型

类　型	定　义	特　点
安全动态 MAC 地址	使能端口安全而未使能 Sticky MAC 功能时转换的 MAC 地址	设备重启后表项会丢失，需要重新学习。默认情况下不会被老化，只有在配置安全 MAC 的老化时间后才可以被老化
安全静态 MAC 地址	使能端口安全时，手工配置的静态 MAC 地址	不会被老化，手动保存配置后重启设备不会丢失
Sticky MAC 地址	使能端口安全后，又同时使能 Sticky MAC 功能后转换到的 MAC 地址	不会被老化，手动保存配置后重启设备不会丢失

安全 MAC 地址通常与安全保护动作结合使用，常见的安全保护动作如下。

（1）Restrict：丢弃源 MAC 地址不存在的报文并上报告警。

（2）Protect：只丢弃源 MAC 地址不存在的报文，不上报告警。

（3）Shutdown：接口状态被置为 error–down 并上报告警。

4.2.4　MAC 地址漂移的防止与检测

1. MAC 地址漂移概述

MAC 地址漂移是指交换机上一个 VLAN 内有两个端口学习到同一个 MAC 地址，后学习到的 MAC 地址表项覆盖原 MAC 地址表项的现象。正常情况下，网络中不会在短时间内出现大量 MAC 地址漂移的情况。如果出现这种情况，一般都意味着网络中存在环路或网络攻击行为。

2. 防止 MAC 地址漂移

如果由环路引起 MAC 地址漂移，则治本的方法是部署防环技术，如 STP，消除二层环路；如果由网络攻击等其他原因引起，则可以使用以下 MAC 地址防漂移特性。

（1）配置接口 MAC 地址学习优先级。当 MAC 地址在交换机的两个接口之间发生漂移时，可以将其中一个接口的 MAC 地址学习优先级提高。高优先级的接口学习到的 MAC 地址表项将覆盖低优先级接口学习到的 MAC 地址表项。

（2）配置不允许相同优先级接口 MAC 地址漂移。当伪造网络设备所连接口的 MAC 地址优先级与安全的网络设备相同时，后学习到的伪造网络设备 MAC 地址表项不会覆盖之前正确的表项。

3. MAC 地址漂移检测

交换机支持 MAC 地址漂移检测机制，主要分为以下两种方式。

（1）基于 VLAN 的 MAC 地址漂移检测。

①配置 VLAN 的 MAC 地址漂移检测功能可以检测指定 VLAN 下的所有 MAC 地址是否发生漂移。

②当 MAC 地址发生漂移后，可以配置指定的动作，如告警、阻断接口或阻断 MAC 地址。

（2）全局 MAC 地址漂移检测。

①该功能可以检测设备上的所有 MAC 地址是否发生了漂移。

②如果发生漂移，设备会上报告警到网管系统。

③用户也可以指定发生漂移后的处理动作。例如，将接口关闭或退出VLAN。

4.2.5 MACsec

1. MACsec 简介

（1）背景。绝大部分数据在局域网链路中都是以明文形式传输的，只在某些安全性要求较高的场景下才存在安全隐患。

（2）MACsec 概述。MACsec 定义了基于以太网的数据安全通信方法，通过逐跳设备之间数据的加密，保证数据传输的安全性，对应的标准为 802.1AE。它的功能如下：

①数据帧的完整性检查。

②用户数据加密。

③数据源的真实性校验。

④重放保护。

（3）MACsec 典型应用场景。如图 4-1 所示，在交换机之间部署 MACsec 保护数据安全。例如，在接入交换机与上联的汇聚或核心交换机之间部署 。当交换机之间存在传输设备时，可部署 MACsec 保护数据安全。

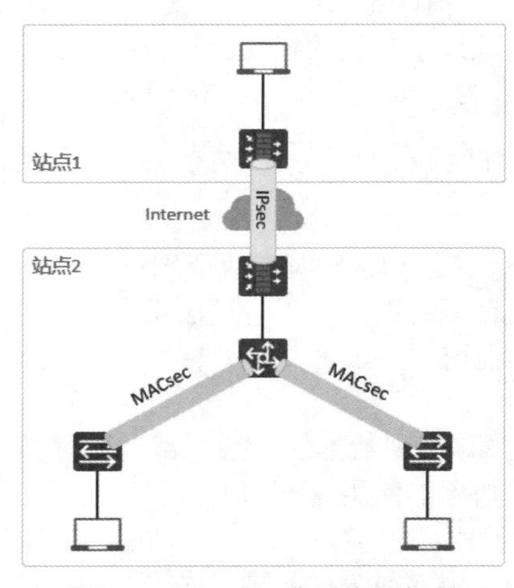

图 4-1　MACsec 典型应用场景

2. MACsec 工作机制

如图 4-2 所示，MACsec 工作机制如下：

（1）网络管理员在两台设备上通过命令行预配置相同的 CAK（Secure Connectivity Association Key，安全连接关联密钥）。

（2）两台设备会通过 MKA（MACsec Key Agreement Protocol，MACsec 数据加密密钥的协商协议）选举出一个 Key Server，该 Key Server 决定加密方案，并且会根据 CAK 等参数使用某种加密算法生成 SAK 数据密钥。

（3）由 Key Server 将 SAK 分发给对端设备，这样两台设备拥有相同的 SAK 数据密钥，可以进行后续 MACsec 数据报文加 / 解密收发。

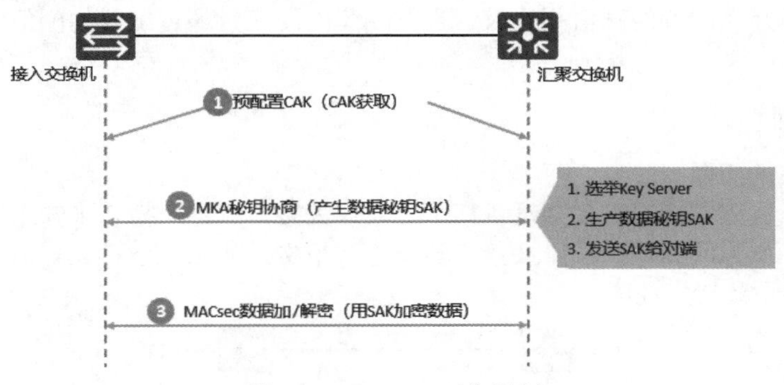

图 4–2　MACsec 工作机制

4.2.6　流量抑制

1. 网络中存在的问题

如图 4–3 所示，正常情况下，当设备某个二层以太接口收到广播、未知组播或未知单播报文时，会向同一 VLAN 内的其他二层以太接口转发这些报文，从而导致流量泛洪，降低设备转发性能；当设备某个以太接口收到已知组播或已知单播报文时，如果某种报文流量过大，则可能会对设备造成冲击，影响其他业务的正常处理。

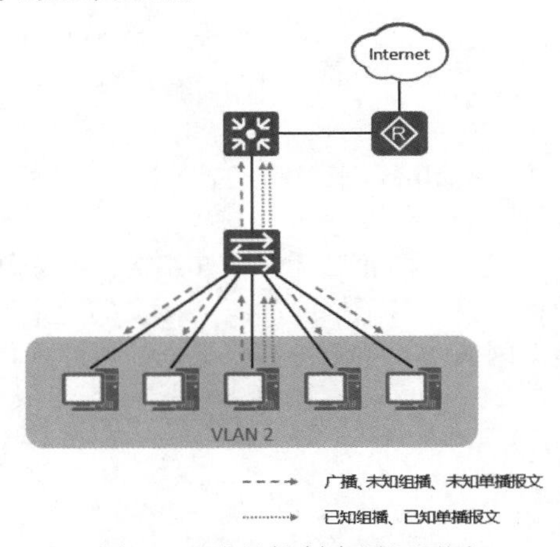

图 4–3　流量抑制应用组网图

2. 可用的解决方案

流量抑制可以通过配置阈值来限制广播、未知组播、未知单播、已知组播和已知单播报文的速率，从而防止广播、未知组播报文和未知单播报文产生流量泛洪，阻止已知组播报文和已知单播报文的大流量冲击。

3. 流量抑制的工作原理

（1）在接口入方向上，设备支持对广播、未知组播、未知单播、已知组播和已知单播报文按百分比、包速率和比特速率进行流量抑制。设备监控接口下的各类报文速率并与配置的阈值相比较，当入口流量超过配置的阈值时，设备会丢弃超额的流量。

（2）在 VLAN 视图下，设备支持对广播报文按比特速率进行流量抑制。设备监控同一 VLAN 内广播报文的速率并与配置的阈值相比较，当 VLAN 内的流量超过配置的阈值时，设备会丢弃超额的流量。

4.2.7 风暴控制

1. 网络中存在的问题

如图 4–4 所示，正常情况下，当设备某个二层以太接口收到广播、未知组播或未知单播报文时，会向同一 VLAN 内的其他二层以太接口转发这些报文，如果网络存在环路，则会导致广播风暴，从而严重降低设备转发性能。

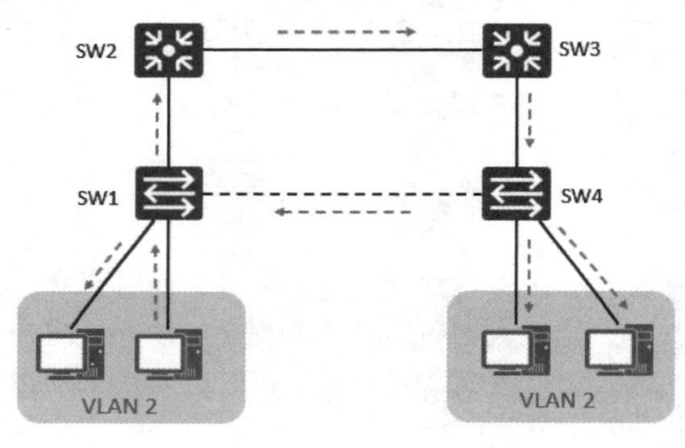

图 4–4 风暴控制应用组网图

2. 可用的解决方案

风暴控制可以通过阻塞报文或关闭端口来阻断广播、未知组播和未知单播报文的流量。

3. 风暴控制的工作原理

如图 4–5 所示，风暴控制可以用于防止广播、未知组播以及未知单播报文产生广播风暴。在风暴控制检测时间间隔内，设备监控接口下接收的三类报文的包平均速率与配置的最大阈值相比较。当报文速率大于配置的最大阈值时，风暴控制将根据配置的动作对接口进行阻塞报文或关闭接口的处理。

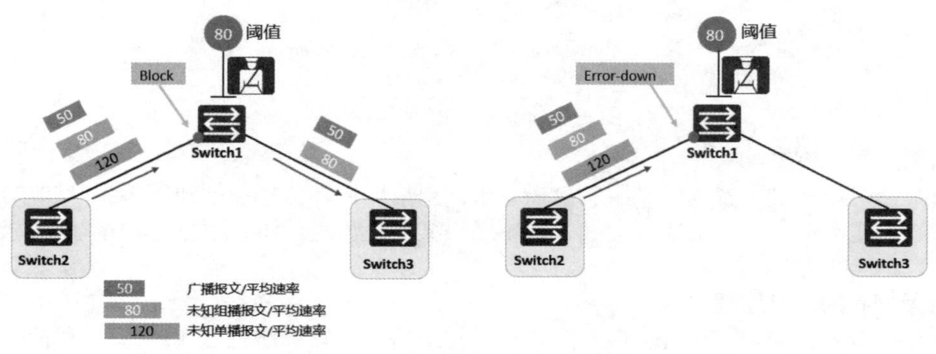

图 4–5 风暴控制的工作原理图

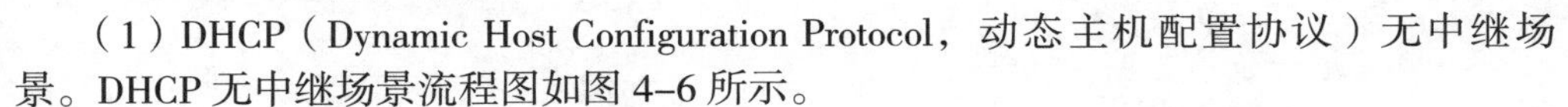

4.2.8 DHCP Snooping

1. DHCP 工作原理概述

（1）DHCP（Dynamic Host Configuration Protocol，动态主机配置协议）无中继场景。DHCP 无中继场景流程图如图 4–6 所示。

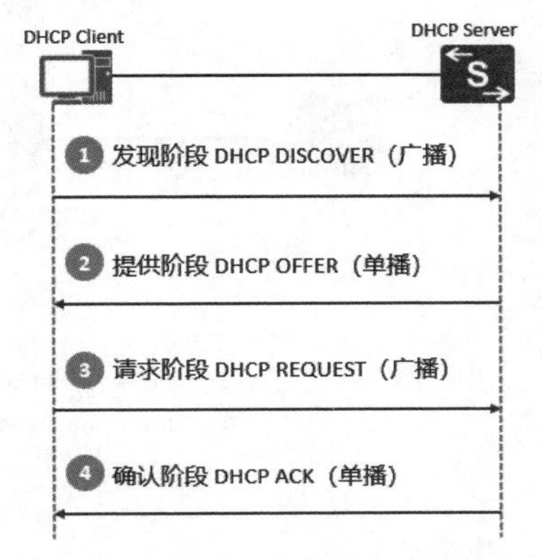

图 4–6　DHCP 无中继场景流程图

DHCP 无中继场景流程如下。

①发现阶段：DHCP 客户端通过发送 DHCP DISCOVER 报文（广播）来发现 DHCP 服务器。DHCP DISCOVER 报文中携带了客户端的 MAC 地址（DHCP DISCOVER 报文中的 CHADDR 字段）、需要请求的参数列表选项（Option55）、广播标志位（DHCP DISCOVER 报文中的 Flags 字段，表示客户端请求服务器以单播或广播形式发送响应报文）等信息。

②提供阶段：服务器接收到 DHCP DISCOVER 报文后，选择跟接收 DHCP DISCOVER 报文接口的 IP 地址处于同一网段的地址池，并且从中选择一个可用的 IP 地址，然后通过 DHCP OFFER 报文将该可用的 IP 地址发送给 DHCP 客户端。

③请求阶段：如果有多个 DHCP 服务器向 DHCP 客户端回应 DHCP OFFER 报文，则 DHCP 客户端一般只接收第 1 个收到的 DHCP OFFER 报文，然后以广播方式发送 DHCP REQUEST 报文，该报文中包含客户端想选择的 DHCP 服务器标识符（Option54）和客户端 IP 地址（Option50，填充了接收的 DHCP OFFER 报文中 yiaddr 字段的 IP 地址）。以广播方式发送 DHCP REQUEST 报文，是为了通知所有的 DHCP 服务器，它将选择某个 DHCP 服务器提供的 IP 地址，其他 DHCP 服务器可以重新将曾经分配给客户端的 IP 地址分配给其他客户端。

④确认阶段：DHCP 客户端收到 DHCP ACK 报文后，会广播发送免费 ARP 报文，以探测本网段是否有其他终端使用服务器分配的 IP 地址。

（2）DHCP 有中继场景。DHCP 有中继场景流程图如图 4–7 所示。

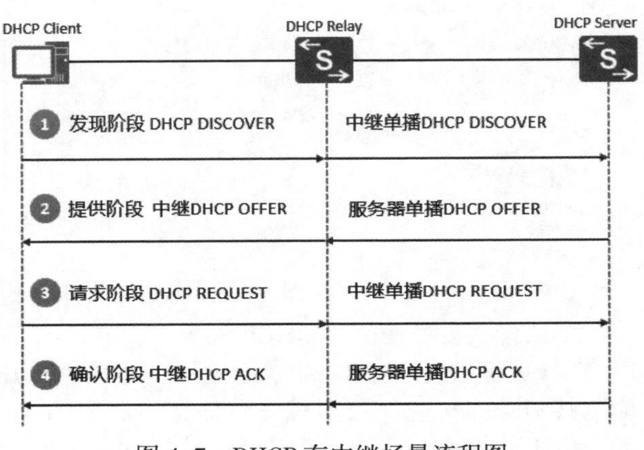

图 4–7　DHCP 有中继场景流程图

2. DHCP Snooping 信任功能

DHCP Snooping 信任功能示意图如图 4-8 所示。

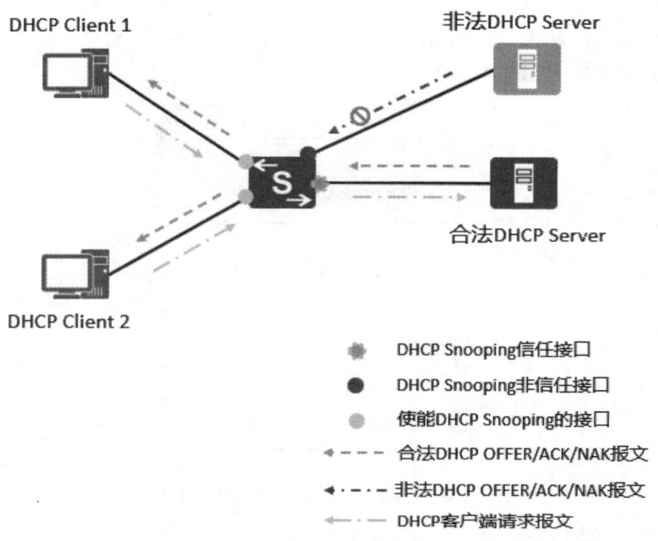

图 4-8　DHCP Snooping 信任功能示意图

DHCP Snooping 是 DHCP 的一种安全特性，其信任功能可以保证 DHCP 客户端从合法的 DHCP 服务器获取 IP 地址。DHCP Snooping 信任功能将接口分为信任接口和非信任接口。

（1）信任接口正常接收 DHCP 服务器响应的 DHCP ACK、DHCP NAK 和 DHCP OFFER 报文。

（2）设备只将 DHCP 客户端的 DHCP 请求报文通过信任接口发送给合法的 DHCP 服务器，不会向非信任接口转发。

（3）非信任接口收到的 DHCP Server 发送的 DHCP OFFER、DHCP ACK、DHCP NAK 报文会被直接丢弃。

3.DHCP Snooping 绑定表

DHCP Snooping 绑定表的 形成方式示意图如图 4-9 所示。

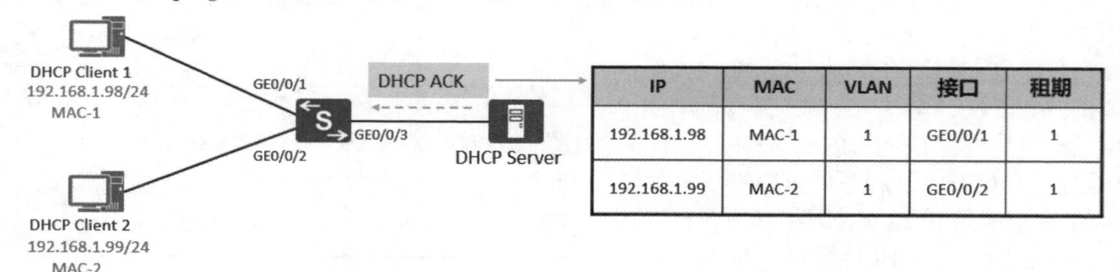

IP	MAC	VLAN	接口	租期
192.168.1.98	MAC-1	1	GE0/0/1	1
192.168.1.99	MAC-2	1	GE0/0/2	1

图 4-9　DHCP Snooping 绑定表的形成方式示意图

二层接入设备使能了 DHCP Snooping 功能后，从收到 DHCP ACK 报文中提取关键信息（包括 PC 的 MAC 地址以及获取到的 IP 地址、地址租期），并获取与 PC 连接的使能了 DHCP Snooping 功能的接口信息（包括接口编号及该接口所属的 VLAN），根据这些信息生成 DHCP Snooping 绑定表。由于 DHCP Snooping 绑定表记录了 DHCP 客户端 IP 地址与 MAC 地址等参数的对应关系，因此通过对报文与 DHCP Snooping 绑定表进行匹配检查，能够有效防范非法用户的攻击。

4. DHCP 饿死攻击

（1）攻击原理：攻击者通过不断伪造与源 MAC 地址不同的 DHCP DISCOVER 报文，向服务器申请 IP 地址，直到耗尽 DHCP Server 地址池中的 IP 地址，导致 DHCP Server 不能给正常的用户进行分配。

（2）漏洞分析：DHCP Server 向申请者分配 IP 地址时，无法区分正常的申请者与恶意的申请者。

（3）解决方法：对于饿死攻击，用户可以通过 DHCP Snooping 的 MAC 地址限制功能来防止。该功能通过限制交换机接口上允许学习到的最多 MAC 地址数，防止通过变换 MAC 地址大量发送 DHCP 请求。

5. 改变 CHADDR 值的 DoS 攻击

（1）攻击原理：攻击者通过不断伪造与 CHADDR 值不同的 DHCP REQUEST 报文，向服务器申请 IP 地址，直到耗尽 DHCP Server 地址池中的 IP 地址，导致 DHCP Server 不能给正常的用户进行分配。

（2）漏洞分析：DHCP Server 向申请者分配 IP 地址时，无法区分正常的申请者与恶意的申请者。

（3）解决方法：为了避免受到攻击者改变 CHADDR 值的攻击，管理员可以在设备上配置 DHCP Snooping 功能，检查 DHCP REQUEST 报文中的 CHADDR 字段。如果该字段与数据帧头部的源 MAC 相匹配，则转发报文；否则，丢弃报文。从而保证合法用户可以正常使用网络服务。

6. DHCP 中间人攻击

（1）攻击原理：攻击者利用 ARP 机制，让 Client 学习到 DHCP Server IP 与 Attacker MAC 的映射关系，又让 Server 学习到 Client IP 与 Attacker MAC 的映射关系。如此一来，Client 与 Server 之间交互的 IP 报文都会经过攻击者中转。

（2）漏洞分析：从本质上讲，中间人攻击是一种 IP/MAC Spoofing 攻击，中间人利用了虚假的 IP 地址与 MAC 地址之间的映射关系来同时欺骗 DHCP 的客户端和服务器。

（3）解决方法：为防御中间人攻击与 IP/MAC Spoofing 攻击，管理员可使用 DHCP Snooping 的绑定表工作模式，当接口接收到 ARP 或者 IP 报文时，使用 ARP 或者 IP 报文中的“源 IP+ 源 MAC”匹配 DHCP Snooping 绑定表。如果匹配，就进行转发；否则，就丢弃。

4.2.9　IPSG

1. IPSG 技术概述

IP 地址欺骗攻击中，攻击者通过伪造合法用户的 IP 地址获取网络访问权限，非法访问网络，甚至造成合法用户无法访问网络，或者信息泄露。IPSG（IP Source Guard，IP 源防攻击）针对 IP 地址欺骗攻击提供了一种防御机制，可以有效阻止此类网络攻击行为。

如图 4-10 所示，IPSG 是一种基于二层接口的源 IP 地址过滤技术。它能够防止恶意主机伪造合法主机的 IP 地址来仿冒合法主机，还能确保非授权主机不能通过自己指定 IP 地址的方式来访问网络或攻击网络。

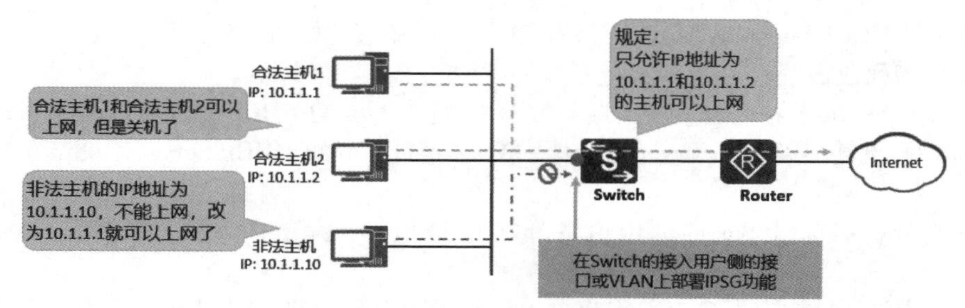

图 4-10　IPSG 应用组网图

2. IPSG 工作原理

IPSG 利用绑定表（源 IP 地址、源 MAC 地址、所属 VLAN、入接口的绑定关系）去匹配检查二层接口上收到的 IP 报文，只有匹配绑定表的报文才允许通过，其他报文将被丢弃。常见的绑定表有静态绑定表和 DHCP Snooping 动态绑定表。

4.3　防火墙的高级特性

4.3.1　防火墙双机热备

1. 防火墙双机热备简介

双机热备需要两台硬件和软件配置均相同的防火墙组成双机热备系统。防火墙之间通过独立的链路连接（心跳线）了解对端的健康状况，向对端备份配置和表项（如会话表、IPSec SA 等）。当一台防火墙出现故障时，业务流量能平滑地切换到另一台设备上处理，使业务不中断。

2. 部署要求

（1）目前只支持两台设备进行双机热备。

（2）主备设备的产品型号和版本必须相同。

（3）主备设备业务板和接口卡的位置、类型和数目都必须相同，否则会出现主用设备备份过去的信息与备用设备的物理配置无法兼容，导致主备切换后出现问题。

3. 防火墙双机热备的关键组件

（1）VRRP（Virtual Router Redundancy Protocol）。VRRP 是一种容错协议，它保证当主机的下一跳路由器（默认网关）出现故障时，备份路由器能自动代替前者完成报文转发任务，从而保持网络通信的连续性和可靠性。

（2）VGMP（VRRP Group Management Protocol）。VGMP 可以将防火墙上的所有 VRRP 组都加入一个 VGMP 组中，由 VGMP 组来集中监控并管理所有的 VRRP 组状态。如果 VGMP 组检测到其中一个 VRRP 组的状态变化，则 VGMP 组会控制组中的所有 VRRP 组统一进行状态切换，保证各 VRRP 备份组状态的一致性。

（3）HRP（Huawei Redundancy Protocol）。HRP 实现防火墙双机之间动态状态数据和关键配置命令的备份。

4. 防火墙双机热备的典型组网场景

（1）双机热备直路部署，连接二层设备。如图 4-11 所示，防火墙的业务接口工作在三层，上下行连接交换机。终端可将默认网关设置为 VRRP VRID1 的虚拟 IP 地址。SW3/SW4 配置回程路由时，可将下一跳设置为 VRRP VRID100 的虚拟 IP 地址。

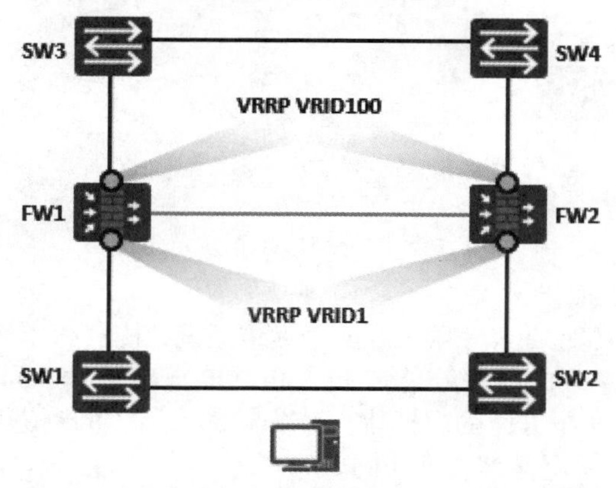

图 4–11　双机热备直路部署，连接二层设备

（2）双机热备直路部署，连接三层设备。如图 4–12 所示，防火墙的业务接口工作在三层，上下行连接路由器。防火墙与路由器之间运行 OSPF。当 FW1 的业务接口故障时，其切换成备用设备，FW2 成为主用设备。FW1 发布的路由 Cost 值自动修改为 65500。路由重新收敛后，流量通过 FW2 转发。

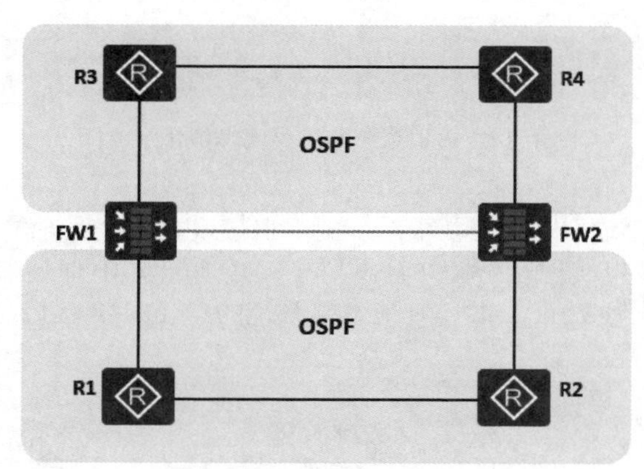

图 4–12　双机热备直路部署，连接三层设备

4.3.2　虚拟系统

1. 虚拟系统的实验目的

如图 4–13 所示，在防火墙上存在两种类型的虚拟系统。

（1）根系统（Public）。默认即存在的一个特殊虚拟系统。默认情况下，管理员对防火墙进行配置等同于对根系统进行配置。在虚拟系统这个特性中，根系统的作用是管理其他虚拟系统，并分别为虚拟系统之间的通信提供服务。

（2）虚拟系统（VSYS）。虚拟系统是在防火墙上划分出来的、独立运行的逻辑设备。

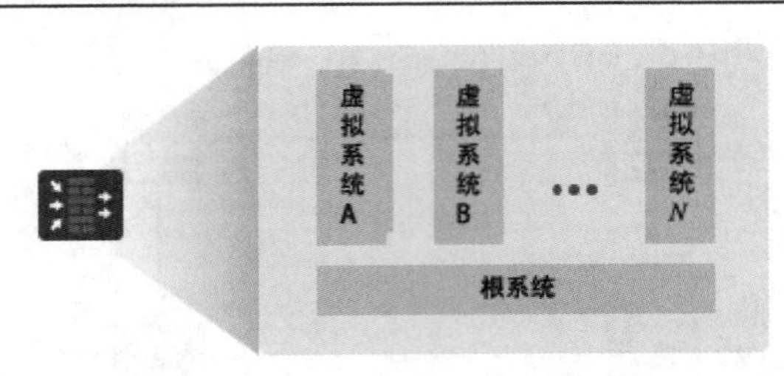

图 4-13　防火墙虚拟系统示意图

2. 虚拟系统与防火墙虚拟化

如图 4-14 所示，为了实现 vsysa 和 vsysb 两个虚拟系统，其中 PC1 属于 vsysa，PC2 属于 vsysb，最终目标是确保每个虚拟系统的业务都能被 PC1 和 PC2 正确转发，同时实现独立管理和相互隔离。为此，FW 主要实现了 4 个方面的虚拟化：资源虚拟化、配置虚拟化、安全功能虚拟化以及路由虚拟化，以确保各虚拟系统间能够相互访问。

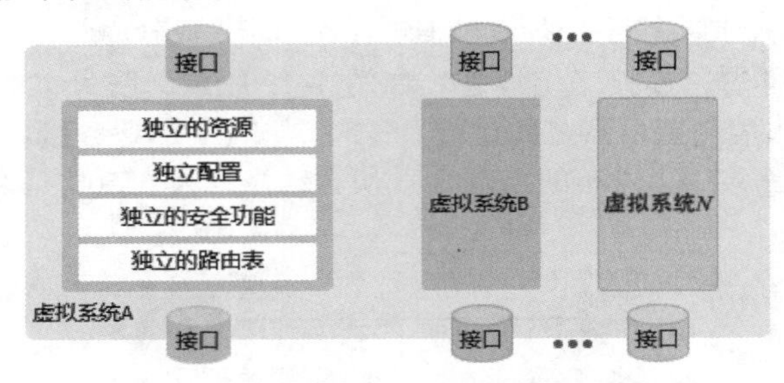

图 4-14　虚拟系统与防火墙虚拟化示意图

3. 虚拟接口

如图 4-15 所示，各个虚拟系统以及根系统的虚拟接口之间默认通过一条“虚拟链路”连接。虚拟系统之间通过虚拟接口实现互访，并且虚拟接口必须配置 IP 地址加入安全区域才能正常工作，根系统的虚拟接口号为 Virtual-if0，其他虚拟系统的 Virtual-if 接口号从 1 开始，根据系统中接口号的占用情况自动分配。

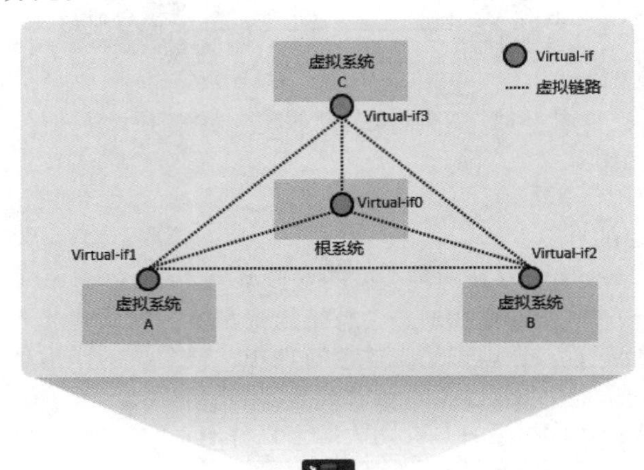

图 4-15　防火墙虚拟接口示意图

4.4　网络安全实验

4.4.1　配置端口隔离

1. 实验目的

要求在 LSW1 中配置端口隔离，实现在 VLAN 10 中 PC1 可以访问 PC2 和 PC3，PC2 和 PC3 不能相互访问。在 VLAN 20 中，PC4 主机存在安全隐患，它向其他主机发送大量的广播报文，通过配置接口间的单向隔离来实现其他主机对该主机报文的隔离。

2. 实验拓扑

配置端口隔离的实验拓扑如图 4-16 所示。

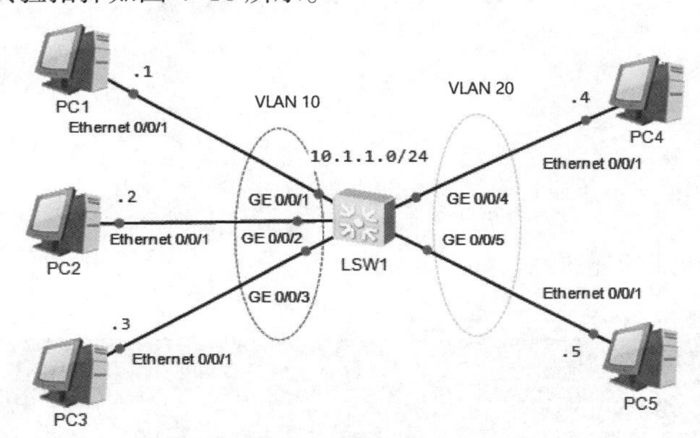

图 4-16　配置端口隔离的实验拓扑

3. 实验步骤

步骤 1： 配置 IP 地址。

配置 IP 地址（此处省略）。

步骤 2： 创建 VLAN，并把接口划入 VLAN。

```
[Huawei]sysname LSW1
[LSW1]vlan 10
[LSW1-vlan10]quit
[LSW1]vlan 20
[LSW1-vlan20]quit
[LSW1]interface g0/0/1
[LSW1-GigabitEthernet0/0/1]port link-type access
[LSW1-GigabitEthernet0/0/1]port default vlan 10
[LSW1-GigabitEthernet0/0/1]quit
[LSW1]interface g0/0/2
[LSW1-GigabitEthernet0/0/2]port link-type access
[LSW1-GigabitEthernet0/0/2]port default vlan 10
[LSW1-GigabitEthernet0/0/2]quit
[LSW1]interface g0/0/3
[LSW1-GigabitEthernet0/0/3]port link-type access
[LSW1-GigabitEthernet0/0/3]port default vlan 10
```

```
[LSW1-GigabitEthernet0/0/3]quit
[LSW1]interface g0/0/4
[LSW1-GigabitEthernet0/0/4]port link-type access
[LSW1-GigabitEthernet0/0/4]port default vlan 20
[LSW1-GigabitEthernet0/0/4]quit
[LSW1]interface g0/0/5
[LSW1-GigabitEthernet0/0/5]port link-type access
[LSW1-GigabitEthernet0/0/5]port default vlan 20
[LSW1-GigabitEthernet0/0/5]quit
```

在 VLAN 10 中，PC1 可以访问 PC2 和 PC3，PC2 和 PC3 不能相互访问。

```
[LSW1]port-isolate mode L2  //配置端口隔离模式为 L2，端口隔离模式为二层隔离、三层互通
[LSW1]interface g0/0/2
[LSW1-GigabitEthernet0/0/2]port-isolate enable group 10 //使能端口隔离功能，加入的
                                                        //端口隔离为 10
[LSW1-GigabitEthernet0/0/2]quit
[LSW1]interface g0/0/3
[LSW1-GigabitEthernet0/0/3]port-isolate enable  group 10
[LSW1-GigabitEthernet0/0/3]quit
```

测试 PC1、PC2、PC3 的连通性，如图 4-17 所示。

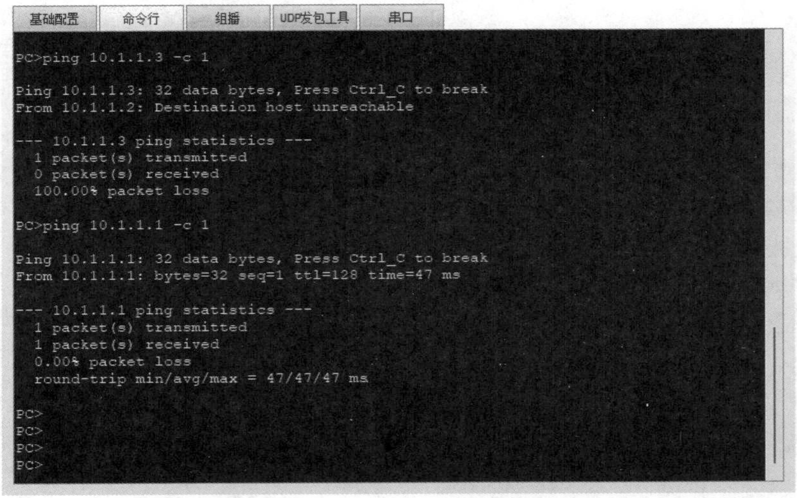

```
基础配置  命令行  组播  UDP发包工具  串口

PC>ping 10.1.1.3 -c 1

Ping 10.1.1.3: 32 data bytes, Press Ctrl_C to break
From 10.1.1.2: Destination host unreachable

--- 10.1.1.3 ping statistics ---
  1 packet(s) transmitted
  0 packet(s) received
  100.00% packet loss

PC>ping 10.1.1.1 -c 1

Ping 10.1.1.1: 32 data bytes, Press Ctrl_C to break
From 10.1.1.1: bytes=32 seq=1 ttl=128 time=47 ms

--- 10.1.1.1 ping statistics ---
  1 packet(s) transmitted
  1 packet(s) received
  0.00% packet loss
  round-trip min/avg/max = 47/47/47 ms

PC>
PC>
PC>
PC>
```

图 4-17　PC2 访问 PC3、PC1 测试图

通过以上输出可以看出，PC2 与 PC3 不能相互访问，但是可以访问 PC1。

在 VLAN 20 中，PC4 主机存在安全隐患，它向其他主机发送大量的广播报文，通过配置接口间的单向隔离来实现其他主机对该主机报文的隔离。

```
[LSW1]interface g0/0/4
[LSW1-GigabitEthernet0/0/4]am isolate GigabitEthernet 0/0/5  //g0/0/4 的报文不能发
                                          //给 g0/0/5，g0/0/5 的报文可以发给 g0/0/4
[LSW1-GigabitEthernet0/0/4]quit
```

抓包查看现象的步骤如下：

（1）PC4 访问 PC5 时，在交换机抓 g0/0/4 和 g0/0/5 的数据包，如图 4-18 所示。

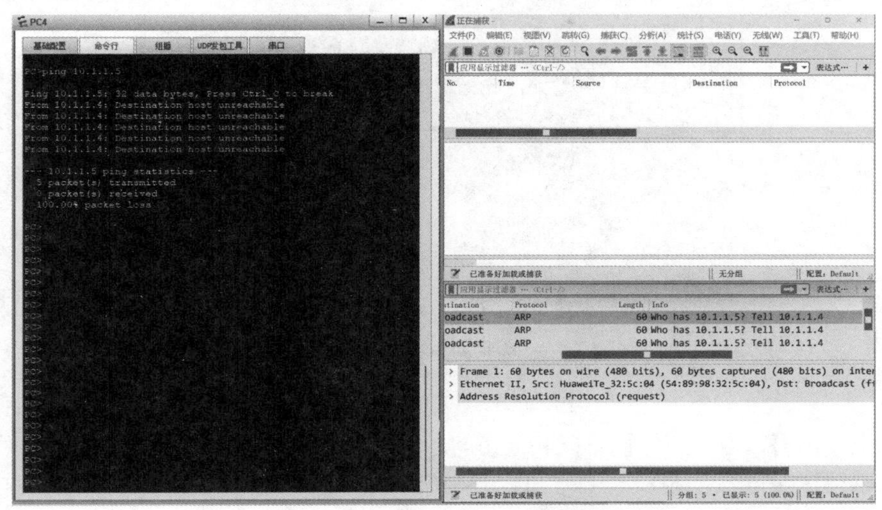

图 4-18　PC4 访问 PC5 测试图

通过以上输出可以看出，g0/0/4 的包文不能到达 g0/0/5。

（2）PC5 访问 PC4 时，在交换机抓 g0/0/4 和 g0/0/5 的数据包，如图 4-19 所示。

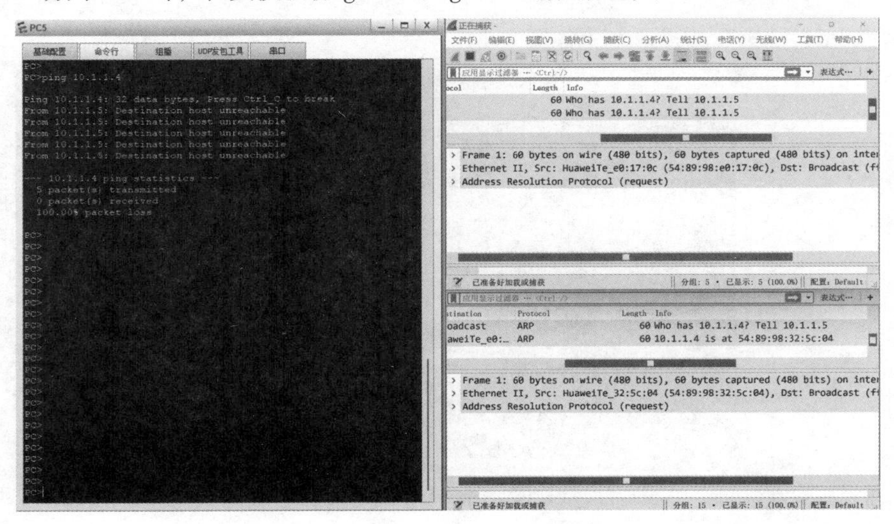

图 4-19　PC5 访问 PC4 测试图

通过以上输出可以看出，g0/0/5 的包文到达了 g0/0/4。

步骤 3：实验调试。

（1）查看端口隔离组中的端口：

```
<LSW1>display port-isolate group all
  The ports in isolate group 10:                    //组的编号为 10
GigabitEthernet0/0/2      GigabitEthernet0/0/3      //组 10 中有两个端口
```

（2）测试 PC2 和 PC3 的连通性，如图 4-20 所示。

图 4-20　PC2 访问 PC3 测试图（1）

通过以上输出可以看出，它们不能相互访问。

（3）创建三层接口，让 PC2 和 PC3 可以相互访问：

```
[LSW1]interface Vlanif 10
[LSW1-Vlanif10]ip address 10.1.1.254 24
[LSW1-Vlanif10]arp-proxy inner-sub-vlan-proxy enable    //使能 VLAN 内的 Proxy ARP 功能
[LSW1-Vlanif10]quit
```

再次测试 PC2 和 PC3 的连通性，如图 4-21 所示。

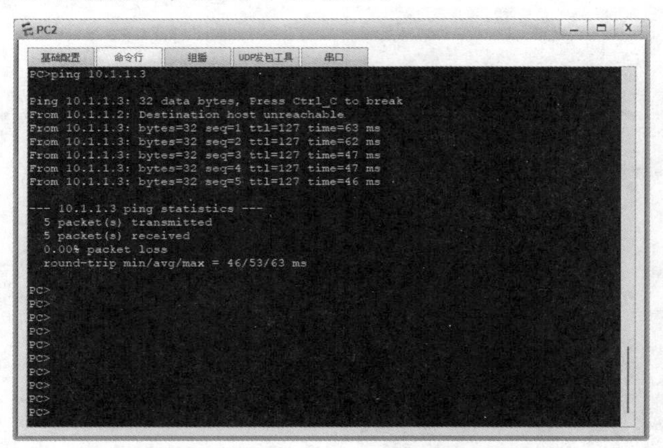

图 4-21　PC2 访问 PC3 测试图（2）

通过以上输出可以看出，PC2 和 PC3 可以相互访问了。

【技术要点】L2（二层隔离、三层互通）

（1）隔离同一VLAN内的广播报文，但是不同端口下的用户还可以进行三层通信。默认情况下，端口隔离模式为二层隔离、三层互通。

（2）采用二层隔离、三层互通的隔离模式时，在VLANIF接口上使能VLAN内的Proxy ARP功能，配置arp-proxy inner-sub-vlan-proxy enable，可以实现同一VLAN内的主机通信。

4.4.2　配置 MAC 地址表安全

1. 实验目的

要求配置 MAC 地址表安全实现以下需求：S1 的 g0/0/1 下面只允许学习到 1 个 MAC 地址，当 MAC 地址学习数达到限制后发出报警信息，将 PC3 的 MAC 地址设置为黑洞 MAC，不允许 PC3 通信。

2. 实验拓扑

配置 MAC 地址表安全的实验拓扑如图 4–22 所示。

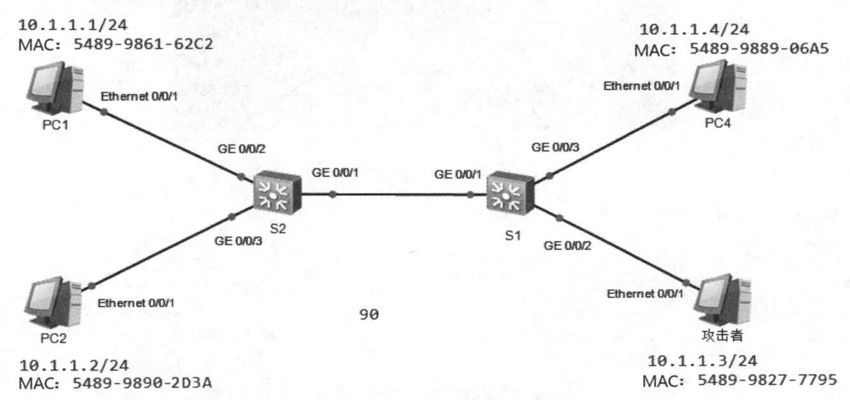

图 4–22　配置 MAC 地址表安全的实验拓扑

3. 实验步骤

步骤 1： 配置 IP 地址（此处省略）。

步骤 2： 配置基于接口限制 MAC 地址学习数，查看配置是否成功。

```
<Huawei>system-view
[Huawei]undo info-center enable
Info: Information center is disabled.
[Huawei]sysname S1
[S1]interface g0/0/1
[S1-GigabitEthernet0/0/1]mac-limit maximum 1 alarm enable  //最大 MAC 地址学习数为 1,
                                                            //达到上限就报警
[S1-GigabitEthernet0/0/1]quit
[S1]display mac-limit                    //查看 MAC 地址学习数限制规则
MAC Limit is enabled
Total MAC Limit rule count : 1
PORT          VLAN/VSI/SI    SLOT    Maximum    Rate(ms)    Action    Alarm
-----------------------------------------------------------------------------
GE0/0/1       -              -       1          -           forward   enable
```

通过以上输出可以看出，g0/0/1 下最多学习到 1 个 MAC 地址，如果超过了数量会报警。

PC1 访问 PC4 测试图如图 4–23 所示。

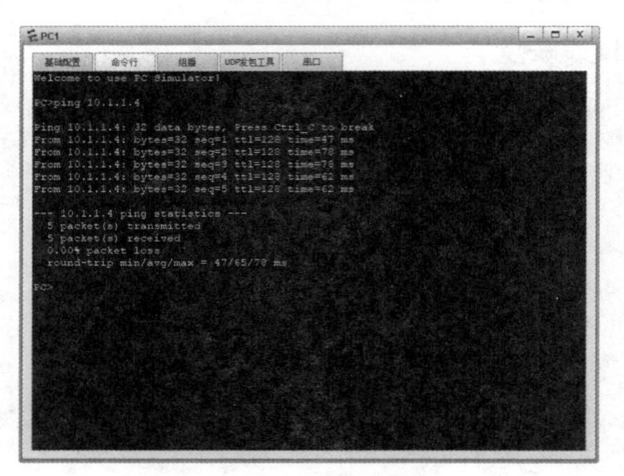

图 4-23　PC1 访问 PC4 测试图

在 S1 上查看 MAC 地址表：

```
[S1]display mac-address                   //查看MAC地址表
MAC address table of slot 0:
-------------------------------------------------------------------------------
MAC Address      VLAN/       PEVLAN     CEVLAN      Port          Type       LSP/LSR-ID
                 VSI/SI                                                      MAC-Tunnel
-------------------------------------------------------------------------------
5489-9889-06a5 1              -          -          GE0/0/3       dynamic     0/-
5489-9861-62c2 1              -          -          GE0/0/1       dynamic     0/-
-------------------------------------------------------------------------------
Total matching items on slot 0 displayed = 2
```

通过以上输出可以看出，g0/0/1 接口学习到了 PC1 的 MAC 地址。

打开信息中心的输出功能，PC2 访问 PC4，然后查看 S1 的 MAC 地址表。PC2 访问 PC4 测试图如图 4-24 所示。

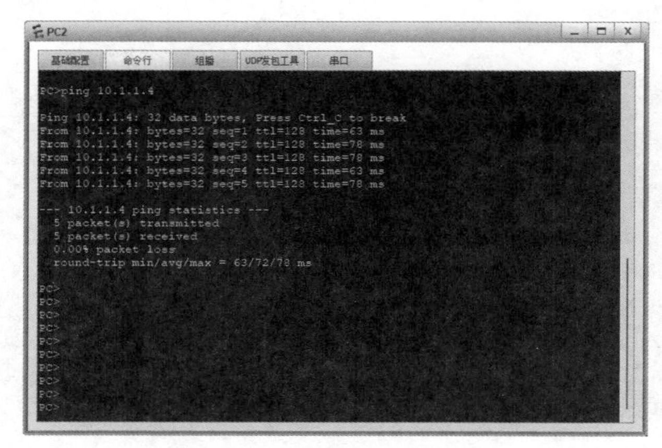

图 4-24　PC2 访问 PC4 测试图

```
[S1]info-center enable
[S1]display mac-address
```

```
MAC address table of slot 0:
-------------------------------------------------------------------------------
MAC Address      VLAN/     PEVLAN    CEVLAN    Port      Type      LSP/LSR-ID
                 VSI/SI                                            MAC-Tunnel
-------------------------------------------------------------------------------
5489-9861-62c2   1         -         -         GE0/0/1   dynamic   0/-
5489-9889-06a5   1         -         -         GE0/0/3   dynamic   0/-
-------------------------------------------------------------------------------
Total matching items on slot 0 displayed = 2
```

通过以上输出可以看出，g0/0/1 接口下的 MAC 地址只有 1 个。

```
Mar 15 2023 15:17:27-08:00 S1 L2IFPPI/4/MAC_LIMIT_ALARM:OID 1.3.6.1.4.1.2011.5.2
5.42.2.1.7.11 MAC address learning reached the limit. (L2IfPort=0,MacLimitVlanId
=0, MacLimitVsiName=,L2IfPort=0, BaseTrapSeverity=4, BaseTrapProbableCause=549,
BaseTrapEventType=1,MacDynAddressLearnNum=1, MacLimitMaxMac=1,L2IfPortName=Gigab
itEthernet0/0/1)
```

通过以上输出可以看出，交换机在报警，因为 MAC 地址超过了上限。

【技术要点】限制MAC地址学习数

（1）模拟器只支持超过MAC地址上限就报警，但是数据还是会转发。

（2）真实设备可以做到超过MAC地址上限可以报警和丢弃数据，配置如下：

①配置基于接口限制MAC地址学习数。

```
[Huawei-GigabitEthernet0/0/1] mac-limit maximum max-num
```

默认情况下，不限制MAC地址学习数。

②配置当MAC地址学习数达到限制后，对报文应采取的动作。

```
[Huawei-GigabitEthernet0/0/1] mac-limit action { discard | forward }
```

默认情况下，对超过MAC地址学习数限制的报文采取丢弃动作。

③配置当MAC地址学习数达到限制后是否进行报警。

```
[Huawei-GigabitEthernet0/0/1] mac-limit alarm { disable | enable }
```

默认情况下，对超过MAC地址学习数限制的报文进行报警。

④配置基于VLAN限制MAC地址学习数。

```
[Huawei-vlan2] mac-limit maximum max-num
```

默认情况下，不限制MAC地址学习数。

管理员发现 PC3 为一个攻击者，设置黑洞 MAC 地址表项，如果收到 PC3 的数据，则直接丢弃。

（1）配置黑洞 MAC 地址表项：

```
[S1]mac-address blackhole 5489-9827-7795  vlan 1 //把 PC3 的 MAC 地址设置为黑洞 MAC 地址
```

（2）查看黑洞 MAC 地址表项：

```
[S1]display mac-address blackhole
MAC address table of slot 0:
-------------------------------------------------------------------------------
MAC Address      VLAN/     PEVLAN    CEVLAN    Port      Type      LSP/LSR-ID
                 VSI/SI                                            MAC-Tunnel
-------------------------------------------------------------------------------
```

```
5489-9827-7795    1      -      -      -      blackhole    -
--------------------------------------------------------------------------
Total matching items on slot 0 displayed = 1
```

通过以上输出可以看出，黑洞 MAC 表项添加成功。

（3）测试 PC3 是否可以访问 PC1、PC2、PC4，如图 4-25 所示。

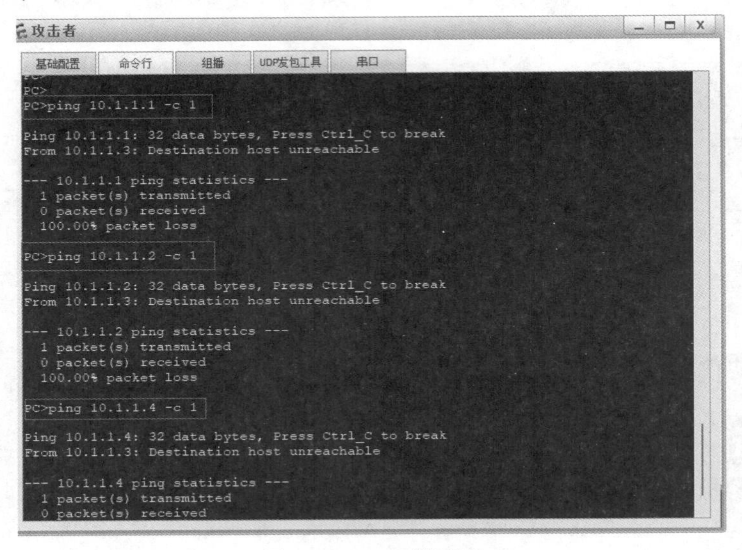

图 4-25　PC3 测试图

通过以上输出可以看出，PC3 不能访问任何设备。

S1 的 g0/0/3 接口接入的是固定用户，配置静态 MAC 地址绑定来保证其安全通信。

（1）配置静态 MAC 地址绑定：

```
[S1]mac-address static 5489-9889-06A5 g0/0/3 vlan 1
```

（2）查看 MAC 地址表：

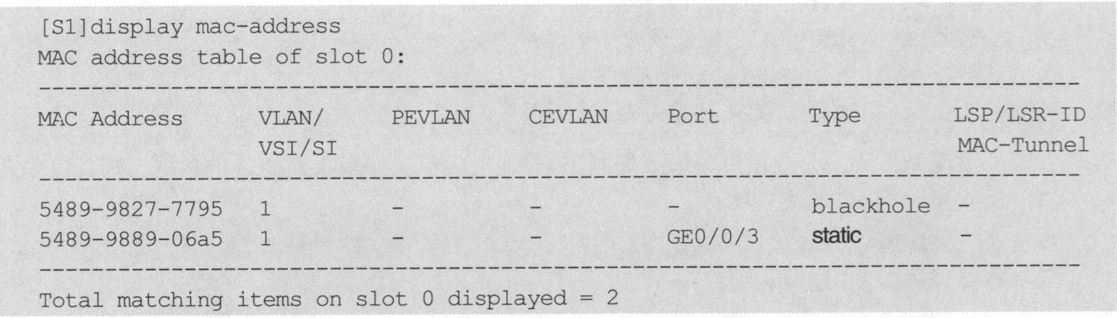

```
[S1]display mac-address
MAC address table of slot 0:
--------------------------------------------------------------------------
MAC Address      VLAN/    PEVLAN   CEVLAN   Port      Type        LSP/LSR-ID
                 VSI/SI                                           MAC-Tunnel
--------------------------------------------------------------------------
5489-9827-7795   1        -        -        -         blackhole   -
5489-9889-06a5   1        -        -        GE0/0/3   static      -
--------------------------------------------------------------------------
Total matching items on slot 0 displayed = 2
```

通过以上输出可以看出，PC4 的 MAC 地址已绑定在 MAC 地址表中。

4.4.3　配置端口安全

1. 实验目的

LSW1 为某公司的接入层设备，为了防止员工私接小路由，需要在设备上配置端口安全，设置端口的 MAC 地址学习数为 1，PC3 和 PC4 为固定工位终端，分别配置安全静态 MAC 和 Sticky MAC 以保证接入的安全性。LSW1 的 g0/0/1 接口只能接入一台终端设备，现在有员工私接小路由，当接口超过 MAC 地址的学习数时，设备需要自动将端口 shutdown。

扫一扫，看视频

2. 实验拓扑

配置端口安全的实验拓扑如图 4–26 所示。

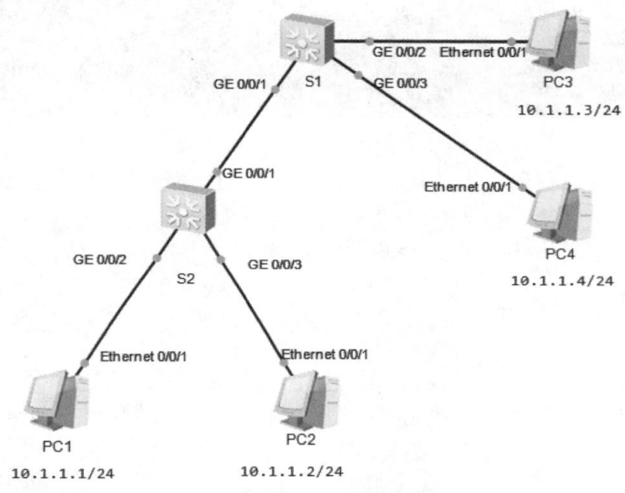

图 4–26　配置端口安全的实验拓扑

3. 实验步骤

步骤 1： 配置 S1 g0/0/1 的端口安全，并且配置保护动作以将接口 shutdown。

```
[S1]interface g0/0/1
[S1-GigabitEthernet0/0/1]port-security enable          //开启端口安全功能
[S1-GigabitEthernet0/0/1]port-security max-mac-num 1 //配置最大的MAC地址学习数为1
[S1-GigabitEthernet0/0/1]port-security protect-action shutdown //配置安全保护动作以
                                                                  //将端口shutdown
```

分别使用 PC1、PC2 访问 PC4，查看 S1 的 g0/0/1 接口超过 MAC 地址学习数是否会 shutdown，如图 4–27 和图 4–28 所示。

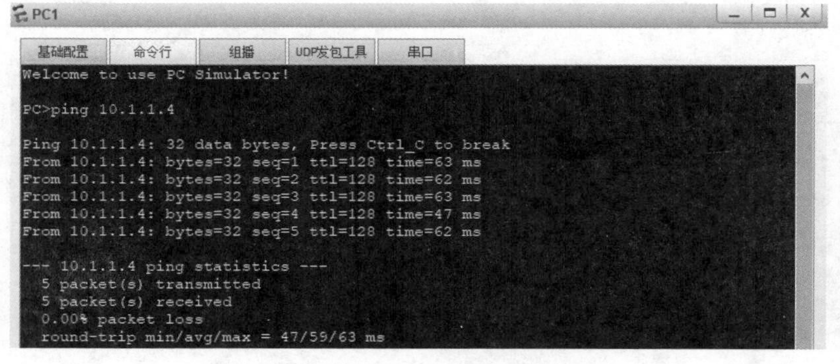

图 4–27　使用 PC1 访问 PC4 测试图

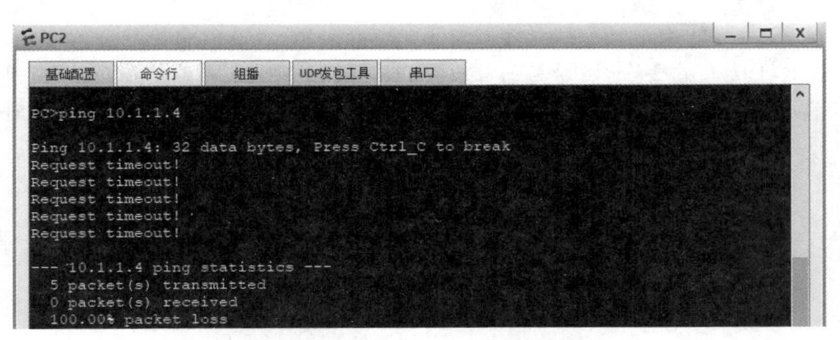

图 4–28　使用 PC2 访问 PC4 测试图

通过以上输出可以看出，PC2 无法访问 PC4，在 S1 上查看日志信息，提示如下：

```
Dec 4 2023 10:23:27-08:00 S1 L2IFPPI/4/PORTSEC_ACTION_ALARM:OID 1.3.6.1.4.1.2011.
5.25.42.2.1.7.6 The number of MAC address on interface (6/6) GigabitEthernet0/0/1
reaches the limit, and the port status is : 3. (1:restrict;2:protect;3:shutdown)
Dec  4 2023 10:23:27-08:00 S1 %%01PHY/1/PHY(l)[0]: GigabitEthernet0/0/1: change
status to down
```

提示由于 g0/0/1 接口 MAC 地址学习数超过限制的数值，接口被 shutdown。

如果需要恢复网络，则需要处理攻击者，并且由管理员手动开启接口或者配置自动恢复功能。

步骤 2：将连接 PC3 的接口配置为安全静态 MAC。

```
[S1]interface g0/0/2
[S1-GigabitEthernet0/0/2]port-security enable                //使能端口安全功能
[S1-GigabitEthernet0/0/2]port-security mac-address sticky   //使能 Sticky MAC 功能
[S1-GigabitEthernet0/0/2]port-security mac-address sticky 5489-98B0-7D44 vlan 1
//配置 VLAN 1 的安全静态 MAC，此处绑定 PC3 的 MAC 地址
[S1-GigabitEthernet0/0/2]port-security max-mac-num 1 //配置该接口的最大 MAC 地址学习
                                                      //数为 1
```

查看 S1 的 MAC 地址表：

```
[S1]display mac-address
MAC address table of slot 0:
-------------------------------------------------------------------------------
MAC Address      VLAN/     PEVLAN    CEVLAN    Port      Type      LSP/LSR-ID
                 VSI/SI                                            MAC-Tunnel
-------------------------------------------------------------------------------
5489-98b0-7d44   1         -         -         GE0/0/2   sticky    -
-------------------------------------------------------------------------------
Total matching items on slot 0 displayed = 1
```

通过以上输出可以看出，即使 PC3 没通信，其 MAC 地址以及静态绑定在该接口上，类型为 sticky 并且不会被老化。

步骤 3：配置 S1 的 g0/0/3 接口为 Sticky MAC。

```
[S1]interface g0/0/3
[S1-GigabitEthernet0/0/3]port-security enable
[S1-GigabitEthernet0/0/3]port-security mac-address sticky
```

```
[S1-GigabitEthernet0/0/3]port-security max-mac-num 1
```

在 PC4 没通信之前，交换机的 MAC 地址表并没有其 MAC 地址对应关系，查看 MAC 地址表。

```
[SW1]display mac-address
MAC address table of slot 0:
-------------------------------------------------------------------------------
MAC Address       VLAN/       PEVLAN     CEVLAN     Port       Type      LSP/LSR-ID
                  VSI/SI                                                 MAC-Tunnel
-------------------------------------------------------------------------------
5489-98b0-7d44    1           -          -          GE0/0/2    sticky    -
-------------------------------------------------------------------------------
Total matching items on slot 0 displayed = 1
```

使用 PC4 访问 PC3：

```
PC>ping 10.1.1.3
Ping 10.1.1.3: 32 data bytes, Press Ctrl_C to break
From 10.1.1.3: bytes=32 seq=1 ttl=128 time=47 ms
From 10.1.1.3: bytes=32 seq=2 ttl=128 time=46 ms
From 10.1.1.3: bytes=32 seq=3 ttl=128 time=47 ms
From 10.1.1.3: bytes=32 seq=4 ttl=128 time=62 ms
From 10.1.1.3: bytes=32 seq=5 ttl=128 time=63 ms
--- 10.1.1.3 ping statistics ---
  5 packet(s) transmitted
  5 packet(s) received
  0.00% packet loss
  round-trip min/avg/max = 46/53/63 ms
```

再次查看 S1 的 MAC 地址表：

```
[S1]display mac-address
MAC address table of slot 0:
-------------------------------------------------------------------------------
MAC Address       VLAN/       PEVLAN     CEVLAN     Port       Type      LSP/LSR-ID
                  VSI/SI                                                 MAC-Tunnel
-------------------------------------------------------------------------------
5489-98b0-7d44    1           -          -          GE0/0/2    sticky    -
5489-9808-15e8    1           -          -          GE0/0/3    sticky    -
-------------------------------------------------------------------------------
Total matching items on slot 0 displayed = 2
```

通过以上输出可以看出，g0/0/3 接口学习到的 MAC 地址为 PC4 的 MAC 地址，并且类型为 sticky。

4.4.4 配置防火墙双机热备

1. 实验目的

PC1 为公司内部网络设备，AR1 为出口设备，在 FW1 和 FW2 上配置双机热备，当网络正常时，PC1 访问 AR1 的路径为 FW1 → AR1，当 FW1 出现故障时，切换路径为 FW2 → AR1。

扫一扫，看视频

2. 实验拓扑

配置防火墙双机热备的实验拓扑如图 4-29 所示。

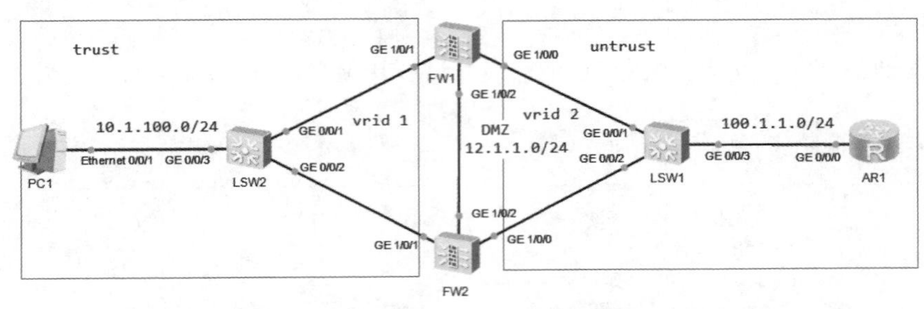

图 4-29　配置防火墙双机热备的实验拓扑

3. 实验步骤

步骤 1：配置 IP 地址。

FW1 的配置：

```
[FW1]interface GigabitEthernet1/0/1
[FW1-GigabitEthernet1/0/1]ip address 10.1.100.1 255.255.255.0
[FW1]interface GigabitEthernet1/0/0
[FW1-GigabitEthernet1/0/0]ip address 100.1.1.1 255.255.255.0
[FW1]interface GigabitEthernet1/0/2
[FW1-GigabitEthernet1/0/2]ip address 12.1.1.1 255.255.255.0
```

FW2 的配置：

```
[FW2]interface GigabitEthernet1/0/1
[FW2-GigabitEthernet1/0/1]ip address 10.1.100.2 255.255.255.0
[FW2]interface GigabitEthernet1/0/0
[FW2-GigabitEthernet1/0/0]ip address 100.1.1.2 255.255.255.0
[FW2]interface GigabitEthernet1/0/2
[FW2-GigabitEthernet1/0/2]ip address 12.1.1.2 255.255.255.0
```

AR1 的配置：

```
[AR1]interface GigabitEthernet0/0/0
[AR1-GigabitEthernet0/0/0]ip address 100.1.1.3 255.255.255.0
[AR1]interface LoopBack0
[AR1-LoopBack0]ip address 1.1.1.1 255.255.255.255
```

步骤 2：将接口加入安全区域。

FW1 的配置：

```
[FW1]firewall zone trust
[FW1-zone-trust]add interface GigabitEthernet1/0/1
[FW1-zone-trust]q
[FW1]firewall zone untrust
[FW1-zone-untrust]add interface GigabitEthernet1/0/0
[FW1-zone-untrust]q
[FW1]firewall zone dmz
[FW1-zone-dmz]add interface GigabitEthernet1/0/2
```

FW2 的配置：

```
[FW2]firewall zone trust
```

```
[FW2-zone-trust]add interface GigabitEthernet1/0/1
[FW2-zone-trust]q
[FW2]firewall zone untrust
[FW2-zone-untrust]add interface GigabitEthernet1/0/0
[FW2-zone-untrust]q
[FW2]firewall zone dmz
[FW2-zone-dmz]add interface GigabitEthernet1/0/2
```

步骤 3: 配置静态路由，实现 AR1 和 PC 的互通。

FW1 的配置:

```
[FW1]ip route-static 1.1.1.1 32 100.1.1.3
```

FW2 的配置:

```
[FW2]ip route-static 1.1.1.1 32 100.1.1.3
```

AR1 的配置:

```
[AR1]IP route-static 10.1.100.0 24 100.1.1.254//AR1 访问内部 PC 的下一跳为 VRRP VRID2 的虚拟 IP 地址
```

步骤 4: 配置 VRRP，将 FW1 设置为主设备，将 FW2 设置为备用设备。

FW1 的配置:

```
[FW1]interface GigabitEthernet1/0/1
[FW1-GigabitEthernet1/0/1]vrrp vrid 1 virtual-ip 10.1.100.254 active
[FW1-GigabitEthernet1/0/1]q
[FW1]interface GigabitEthernet1/0/0
[FW1-GigabitEthernet1/0/0]vrrp vrid 2 virtual-ip 100.1.1.254 active
```

FW2 的配置:

```
[FW2]interface GigabitEthernet1/0/1
[FW2-GigabitEthernet1/0/1]vrrp vrid 1 virtual-ip 10.1.100.254 standby
[FW2-GigabitEthernet1/0/1]q
[FW2]interface GigabitEthernet1/0/0
[FW2-GigabitEthernet1/0/0]vrrp vrid 2 virtual-ip 100.1.1.254 standby
```

步骤 5: 配置安全策略，允许心跳接口之间交互 HRP 报文。

FW1 的配置:

```
[FW1]security-policy
[FW1-policy-security]rule name local_dmz
[FW1-policy-security-rule-local_dmz]source-zone dmz
[FW1-policy-security-rule-local_dmz]source-zone local
[FW1-policy-security-rule-local_dmz]destination-zone dmz
[FW1-policy-security-rule-local_dmz]destination-zone local
[FW1-policy-security-rule-local_dmz]service protocol udp destination-port 18514
[FW1-policy-security-rule-local_dmz]action permit
```

FW2 的配置:

```
[FW2]security-policy
[FW2-policy-security]rule name local_dmz
[FW2-policy-security-rule-local_dmz]source-zone dmz
[FW2-policy-security-rule-local_dmz]source-zone local
```

```
[FW2-policy-security-rule-local_dmz]destination-zone dmz
[FW2-policy-security-rule-local_dmz]destination-zone local
[FW2-policy-security-rule-local_dmz]service protocol udp destination-port 18514
[FW2-policy-security-rule-local_dmz]action permit
```

步骤 6： 配置 HRP（配置 HRP 是为了实现主设备故障后，备用设备能够平滑接替主设备的工作）。

FW1 的配置：

```
[FW1]hrp interface g1/0/2 remote 12.1.1.2
[FW1]hrp enable
```

FW2 的配置：

```
[FW2]hrp interface g1/0/2 remote 12.1.1.1
[FW2]hrp enable
```

在 FW1 上查看 HRP 的状态：

```
HRP_M[FW1]display hrp state
Role: active, peer: standby //此处表示本设备状态为主，邻居为备用设备
 Running priority: 45000, peer: 45000
 Backup channel usage: 0.00%
 Stable time: 0 days, 0 hours, 6 minutes
 Last state change information: 2023-03-28 7:13:26 HRP link changes to up.
```

查看 FW1 的 VRRP 状态：

```
HRP_M[FW1]display vrrp brief
2023-03-28 07:23:32.950
Total:2     Master:2     Backup:0     Non-active:0
VRID  State        Interface                Type     Virtual IP
----------------------------------------------------------------
1     Master       GE1/0/1                  Vgmp     10.1.100.254
2     Master       GE1/0/0                  Vgmp     100.1.1.254
```

通过以上输出可以看出，FW1 的上下行接口都为 VRRP 的主设备。

步骤 7： 配置放行 PC 访问 AR1 流量的安全策略。

FW1 的配置：

```
HRP_M[FW1]security-policy
HRP_M[FW1-policy-security]rule name trust_untrust
HRP_M[FW1-policy-security-rule-trust_untrust]source-zone trust
HRP_M[FW1-policy-security-rule-trust_untrust]destination-zone untrust
HRP_M[FW1-policy-security-rule-trust_untrust]action permit
```

在 FW1 上配置了安全策略，设备会自动通过 HRP 将对应的安全策略配置同步到备用设备上。在 FW2 上查看设备的安全策略：

```
HRP_S[FW2]display security-policy rule name trust_untrust
2023-03-28 07:25:17.240
 (0 times matched)
 rule name trust_untrust
  source-zone trust
```

```
  destination-zone untrust
  action permit
```

通过以上输出可以看出，FW2 通过 HRP 自动备份了 FW1 的安全策略。

步骤 8: 测试。

使用 PC1 访问 AR1 的环回口测试图如图 4–30 所示。

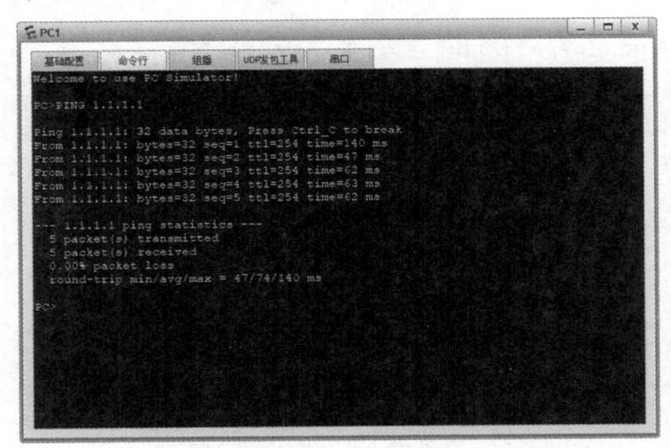

图 4–30　使用 PC1 访问 AR1 的环回口测试图

查看 FW1 的会话表：

```
HRP_M[FW1]display firewall session table
2023-03-28 07:26:30.340
 Current Total Sessions : 8
 icmp  VPN: public --> public  10.1.100.100:20121 --> 1.1.1.1:2048
 udp  VPN: public --> public  12.1.1.2:16384 --> 12.1.1.1:18514
 udp  VPN: public --> public  12.1.1.1:49152 --> 12.1.1.2:18514
 icmp  VPN: public --> public  10.1.100.100:19353 --> 1.1.1.1:2048
 icmp  VPN: public --> public  10.1.100.100:20377 --> 1.1.1.1:2048
 icmp  VPN: public --> public  10.1.100.100:19609 --> 1.1.1.1:2048
 icmp  VPN: public --> public  10.1.100.100:19865 --> 1.1.1.1:2048
 udp  VPN: public --> public  12.1.1.2:49152 --> 12.1.1.1:18514
```

流量通过 FW1 访问 AR1，产生了 ICMP 的会话表项。

查看 FW2 的会话表：

```
HRP_S[FW2]display firewall session table
2023-03-28 07:27:29.090
 Current Total Sessions : 8
 icmp  VPN: public --> public  Remote 10.1.100.100:40089 --> 1.1.1.1:2048
 udp  VPN: public --> public  12.1.1.1:16384 --> 12.1.1.2:18514
 udp  VPN: public --> public  12.1.1.1:49152 --> 12.1.1.2:18514
 icmp  VPN: public --> public  Remote 10.1.100.100:39833 --> 1.1.1.1:2048
 icmp  VPN: public --> public  Remote 10.1.100.100:40857 --> 1.1.1.1:2048
 icmp  VPN: public --> public  Remote 10.1.100.100:40601 --> 1.1.1.1:2048
 icmp  VPN: public --> public  Remote 10.1.100.100:41113 --> 1.1.1.1:2048
 udp  VPN: public --> public  12.1.1.2:49152 --> 12.1.1.1:18514
```

流量并不会经过 FW2，但是 FW2 可以通过 HRP 备份会话表项信息来实现主备的平滑切换。

4.4.5 配置防火墙虚拟系统

1. 实验目的

在防火墙上创建两个虚拟系统，分别命名为 vsysa、vsysb。其中，PC1 属于 vsysa；PC2 属于 vsysb，最终实现 PC1 和 PC2 能够互相访问。

2. 实验拓扑

配置防火墙虚拟系统的实验拓扑如图 4-31 所示。

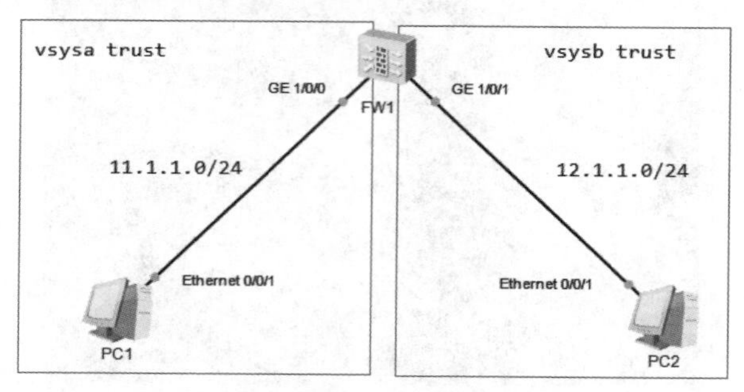

图 4-31 配置防火墙虚拟系统的实验拓扑

3. 实验步骤

步骤 1：创建虚拟系统，并且将虚拟系统、根系统都视为独立的设备，通过划分到对应的虚拟系统，将虚拟接口视为设备之间通信对应的接口。

FW1 的配置：

```
[FW1]vsys enable
[FW1]vsys name vsysa      //创建虚拟系统，命名为 vsysa
[FW1-vsys-vsysa]assign interface GigabitEthernet1/0/0//将 g0/0/0 接口划分到 vsysa
[FW1-vsys-vsysa]quit
[FW1]vsys name vsysb      //创建虚拟系统，命名为 vsysb
[FW1-vsys-vsysb]assign interface GigabitEthernet1/0/1//将虚拟接口加入 g0/0/1 接口划分到 vsysb
[FW1-vsys-vsysb]quit
```

步骤 2：进入虚拟系统视图，配置接口 IP 地址，并且将接口划分到安全区域。

（1）配置 vsysa 的 IP 地址：

```
[FW1]switch vsys vsysa
[FW1-vsysa]display inter br
Interface              PHY    Protocol    InUti    OutUti    inErrors    outErrors
GigabitEthernet1/0/0   up     up          0%       0%        0           0
Virtual-if1            up     up(s)       --       --        0           0
```

通过以上输出可以看出，设备为 vsysa 分配了一个虚拟接口 Virtual-if1，用于与其他虚拟系统进行通信。

```
[FW1-vsysa]interface GigabitEthernet1/0/0
[FW1-vsysa-GigabitEthernet1/0/0]ip address 11.1.1.1 255.255.255.0
[FW1-vsysa-GigabitEthernet1/0/0]q
[FW1-vsysa]interface Virtual-if1
[FW1-vsysa-Virtual-if1]ip address 172.16.1.1 255.255.255.255
```

（2）将 vsysa 的接口划分到安全区域：

```
[FW1-vsysa]firewall zone trust
[FW1-vsysa-zone-trust]add interface GigabitEthernet1/0/0
[FW1-vsysa-zone-trust]q
[FW1-vsysa]firewall zone dmz
[FW1-vsysa-zone-dmz]add interface Virtual-if1
```

（3）配置一般设备间互访 vsysb 的 IP 地址（配置前注意退出到根系统）：

```
[FW1]switch vsys vsysb
[FW1-vsysb]display int br
Interface               PHY    Protocol    InUti    OutUti    inErrors    outErrors
GigabitEthernet1/0/1    up     up          0%       0%        0           0
Virtual-if2             up     up(s)       --       --        0           0
```

通过以上输出可以看出，设备为 vsysb 分配了一个虚拟接口 Virtual-if2，用于与其他虚拟系统进行通信。

```
[FW1-vsysb]interface GigabitEthernet1/0/1
[FW1-vsysb-GigabitEthernet1/0/1]ip address 12.1.1.1 255.255.255.0
[FW1-vsysb-GigabitEthernet1/0/1]q
[FW1-vsysb]interface Virtual-if2
[FW1-vsysb-Virtual-if2]ip address 172.16.2.1 255.255.255.255
```

（4）将 vsysb 的接口划分到安全区域：

```
[FW1-vsysb]firewall zone trust
[FW1-vsysb-zone-trust]add interface GigabitEthernet1/0/1
[FW1-vsysb-zone-trust]q
[FW1-vsysb]firewall zone dmz
[FW1-vsysb-zone-dmz]add interface Virtual-if2
```

查看设备的 VPN 实例：

```
[FW1]display ip vpn-instance
2023-12-02 03:21:57.000
 Total VPN-Instances configured      : 3
 Total IPv4 VPN-Instances configured : 3
 Total IPv6 VPN-Instances configured : 2
  VPN-Instance Name                 RD                     Address-family
  default                                                  IPv4
  vsysa                                                    IPv4
  vsysa                                                    IPv6
  vsysb                                                    IPv4
  vsysb                                                    IPv6
```

通过以上输出可以看出，在创建虚拟系统 vsysa 和 vsysb 的同时，设备上会自动创建两个同名的 VPN 实例。虚拟系统之间的通信则查询对应的 VPN 实例路由表即可。

步骤 3：配置静态路由和策略，就能实现路由可达，不同的虚拟系统需要通过 Virtual-if 接口进行通信，因此静态路由的下一跳写虚拟系统的 VPN 实例即可。

```
[FW1]ip route-static vpn-instance vsysa 12.1.1.0 24 vpn-instance vsysb
[FW1]ip route-static vpn-instance vsysb 11.1.1.0 24 vpn-instance vsysa
```

查询路由表：

```
[FW1]display ip routing-table vpn-instance vsysa
2023-12-02 03:23:27.090
Route Flags: R - relay, D - download to fib
------------------------------------------------------------------------------
Routing Tables: vsysa
         Destinations : 4        Routes : 4
Destination/Mask    Proto    Pre    Cost    Flags    NextHop      Interface
      11.1.1.0/24   Direct   0      0       D        11.1.1.1     GigabitEthernet1/0/0
      11.1.1.1/32   Direct   0      0       D        127.0.0.1    GigabitEthernet1/0/0
      12.1.1.0/24   Static   60     0       D        172.16.2.1   Virtual-if2
    172.16.1.1/32   Direct   0      0       D        127.0.0.1    Virtual-if1

[FW1]display ip routing-table vpn-instance vsysb
2023-12-02 03:23:29.040
Route Flags: R - relay, D - download to fib
------------------------------------------------------------------------------
Routing Tables: vsysb
         Destinations : 4        Routes : 4
Destination/Mask    Proto    Pre    Cost    Flags    NextHop      Interface
      11.1.1.0/24   Static   60     0       D        172.16.1.1   Virtual-if1
      12.1.1.0/24   Direct   0      0       D        12.1.1.1     GigabitEthernet1/0/1
      12.1.1.1/32   Direct   0      0       D        127.0.0.1    GigabitEthernet1/0/1
    172.16.2.1/32   Direct   0      0       D        127.0.0.1    Virtual-if2
```

通过以上输出可以看出，vsysa、vsysb 的两个 VPN 实例有去往对端业务网段的路由，并且下一跳为对端虚拟系统的 Vritual-if 接口。

步骤 4：配置两个虚拟系统和根系统的安全策略。

配置 vsysa 的安全策略，放行 PC1 访问 PC2 的流量以及 PC2 访问 PC1 的流量：

```
[FW1]switch vsys vsysa
<FW1-vsysa>sys
Enter system view, return user view with Ctrl+Z.
[FW1-vsysa]security-policy
[FW1-vsysa-policy-security]rule name trust_dmz
[FW1-vsysa-policy-security-rule-trust_dmz]source-zone dmz
[FW1-vsysa-policy-security-rule-trust_dmz]source-zone trust
[FW1-vsysa-policy-security-rule-trust_dmz]destination-zone dmz
[FW1-vsysa-policy-security-rule-trust_dmz]destination-zone trust
[FW1-vsysa-policy-security-rule-trust_dmz]action permit
[FW1]switch vsys vsysb
<FW1-vsysb>sys
Enter system view, return user view with Ctrl+Z.
[FW1-vsysb]security-policy
[FW1-vsysb-policy-security]rule name trust_dmz
[FW1-vsysb-policy-security-rule-trust_dmz]source-zone dmz
[FW1-vsysb-policy-security-rule-trust_dmz]source-zone trust
[FW1-vsysb-policy-security-rule-trust_dmz]destination-zone dmz
```

```
[FW1-vsysb-policy-security-rule-trust_dmz]destination-zone trust
[FW1-vsysb-policy-security-rule-trust_dmz]action permit
```

步骤 5： 测试。

PC1 访问 PC2，如图 4–32 所示。

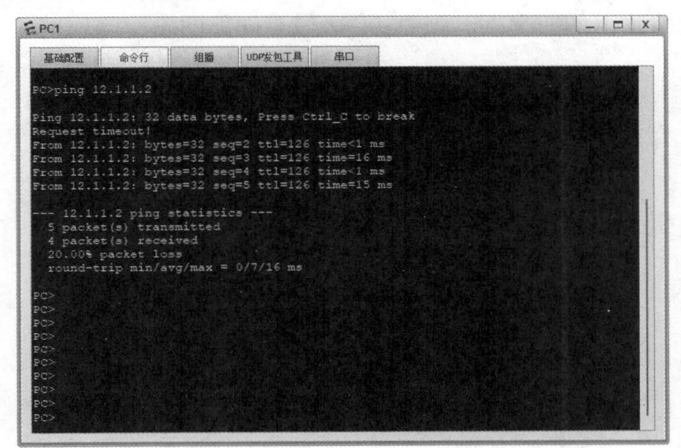

图 4–32 PC1 访问 PC2 测试图

PC2 访问 PC1，如图 4–33 所示。

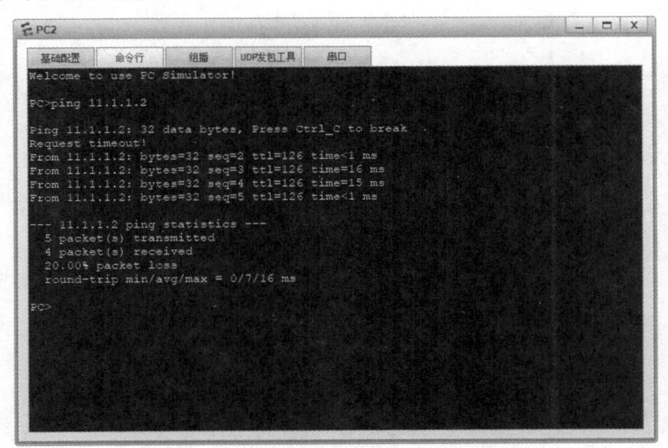

图 4–33 PC2 访问 PC1 测试图

4.4.6 防火墙综合应用

1. 实验目的

（1）在 SW1 上通过部署两个 VPN 实例，实现内部 RD、访客网络隔离。PC1 属于 guest（访客部门），PC2 属于 RD（研发部门）。

（2）FW1 使用虚拟系统与 SW1 上的两个 VPN 实例 RD 和 guest 对接，系统与其对接的 VPN 实例名称保持一致。

（3）AR1 为出口路由器，SW1 与 AR1 使用 VLAN 104 对接，互联 IP 为 10.1.104.0/24，并且运行在 OSPF 的区域 0 中。

（4）FW1 的虚拟系统 guest 与 SW1 分别使用 VLAN 100 和 VLAN 102 与 SW1 对接。VLAN 100 互联 IP 10.1.100.0/24，VLAN 102 互联 IP 10.1.102.0/24，并且运行在 OSPF 的区域 1 中。

（5）FW1 的虚拟系统 RD 与 SW1 分别使用 VLAN 101 和 VLAN 103 与 SW1 对接。VLAN 101 互联 IP 10.1.101.0/24，VLAN 103 互联 IP 10.1.103.0/24，并且运行在 OSPF 的区域 2 中。

（6）在不使用策略路由的前提下，跨 OSPF 区域（内部网络访问公网或者内部互访）的流量必须经过防火墙。

2. 实验拓扑

防火墙综合应用的实验拓扑如图 4-34 所示。

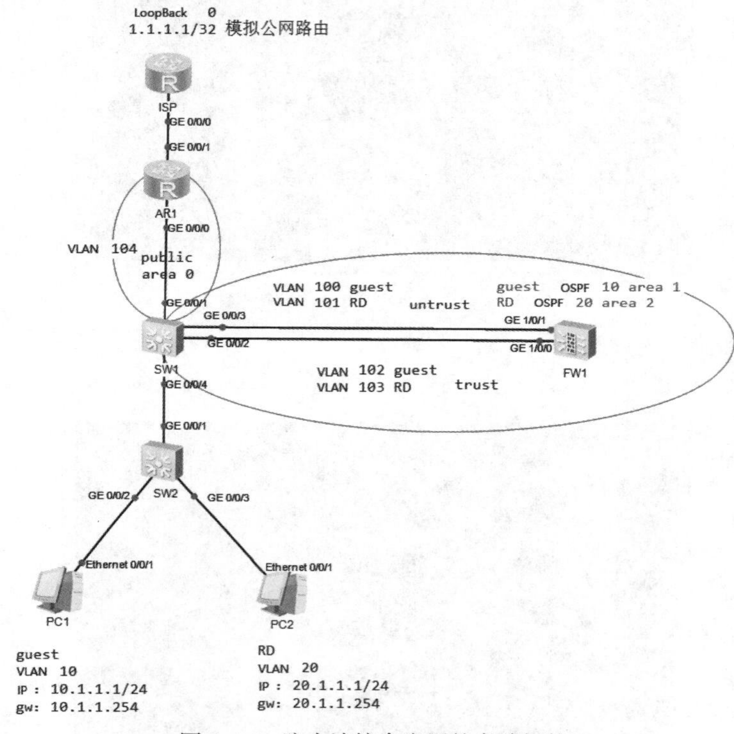

图 4-34 防火墙综合应用的实验拓扑

3. 逻辑拓扑

防火墙 guest 虚拟网络的逻辑拓扑如图 4-35 所示。

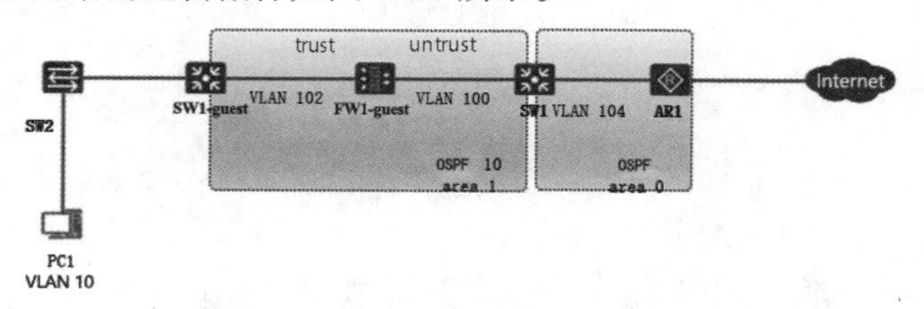

图 4-35 防火墙 guest 虚拟网络的逻辑拓扑

防火墙 RD 虚拟网络的逻辑拓扑如图 4-36 所示。

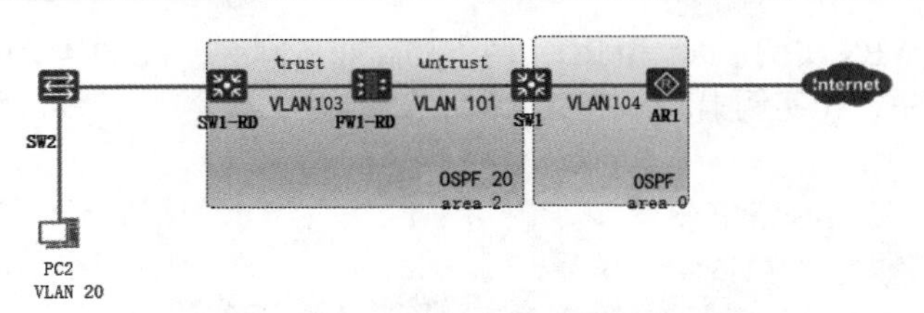

图 4-36 防火墙 RD 虚拟网络的逻辑拓扑

4. 实验步骤

步骤 1：在 SW1 上通过部署两个 VPN 实例，实现内部 RD、访客网络隔离。其中，PC1 属于 guest（访客部门），PC2 属于 RD（研发部门）。

（1）在 SW1 和 SW2 上创建对应的 VLAN，并且配置链路类型。

SW1 的配置：

```
[SW1]vlan batch 10 20
[SW1]vlan batch 100 to 104
[SW1]interface GigabitEthernet0/0/4
[SW1-GigabitEthernet0/0/4]port link-type trunk
[SW1-GigabitEthernet0/0/4]port trunk allow-pass vlan 10 20  //连接SW2的接口，用于
                                                            //接收用户流量
[SW1-GigabitEthernet0/0/4]q
[SW1]interface GigabitEthernet0/0/1
[SW1-GigabitEthernet0/0/1]port link-type access
[SW1-GigabitEthernet0/0/1]port default vlan 104 //对接AR1，使用VLAN 104三层互联
[SW1-GigabitEthernet0/0/1]q
[SW1]interface GigabitEthernet0/0/2
[SW1-GigabitEthernet0/0/2]port link-type trunk
[SW1-GigabitEthernet0/0/2]port trunk allow-pass vlan 102 103//对接FW1，guest网络
                                        //使用VLAN 102互联，RD网络使用VLAN 103互联
[SW1-GigabitEthernet0/0/2]q
[SW1]interface GigabitEthernet0/0/3
[SW1-GigabitEthernet0/0/3]port link-type trunk
[SW1-GigabitEthernet0/0/3]port trunk allow-pass vlan 100 101//对接FW1，guest网络
                                        //使用VLAN 100互联，RD网络使用VLAN 101互联
```

SW2 的配置：

```
[SW2]vlan batch 10 20
[SW2]interface GigabitEthernet0/0/1
[SW2-GigabitEthernet0/0/1]port link-type trunk
[SW2-GigabitEthernet0/0/1]port trunk allow-pass vlan 10 20
[SW2-GigabitEthernet0/0/1]q
[SW2]interface GigabitEthernet0/0/2
[SW2-GigabitEthernet0/0/2]port link-type access
[SW2-GigabitEthernet0/0/2]port default vlan 10
[SW2-GigabitEthernet0/0/2]q
[SW2]interface GigabitEthernet0/0/3
[SW2-GigabitEthernet0/0/3]port link-type access
[SW2-GigabitEthernet0/0/3]port default vlan 20
```

（2）配置SW1的VPN实例guest和RD，用于隔离业务网络，将VLAN 10和VLAN 20的Vlanif接口分别绑定到对应的VPN实例中。

SW1的配置：

```
[SW1]ip vpn-instance guest
[SW1-vpn-instance-guest]ipv4-family
[SW1-vpn-instance-guest-af-ipv4]route-distinguisher 100:1
[SW1-vpn-instance-guest-af-ipv4]q
[SW1-vpn-instance-guest]q
[SW1]ip vpn-instance RD
[SW1-vpn-instance-RD]ipv4-family
[SW1-vpn-instance-RD-af-ipv4]route-distinguisher 100:2
[SW1-vpn-instance-RD-af-ipv4]q
[SW1-vpn-instance-RD]q
[SW1]interface Vlanif10
[SW1-Vlanif10]ip address 10.1.1.254 255.255.255.0
[SW1]interface Vlanif20
[SW1-Vlanif20]ip binding vpn-instance RD
[SW1-Vlanif20]ip address 20.1.1.254 255.255.255.0
```

步骤2：FW1使用虚拟系统与SW1上的两个VPN实例RD和guest对接，系统与其对接的VPN实例名称保持一致（在防火墙上创建两个虚拟系统，分别为RD和guest）。

FW1的配置：

```
[FW1]vsys enable                                    //开启虚拟系统功能
[FW1]vlan batch 100 to 103                          //创建与SW1的互联VLAN
[FW1]interface GigabitEthernet1/0/0
[FW1-GigabitEthernet1/0/0]portswitch                //设置端口为二层端口
[FW1-GigabitEthernet1/0/0]port link-type trunk
[FW1-GigabitEthernet1/0/0]port trunk allow-pass vlan 102 to 103
[FW1-GigabitEthernet1/0/0]q
[FW1]interface GigabitEthernet1/0/1
[FW1-GigabitEthernet1/0/1]portswitch.
[FW1-GigabitEthernet1/0/1]port link-type trunk
[FW1-GigabitEthernet1/0/1]port trunk allow-pass vlan 100 to 101
```

创建虚拟系统：

```
[FW1]vsys name guest                    //创建虚拟系统guest
[FW1-vsys-guest]assign vlan 100
[FW1-vsys-guest]assign vlan 102         //将互联VLAN 100和VLAN 102加入虚拟系统中
[FW1-vsys-guest]quit
[FW1]vsys name RD                       //创建虚拟系统RD
[FW1-vsys-RD]assign vlan 101
[FW1-vsys-RD]assign vlan 103            //将互联VLAN 101和VLAN 103加入虚拟系统中
```

步骤3：配置SW1和FW1以及SW1和AR1之间互联VLAN的IP地址，并且将防火墙的接口划分到安全区域。

SW1的配置：

```
[SW1]interface Vlanif100
```

```
[SW1-Vlanif100]ip address 10.1.100.1 255.255.255.0
[SW1-Vlanif100]
[SW1]interface Vlanif101
[SW1-Vlanif101]ip address 10.1.101.1 255.255.255.0
[SW1-Vlanif101]q
[SW1]interface Vlanif102
[SW1-Vlanif102]ip binding vpn-instance guest
[SW1-Vlanif102]ip address 10.1.102.1 255.255.255.0
[SW1-Vlanif102]q
[SW1]interface Vlanif103
[SW1-Vlanif103]ip binding vpn-instance RD
[SW1-Vlanif103]ip address 10.1.103.1 255.255.255.0
[SW1-Vlanif103]q
[SW1]interface Vlanif104
[SW1-Vlanif104]ip address 10.1.104.1 255.255.255.0
```

需要注意的是，SW1 的 VLAN 102 和 VLAN 103 需要分别加入 VPN 实例 guest 和 RD 中，用于将实例路由传递给 FW1，VLAN 100 和 VLAN 101 无须加入 VPN 实例中，防火墙通过 VLAN 102 和 VLAN 103 收集到 VPN 实例路由后，可以通过 VLAN 100 和 VLAN 101 将实例路由传递给 SW1 的全局路由表，从而实现内部网络访问外部的流量走向为 PC → SW2 → SW1 → FW1 → SW1 → AR1。

FW1 的配置：

```
[FW1]interface Vlanif100
[FW1-Vlanif100]ip address 10.1.100.2 255.255.255.0
[FW1-Vlanif100]q
[FW1]interface Vlanif101
[FW1-Vlanif101]ip address 10.1.101.2 255.255.255.0
[FW1-Vlanif101]q
[FW1]interface Vlanif102
[FW1-Vlanif102]ip address 10.1.102.2 255.255.255.0
[FW1-Vlanif102]q
[FW1]interface Vlanif103
[FW1-Vlanif103]ip address 10.1.103.2 255.255.255.0
```

配置 FW1 guset 虚拟系统：

```
[FW1]switch vsys guest //进入虚拟系统 guest，将 Vlanif100 划分到 untrust 区域，将 Vlanif102
                       //划分到 trust 区域
<FW1-guest>system-view
[FW1-guest]firewall zone trust
[FW1-guest-zone-trust]add interface Vlanif102
[FW1-guest-zone-trust]q
[FW1-guest]firewall zone untrust
[FW1-guest-zone-untrust]add interface Vlanif100
```

配置 FW1 RD 虚拟系统：

```
[FW1]switch vsys RD //进入虚拟系统 guest，将 Vlanif101 划分到 untrust 区域，将 Vlanif103
                    //划分到 trust 区域
<FW1-RD>system-view
```

```
[FW1-RD]firewall zone trust
[FW1-RD-zone-trust]add interface Vlanif103
[FW1-RD-zone-trust]q
[FW1-RD]firewall zone untrust
[FW1-RD-zone-untrust]add interface Vlanif101
```

AR1 的配置：

```
[AR1]interface GigabitEthernet0/0/0
[AR1-GigabitEthernet0/0/0]ip address 10.1.104.2 255.255.255.0
[AR1]interface  LoopBack 0
[AR1-LoopBack0]ip address  1.1.1.1 32
```

步骤 4：配置 OSPF 实现路由互通。

AR1 的配置：

```
[AR1]ospf 1
[AR1-ospf-1]area 0
[AR1-ospf-1-area-0.0.0.0]network 10.1.104.0 0.0.0.255
```

SW1 public 网络的 OSPF 配置：

```
[SW1]ospf 1
[SW1-ospf-1]area 0
[SW1-ospf-1-area-0.0.0.0]network 10.1.104.0 0.0.0.255
[SW1-ospf-1-area-0.0.0.0]area 1
[SW1-ospf-1-area-0.0.0.1]network 10.1.100.0 0.0.0.255
[SW1-ospf-1-area-0.0.0.1]area 2
[SW1-ospf-1-area-0.0.0.2]network 10.1.101.0 0.0.0.255
```

SW1 guest 网络的 OSPF 的配置：

```
[SW1]ospf 10 vpn-instance guest
[SW1-ospf-10]vpn-instance-capability simple // 互联 3 类 LSA 的 DN 检测，直接计算路由
[SW1-ospf-10]area 1
[SW1-ospf-10-area-0.0.0.1]network 10.1.102.0 0.0.0.255
[SW1-ospf-10-area-0.0.0.1]network  10.1.1.0 0.0.0.255
```

SW1 RD 网络的 OSPF 配置：

```
[SW1]ospf 20 vpn-instance RD
[SW1-ospf-20]vpn-instance-capability simple
[SW1-ospf-20]area 2
[SW1-ospf-20-area-0.0.0.2]network 10.1.103.0 0.0.0.255
[SW1-ospf-20-area-0.0.0.2]network 20.1.1.0 0.0.0.255
```

FW1 的配置：

```
[FW1]ospf 10 vpn-instance guest
[FW1-ospf-10]vpn-instance-capability simple
[FW1-ospf-10]area 1
[FW1-ospf-10-area-0.0.0.1]network 10.1.100.0 0.0.0.255
[FW1-ospf-10-area-0.0.0.1]network 10.1.102.0 0.0.0.255
[FW1]ospf 20 vpn-instance RD
[FW1-ospf-20]vpn-instance-capability simple
```

```
[FW1-ospf-20]area 2
[FW1-ospf-20-area-0.0.0.2]network 10.1.101.0 0.0.0.255
[FW1-ospf-20-area-0.0.0.2]network 10.1.103.0 0.0.0.255
```

查看 OSPF 的邻居建立情况：

```
[SW1]display  ospf peer  brief
        OSPF Process 1 with Router ID 10.1.100.1
                 Peer Statistic Information
 ----------------------------------------------------------------------------
 Area Id          Interface                       Neighbor id      State
 0.0.0.0          Vlanif104                       10.1.104.2       Full
 0.0.0.1          Vlanif100                       10.1.100.2       Full
 0.0.0.2          Vlanif101                       10.1.101.2       Full
 ----------------------------------------------------------------------------
        OSPF Process 10 with Router ID 10.1.1.254
                 Peer Statistic Information
 ----------------------------------------------------------------------------
 Area Id          Interface                       Neighbor id      State
 0.0.0.1          Vlanif102                       10.1.100.2       Full
 ----------------------------------------------------------------------------
        OSPF Process 20 with Router ID 20.1.1.254
                 Peer Statistic Information
 ----------------------------------------------------------------------------
 Area Id          Interface                       Neighbor id      State
 0.0.0.2          Vlanif103                       10.1.101.2       Full
 ----------------------------------------------------------------------------

[FW1]display  ospf peer  brief
        OSPF Process 10 with Router ID 10.1.100.2
                 Peer Statistic Information
 ----------------------------------------------------------------------------
 Area Id          Interface                       Neighbor id      State
 0.0.0.1          Vlanif100                       10.1.100.1       Full
 0.0.0.1          Vlanif102                       10.1.1.254       Full
 ----------------------------------------------------------------------------
 Total Peer(s):      2
        OSPF Process 20 with Router ID 10.1.101.2
                 Peer Statistic Information
 ----------------------------------------------------------------------------
 Area Id          Interface                       Neighbor id      State
 0.0.0.2          Vlanif101                       10.1.100.1       Full
 0.0.0.2          Vlanif103                       20.1.1.254       Full
 ----------------------------------------------------------------------------
 Total Peer(s):      2
```

通过以上输出可以看出，guest 实例（OSPF 10）通过 VLAN 102 与防火墙建立邻居，将本实例的路由发布给防火墙；RD 实例（OSPF 20）通过 VLAN 103 和防火墙建立邻居，将本实例的路由发布给防火墙。

SW1 的 public 实例（OSPF 1）与 AR1 和 FW1 建立 OSPF 邻居，用于通过与防火墙建立邻居关系，将 VPN 实例的路由接收放入全局路由表中。

查看 SW1 的路由表。全局路由表：

```
[SW1]display ip routing-table
Route Flags: R - relay, D - download to fib
------------------------------------------------------------------------------
Routing Tables: Public
         Destinations : 12       Routes : 12
Destination/Mask      Proto    Pre    Cost     Flags    NextHop        Interface
     10.1.1.0/24      OSPF     10     3        D        10.1.100.2     Vlanif100
   10.1.100.0/24      Direct   0      0        D        10.1.100.1     Vlanif100
   10.1.100.1/32      Direct   0      0        D        127.0.0.1      Vlanif100
   10.1.101.0/24      Direct   0      0        D        10.1.101.1     Vlanif101
   10.1.101.1/32      Direct   0      0        D        127.0.0.1      Vlanif101
   10.1.102.0/24      OSPF     10     2        D        10.1.100.2     Vlanif100
   10.1.103.0/24      OSPF     10     2        D        10.1.101.2     Vlanif101
   10.1.104.0/24      Direct   0      0        D        10.1.104.1     Vlanif104
   10.1.104.1/32      Direct   0      0        D        127.0.0.1      Vlanif104
     20.1.1.0/24      OSPF     10     3        D        10.1.101.2     Vlanif101
     127.0.0.0/8      Direct   0      0        D        127.0.0.1      InLoopBack0
    127.0.0.1/32      Direct   0      0        D        127.0.0.1      InLoopBack0
```

全局路由表可以通过防火墙和后续流量的转发学习到 VPN 实例 guest 和 RD 的路由。

VPN 实例 guest 的路由表：

```
[SW1]display ip routing-table vpn-instance guest
Route Flags: R - relay, D - download to fib
------------------------------------------------------------------------------
Routing Tables: guest
         Destinations : 9        Routes : 9
Destination/Mask      Proto    Pre    Cost     Flags    NextHop        Interface
     10.1.1.0/24      Direct   0      0        D        10.1.1.254     Vlanif10
   10.1.1.254/32      Direct   0      0        D        127.0.0.1      Vlanif10
   10.1.100.0/24      OSPF     10     2        D        10.1.102.2     Vlanif102
   10.1.101.0/24      OSPF     10     3        D        10.1.102.2     Vlanif102
   10.1.102.0/24      Direct   0      0        D        10.1.102.1     Vlanif102
   10.1.102.1/32      Direct   0      0        D        127.0.0.1      Vlanif102
   10.1.103.0/24      OSPF     10     4        D        10.1.102.2     Vlanif102
   10.1.104.0/24      OSPF     10     3        D        10.1.102.2     Vlanif102
     20.1.1.0/24      OSPF     10     5        D        10.1.102.2     Vlanif102
```

VPN 实例 RD 的路由表：

```
[SW1]display ip routing-table vpn-instance RD
Route Flags: R - relay, D - download to fib
------------------------------------------------------------------------------
Routing Tables: RD
         Destinations : 9        Routes : 9
Destination/Mask      Proto    Pre    Cost     Flags    NextHop        Interface
```

```
      10.1.1.0/24      OSPF     10     5      D        10.1.103.2     Vlanif103
    10.1.100.0/24      OSPF     10     3      D        10.1.103.2     Vlanif103
    10.1.101.0/24      OSPF     10     2      D        10.1.103.2     Vlanif103
    10.1.102.0/24      OSPF     10     4      D        10.1.103.2     Vlanif103
    10.1.103.0/24      Direct   0      0      D        10.1.103.1     Vlanif103
    10.1.103.1/32      Direct   0      0      D        127.0.0.1      Vlanif103
    10.1.104.0/24      OSPF     10     3      D        10.1.103.2     Vlanif103
      20.1.1.0/24      Direct   0      0      D        20.1.1.254     Vlanif20
    20.1.1.254/32      Direct   0      0      D        127.0.0.1      Vlanif20
```

通过以上输出可以看出，guest 和 RD 可以相互学习到各自的路由，原因是 SW1 的 OSPF 1 接收到 area 1 和 area 2 的路由后会生成 3 类 LSA 发布给 FW1。如果在防火墙上放行对应流量的安全策略，则 PC1 和 PC2 可以互访，并且虚拟系统之间也可以互访。如果无此类需求，则需要在 SW1 上使用路由过滤，让两个 VPN 实例的路由相互独立。

在 SW1 上配置路由过滤：

```
[SW1]ip ip-prefix guest deny 10.1.1.0 24
[SW1]ip ip-prefix guest permit  0.0.0.0 0 less-equal 32 //使用前缀列表 guest 匹配过滤
                                                         //VLAN 10 的路由
[SW1]ip ip-prefix RD deny  20.1.1.0 24
[SW1]ip ip-prefix RD permit 0.0.0.0 0 less-equal  32    //使用前缀列表 RD 匹配过滤 VLAN 20
                                                         //的路由
```

在 SW1 上配置区域间的路由过滤：

```
[SW1]ospf 1
[SW1-ospf-1]area 1
[SW1-ospf-1-area-0.0.0.1]filter ip-prefix RD import
[SW1-ospf-1-area-0.0.0.1]area 2
[SW1-ospf-1-area-0.0.0.2]filter ip-prefix guest import
```

查看过滤后的 VPN 实例路由表：

```
[SW1]display IP routing-table vpn-instance guest
Route Flags: R - relay, D - download to fib
------------------------------------------------------------------------------
Routing Tables: guest
         Destinations : 8          Routes : 8
Destination/Mask     Proto    Pre    Cost     Flags    NextHop        Interface
     10.1.1.0/24     Direct   0      0        D        10.1.1.254     Vlanif10
   10.1.1.254/32     Direct   0      0        D        127.0.0.1      Vlanif10
   10.1.100.0/24     OSPF     10     2        D        10.1.102.2     Vlanif102
   10.1.101.0/24     OSPF     10     3        D        10.1.102.2     Vlanif102
   10.1.102.0/24     Direct   0      0        D        10.1.102.1     Vlanif102
   10.1.102.1/32     Direct   0      0        D        127.0.0.1      Vlanif102
   10.1.103.0/24     OSPF     10     4        D        10.1.102.2     Vlanif102
   10.1.104.0/24     OSPF     10     3        D        10.1.102.2     Vlanif102
[SW1]display IP routing-table vpn-instance RD
Route Flags: R - relay, D - download to fib
------------------------------------------------------------------------------
Routing Tables: RD
```

```
        Destinations : 8        Routes : 8
Destination/Mask    Proto   Pre  Cost    Flags  NextHop        Interface
   10.1.100.0/24    OSPF    10   3       D      10.1.103.2     Vlanif103
   10.1.101.0/24    OSPF    10   2       D      10.1.103.2     Vlanif103
   10.1.102.0/24    OSPF    10   4       D      10.1.103.2     Vlanif103
   10.1.103.0/24    Direct  0    0       D      10.1.103.1     Vlanif103
   10.1.103.1/32    Direct  0    0       D      127.0.0.1      Vlanif103
   10.1.104.0/24    OSPF    10   3       D      10.1.103.2     Vlanif103
     20.1.1.0/24    Direct  0    0       D      20.1.1.254     Vlanif20
   20.1.1.254/32    Direct  0    0       D      127.0.0.1      Vlanif20
```

步骤 5： 在防火墙上放行 PC 访问公网的流量。

```
[FW1]switch vsys guest
<FW1-guest>system-view
[FW1-guest]security-policy
[FW1-guest-policy-security]rule name trust_untrust
[FW1-guest-policy-security-rule-trust_untrust]source-zone trust
[FW1-guest-policy-security-rule-trust_untrust]destination-zone untrust
[FW1-guest-policy-security-rule-trust_untrust]action permit
[FW1]switch vsys RD
<FW1-RD>system-view
[FW1-RD]security-policy
[FW1-RD-policy-security]rule name trust_untrust
[FW1-RD-policy-security-rule-trust_untrust]source-zone trust
[FW1-RD-policy-security-rule-trust_untrust]destination-zone untrust
[FW1-RD-policy-security-rule-trust_untrust]action permit
```

步骤 6： 在 AR1 配置访问公网的路由，并且配置 NAT。

AR1 的配置：

```
[AR1]interface GigabitEthernet0/0/1
[AR1-GigabitEthernet0/0/1]ip address 100.1.1.1 255.255.255.0
[AR1]ip route-static 0.0.0.0 0 100.1.1.2
```

ISP 的配置：

```
<ISP>system-view
[ISP]interface GigabitEthernet0/0/0
[ISP-GigabitEthernet0/0/0]ip address 100.1.1.2 255.255.255.0
[ISP-GigabitEthernet0/0/0]interface LoopBack0
[ISP-LoopBack0]ip addess 1.1.1.1 255.255.255.255
```

配置 NAT：

```
[AR1]acl number 2000
[AR1-acl-basic-2000]rule 5 permit
[AR1-acl-basic-2000]interface GigabitEthernet0/0/1
[AR1-GigabitEthernet0/0/1]nat outbound 2000
```

步骤 7： 下发默认路由给下面所有的 3 层设备。

```
[AR1]ospf 1
[AR1-ospf-1]default-route-advertise  always
```

步骤 8：测试。

使用 PC1 和 PC2 访问公网（1.1.1.1/32），PC1 和 PC2 访问公网的路径为 PC → SW2 → SW1 → FW1 → SW1 → AR1 → ISP，如图 4-37 和图 4-38 所示。

```
PC1
基础配置 命令行 组播 UDP发包工具 串口
PC>tracert 1.1.1.1

traceroute to 1.1.1.1, 8 hops max
(ICMP), press Ctrl+C to stop
 1  10.1.1.254   31 ms  47 ms  47 ms
 2    *  *  *
 3  10.1.100.1   78 ms  63 ms  62 ms
 4    *  *  *
 5  1.1.1.1   110 ms  109 ms  109 ms

PC>
PC>
PC>
PC>
PC>
PC>
PC>
PC>
PC>
PC>
PC>
PC>
PC>
PC>
PC>
```

图 4-37　PC1 访问 1.1.1.1 测试图

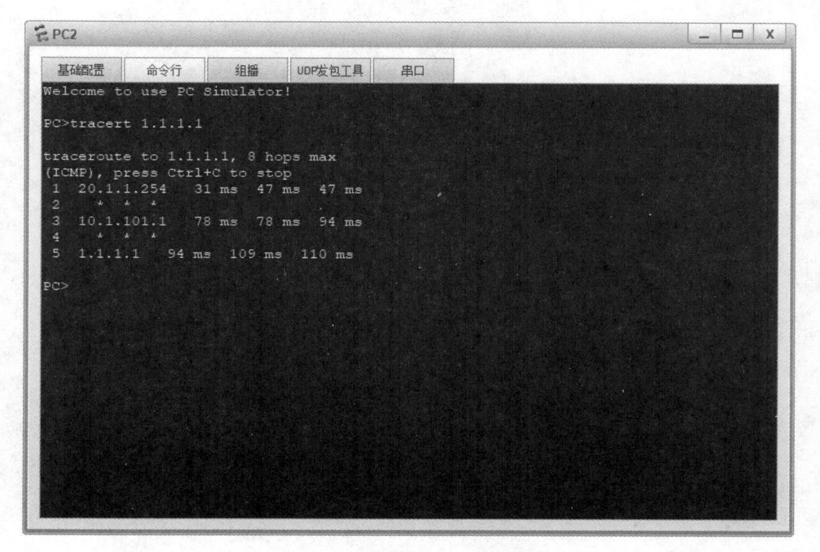

图 4-38　PC2 访问 1.1.1.1 测试图

第5章

MPLS 技术

5.1 MPLS 和 MPLS LDP 概述

MPLS（Multiprotocol Label Switching，多协议标签交换）是一种根据报文中携带的标签来转发数据的技术。

MPLS 的一个基本概念就是两台 LSR（Label Switching Router，标签交换路由器）必须对在它们之间转发的数据的标签使用上“达成共识”。LSR 之间可以运行 LDP（Label Distribution Protocol，标签分发协议）来告知其他 LSR 本设备上的标签绑定信息，从而实现标签报文的正确转发。

5.1.1 MPLS 基本概念

MPLS 是一种 IP（Internet Protocol）骨干网技术。其在无连接的 IP 网络上引入面向连接的标签交换概念，将第三层路由技术和第二层交换技术相结合，充分发挥了 IP 路由的灵活性和二层交换的简捷性。

1. 网络结构

MPLS 网络的典型结构图如图 5-1 所示。MPLS 基于标签进行转发，图 5-1 中进行 MPLS 标签交换和报文转发的网络设备称为 LSR；由 LSR 构成的网络区域称为 MPLS 域（MPLS Domain）。位于 MPLS 域边缘、连接其他网络的 LSR 称为 LER（Label Edge Router，边缘路由器），区域内部的 LSR 称为核心 LSR（Core LSR）。

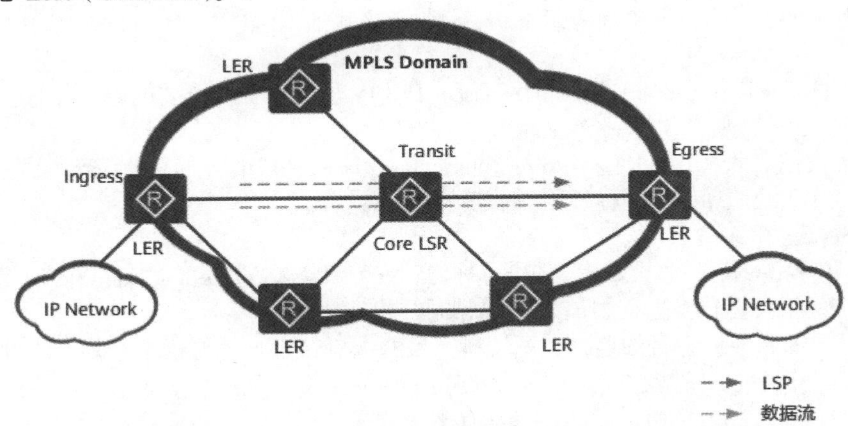

图 5-1　MPLS 网络的典型结构图

IP 报文进入 MPLS 网络时，MPLS 入口的 LER 分析 IP 报文的内容并且为这些 IP 报文添加合适的标签，所有 MPLS 网络中的 LSR 根据标签转发数据。当该 IP 报文离开 MPLS 网络时，标签由出口 LER 弹出。

IP 报文在 MPLS 网络中经过的路径称为 LSP。这是一个单向路径，与数据流的方向一致。

LSP 的入口 LER 称为入节点（Ingress）；位于 LSP 中间的 LSR 称为中间节点（Transit）；LSP 的出口 LER 称为出节点（Egress）。一条 LSP 可以有 0 个、1 个或多个中间节点，但有且只有一个入节点和一个出节点。

根据 LSP 的方向，如果 MPLS 报文由 Ingress 发往 Egress，则 Ingress 是 Transit 的上游节点，Transit 是 Ingress 的下游节点。同理，Transit 是 Egress 的上游节点，Egress 是 Transit 的下游节点。

2. 体系结构

MPLS 的体系结构如图 5-2 所示，它由控制平面（Control Plane）和转发平面（Forwarding Plane）组成。

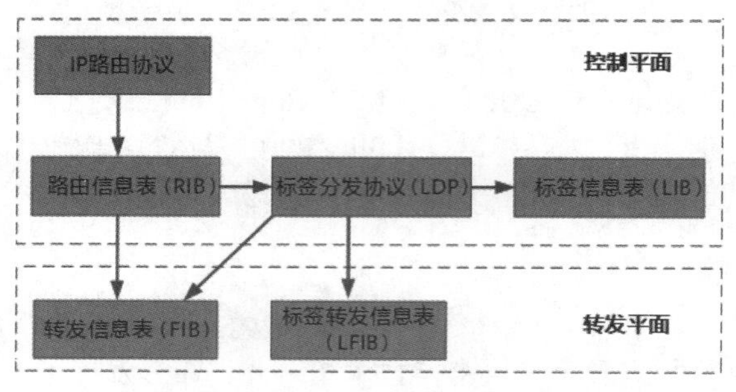

图 5-2 MPLS 的体系结构

（1）控制平面：负责产生和维护路由信息以及标签信息。

①路由信息表（Routing Information Base，RIB）：由 IP 路由协议（IP Routing Protocol）生成，用于选择路由。

②标签分发协议：负责标签的分配、标签转发信息表的建立、标签交换路径的建立、拆除等工作。

③标签信息表（Label Information Base，LIB）：由 LDP 生成，用于管理标签信息。

（2）转发平面：又称数据平面（Data Plane），负责普通 IP 报文的转发以及带 MPLS 标签报文的转发。

①转发信息表（Forwarding Information Base，FIB）：从 RIB 提取必要的路由信息生成，负责普通 IP 报文的转发。

②标签转发信息表（Label Forwarding Information Base，LFIB）：简称标签转发表，由 LDP 在 LSR 上建立 LFIB，负责带 MPLS 标签报文的转发。

3. MPLS 标签

MPLS 将具有相同特征的报文归为一类，称为转发等价类（Forwarding Equivalence Class，FEC）。属于相同 FEC 的报文在转发过程中被 LSR 以相同方式处理。

FEC 可以根据源地址、目的地址、源端口、目的端口、VPN 等要素进行划分。例如，在传统的采用最长匹配算法的 IP 转发中，到同一条路由的所有报文就是一个 FEC。

标签（Label）是一个短而定长的、只具有本地意义的标识符，用于唯一标识一个分组所属的 FEC。例如，在某些情况下要进行负载分担，对应一个 FEC 可能会有多个入标签，但是一台设备上，一个标签只能代表一个 FEC。

MPLS 报文与普通的 IP 报文相比增加了 MPLS 标签信息，MPLS 标签的长度为 4 字节。MPLS 标签封装在链路层和网络层之间，可以支持任意的链路层协议。MPLS 标签的封装结构如图 5-3 所示。

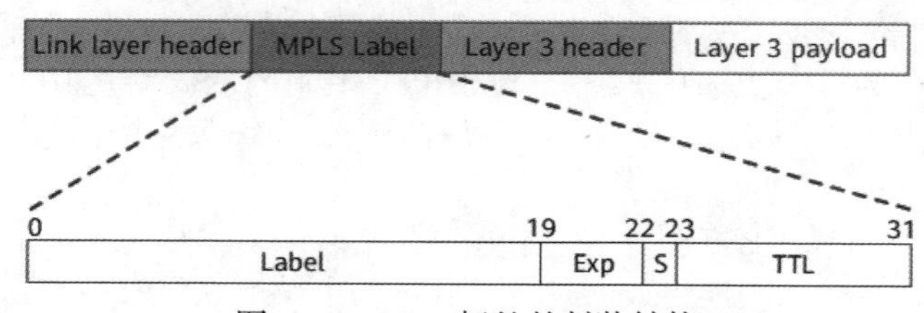

图 5-3 MPLS 标签的封装结构

MPLS 标签共有以下 4 个字段。

（1）Label：20bit，标签值域。

（2）Exp：3bit，用于扩展。现在通常用作 CoS（Class of Service），当设备阻塞时，优先发送优先级高的报文。

（3）S：1bit，栈底标识。MPLS 支持多层标签，即标签嵌套。当 S 值为 1 时，表示最底层标签。

（4）TTL：8bit，和 IP 报文中的 TTL 意义相同。

标签栈（Label Stack）是指标签的排序集合。在图 5-3 中，靠近二层首部的标签称为栈顶 MPLS 标签或外层 MPLS 标签（Outer MPLS Label）；靠近 IP 首部的标签称为栈底 MPLS 标签或内层 MPLS 标签（Inner MPLS Label）。理论上，MPLS 标签可以无限嵌套。目前 MPLS 标签嵌套主要应用在 MPLS VPN、TE FRR（Traffic Engineering Fast ReRoute）中。

5.1.2 MPLS LDP 基本概念

LDP 规定了标签分发过程中的各种消息以及相关的处理过程。通过 LDP，LSR 可以把网络层的路由信息映射到数据链路层的交换路径上，进而建立起 LSP。

1. LDP 消息类型

LDP 主要使用以下 4 类消息。

（1）发现（Discovery）消息：用于通告和维护网络中 LSR 的存在，如 Hello 消息。

（2）会话（Session）消息：用于建立、维护和终止 LDP 对等体之间的会话，如 Initialization 消息、Keepalive 消息。

（3）通告（Advertisement）消息：用于创建、改变和删除 FEC 的标签映射。

（4）通知（Notification）消息：用于提供建议性的消息和差错通知。

为保证 LDP 消息的可靠发送，除了 Discovery 消息使用 UDP（User Datagram Protocol）传输外，Session 消息、Advertisement 消息和 Notification 消息都使用 TCP（Transmission Control Protocol）传输。

LDP 的工作过程主要分为以下两个阶段。

（1）LDP 会话的建立。通过 Hello 消息发现邻居后，LSR 之间开始建立 LDP 会话。会话建立后，LDP 对等体之间通过不断发送 Hello 消息和 Keepalive 消息来维护这个会话。

LDP 对等体之间通过周期性发送 Hello 消息表明自己希望继续维持这种邻接关系。如果 Hello 保持定时器超时仍没有收到新的 Hello 消息，则删除 Hello 邻接关系。邻接关系被删除后，本端 LSR 将发送 Notification 消息，结束该 LDP 会话。

LDP 对等体之间通过 LDP 会话连接上传送的 Keepalive 消息来维持 LDP 会话。如果会话保持定时器（Keepalive 保持定时器）超时仍没有收到任何 Keepalive 消息，则关闭 TCP 连接，本端 LSR 将发送 Notification 消息，结束 LDP 会话。

（2）LDP LSP 的建立。LDP 会话建立后，LDP 通过发送标签请求和标签映射消息，在 LDP 对等体之间通告 FEC 和标签的绑定关系，从而建立 LSP。

2. MPLS LDP 的工作机制

（1）LDP 发现机制。LDP 发现机制用于 LSR 发现潜在的 LDP 对等体。LDP 有以下两种发现机制。

①基本发现机制：用于发现链路上直连的 LSR。LSR 通过周期性地发送 LDP 链路 Hello 消息（LDP Link Hello），实现 LDP 基本发现机制，建立本地 LDP 会话。LDP 链路 Hello 消息使用 UDP 报文，目的地址是组播地址 224.0.0.2。如果 LSR 在特定接口接收到 LDP 链路 Hello 消息，表明该

接口存在 LDP 对等体。

②扩展发现机制：用于发现链路上非直连的 LSR。LSR 周期性地发送 LDP 目标 Hello 消息（LDP Targeted Hello）到指定 IP 地址，实现 LDP 扩展发现机制，建立远端 LDP 会话。LDP 目标 Hello 消息使用 UDP 报文，目的地址是指定 IP 地址。如果 LSR 接收到 LDP 目标 Hello 消息，表明该 LSR 存在 LDP 对等体。

（2）LDP 会话的建立过程。两台 LSR 之间交换 Hello 消息触发 LDP 会话的建立。

LDP 会话的建立过程如图 5-4 所示。

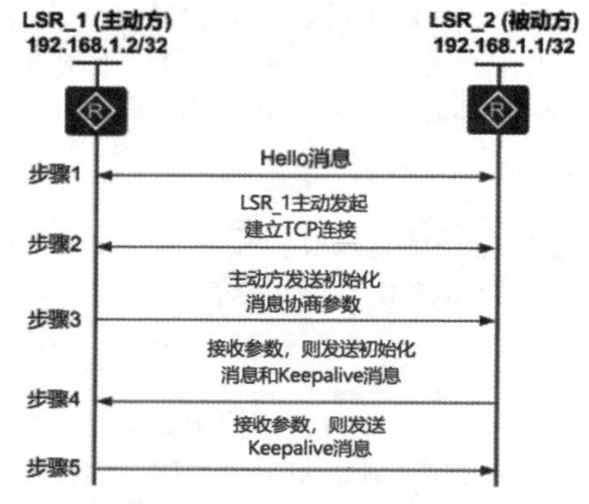

图 5-4　LDP 会话的建立过程

①两台 LSR 之间互相发送 Hello 消息。

② Hello 消息中携带传输地址（设备的 IP 地址），双方使用传输地址建立 LDP 会话。

③传输地址较大的一方作为主动方，发起建立 TCP 连接。

④ LSR_1 作为主动方发起建立 TCP 连接，LSR_2 作为被动方等待对方发起连接。

⑤ TCP 连接建立成功后，由主动方 LSR_1 发送初始化消息，协商建立 LDP 会话的相关参数。

⑥ LDP 会话的相关参数包括 LDP 协议版本、标签分发方式、Keepalive 保持定时器的值、最大 PDU 长度和标签空间等。

⑦如果被动方 LSR_2 收到初始化消息后能够接收相关参数，则发送初始化消息，同时发送 Keepalive 消息给主动方 LSR_1。

⑧如果被动方 LSR_2 不能接收相关参数，则发送 Notification 消息终止 LDP 会话的建立。

⑨初始化消息中包括 LDP 协议版本、标签分发方式、Keepalive 保持定时器的值、最大 PDU 长度和标签空间等。

⑩如果主动方 LSR_1 收到初始化消息后能够接收相关参数，则发送 Keepalive 消息给被动方 LSR_2。

⑪ 如果主动方 LSR_1 不能接收相关参数，则发送 Notification 消息给被动方 LSR_2 终止 LDP 会话的建立。

（3）标签的发布和管理。在 MPLS 网络中，下游 LSR 决定标签和 FEC 的绑定关系，并将这种绑定关系发布给上游 LSR。LDP 通过发送标签请求和标签映射消息，在 LDP 对等体之间通告 FEC 和标签的绑定关系来建立 LSP。标签的发布和管理由标签发布方式、标签分配控制方式和标签保持方式来决定，见表 5-1。

表 5-1 标签的发布和管理方式

内 容	名 称	默 认	含 义
标签的发布方式（Label Advertisement Mode）	下游自主方式（DU）	是	对于一个特定的 FEC，LSR 无须从上游获得标签请求消息即可进行标签分配与分发
	下游按需方式（DoD）	否	对于一个特定的 FEC，LSR 获得标签请求消息之后才进行标签分配与分发
标签的分配控制方式（Label Distribution Control Mode）	独立方式（Independent）	否	本地 LSR 可以自主地分配一个标签绑定到某个 FEC，并通告给上游 LSR，而无须等待下游的标签
	有序方式（Ordered）	是	对于 LSR 上某个 FEC 的标签映射，只有当该 LSR 已经具有此 FEC 下一跳的标签映射消息，或者该 LSR 就是此 FEC 的出节点时，该 LSR 才可以向上游发送此 FEC 的标签映射
标签的保持方式（Label Retention Mode）	自由方式（Liberal）	是	对于从邻居 LSR 收到的标签映射，无论邻居 LSR 是不是自己的下一跳都保留
	保守方式（Conservative）	否	对于从邻居 LSR 收到的标签映射，只有当邻居 LSR 是自己的下一跳时才保留

5.2 MPLS 和 MPLS LDP 实验

5.2.1 配置 MPLS 的静态 LSP

1. 实验目的

全网的 IGP 都运行 OSPF，并且按照 FEC 的标签规划配置静态 LSP，从而实现 AR1 的 1.1.1.1/32 能够使用 MPLS LSP 隧道访问 AR3 的 3.3.3.3/32。

扫一扫，看视频

2. 实验拓扑

配置 MPLS 的静态 LSP 的实验拓扑如图 5-5 所示。

FEC:3.3.3.3/32
out label: 201
FEC:3.3.3.3/32
in label: 302
out label:201
FEC:3.3.3.3/32
in label: 302
10.0.12.0/24
10.0.23.0/24
GE 0/0/0
GE 0/0/0
GE 0/0/1
GE 0/0/0
AR1
AR2
AR3
FEC:1.1.1.1/32
in label: 102
FEC:1.1.1.1/32
in label: 203
out label:102
FEC:1.1.1.1/32
out label: 203

图 5-5 配置 MPLS 的静态 LSP 的实验拓扑

3. 实验步骤

步骤 1： 配置 IP 地址，MPLS 静态 LSP 实验 IP 规划表见表 5-2。

表 5-2　MPLS 静态 LSP 实验 IP 规划表

设备名称	接口编号	IP 地址
AR1	G0/0/0	10.0.12.1/24
	LoopBack 0	1.1.1.1/32
AR2	G0/0/0	10.0.12.2/24
	G0/0/1	10.0.23.2/24
	LoopBack 0	2.2.2.2/32
AR3	G0/0/0	10.0.23.3/24
	LoopBack 0	3.3.3.3/24

步骤 2：配置 OSPF。

AR1 的配置：

```
[AR1]ospf
[AR1-ospf-1]area 0
[AR1-ospf-1-area-0.0.0.0]network 10.0.12.0 0.0.0.255
[AR1-ospf-1-area-0.0.0.0]network 1.1.1.1 0.0.0.0
```

AR2 的配置：

```
[AR2]ospf
[AR2-ospf-1]area 0
[AR2-ospf-1-area-0.0.0.0]network 10.0.12.0 0.0.0.255
[AR2-ospf-1-area-0.0.0.0]network 10.0.23.0 0.0.0.255
[AR2-ospf-1-area-0.0.0.0]network 2.2.2.2 0.0.0.0
```

AR3 的配置：

```
[AR3]ospf
[AR3-ospf-1]area 0
[AR3-ospf-1-area-0.0.0.0]network 10.0.23.0 0.0.0.255
[AR3-ospf-1-area-0.0.0.0]network 3.3.3.3 0.0.0.0
```

步骤 3：配置静态 LSP。使能接口以及全局的 MPLS 功能。

AR1 的配置：

```
[AR1]mpls lsr-id 1.1.1.1          //配置设备的 LSR-ID 为环回口的地址 1.1.1.1
[AR1]mpls                         //全局开启 MPLS 功能
[AR1]interface g0/0/0
[AR1-GigabitEthernet0/0/0]mpls    //接口开启 MPLS 功能
```

AR2 的配置：

```
[AR2]mpls  lsr-id  2.2.2.2
[AR2]mpls
[AR2]interface g0/0/0
[AR2-GigabitEthernet0/0/0]mpls
[AR2]interface g0/0/1
[AR2-GigabitEthernet0/0/1]mpls
```

AR3 的配置：

```
[AR3]mpls lsr-id 3.3.3.3
[AR3]mpls
[AR3]interface g0/0/0
[AR3-GigabitEthernet0/0/0]mpls
```

配置 FEC 为 3.3.3.3 的静态 LSP。

AR1 的配置：

```
[AR1]static-lsp ingress 1to3 destination 3.3.3.3 32 nexthop 10.0.12.2 outgoing-
interface g0/0/0 out-label 201//配置AR1为去往FEC 3.3.3.3/32的ingress（入站LSR），
          //静态LSP命名为1to3，下一跳地址为10.0.12.2，出接口为G0/0/0，出标签为201
```

查看 MPLS LSP：

```
<AR1>display mpls lsp
-------------------------------------------------------------------------------
                 LSP Information: STATIC LSP
-------------------------------------------------------------------------------
FEC                 In/Out Label           In/Out IF             Vrf Name
3.3.3.3/32          NULL/201               -/GE0/0/0
```

通过以上输出可以看出，LSP Information: STATIC LSP 表示此 LSP 为静态 LSP，当设备在发送目标网段为 3.3.3.3/32 的数据时，从 G0/0/0 接口转发，出标签为 201。

AR2 的配置：

```
[AR2]static-lsp transit 1to3 incoming-interface g0/0/0 in-label 201 nexthop 10.0.23.3
out-label 302 //配置AR2为去往FEC 3.3.3.3/32的transit（中转LSR），静态
          //LSP命名为1-3，入接口为G0/0/0，入标签为201，下一跳地址为10.0.32.3，出标签为302
[AR2]display mpls lsp
------------------------------------
---------------------------------------
                 LSP Information: STATIC LSP
-------------------------------------------------------------------------------
FEC                 In/Out Label           In/Out IF             Vrf Name
-/-                 201/302                GE0/0/0/GE0/0/1
```

以上信息表示，如果 AR2 在 G0/0/0 接口收到标签为 302 的数据，则发往 G0/0/1 接口，并打上标签 201。

AR3 的配置：

```
[AR3]static-lsp egress 1to3 incoming-interface g0/0/0 in-label 302 //配置AR3为
//FEC 3.3.3.3/32的egress（出站LSR），静态LSP命名为1to3，入接口为G0/0/0，入标签为302
[AR3]display mpls lsp
-------------------------------------------------------------------------------
                 LSP Information: STATIC LSP
-------------------------------------------------------------------------------
FEC                 In/Out Label           In/Out IF                   Vrf Name
-/-                 302/NULL               GE0/0/0/-
```

以上信息表示，如果 AR3 在 G0/0/0 接口收到标签为 302 的数据，则剥离标签。

在 AR1 上测试，并且在 G0/0/0 接口抓包查看数据特征。

图 5-6 所示为 1.1.1.1 发送 3.3.3.3 的抓包结果，可以看到发送时打上了标签 201。

```
> Frame 6: 102 bytes on wire (816 bits), 102 bytes captured (816 bits) on interface -, id 0
> Ethernet II, Src: HuaweiTe_4f:36:ee (00:e0:fc:4f:36:ee), Dst: HuaweiTe_e5:0c:69 (00:e0:fc:e5:0c:69)
> MultiProtocol Label Switching Header, Label: 201, Exp: 0, S: 1, TTL: 255
> Internet Protocol Version 4, Src: 1.1.1.1, Dst: 3.3.3.3
> Internet Control Message Protocol
```

图 5-6　AR1 的 G0/0/0 接口的抓包结果（1）

图 5-7 所示为 3.3.3.3 回复 1.1.1.1 的抓包结果，可以看到回复的报文并没有打上标签。

```
> Frame 7: 98 bytes on wire (784 bits), 98 bytes captured (784 bits) on interface -, id 0
> Ethernet II, Src: HuaweiTe_e5:0c:69 (00:e0:fc:e5:0c:69), Dst: HuaweiTe_4f:36:ee (00:e0:fc:4f:36:ee)
> Internet Protocol Version 4, Src: 3.3.3.3, Dst: 1.1.1.1
> Internet Control Message Protocol
```

图 5-7　AR1 的 G0/0/0 接口的抓包结果（2）

因此现在还只是一个单向的隧道。

步骤 4：配置 FEC 为 1.1.1.1 的静态 LSP。

AR3 的配置：

```
[AR3]static-lsp ingress 3to1 destination 1.1.1.1 32 nexthop 10.0.23.2 out-label
203
```

AR2 的配置：

```
[AR2]static-lsp transit 3to1 incoming-interface g0/0/1 in-label 203 nexthop
10.0.12.1 out-label 102
```

AR1 的配置：

```
[AR1]static-lsp egress  3to1 incoming-interface g0/0/0 in-label 102
```

图 5-8 所示为再次在 AR1 上 ping 测试 3.3.3.3，查看抓包结果。

```
> Frame 242: 102 bytes on wire (816 bits), 102 bytes captured (816 bits) on interface -, id 0
> Ethernet II, Src: HuaweiTe_4f:36:ee (00:e0:fc:4f:36:ee), Dst: HuaweiTe_e5:0c:69 (00:e0:fc:e5:0c:69)
> MultiProtocol Label Switching Header, Label: 201, Exp: 0, S: 1, TTL: 255
> Internet Protocol Version 4, Src: 1.1.1.1, Dst: 3.3.3.3
> Internet Control Message Protocol
```

图 5-8　AR1 的 G0/0/0 接口抓包结果（3）

如图 5-9 所示，可以看到来回的报文都打上对应的标签，迭代进入了静态的 LSP 隧道。

```
> Frame 243: 102 bytes on wire (816 bits), 102 bytes captured (816 bits) on interface -, id 0
> Ethernet II, Src: HuaweiTe_e5:0c:69 (00:e0:fc:e5:0c:69), Dst: HuaweiTe_4f:36:ee (00:e0:fc:4f:36:ee)
> MultiProtocol Label Switching Header, Label: 102, Exp: 0, S: 1, TTL: 254
> Internet Protocol Version 4, Src: 3.3.3.3, Dst: 1.1.1.1
> Internet Control Message Protocol
```

图 5-9　AR1 的 G0/0/0 接口抓包结果（4）

5.2.2　配置 MPLS LDP

1. 实验目的

4 台路由器都运行 OSPF 协议，通过配置 MPLS LDP ，使设备的环回口都通过 MPLS 通信。

2. 实验拓扑

配置 MPLS LDP 的实验拓扑如图 5–10 所示。

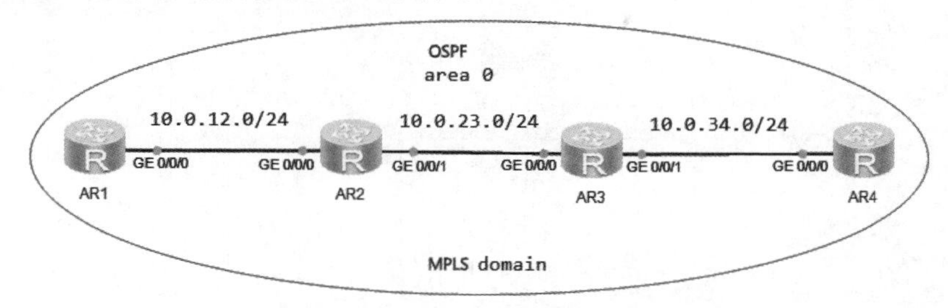

图 5–10　配置 MPLS LDP 的实验拓扑

3. 实验步骤

步骤 1：配置 IP 地址，MPLS LDP 实验 IP 规划表见表 5–3。

表 5–3　MPLS LDP 实验 IP 规划表

设备名称	接口编号	IP 地址
AR1	G0/0/0	10.0.12.1/24
	LoopBack 0	1.1.1.1/32
AR2	G0/0/0	10.0.12.2/24
	G0/0/1	10.0.23.2/24
	LoopBack 0	2.2.2.2/32
AR3	G0/0/0	10.0.23.3/24
	G0/0/1	10.0.34.3/24
	LoopBack 0	3.3.3.3/32
AR4	G0/0/0	10.0.34.4/24
	LoopBack 0	4.4.4.4/32

步骤 2：全网运行 OSPF 协议。

AR1 的配置：

```
[AR1]ospf
[AR1-ospf-1]area 0
[AR1-ospf-1-area-0.0.0.0]network 1.1.1.1 0.0.0.0
[AR1-ospf-1-area-0.0.0.0]network 10.0.12.0 0.0.0.255
```

AR2 的配置：

```
[AR2]ospf
[AR2-ospf-1]area 0
[AR2-ospf-1-area-0.0.0.0]network 2.2.2.2 0.0.0.0
[AR2-ospf-1-area-0.0.0.0]network 10.0.12.0 0.0.0.255
[AR2-ospf-1-area-0.0.0.0]network 10.0.23.0 0.0.0.255
```

AR3 的配置：

```
[AR3]ospf
[AR3-ospf-1]area 0
[AR3-ospf-1-area-0.0.0.0]network 3.3.3.3 0.0.0.0
[AR3-ospf-1-area-0.0.0.0]network 10.0.23.0 0.0.0.255
[AR3-ospf-1-area-0.0.0.0]network 10.0.34.0 0.0.0.255
```

AR4 的配置：

```
[AR4]ospf
[AR4-ospf-1]area 0
[AR4-ospf-1-area-0.0.0.0]network 4.4.4.4 0.0.0.0
[AR4-ospf-1-area-0.0.0.0]network 10.0.34.0 0.0.0.255
```

步骤 3： 配置 MPLS LDP。

AR1 的配置：

```
[AR1]mpls lsr-id 1.1.1.1                        //配置 MPLS LSR-ID
[AR1]mpls                                       //全局开启 MPLS
[AR1-mpls]q
[AR1]mpls ldp                                   //全局开启 MPLS LDP 协议
[AR1]interface g0/0/0
[AR1-GigabitEthernet0/0/0]mpls                  //接口开启 MPLS
[AR1-GigabitEthernet0/0/0]mpls ldp              //接口开启 MPLS LDP 功能
```

AR2 的配置：

```
[AR2]mpls lsr-id 2.2.2.2
[AR2]mpls
[AR2-mpls]mpls ldp
[AR2]interface g0/0/0
[AR2-GigabitEthernet0/0/0]mpls
[AR2-GigabitEthernet0/0/0]mpls ldp
[AR2]interface g0/0/1
[AR2-GigabitEthernet0/0/1]mpls
[AR2-GigabitEthernet0/0/1]mpls ldp
```

AR3 的配置：

```
[AR3]mpls lsr-id 3.3.3.3
[AR3]mpls
[AR3-mpls]mpls ldp
[AR3]interface g0/0/0
[AR3-GigabitEthernet0/0/0]mpls
[AR3-GigabitEthernet0/0/0]mpls ldp
[AR3]interface g0/0/1
[AR3-GigabitEthernet0/0/1]mpls
[AR3-GigabitEthernet0/0/1]mpls ldp
```

AR4 的配置：

```
[AR4]mpls lsr-id 4.4.4.4
[AR4]mpls
```

```
[AR4-mpls]mpls ldp
[AR4]interface g0/0/0
[AR4-GigabitEthernet0/0/0]mpls
[AR4-GigabitEthernet0/0/0]mpls ldp
```

查看 AR1 的 LDP 会话建立情况：

```
[AR1]display mpls ldp session
 LDP Session(s) in Public Network
 Codes: LAM(Label Advertisement Mode), SsnAge Unit(DDDD:HH:MM)
 A '*' before a session means the session is being deleted.
 ------------------------------------------------------------------------------
 PeerID          Status          LAM      SsnRole      SsnAge           KASent/Rcv
 ------------------------------------------------------------------------------
 2.2.2.2:0       Operational     DU       Passive      0000:00:03       13/13
 ------------------------------------------------------------------------------
 TOTAL: 1 session(s) Found.
```

通过以上输出可以看出，AR1 和 2.2.2.2（AR2）建立了 LDP 的会话关系。其中，PeerID 表示对等体的 LDP 标识符，格式为 <LSR ID>:< 标签空间 >。标签空间取值如下：0 表示全局标签空间，1 表示接口标签空间。Status 为 Operational 表示 LDP 会话建立成功。LAM 为 DU 表示标签的分发方式为下游自主。

查看 LDP 动态建立的 LSP：

```
[AR1]display mpls lsp
-------------------------------------------------------------------------------
                 LSP Information: LDP LSP
-------------------------------------------------------------------------------
FEC                 In/Out Label            In/Out IF               Vrf Name
1.1.1.1/32          3/NULL                  -/-
2.2.2.2/32          NULL/3                  -/GE0/0/0
2.2.2.2/32          1024/3                  -/GE0/0/0
3.3.3.3/32          NULL/1025               -/GE0/0/0
3.3.3.3/32          1025/1025               -/GE0/0/0
4.4.4.4/32          NULL/1026               -/GE0/0/0
4.4.4.4/32          1026/1026               -/GE0/0/0
```

通过以上输出可以看出，设备为每一个 32 位的主机地址分配了标签，并且动态建立了 LSP 隧道。以 3.3.3.3/32 这条 FEC 为例，FEC 为 3.3.3.3/32，In/Out Label 为 1025/1025，In/Out IF 为 –/GE0/0/0，表示当设备收到目标 IP 地址为 3.3.3.3 的数据时，如果入标签为 1025，则将标签交换 1025 并且从 G0/0/0 接口转发出去。

使用 AR1 测试 3.3.3.3 的连通性，并且在 AR1 的 G0/0/0 接口抓包查看现象。

```
[AR1]ping -a 1.1.1.1 3.3.3.3
  PING 3.3.3.3: 56  data bytes, press CTRL_C to break
    Reply from 3.3.3.3: bytes=56 Sequence=1 ttl=254 time=40 ms
    Reply from 3.3.3.3: bytes=56 Sequence=2 ttl=254 time=20 ms
    Reply from 3.3.3.3: bytes=56 Sequence=3 ttl=254 time=20 ms
    Reply from 3.3.3.3: bytes=56 Sequence=4 ttl=254 time=20 ms
    Reply from 3.3.3.3: bytes=56 Sequence=5 ttl=254 time=30 ms
```

```
--- 3.3.3.3 ping statistics ---
  5 packet(s) transmitted
  5 packet(s) received
  0.00% packet loss
  round-trip min/avg/max = 20/26/40 ms
```

如图 5-11 所示，通过抓包结果可以发现，AR1 访问 3.3.3.3 时，设备会查看 MPLS LSP，MPLS LSP 中的出标签为 1025，因此设备在发送数据时会为数据包封装一层 MPLS 头部，并且携带标签为 1025，当下一跳设备收到该报文时，就可以直接通过标签转发，而不需要再查询路由表。

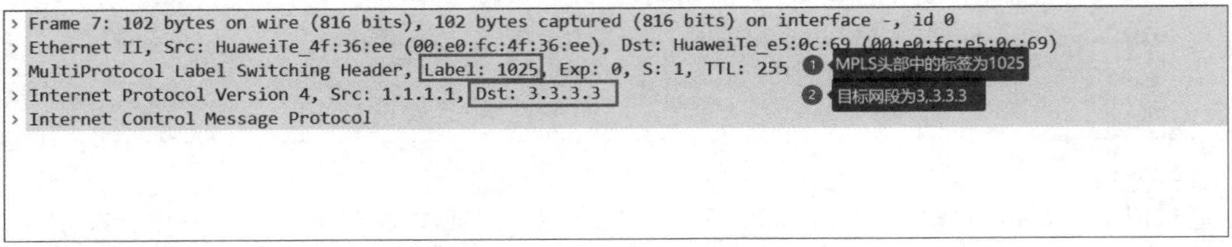

> Frame 7: 102 bytes on wire (816 bits), 102 bytes captured (816 bits) on interface -, id 0
> Ethernet II, Src: HuaweiTe_4f:36:ee (00:e0:fc:4f:36:ee), Dst: HuaweiTe_e5:0c:69 (00:e0:fc:e5:0c:69)
> MultiProtocol Label Switching Header, Label: 1025, Exp: 0, S: 1, TTL: 255
> Internet Protocol Version 4, Src: 1.1.1.1, Dst: 3.3.3.3
> Internet Control Message Protocol
① MPLS头部中的标签为1025
② 目标网段为3,.3.3.3

图 5-11　AR1 的 G0/0/0 接口的抓包结果

第 6 章

MPLS VPN 技术

6.1 MPLS VPN 概述

BGP/MPLS IP VPN 是一种 L3VPN（Layer 3 Virtual Private Network）。它使用 BGP 在服务提供商骨干网上发布 VPN 路由，使用 MPLS 在服务提供商骨干网上转发 VPN 报文。这里的 IP 是指 VPN 承载的 IP 报文。

6.1.1 MPLS VPN 模型

BGP/MPLS IP VPN 的基本模型由三部分组成：CE（Customer Edge）、PE（Provider Edge）和 P（Provider），如图 6-1 所示。

（1）CE 是指用户网络中的边缘设备，有接口直接与服务提供商网络相连。CE 可以是路由器或交换机，也可以是一台主机。通常情况下，CE“感知”不到 VPN 的存在，也不需要支持 MPLS。

（2）PE 是指服务提供商网络中的边缘设备，与 CE 直接相连。在 MPLS 网络中，对 VPN 的所有处理都发生在 PE 上，对 PE 性能要求较高。

（3）P 是指服务提供商网络中的骨干设备，不与 CE 直接相连。P 设备只需具备基本 MPLS 转发能力，不维护 VPN 信息。

PE 和 P 设备仅由服务提供商管理；CE 设备仅由用户管理，除非用户把管理权委托给服务提供商。一台 PE 设备可以接入多台 CE 设备。一台 CE 设备也可以连接属于相同或不同服务提供商的多台 PE 设备。

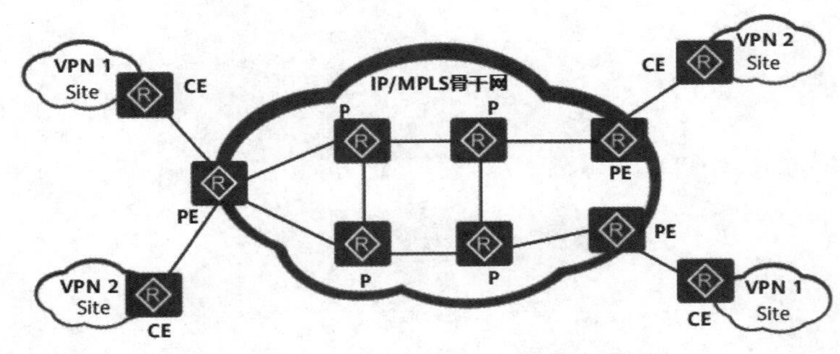

图 6-1　BGP/MPLS IP VPN 的基本模型

6.1.2 BGP/MPLS IP VPN 基本原理

1. 私网标签分配

在 BGP/MPLS IP VPN 中，PE 通过 MP-BGP 发布私网路由给骨干网的其他相关的 PE 前，需要为私网路由分配 MPLS 标签（私网标签）。当数据包在骨干网传输时，携带私网标签。

在 PE 上分配私网标签的方法有以下两种。

（1）基于路由的 MPLS 标签分配：为 VPN 路由表的每一条路由分配一个标签（One Label Per Route）。这种方式的缺点：当路由数量比较多时，设备入标签映射表（Incoming Label Map，ILM）需要维护的表项也会增多，从而提高了对设备容量的要求。

（2）基于 VPN 实例的 MPLS 标签分配：为整个 VPN 实例分配一个标签，该 VPN 实例中的所有路由都共享一个标签。使用这种分配方法的好处是节约了标签。

2. 私网路由交叉

两台 PE 之间通过 MP-BGP 传播的路由是 VPNv4 路由。当接收到 VPNv4 路由时，PE 先进行以下处理。

（1）检查其下一跳是否可达。如果下一跳不可达，则该路由被丢弃。

（2）对于 RR 发送过来的 VPNv4 路由，如果收到的路由中的 cluster_list 包含自己的 cluster_id，则丢弃这条路由。

（3）进行 BGP 的路由策略过滤，如果不通过，则丢弃该路由。

（4）PE 把没有丢弃的路由与本地的各个 VPN 实例的 Import Target 属性匹配。VPNv4 路由与本地 VPN 实例的 Import VPN-Target 进行匹配的过程称为私网路由交叉。

PE 上有种特殊的路由，即来自本地 CE 的属于不同 VPN 的路由。对于这种路由，如果其下一跳直接可达或可迭代成功，则 PE 也将其与本地的其他 VPN 实例的 Import Target 属性匹配，该过程称为本地交叉。例如，CE1 所在的 Site 属于 VPN1，CE2 所在的 Site 属于 VPN2，并且 CE1 和 CE2 同时接入 PE1。当 PE1 收到来自 CE1 的 VPN1 的路由时，也会与 VPN2 对应的 VPN 实例的 Import Target 属性匹配。

3. 公网隧道迭代

为了将私网流量通过公网传递到另一端，需要有一条公网隧道承载这个私网流量。因此，私网路由交叉完成后，需要根据目的 IPv4 前缀进行路由迭代，查找合适的隧道（本地交叉的路由除外）；只有隧道迭代成功，该路由才被放入对应的 VPN 实例路由表。将路由迭代到相应的隧道的过程称为隧道迭代。

隧道迭代成功后，保留该隧道的标识符（Tunnel ID），以供后续转发报文时使用。Tunnel ID 用于唯一标识一条隧道。VPN 报文转发时根据 Tunnel ID 查找对应的隧道，然后从隧道上发送出去。

4. 私网路由的选择规则

经过路由交叉和隧道迭代的路由并不是全部被放入 VPN 实例路由表中。从本地 CE 收到的路由和本地交叉路由也不是全部被放入 VPN 实例路由表中。

对于到同一目的地址的多条路由，如果不进行路由的负载分担，则按以下规则选择其中的一条。

（1）如果同时存在直接从 CE 收到的路由和交叉成功后的同一目的地址路由，则优选从 CE 收到的路由。

（2）如果同时存在本地交叉路由和从其他 PE 接收并交叉成功后的同一目的地址路由，则优选本地交叉路由。

（3）对于到同一目的地址的多条路由，如果进行路由的负载分担，则按以下规则选择一条。

①优先选择从本地 CE 收到的路由。在只有一条从本地 CE 收到的路由而有多条交叉路由的情况下，也只选择从本地 CE 收到的路由。

②只在从本地 CE 收到的路由之间分担或只在交叉路由之间分担，不会在本地 CE 收到的路由和交叉路由之间分担。

需要注意的是，负载分担的 AS_PATH 属性必须完全相同。

5. BGP/MPLS IP VPN 的路由发布

在基本 BGP/MPLS IP VPN 组网中，VPN 路由信息的发布涉及 CE 和 PE，P 设备只维护骨干网的路由，不需要了解任何 VPN 路由信息。PE 设备一般维护所有 VPN 路由。

VPN 路由信息的发布过程包括三部分：本地 CE 到入口 PE、入口 PE 到出口 PE、出口 PE 到远端 CE。完成这三部分后，本地 CE 与远端 CE 之间建立可达路由，VPN 路由信息能够在骨干网上发布。

下面分别对这三部分进行介绍。

（1）本地 CE 到入口 PE 的路由信息交换。

CE 与直接相连的 PE 建立邻居或对等体关系后，把本站点的 IPv4 路由发布给 PE。CE 与 PE 之间可以使用静态路由、RIP（Routing Information Protocol）、OSPF、IS-IS 或 BGP。无论使用哪种路由协议，CE 发布给 PE 的都是标准的 IPv4 路由。

（2）入口 PE 到出口 PE 的路由信息交换。

PE 从 CE 学到 VPN 路由信息后，存放到 VPN 实例中。同时，为这些标准 IPv4 路由增加 RD，形成 VPN-IPv4（以下简写为 VPNv4）路由。

入口 PE 通过 MP-BGP 的 Update 报文把 VPNv4 路由发布给出口 PE。Update 报文中携带 Export VPN Target 属性及 MPLS 标签。

出口 PE 收到 VPNv4 路由后，在下一跳可达的情况下进行路由交叉、隧道迭代和路由优选，决定是否将该路由加入到 VPN 实例的路由表。被加入到 VPN 路由表的路由，本地 PE 为其保留 MP-BGP Update 消息中携带的 MPLS 标签值和 Tunnel ID，以供后续转发报文时使用。

（3）出口 PE 到远端 CE 的路由信息交换。

远端 CE 有多种方式可以从出口 PE 学习 VPN 路由，包括静态路由、RIP、OSPF、IS-IS 和 BGP，与本地 CE 到入口 PE 的路由信息交换相同，此处不再赘述。需要注意的是，出口 PE 发布给远端 CE 的路由是普通 IPv4 路由。

6.1.3 MPLS VPN 的路由交互

MPLS VPN 的路由交互过程如图 6-2 所示。

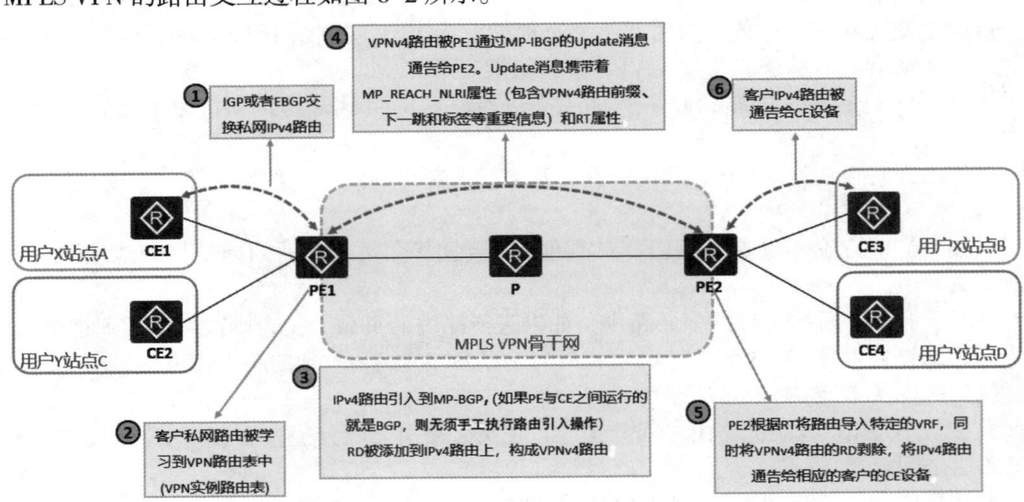

图 6-2　MPLS VPN 的路由交互过程

（1）CE 与 PE 之间：CE 与 PE 之间可以使用静态路由、OSPF、IS-IS 或 BGP 交换路由信息。无论使用哪种路由协议，CE 与 PE 之间交换的都是标准的 IPv4 路由。

（2）PE 与 PE 之间：PE 与 PE 之间使用 MP-BGP 传递 VPNv4 路由信息，并且需要注意以下几个问题。

①不同站点之间需要把私网路由传递给 PE，可能有地址重叠的问题，需要使用 VRF 技术将 CE 发送过来的路由存储到不同的 VPN 实例路由表中，以解决地址重叠问题。

② PE 收到不同 VPN 的 CE 发来的 IPv4 地址前缀，本地根据 VPN 实例配置来区分这些地址前缀。但是 VPN 实例只是一个本地的概念，PE 无法将 VPN 实例信息传递到对端 PE，需要使用 RD（路由标识符）区分不同 CE 发送的路由，IPv4 路由前缀 +RD=VPNv4 路由。

③ MP-BGP 将 VPNv4 传递到远端 PE 之后，远端 PE 需要将 VPNv4 路由导入正确的 VPN 实例。

MPLS VPN 使用 BGP 扩展团体属性 VPN Target（也称为 Route Target）来控制 VPN 路由信息的发布与接收。本地 PE 在发布 VPNv4 路由前附上 RT 属性，对端 PE 在接收到 VPNv4 路由后根据 RT 将路由导入对应的 VPN 实例。

在 PE 上，每个 VPN 实例都会与一个或多个 VPN Target 属性绑定，有以下两类 VPN Target 属性。

（1）Export Target（ERT）：本地 PE 从直接相连站点学到 IPv4 路由后，转换为 VPN IPv4 路由，并为这些路由添加 Export Target 属性。Export Target 属性作为 BGP 的扩展团体属性随路由发布。

（2）Import Target（IRT）：PE 收到其他 PE 发布的 VPNv4 路由时，检查其 Export Target 属性。当此属性与 PE 上某个 VPN 实例的 Import Target 属性匹配时，PE 就把路由加入该 VPN 实例的路由表中。

6.1.4　MPLS VPN 报文的转发

MPLS VPN 的报文转发过程如图 6-3 所示。

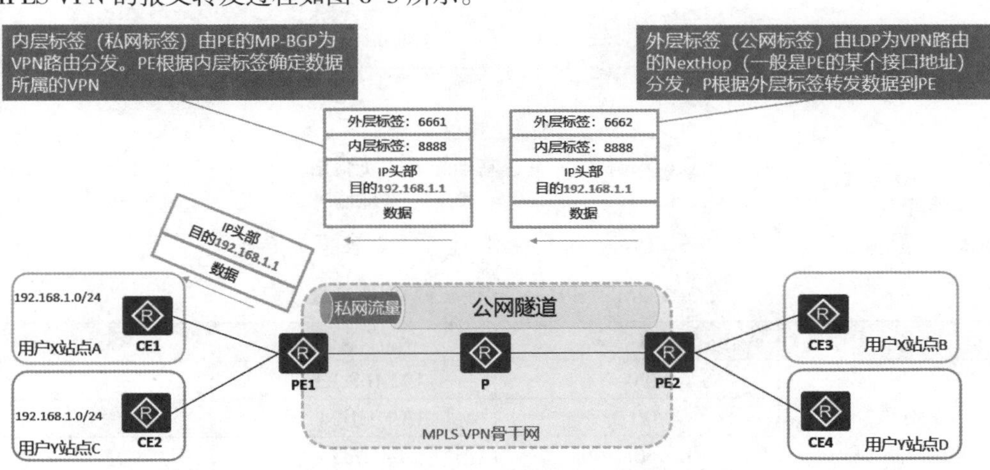

图 6-3　MPLS VPN 的报文转发过程

（1）CE 给本端 PE 发送普通的 IPv4 报文。

（2）本端 PE 发给 P 设备。

①根据报文入接口找到 VPN 实例，查找对应 VPN 的转发表。

②匹配目的 IPv4 前缀，并打上对应的内层标签（由 MP-BGP 分配）。

③根据下一跳地址查找对应的 Tunnel ID。

④将报文从隧道发送出去，即打上外层标签（由 LDP 分配）。

（3）P 设备转发报文：查看外层标签并且进行标签交换，把报文交给远端 PE。

（4）远端 PE 收到报文。

①收到该携带两层标签的报文，交给 MPLS 处理，MPLS 协议将去掉外层标签。

②继续处理内层标签：根据内层标签确定对应的下一跳，并将内层标签剥离后发送纯 IPv4 报文给远端 CE。

6.2 MPLS VPN 实验

6.2.1 MPLS VPN 基础配置

扫一扫，看视频

扫一扫，看视频

1. 实验目的

CE1 和 CE2 属于同一个公司的不同站点的出口设备，G0/0/0 连接运营商的 MPLS VPN 专线，G0/0/1 接口为公司租用的物理专线，直连对端 CE 设备（带宽较小），要求站点之间互相访问主链路使用 MPLS VPN，备用链路使用物理专线。

CE 与 PE、CE 与 CE 之间的 IGP 协议选择 OSPF，运营商内部 AS 号为 AS 100，IGP 协议也选择 OSPF。

2. 实验拓扑

MPLS VPN 基础配置的实验拓扑如图 6-4 所示。

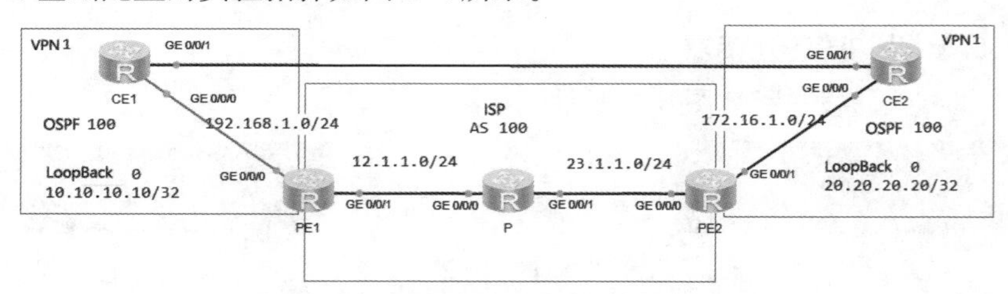

图 6-4　MPLS VPN 基础配置的实验拓扑

3. 实验步骤

步骤 1： 配置 IP 地址，MPLS VPN 基础配置 IP 地址规划表见表 6-1。

表 6-1　MPLS VPN 基础配置 IP 地址规划表

设备名称	接口编号	IP 地址	所属 VPN 实例
PE1	G0/0/0	192.168.1.1/24	A
	G0/0/1	12.1.1.1/24	
	LoopBack 0	1.1.1.1/32	
P	G0/0/0	12.1.1.2/24	
	G0/0/1	23.1.1.1/24	
	LoopBack 0	2.2.2.2/32	
PE2	G0/0/0	23.1.1.2/24	
	G0/0/1	172.16.1.1/24	A
	LoopBack 0	3.3.3.3/32	

续表

设备名称	接口编号	IP 地址	所属 VPN 实例
CE1	G0/0/0	192.168.1.2/24	
	G0/0/1	10.0.12.1/24	
	LoopBack 0	10.10.10.10/32	
CE2	G0/0/0	172.16.1.2/24	
	G0/0/1	10.0.12.2/24	
	LoopBack 0	20.20.20.20/32	

步骤 2: 配置 ISP 内部的 IGP 协议。

PE1 的配置:

```
[PE1]ospf 1
[PE1-ospf-1]area 0.0.0.0
[PE1-ospf-1-area-0.0.0.0]network 1.1.1.1 0.0.0.0
[PE1-ospf-1-area-0.0.0.0]network 12.1.1.0 0.0.0.255
```

P 的配置:

```
[P]ospf 1
[P-ospf-1]area 0.0.0.0
[P-ospf-1-area-0.0.0.0]network 2.2.2.2 0.0.0.0
[P-ospf-1-area-0.0.0.0]network 12.1.1.0 0.0.0.255
[P-ospf-1-area-0.0.0.0]network 23.1.1.0 0.0.0.255
```

PE2 的配置:

```
[PE2]ospf 1
[PE2-ospf-1]area 0.0.0.0
[PE2-ospf-1-area-0.0.0.0]network 3.3.3.3 0.0.0.0
[PE2-ospf-1-area-0.0.0.0]network 23.1.1.0 0.0.0.255
```

步骤 3: 配置 IGP 内部的 MPLS LDP。

PE1 的配置:

```
[PE1]mpls lsr-id 1.1.1.1
[PE1]mpls
[PE1-mpls]mpls ldp
[PE1-mpls-ldp]interface GigabitEthernet0/0/1
[PE1-GigabitEthernet0/0/1]mpls
[PE1-GigabitEthernet0/0/1]mpls ldp
```

P 的配置:

```
[P]mpls lsr-id 2.2.2.2
[P]mpls
[P-mpls]mpls ldp
[P-mpls-ldp]interface GigabitEthernet0/0/0
[P-GigabitEthernet0/0/0]mpls
[P-GigabitEthernet0/0/0]mpls ldp
[P-GigabitEthernet0/0/0]interface GigabitEthernet0/0/1
```

```
[P-GigabitEthernet0/0/1]mpls
[P-GigabitEthernet0/0/1]mpls ldp
```

PE2 的配置：

```
[PE2]mpls lsr-id 3.3.3.3
[PE2]mpls
[PE2-mpls]mpls ldp
[PE2-mpls-ldp]interface GigabitEthernet0/0/0
[PE2-GigabitEthernet0/0/0]mpls
[PE2-GigabitEthernet0/0/0]mpls ldp
```

查看 LDP 的隧道建立情况，可以看到每一个 32 位的主机路由都建立了对应的 LDP LSP。

```
[PE1]display mpls lsp
                 LSP Information: LDP LSP
-------------------------------------------------------------------------------
FEC                 In/Out Label           In/Out IF               Vrf Name
2.2.2.2/32          NULL/3                  -/GE0/0/1
2.2.2.2/32          1024/3                  -/GE0/0/1
3.3.3.3/32          NULL/1024               -/GE0/0/1
3.3.3.3/32          1025/1024               -/GE0/0/1
1.1.1.1/32          3/NULL                  -/-
```

步骤 4: 配置 VPN 实例，将接口加入 VPN 实例。

PE1 的配置：

```
[PE1]ip vpn-instance vpn1            //创建VPN实例，命名为vpn1
[PE1-vpn-instance-vpn1]ipv4-family   //进入IPv4地址族视图
[PE1-vpn-instance-vpn1-af-ipv4]route-distinguisher 100:1 //配置RD为100: 1
                                     //配置import 、export RT都为100: 1
[PE1-vpn-instance-vpn1-af-ipv4]vpn-target 100:1 export-extcommunity
[PE1-vpn-instance-vpn1-af-ipv4]vpn-target 100:1 import-extcommunity
[PE1]interface GigabitEthernet0/0/0
[PE1-GigabitEthernet0/0/0]ip binding vpn-instance vpn1 //接口绑定到VPN实例vpn1
[PE1-GigabitEthernet0/0/0]ip address 192.168.1.1 255.255.255.0
```

PE2 的配置：

```
[PE2]ip vpn-instance vpn1
[PE2-vpn-instance-vpn1]ipv4-family
[PE2-vpn-instance-vpn1-af-ipv4]route-distinguisher 100:1
[PE2-vpn-instance-vpn1-af-ipv4]vpn-target 100:1 export-extcommunity
[PE2-vpn-instance-vpn1-af-ipv4]vpn-target 100:1 import-extcommunity
[PE2]interface GigabitEthernet0/0/1
[PE2-GigabitEthernet0/0/1]ip binding vpn-instance vpn1
[PE2-GigabitEthernet0/0/1]ip address 172.16.1.1 255.255.255.0
```

RD 用于标记 VPNv4 路由，BGP 传递 VPNv4 路由时会携带 RD 值，代表这是一条唯一的 VPNv4 路由。

RT 用于控制 VPNv4 路由的接收，如果出方向 RT 等于对端设备入方向 RT，则接收路由，并且将路由加入对应的 VPN 实例路由表。

步骤 5: 配置 CE 和 PE 之间的路由协议。

PE1 的配置:

```
[PE1]ospf 100 vpn-instance vpn1
[PE1-ospf-100]area 0.0.0.0
[PE1-ospf-100-area-0.0.0.0]network 192.168.1.0 0.0.0.255
```

CE1 的配置:

```
[ce1]ospf 100
[ce1-ospf-100]area 0.0.0.0
[ce1-ospf-100-area-0.0.0.0]network 10.10.10.10 0.0.0.0
[ce1-ospf-100-area-0.0.0.0]network 192.168.1.0 0.0.0.255
```

查看 PE1 和 CE1 的邻居关系以及 PE1 的 VPN1 实例的路由表:

```
[PE1]display ospf peer brief
        OSPF Process 1 with Router ID 12.1.1.1
                 Peer Statistic Information
 ----------------------------------------------------------------------------
 Area Id          Interface                        Neighbor id      State
 0.0.0.0          GigabitEthernet0/0/1             12.1.1.2         Full
 ----------------------------------------------------------------------------
        OSPF Process 100 with Router ID 192.168.1.1
                 Peer Statistic Information
 ----------------------------------------------------------------------------
 Area Id          Interface                        Neighbor id      State
 0.0.0.0          GigabitEthernet0/0/0             192.168.1.2      Full
 ----------------------------------------------------------------------------
[PE1]display  ip routing-table  vpn-instance vpn1
 Destination/Mask     Proto    Pre   Cost   Flags   NextHop      Interface
    10.10.10.10/32    OSPF     10    1      D       192.168.1.2  GigabitEthernet0/0/0
    192.168.1.0/24    Direct   0     0      D       192.168.1.1  GigabitEthernet0/0/0
    192.168.1.1/32    Direct   0     0      D       127.0.0.1    GigabitEthernet0/0/0
  192.168.1.255/32    Direct   0     0      D       127.0.0.1    GigabitEthernet0/0/0
255.255.255.255/32    Direct   0     0      D       127.0.0.1    InLoopBack0
```

通过以上输出可以看出，PE1 可以通过 OSPF 100 学习到 CE1 的环回口 10.10.10.10/32 的路由（PE2 和 CE2 的现象同理）。

PE2 的配置:

```
[PE2]ospf 100 vpn-instance vpn1
[PE2-ospf-100]area 0.0.0.0
[PE2-ospf-100-area-0.0.0.0]network 172.16.1.0 0.0.0.255
```

CE2 的配置:

```
[ce2]ospf 100
[ce2-ospf-100]area 0
[ce2-ospf-100-area-0.0.0.0]network 172.16.1.0 0.0.0.255
[ce2-ospf-100-area-0.0.0.0]network 20.20.20.20 0.0.0.0
```

步骤 6: 配置 PE1 和 PE2 之间的 MP-BGP 协议。

PE1 的配置:

```
[PE1]bgp 100
[PE1-bgp]peer 3.3.3.3 as-number 100
[PE1-bgp]peer 3.3.3.3 connect-interface LoopBack0
[PE1-bgp]ipv4-family vpnv4
[PE1-bgp-af-vpnv4]peer 3.3.3.3 enable
```

PE2 的配置:

```
[PE2]bgp 100
[PE2-bgp]peer 1.1.1.1 as-number 100
[PE2-bgp]peer 1.1.1.1 connect-interface LoopBack0
[PE2-bgp]ipv4-family vpnv4
[PE2-bgp-af-vpnv4]peer 1.1.1.1 enable
```

查看 PE1 和 PE2 的 MP-BGP 邻居关系:

```
[PE1]display  bgp vpnv4 all  peer
 BGP local router ID : 12.1.1.1
 Local AS number : 100
 Total number of peers : 2              Peers in established state : 1
 Peer        V    AS     MsgRcvd    MsgSent    OutQ    Up/Down      State        PrefRcv
 3.3.3.3     4    100    134        132        0       02:07:20     Established   0
```

步骤 7: 将 PE1、PE2 的 OSPF 100 的路由引入 BGP 中，把 VPN 实例 VPN1 的路由变为 VPNv4 路由，使用 MP-BGP 传递给对端 PE，并且将 BGP 的路由引入 OSPF 100 中。

PE1 的配置:

```
[PE1]bgp 100
[PE1-bgp]ipv4-family vpn-instance vpn1
[PE1-bgp-vpn1]import-route ospf 100
[PE1]ospf 100
[PE1-ospf-100]import-route bgp
```

PE2 的配置:

```
[PE2]bgp 100
[PE2-bgp]ipv4-family vpn-instance vpn1
[PE2-bgp-vpn1]import-route ospf 100
[PE2]ospf 100
[PE2-ospf-100]import-route bgp
```

查看 PE2 的 VPNv4 路由，可以看到能够学习到对端 CE 和本端 CE 的路由:

```
[PE2]display  bgp vpnv4 all  routing-table
 BGP Local router ID is 23.1.1.2
 Status codes: * - valid, > - best, d - damped,
               h - history,  i - internal, s - suppressed, S - Stale
               Origin : i - IGP, e - EGP, ? - incomplete
 Total number of routes from all PE: 4
 Route Distinguisher: 100:1
       Network             NextHop         MED         LocPrf      PrefVal      Path/Ogn
```

```
 *>i   10.10.10.10/32     1.1.1.1         2          100         0            ?
 *>    20.20.20.20/32     0.0.0.0         2                      0            ?
 *>    172.16.1.0/24      0.0.0.0         0                      0            ?
 *>i   192.168.1.0        1.1.1.1         0          100         0            ?
 VPN-Instance vpn1, Router ID 23.1.1.2:
 Total Number of Routes: 4
       Network            NextHop        MED        LocPrf      PrefVal      Path/Ogn
 *>i   10.10.10.10/32     1.1.1.1         2          100         0            ?
 *>    20.20.20.20/32     0.0.0.0         2                      0            ?
 *>    172.16.1.0/24      0.0.0.0         0                      0            ?
 *>i   192.168.1.0        1.1.1.1         0          100         0            ?
```

查看 CE2 的路由表：

```
[CE2]display  ip routing-table
Destination/Mask     Proto    Pre   Cost  Flags    NextHop       Interface
     10.0.12.0/24    Direct   0     0     D        10.0.12.2     GigabitEthernet0/0/1
     10.0.12.2/32    Direct   0     0     D        127.0.0.1     GigabitEthernet0/0/1
   10.0.12.255/32    Direct   0     0     D        127.0.0.1     GigabitEthernet0/0/1
   10.10.10.10/32    OSPF     10    3     D        172.16.1.1    GigabitEthernet0/0/0
   20.20.20.20/32    Direct   0     0     D        127.0.0.1     LoopBack0
      127.0.0.0/8    Direct   0     0     D        127.0.0.1     InLoopBack0
     127.0.0.1/32    Direct   0     0     D        127.0.0.1     InLoopBack0
127.255.255.255/32   Direct   0     0     D        127.0.0.1     InLoopBack0
    172.16.1.0/24    Direct   0     0     D        172.16.1.2    GigabitEthernet0/0/0
    172.16.1.2/32    Direct   0     0     D        127.0.0.1     GigabitEthernet0/0/0
  172.16.1.255/32    Direct   0     0     D        127.0.0.1     GigabitEthernet0/0/0
   192.168.1.0/24    O_ASE    150   1     D        172.16.1.1    GigabitEthernet0/0/0
255.255.255.255/32   Direct   0     0     D        127.0.0.1     InLoopBack0
```

使用 CE2 测试 CE1 的连通性：

```
[CE2]ping 10.10.10.10
  PING 10.10.10.10: 56  data bytes, press CTRL_C to break
    Reply from 10.10.10.10: bytes=56 Sequence=1 ttl=252 time=70 ms
    Reply from 10.10.10.10: bytes=56 Sequence=2 ttl=252 time=50 ms
    Reply from 10.10.10.10: bytes=56 Sequence=3 ttl=252 time=40 ms
    Reply from 10.10.10.10: bytes=56 Sequence=4 ttl=252 time=40 ms
    Reply from 10.10.10.10: bytes=56 Sequence=5 ttl=252 time=40 ms
  --- 10.10.10.10 ping statistics ---
    5 packet(s) transmitted
    5 packet(s) received
    0.00% packet loss
round-trip min/avg/max = 40/48/70 ms
```

步骤 8：配置 CE1 和 CE2 之间的物理专线。

CE1 的配置：

```
[CE1]ospf 100
[CE1-ospf-100]area 0
[CE1-ospf-100-area-0.0.0.0]network  10.0.12.0 0.0.0.255
```

CE2 的配置：

```
[CE2]ospf 100
[CE2-ospf-100]area 0
[CE2-ospf-100-area-0.0.0.0]network  10.0.12.0 0.0.0.255
```

查看 CE2 的路由表：

```
[CE2]display  ip routing-table
Route Flags: R - relay, D - download to fib
------------------------------------------------------------------------------
Routing Tables: Public
         Destinations : 13        Routes : 13

 Destination/Mask    Proto   Pre  Cost  Flags   NextHop       Interface

      10.0.12.0/24   Direct  0    0     D       10.0.12.2     GigabitEthernet0/0/1
      10.0.12.2/32   Direct  0    0     D       127.0.0.1     GigabitEthernet0/0/1
    10.0.12.255/32   Direct  0    0     D       127.0.0.1     GigabitEthernet0/0/1
    10.10.10.10/32   OSPF    10   1     D       10.0.12.1     GigabitEthernet0/0/1
    20.20.20.20/32   Direct  0    0     D       127.0.0.1     LoopBack0
       127.0.0.0/8   Direct  0    0     D       127.0.0.1     InLoopBack0
       127.0.0.1/32  Direct  0    0     D       127.0.0.1     InLoopBack0
127.255.255.255/32   Direct  0    0     D       127.0.0.1     InLoopBack0
     172.16.1.0/24   Direct  0    0     D       172.16.1.2    GigabitEthernet0/0/0
     172.16.1.2/32   Direct  0    0     D       127.0.0.1     GigabitEthernet0/0/0
   172.16.1.255/32   Direct  0    0     D       127.0.0.1     GigabitEthernet0/0/0
    192.168.1.0/24   OSPF    10   2     D       10.0.12.1     GigabitEthernet0/0/1
255.255.255.255/32   Direct  0    0     D       127.0.0.1     InLoopBack0
```

通过以上输出可以看出，CE2 此时去往 10.10.10.10/32 的路由下一跳为 10.0.12.1（物理专线），原因是 CE2 从 PE2 收到的 10.10.10.10/32 的路由为 3 类 LSA，通过物理专线收到的 10.10.10.10/32 的路由为 1 类 LSA，因此会优选带宽较小的物理专线。

步骤 9： 配置站点之间的 sham-link，让 CE 从 PE 通过 1 类 LSA 学习到路由，再通过修改 OSPF 开销的方式实现选路。

（1）在 PE1 和 PE2 上创建环回口，用于建立 sham-link，并且加入到对应的 VPN 实例。

PE1 的配置：

```
[PE1]interface LoopBack 1
[PE1-LoopBack1]ip binding vpn-instance vpn1
[PE1-LoopBack1]ip address 100.1.1.1 32
```

PE2 的配置：

```
[PE2]interface LoopBack 1
[PE2-LoopBack1]ip binding vpn-instance vpn1
[PE2-LoopBack1]ip address 100.1.1.2 32
```

（2）将建立 sham-link 的接口通告在 BGP 中，让对端 PE 能够学习到此路由。

PE1 的配置：

```
[PE1]bgp 100
```

```
[PE1-bgp]ipv4-family vpn-instance vpn1
[PE1-bgp-vpn1]network 100.1.1.1 32
```

PE2 的配置：

```
[PE2]bgp 100
[PE2-bgp]ipv4-family vpn-instance vpn1
[PE2-bgp-vpn1]network 100.1.1.2 32
```

查看 VPNv4 路由表，可以看到 100.1.1.1/32 的路由已经被学习到。

```
[PE2]display bgp vpnv4 all routing-table
 BGP Local router ID is 23.1.1.2
 Status codes: * - valid, > - best, d - damped,
               h - history,  i - internal, s - suppressed, S - Stale
               Origin : i - IGP, e - EGP, ? - incomplete
 Total number of routes from all PE: 12
 Route Distinguisher: 100:1
       Network            NextHop        MED       LocPrf      PrefVal      Path/Ogn
 *>    10.0.12.0/24       0.0.0.0        3                     0            ?
 * i                      1.1.1.1        3         100         0            ?
 *>    10.10.10.10/32     0.0.0.0        3         0           ?
 * i                      1.1.1.1        2         100         0            ?
 *>    20.20.20.20/32     0.0.0.0        2         0           ?
 * i                      1.1.1.1        3         100         0            ?
 *>i   100.1.1.1/32       1.1.1.1        0         100         0            i
 *>    100.1.1.2/32       0.0.0.0        0         0           i
 *>    172.16.1.0/24      0.0.0.0        0         0           ?
 * i                      1.1.1.1        4         100         0            ?
 *>    192.168.1.0        0.0.0.0        4         0           ?
 * i                      1.1.1.1        0         100         0            ?
 VPN-Instance vpn1, Router ID 23.1.1.2:
 Total Number of Routes: 12
       Network            NextHop        MED       LocPrf      PrefVal      Path/Ogn
 *>    10.0.12.0/24       0.0.0.0        3                     0            ?
 * i                      1.1.1.1        3         100         0            ?
 *>    10.10.10.10/32     0.0.0.0        3                     0            ?
 * i                      1.1.1.1        2         100         0            ?
 *>    20.20.20.20/32     0.0.0.0        2                     0            ?
 * i                      1.1.1.1        3         100         0            ?
 *>i   100.1.1.1/32       1.1.1.1        0         100         0            i
 *>    100.1.1.2/32       0.0.0.0        0                     0            i
 *>    172.16.1.0/24      0.0.0.0        0                     0            ?
 * i                      1.1.1.1        4         100         0            ?
 *>    192.168.1.0        0.0.0.0        4                     0            ?
 * i                      1.1.1.1        0         100         0            ?
```

（3）配置 sham-link。

PE1 的配置：

```
[PE1]ospf 100
```

```
[PE1-ospf-100]area 0
[PE1-ospf-100-area-0.0.0.0]sham-link 100.1.1.1 100.1.1.2
```

PE2 的配置：

```
[PE2]ospf 100
[PE2-ospf-100]area 0
[PE2-ospf-100-area-0.0.0.0]sham-link 100.1.1.2 100.1.1.1
```

查看 sham-link 的建立情况，可以看到 sham-link 已经建立好了：

```
<PE2>display  ospf sham-link
         OSPF Process 1 with Router ID 23.1.1.2
 Sham Link:
 Area            NeighborId      Source-IP       Destination-IP  State Cost
         OSPF Process 100 with Router ID 172.16.1.1
 Sham Link:
 Area            NeighborId      Source-IP       Destination-IP  State Cost
 0.0.0.0         192.168.1.1     100.1.1.2       100.1.1.1       P-2-P 1
```

查看 CE2 的路由表，可以看到 10.10.10.10/32 的路由还是所选物理链路，原因是通过 PE 设备传递过来的路由 Cost 值比较大，因此接下来修改物理链路的 OSPF Cost：

```
<CE2>display  ip routing-table
Route Flags: R - relay, D - download to fib
------------------------------------------------------------------------------
Routing Tables: Public
         Destinations : 15       Routes : 16
 Destination/Mask    Proto   Pre   Cost  Flags    NextHop        Interface
      10.0.12.0/24   Direct  0     0     D        10.0.12.2      GigabitEthernet0/0/1
      10.0.12.2/32   Direct  0     0     D        127.0.0.1      GigabitEthernet0/0/1
    10.0.12.255/32   Direct  0     0     D        127.0.0.1      GigabitEthernet0/0/1
    10.10.10.10/32   OSPF    10    1     D        10.0.12.1      GigabitEthernet0/0/1
    20.20.20.20/32   Direct  0     0     D        127.0.0.1      LoopBack0
      100.1.1.1/32   O_ASE   150   1     D        172.16.1.1     GigabitEthernet0/0/0
      100.1.1.2/32   O_ASE   150   1     D        10.0.12.1      GigabitEthernet0/0/1
                     O_ASE   150   1     D        172.16.1.1     GigabitEthernet0/0/0
       127.0.0.0/8   Direct  0     0     D        127.0.0.1      InLoopBack0
      127.0.0.1/32   Direct  0     0     D        127.0.0.1      InLoopBack0
127.255.255.255/32   Direct  0     0     D        127.0.0.1      InLoopBack0
     172.16.1.0/24   Direct  0     0     D        172.16.1.2     GigabitEthernet0/0/0
     172.16.1.2/32   Direct  0     0     D        127.0.0.1      GigabitEthernet0/0/0
   172.16.1.255/32   Direct  0     0     D        127.0.0.1      GigabitEthernet0/0/0
    192.168.1.0/24   OSPF    10    2     D        10.0.12.1      GigabitEthernet0/0/1
255.255.255.255/32   Direct  0     0     D        127.0.0.1      InLoopBack0
```

（4）配置 OSPF 的 Cost 值，实现路由选路。

CE1 的配置：

```
[CE1]interface g0/0/1
[CE1-GigabitEthernet0/0/1]ospf cost 100
```

CE2 的配置：

```
[CE2]interface g0/0/1
[CE2-GigabitEthernet0/0/1]ospf cost 100
```

再次查看 CE2 的路由表，下一跳变为 172.16.1.1（MPLS VPN 专线）：

```
[CE2]display ip routing-table  10.10.10.10
Route Flags: R - relay, D - download to fib
------------------------------------------------------------------------------
Routing Table : Public
Summary Count : 1
Destination/Mask    Proto   Pre   Cost    Flags  NextHop       Interface

10.10.10.10/32      OSPF    10    3       D      172.16.1.1   GigabitEthernet0/0/0
```

测试结果表明，CE 之间的互访流量选择的是 MPLS VPN 专线：

```
[CE2]tracert 10.10.10.10
 traceroute to 10.10.10.10(10.10.10.10), max hops: 30 ,packet length: 40,press
 CTRL_C to break
 1 172.16.1.1  20 ms  20 ms  20 ms
 2 23.1.1.1    20 ms  30 ms  20 ms
 3 192.168.1.1 20 ms  20 ms  30 ms
 4 192.168.1.2 40 ms  40 ms  40 ms
```

6.2.2　MCE + Hub Spoke 组网

1. 实验目的

（1）企业的 CE 设备与运营商之间运行 BGP 协议。

（2）企业内部有销售和财务两个部门，要求实现销售部门和财务部门的业务隔离，公司总部使用 OSPF 进行内部组网，销售部门的 OSPF 进程为 100，财务部门的 OSPF 进程为 200。

（3）企业分支同部门之间的互访流量一定需要通过企业总部。

2. 实验拓扑

MCE and Hub Spoke 组网的实验拓扑如图 6-5 所示。

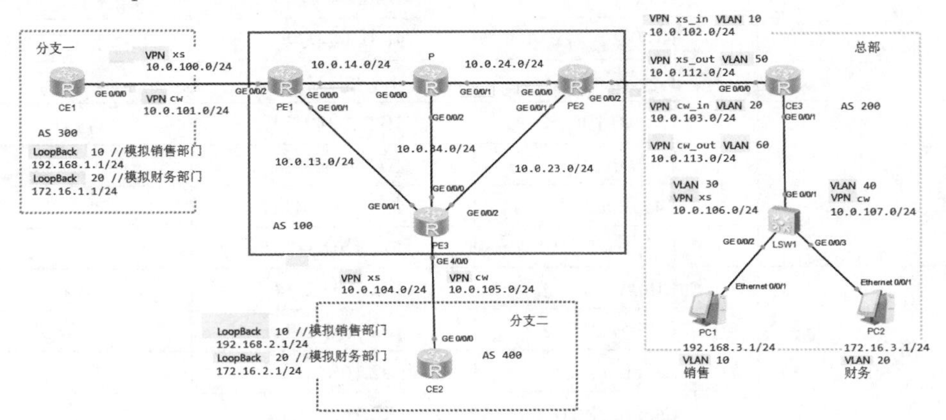

图 6-5　MCE and Hub Spoke 组网的实验拓扑

3. 实验步骤

步骤 1：配置接口 IP 地址，MCE and Hub Spoke 组网 IP 地址规划表见表 6-2。

表 6-2　MCE and Hub Spoke 组网 IP 地址规划表

设备名称	接口编号	IP 地址	所属 VPN 实例
PE1	G0/0/0	10.0.14.1/24	
	G0/0/1	10.0.13.1/24	
	LoopBack 0	1.1.1.1/32	
	G0/0/2.10	10.0.100.1/24	xs
	G0/0/2.20	10.0.101.1/24	cw
PE2	G0/0/0	10.0.24.2/24	
	G0/0/1	10.0.23.2/24	
	LoopBack 0	2.2.2.2/32	
	G0/0/2.10	10.0.102.1/24	xs_in
	G0/0/2.20	10.0.103.1/24	cw_in
PE2	G0/0/2.50	10.0.112.1/24	xs_out
	G0/0/2.60	10.0.113.1/24	cw_out
PE3	G0/0/0	10.0.34.3/24	
	G0/0/1	10.0.13.3/24	
	G0/0/2	10.0.23.3/24	
	LoopBack 0	3.3.3.3/32	
	G0/0/4.10	10.0.104.1/24	xs
	G0/0/4.20	10.0.105.1/24	cw
P	G0/0/0	10.0.14.4/24	
	G0/0/1	10.0.24.4/24	
	G0/0/2	10.0.34.4/24	
	LoopBack 0	4.4.4.4/32	
CE1	G0/0/0.10	10.0.100.2/24	xs
	G0/0/0.20	10.0.101.2/24	cw
	LoopBack 10	192.168.1.1/24	
	LoopBack 20	172.16.1.1/24	
CE2	G0/0/0.10	10.0.104.2/24	xs
	G0/0/0.20	10.0.105.2/24	cw
	LoopBack 10	192.168.2.1/24	
	LoopBack 20	172.16.2.1/24	

续表

设备名称	接口编号	IP 地址	所属 VPN 实例
CE3	G0/0/0.10	10.0.102.2/24	xs-in
	G0/0/0.20	10.0.103.2/24	cw-in
	G0/0/0.50	10.0.112.2/24	xs-out
	G0/0/0.60	10.0.113.2/24	cw-out
	G0/0/1.10	10.0.106.1/24	xs
	G0/0/1.20	10.0.107.1/24	cw
LSW1	Vlanif30	10.0.106.2/24	xs
	Vlanif40	10.0.107.2/24	cw
	Vlanif10	192.168.3.254/24	xs
	Vlanif20	172.16.3.254/24	cw

步骤 2：配置运营商内部的 IGP 协议。

PE1 的配置：

```
[PE1]ospf 1
[PE1-ospf-1]area 0.0.0.0
[PE1-ospf-1-area-0.0.0.0]network 1.1.1.1 0.0.0.0
[PE1-ospf-1-area-0.0.0.0]network 10.0.13.0 0.0.0.255
[PE1-ospf-1-area-0.0.0.0]network 10.0.14.0 0.0.0.255
```

PE2 的配置：

```
[PE2]ospf 1
[PE2-ospf-1]area 0.0.0.0
[PE2-ospf-1-area-0.0.0.0]network 2.2.2.2 0.0.0.0
[PE2-ospf-1-area-0.0.0.0]network 10.0.23.0 0.0.0.255
[PE2-ospf-1-area-0.0.0.0]network 10.0.24.0 0.0.0.255
```

PE3 的配置：

```
[PE3]ospf 1
[PE3-ospf-1]
[PE3-ospf-1]area 0.0.0.0
[PE3-ospf-1-area-0.0.0.0]network 3.3.3.3 0.0.0.0
[PE3-ospf-1-area-0.0.0.0]network 10.0.13.0 0.0.0.255
[PE3-ospf-1-area-0.0.0.0]network 10.0.23.0 0.0.0.255
[PE3-ospf-1-area-0.0.0.0]network 10.0.34.0 0.0.0.255
```

P 的配置：

```
[P]ospf 1
[P-ospf-1]area 0.0.0.0
[P-ospf-1-area-0.0.0.0]network 4.4.4.4 0.0.0.0
[P-ospf-1-area-0.0.0.0]network 10.0.14.0 0.0.0.255
[P-ospf-1-area-0.0.0.0]network 10.0.24.0 0.0.0.255
```

```
[P-ospf-1-area-0.0.0.0]network 10.0.34.0 0.0.0.255
```

步骤 3：配置运营商内部的 MPLS LDP。

PE1 的配置：

```
[PE1]mpls lsr-id 1.1.1.1
[PE1]mpls
Info: Mpls starting, please wait... OK!
[PE1-mpls]mpls ldp
[PE1-mpls-ldp]interface GigabitEthernet0/0/0
[PE1-GigabitEthernet0/0/0]mpls
[PE1-GigabitEthernet0/0/0]mpls ldp
[PE1-GigabitEthernet0/0/0]interface GigabitEthernet0/0/1
[PE1-GigabitEthernet0/0/1]mpls
[PE1-GigabitEthernet0/0/1]mpls ldp
```

PE2 的配置：

```
[PE2]mpls lsr-id 2.2.2.2
[PE2]mpls
Info: Mpls starting, please wait... OK!
[PE2-mpls]mpls ldp
[PE2-mpls-ldp]interface GigabitEthernet0/0/0
[PE2-GigabitEthernet0/0/0]mpls
[PE2-GigabitEthernet0/0/0]mpls ldp
[PE2-GigabitEthernet0/0/0]interface GigabitEthernet0/0/1
[PE2-GigabitEthernet0/0/1]mpls
[PE2-GigabitEthernet0/0/1]mpls ldp
```

PE3 的配置：

```
[PE3]mpls lsr-id 3.3.3.3
[PE3]mpls
Info: Mpls starting, please wait... OK!
[PE3-mpls]mpls ldp
[PE3-mpls-ldp]interface GigabitEthernet0/0/0
[PE3-GigabitEthernet0/0/0]mpls
[PE3-GigabitEthernet0/0/0]mpls ldp
[PE3-GigabitEthernet0/0/0]interface GigabitEthernet0/0/1
[PE3-GigabitEthernet0/0/1]mpls
[PE3-GigabitEthernet0/0/1]mpls ldp
[PE3-GigabitEthernet0/0/1]interface GigabitEthernet0/0/2
[PE3-GigabitEthernet0/0/2]mpls
[PE3-GigabitEthernet0/0/2]mpls ldp
```

P 的配置：

```
[P]mpls lsr-id 4.4.4.4
[P]mpls
Info: Mpls starting, please wait... OK!
[P-mpls]mpls ldp
[P-mpls-ldp]
```

```
[P-mpls-ldp]interface GigabitEthernet0/0/0
[P-GigabitEthernet0/0/0]mpls
[P-GigabitEthernet0/0/0]mpls ldp
[P-GigabitEthernet0/0/0]interface GigabitEthernet0/0/1
[P-GigabitEthernet0/0/1]mpls
[P-GigabitEthernet0/0/1]mpls ldp
[P-GigabitEthernet0/0/1]interface GigabitEthernet0/0/2
[P-GigabitEthernet0/0/2]mpls
[P-GigabitEthernet0/0/2]mpls ldp
```

查看隧道的建立情况：

```
[PE1]display mpls lsp
-------------------------------------------------------------------------------
                 LSP Information: LDP LSP
-------------------------------------------------------------------------------
FEC                In/Out Label           In/Out IF                  Vrf Name
4.4.4.4/32         NULL/3                 -/GE0/0/0
4.4.4.4/32         1024/3                 -/GE0/0/0
1.1.1.1/32         3/NULL                 -/-
2.2.2.2/32         NULL/1025              -/GE0/0/0
2.2.2.2/32         1025/1025              -/GE0/0/0
2.2.2.2/32         NULL/1026              -/GE0/0/1
2.2.2.2/32         1025/1026              -/GE0/0/1
3.3.3.3/32         NULL/3                 -/GE0/0/1
3.3.3.3/32         1026/3                 -/GE0/0/1
```

步骤4：配置PE的VPN实例，并配置PE和CE的直连网段IP地址。

PE1的配置：

```
[PE1]ip vpn-instance xs                          //销售部门VPN实例
[PE1-vpn-instance-xs]ipv4-family
[PE1-vpn-instance-xs-af-ipv4]route-distinguisher 100:1
[PE1-vpn-instance-xs-af-ipv4]vpn-target 1:1 export-extcommunity
[PE1-vpn-instance-xs-af-ipv4]vpn-target 1:2 import-extcommunity
[PE1-vpn-instance-xs-af-ipv4]ip vpn-instance cw  //财务部门VPN实例
[PE1-vpn-instance-cw]ipv4-family
[PE1-vpn-instance-cw-af-ipv4]route-distinguisher 100:2
[PE1-vpn-instance-cw-af-ipv4]vpn-target 2:1 export-extcommunity
[PE1-vpn-instance-cw-af-ipv4]vpn-target 2:2 import-extcommunity
[PE1]interface GigabitEthernet0/0/2.10           //对接销售部门
[PE1-GigabitEthernet0/0/2.10]dot1q termination vid 10
[PE1-GigabitEthernet0/0/2.10]ip binding vpn-instance xs
[PE1-GigabitEthernet0/0/2.10]ip address 10.0.100.1 255.255.255.0
[PE1-GigabitEthernet0/0/2.10]arp broadcast enable
[PE1]interface GigabitEthernet0/0/2.20           //对接财务部门
[PE1-GigabitEthernet0/0/2.20]dot1q termination vid 20
[PE1-GigabitEthernet0/0/2.20]ip binding vpn-instance cw
[PE1-GigabitEthernet0/0/2.20]ip address 10.0.101.1 255.255.255.0
[PE1-GigabitEthernet0/0/2.20]arp broadcast enable
```

CE1 的配置：

```
[CE1]ip vpn-instance cw
[CE1-vpn-instance-cw]ipv4-family
[CE1-vpn-instance-cw-af-ipv4]route-distinguisher 100:7
[CE1]ip vpn-instance xs
[CE1-vpn-instance-xs]ipv4-family
[CE1-vpn-instance-xs-af-ipv4]route-distinguisher 100:8
[CE1-]interface GigabitEthernet0/0/0.10
[CE1-GigabitEthernet0/0/0.10]dot1q termination vid 10
[CE1-GigabitEthernet0/0/0.10]ip binding vpn-instance xs
[CE1-GigabitEthernet0/0/0.10]ip address 10.0.100.2 255.255.255.0
[CE1-GigabitEthernet0/0/0.10]arp broadcast enable
[CE1-GigabitEthernet0/0/0.10]q
[CE1]interface GigabitEthernet0/0/0.20
[CE1-GigabitEthernet0/0/0.20]dot1q termination vid 20
[CE1-GigabitEthernet0/0/0.20]ip binding vpn-instance cw
[CE1-GigabitEthernet0/0/0.20]ip address 10.0.101.2 255.255.255.0
[CE1-GigabitEthernet0/0/0.20]arp broadcast enable
[CE1-GigabitEthernet0/0/0.20]q
[CE1]interface LoopBack10
[CE1-LoopBack10]ip binding vpn-instance xs
[CE1-LoopBack10]ip address 192.168.1.1 255.255.255.0
[CE1-LoopBack10]q
[CE1]interface LoopBack20
[CE1-LoopBack20]ip binding vpn-instance cw
[CE1-LoopBack20]ip address 172.16.1.1 255.255.255.0
```

PE2 的配置：

```
[PE2]ip vpn-instance xs_in
[PE2-vpn-instance-xs_in]ipv4-family
[PE2-vpn-instance-xs_in-af-ipv4]route-distinguisher 100:3
[PE2-vpn-instance-xs_in-af-ipv4]vpn-target 1:1 import-extcommunity
[PE2]ip vpn-instance xs_out
[PE2-vpn-instance-xs_out]ipv4-family
[PE2-vpn-instance-xs_out-af-ipv4]route-distinguisher 100:4
[PE2-vpn-instance-xs_out-af-ipv4]vpn-target 1:2 export-extcommunity
[PE2]ip vpn-instance cw_in
[PE2-vpn-instance-cw_in]ipv4-family
[PE2-vpn-instance-cw_in-af-ipv4]route-distinguisher 100:5
[PE2-vpn-instance-cw_in-af-ipv4]vpn-target 2:1 import-extcommunity
[PE2]ip vpn-instance cw_out
[PE2-vpn-instance-cw_out]ipv4-family
[PE2-vpn-instance-cw_out-af-ipv4]route-distinguisher 100:6
[PE2-vpn-instance-cw_out-af-ipv4]vpn-target 2:2 export-extcommunity
[PE2]interface GigabitEthernet0/0/2.10
[PE2-GigabitEthernet0/0/2.10]dot1q termination vid 10
[PE2-GigabitEthernet0/0/2.10]ip binding vpn-instance xs_in
```

```
[PE2-GigabitEthernet0/0/2.10]ip address 10.0.102.1 255.255.255.0
[PE2-GigabitEthernet0/0/2.10]arp broadcast enable
[PE2]interface GigabitEthernet0/0/2.20
[PE2-GigabitEthernet0/0/2.20]dot1q termination vid 20
[PE2-GigabitEthernet0/0/2.20]ip binding vpn-instance cw_in
[PE2-GigabitEthernet0/0/2.20]ip address 10.0.103.1 255.255.255.0
[PE2-GigabitEthernet0/0/2.20]arp broadcast enable
[PE2]interface GigabitEthernet0/0/2.30
[PE2-GigabitEthernet0/0/2.30]dot1q termination vid 50
[PE2-GigabitEthernet0/0/2.30]ip binding vpn-instance xs_out
[PE2-GigabitEthernet0/0/2.30]ip address 10.0.112.1 255.255.255.0
[PE2-GigabitEthernet0/0/2.30]arp broadcast enable
[PE2]interface GigabitEthernet0/0/2.40
[PE2-GigabitEthernet0/0/2.40]dot1q termination vid 60
[PE2-GigabitEthernet0/0/2.40]ip binding vpn-instance cw_out
[PE2-GigabitEthernet0/0/2.40]ip address 10.0.113.1 255.255.255.0
[PE2-GigabitEthernet0/0/2.40]arp broadcast enable
```

CE3 的配置：

```
[CE3]ip vpn-instance cw
[CE3-vpn-instance-cw]ipv4-family
[CE3-vpn-instance-cw-af-ipv4]route-distinguisher 100:7
[CE3]ip vpn-instance xs
[CE3-vpn-instance-xs]ipv4-family
[CE3-vpn-instance-xs-af-ipv4]route-distinguisher 100:8
[CE3]interface GigabitEthernet0/0/0.10
[CE3-GigabitEthernet0/0/0.10]dot1q termination vid 10
[CE3-GigabitEthernet0/0/0.10]ip binding vpn-instance xs
[CE3-GigabitEthernet0/0/0.10]ip address 10.0.102.2 255.255.255.0
[CE3-GigabitEthernet0/0/0.10]arp broadcast enable
[CE3]interface GigabitEthernet0/0/0.20
[CE3-GigabitEthernet0/0/0.20]dot1q termination vid 20
[CE3-GigabitEthernet0/0/0.20]ip binding vpn-instance cw
[CE3-GigabitEthernet0/0/0.20]ip address 10.0.103.2 255.255.255.0
[CE3-GigabitEthernet0/0/0.20]arp broadcast enable
[CE3]interface GigabitEthernet0/0/0.30
[CE3-GigabitEthernet0/0/0.30]dot1q termination vid 50
[CE3-GigabitEthernet0/0/0.30]ip binding vpn-instance xs
[CE3-GigabitEthernet0/0/0.30]ip address 10.0.112.2 255.255.255.0
[CE3-GigabitEthernet0/0/0.30]arp broadcast enable
[CE3]interface GigabitEthernet0/0/0.40
[CE3-GigabitEthernet0/0/0.40]dot1q termination vid 60
[CE3-GigabitEthernet0/0/0.40]ip binding vpn-instance cw
[CE3-GigabitEthernet0/0/0.40]ip address 10.0.113.2 255.255.255.0
[CE3-GigabitEthernet0/0/0.40]arp broadcast enable
```

PE3 的配置：

```
[PE3]ip vpn-instance xs
```

```
[PE3-vpn-instance-xs]ipv4-family
[PE3-vpn-instance-xs-af-ipv4]route-distinguisher 100:1
[PE3-vpn-instance-xs-af-ipv4]vpn-target 1:1 export-extcommunity
[PE3-vpn-instance-xs-af-ipv4]vpn-target 1:2 import-extcommunity
[PE3]ip vpn-instance cw
[PE3-vpn-instance-cw]ipv4-family
[PE3-vpn-instance-cw-af-ipv4]route-distinguisher 100:2
[PE3-vpn-instance-cw-af-ipv4]vpn-target 2:1 export-extcommunity
[PE3-vpn-instance-cw-af-ipv4]vpn-target 2:2 import-extcommunity
[PE3]interface GigabitEthernet 4/0/0.10
[PE3-GigabitEthernet4/0/0.10]dot1q termination vid 10
[PE3-GigabitEthernet4/0/0.10]ip binding vpn-instance xs
[PE3-GigabitEthernet4/0/0.10]ip address 10.0.104.1 255.255.255.0
[PE3-GigabitEthernet4/0/0.10]arp broadcast enable
[PE3]interface GigabitEthernet 4/0/0.20
[PE3-GigabitEthernet4/0/0.20]dot1q termination vid 20
[PE3-GigabitEthernet4/0/0.20]ip binding vpn-instance cw
[PE3-GigabitEthernet4/0/0.20]ip address 10.0.105.1 255.255.255.0
[PE3-GigabitEthernet4/0/0.20]arp broadcast enable
```

CE2 的配置：

```
[CE2]ip vpn-instance cw
[CE2-vpn-instance-cw]ipv4-family
[CE2-vpn-instance-cw-af-ipv4]route-distinguisher 100:7
[CE2]ip vpn-instance xs
[CE2-vpn-instance-xs]ipv4-family
[CE2-vpn-instance-xs-af-ipv4]route-distinguisher 100:8
[CE2]interface GigabitEthernet0/0/0.10
[CE2-GigabitEthernet0/0/0.10]dot1q termination vid 10
[CE2-GigabitEthernet0/0/0.10]ip binding vpn-instance xs
[CE2-GigabitEthernet0/0/0.10]ip address 10.0.104.2 255.255.255.0
[CE2-GigabitEthernet0/0/0.10]arp broadcast enable
[CE2]interface GigabitEthernet0/0/0.20
[CE2-GigabitEthernet0/0/0.20]dot1q termination vid 20
[CE2-GigabitEthernet0/0/0.20]ip binding vpn-instance cw
[CE2-GigabitEthernet0/0/0.20]ip address 10.0.105.2 255.255.255.0
[CE2-GigabitEthernet0/0/0.20]arp broadcast enable
[CE2]interface LoopBack10
[CE2-LoopBack10]ip binding vpn-instance xs
[CE2-LoopBack10]ip address 192.168.2.1 255.255.255.0
[CE2]interface LoopBack20
[CE2-LoopBack20]ip binding vpn-instance cw
[CE2-LoopBack20]ip address 172.16.2.1 255.255.255.0
```

步骤 5：配置总部内部网络。

配置 CE3 的互联 IP 地址：

```
[CE3]interface GigabitEthernet0/0/1.10
[CE3-GigabitEthernet0/0/1.10]dot1q termination vid 30
```

```
[CE3-GigabitEthernet0/0/1.10]ip binding vpn-instance xs
[CE3-GigabitEthernet0/0/1.10]ip address 10.0.106.1 255.255.255.0
[CE3-GigabitEthernet0/0/1.10]arp broadcast enable
[CE3]interface GigabitEthernet0/0/1.20
[CE3-GigabitEthernet0/0/1.20]dot1q termination vid 40
[CE3-GigabitEthernet0/0/1.20]ip binding vpn-instance cw
[CE3-GigabitEthernet0/0/1.20]ip address 10.0.107.1 255.255.255.0
[CE3-GigabitEthernet0/0/1.20]arp broadcast enable
```

配置 LSW1 的互联 IP 地址：

```
[LSW1]vlan batch 10 20 30 40
[LSW1]ip vpn-instance cw
[LSW1-vpn-instance-cw]ipv4-family
[LSW1]ip vpn-instance xs
[LSW1-vpn-instance-xs]ipv4-family
[LSW1]interface Vlanif10
[LSW1-Vlanif10]ip binding vpn-instance xs
[LSW1-Vlanif10]ip address 192.168.3.254 255.255.255.0
[LSW1]interface Vlanif20
[LSW1-Vlanif20]ip binding vpn-instance cw
[LSW1-Vlanif20]ip address 172.16.3.254 255.255.255.0
[LSW1]interface Vlanif30
[LSW1-Vlanif30]ip binding vpn-instance xs
[LSW1-Vlanif30]ip address 10.0.106.2 255.255.255.0
[LSW1]interface Vlanif40
[LSW1-Vlanif40]ip binding vpn-instance cw
[LSW1-Vlanif40]ip address 10.0.107.2 255.255.255.0
```

配置 LSW1 接口类型：

```
[LSW1]interface GigabitEthernet0/0/1
[LSW1-GigabitEthernet0/0/1]port link-type trunk
[LSW1-GigabitEthernet0/0/1]port trunk allow-pass vlan 10 20 30 40
[LSW1]interface GigabitEthernet0/0/2
[LSW1-GigabitEthernet0/0/2]port link-type access
[LSW1-GigabitEthernet0/0/2]port default vlan 10
[LSW1]interface GigabitEthernet0/0/3
[LSW1-GigabitEthernet0/0/3]port link-type access
[LSW1-GigabitEthernet0/0/3]port default vlan 20
```

配置内部的 OSPF 协议，销售部门为 OSPF 100 ，财务部门为 OSPF 200。

CE3 的配置：

```
[CE3]ospf 100 vpn-instance xs
[CE3-ospf-100]area 0.0.0.0
[CE3-ospf-100-area-0.0.0.0]network 10.0.106.0 0.0.0.255
[CE3]ospf 200 vpn-instance cw
[CE3-ospf-200]area 0.0.0.0
[CE3-ospf-200-area-0.0.0.0]network 10.0.107.0 0.0.0.255
```

LSW1 的配置：

```
[LSW1]ospf 100 vpn-instance xs
[LSW1-ospf-100]area 0.0.0.0
[LSW1-ospf-100-area-0.0.0.0]network 10.0.106.0 0.0.0.255
[LSW1-ospf-100-area-0.0.0.0]network 192.168.3.0 0.0.0.255
[LSW1]ospf 200 vpn-instance cw
[LSW1-ospf-200]area 0.0.0.0
[LSW1-ospf-200-area-0.0.0.0]network 10.0.107.0 0.0.0.255
[LSW1-ospf-200-area-0.0.0.0]network 172.16.3.0 0.0.0.255
```

查看邻居关系：

```
[SW1]display ospf peer brief
         OSPF Process 100 with Router ID 192.168.3.254
                  Peer Statistic Information
 ----------------------------------------------------------------------------
 Area Id          Interface                        Neighbor id      State
 0.0.0.0          Vlanif30                         10.0.102.2       Full
 ----------------------------------------------------------------------------
         OSPF Process 200 with Router ID 172.16.3.254
                  Peer Statistic Information
 ----------------------------------------------------------------------------
 Area Id          Interface                        Neighbor id      State
 0.0.0.0          Vlanif40                         10.0.103.2       Full
 ----------------------------------------------------------------------------
```

步骤 6: 配置 PE 设备的 MP-BGP 邻居关系。

PE1 的配置：

```
[PE1]bgp 100
[PE1-bgp]peer 2.2.2.2 as-number 100
[PE1-bgp]peer 2.2.2.2 connect-interface LoopBack0
[PE1-bgp]ipv4-family vpnv4
[PE1-bgp-af-vpnv4]policy vpn-target
[PE1-bgp-af-vpnv4]peer 2.2.2.2 enable
```

PE2 的配置：

```
[PE2]bgp 100
[PE2-bgp]peer 1.1.1.1 as-number 100
[PE2-bgp]peer 1.1.1.1 connect-interface LoopBack0
[PE2-bgp]peer 3.3.3.3 as-number 100
[PE2-bgp]peer 3.3.3.3 connect-interface LoopBack0
[PE2-bgp]ipv4-family vpnv4
[PE2-bgp-af-vpnv4]policy vpn-target
[PE2-bgp-af-vpnv4]peer 1.1.1.1 enable
[PE2-bgp-af-vpnv4]peer 3.3.3.3 enable
```

PE3 的配置：

```
[PE3]bgp 100
[PE3-bgp]peer 2.2.2.2 as-number 100
```

```
[PE3-bgp]peer 2.2.2.2 connect-interface LoopBack0
[PE3-bgp]ipv4-family vpnv4
[PE3-bgp-af-vpnv4]policy vpn-target
[PE3-bgp-af-vpnv4]peer 2.2.2.2 enable
```

步骤 7: 配置 PE 和 CE 的 BGP 邻居。

PE2 的配置:

```
[PE2]bgp 100
[PE2-bgp]ipv4-family vpn-instance cw_in
[PE2-bgp-cw_in]peer 10.0.103.2 as-number 200
[PE2-bgp-cw_in]q
[PE2-bgp]ipv4-family vpn-instance cw_out
[PE2-bgp-cw_out]peer 10.0.113.2 as-number 200
[PE2-bgp-cw_out]q
[PE2-bgp]ipv4-family vpn-instance xs_in
[PE2-bgp-xs_in]peer 10.0.102.2 as-number 200
[PE2-bgp-xs_in]q
[PE2-bgp]ipv4-family vpn-instance xs_out
[PE2-bgp-xs_out]peer 10.0.112.2 as-number 200
```

CE3 的配置:

```
[CE3]bgp 200
[CE3-bgp]router-id 30.30.30.30
[CE3-bgp]
[CE3-bgp]ipv4-family vpn-instance cw
[CE3-bgp-cw]peer 10.0.103.1 as-number 100
[CE3-bgp-cw]peer 10.0.113.1 as-number 100
[CE3-bgp-cw]q
[CE3-bgp]ipv4-family vpn-instance xs
[CE3-bgp-xs]peer 10.0.102.1 as-number 100
[CE3-bgp-xs]peer 10.0.112.1 as-number 100
```

PE1 的配置:

```
[PE1]bgp 100
[PE1-bgp]ipv4-family vpn-instance cw
[PE1-bgp-cw]peer 10.0.101.2 as-number 300
[PE1-bgp-cw]q
[PE1-bgp]ipv4-family vpn-instance xs
[PE1-bgp-xs]peer 10.0.100.2 as-number 300
```

CE1 的配置:

```
[CE1]bgp 300
[CE1-bgp]router-id 10.10.10.10
[CE1-bgp]ipv4-family vpn-instance cw
[CE1-bgp-cw]peer 10.0.101.1 as-number 100
[CE1-bgp-cw]network 172.16.1.0 255.255.255.0
[CE1-bgp-cw]ipv4-family vpn-instance xs
[CE1-bgp-xs]peer 10.0.100.1 as-number 100
```

```
[CE1-bgp-xs]network 192.168.1.0 255.255.255.0
```

PE3 的配置：

```
[PE3]bgp 100
[PE3-bgp]ipv4-family vpn-instance cw
[PE3-bgp-cw]peer 10.0.105.2 as-number 400
[PE3-bgp-cw]q
[PE3-bgp]ipv4-family vpn-instance xs
[PE3-bgp-xs]peer 10.0.104.2 as-number 400
```

CE2 的配置：

```
[CE2]bgp 400
[CE2-bgp]router-id 20.20.20.20
[CE2-bgp]ipv4-family vpn-instance cw
[CE2-bgp-cw]network 172.16.2.0 255.255.255.0
[CE2-bgp-cw]peer 10.0.105.1 as-number 100
[CE2-bgp-cw]ipv4-family vpn-instance xs
[CE2-bgp-xs]network 192.168.2.0 255.255.255.0
[CE2-bgp-xs]peer 10.0.104.1 as-number 100
```

步骤 8： 在 CE3 上引入内网路由。

```
[CE3]ospf 100 vpn-instance xs
[CE3-ospf-100]import-route bgp
[CE3]ospf 200 vpn-instance cw
[CE3-ospf-200]import-route bgp
[CE3]bgp 200
[CE3-bgp]ipv4-family vpn-instance cw
[CE3-bgp-cw]import-route ospf 200
[CE3-bgp-cw]ipv4-family vpn-instance xs
[CE3-bgp-xs]import-route ospf 100
```

步骤 9： 配置连接总部的 PE 节点 xs_out 、cw_out 允许 AS 号重复，否则这两个实例由于 AS 的防环无法从总部学习到其他站点的路由。

PE2 的配置：

```
[PE2]bgp 100
[PE2-bgp]ipv4-family vpn-instance cw_out
[PE2-bgp-cw_out]peer 10.0.113.2 allow-as-loop
[PE2-bgp-cw_out]ipv4-family vpn-instance xs_out
[PE2-bgp-xs_out]peer 10.0.112.2 allow-as-loop
```

步骤 10： 在 LSW1 上配置忽略 VPN 实例的防环规则。

LSW1 的配置：

```
[LSW1]ospf 100 vpn-instance xs
[LSW1-ospf-100]vpn-instance-capability simple
[LSW1-ospf-100]ospf 200 vpn-instance cw
[LSW1-ospf-200]vpn-instance-capability simple
```

查看 CE 设备的路由表：

```
<CE1>display bgp vpnv4 all routing-table
```

```
 BGP Local router ID is 10.10.10.10
 Status codes: * - valid, > - best, d - damped,
               h - history,  i - internal, s - suppressed, S - Stale
               Origin : i - IGP, e - EGP, ? - incomplete
 Total number of routes from all PE: 8
 Route Distinguisher: 100:7
       Network            NextHop        MED      LocPrf        PrefVal  Path/Ogn
 *>    10.0.106.0/24      10.0.100.1                            0        100 200?
 *>    192.168.1.0        0.0.0.0        0                      0        i
 *>    192.168.2.0        10.0.100.1                            0        100 200 100 400i
 *>    192.168.3.0        10.0.100.1                            0        100 200?
 Route Distinguisher: 100:8
       Network            NextHop        MED      LocPrf        PrefVal  Path/Ogn
 *>    10.0.107.0/24      10.0.101.1                            0        100 200?
 *>    172.16.1.0/24      0.0.0.0        0                      0        i
 *>    172.16.2.0/24      10.0.101.1                            0        100 200 100 400i
 *>    172.16.3.0/24      10.0.101.1                            0        100 200?
 VPN-Instance cw, Router ID 10.10.10.10:
 Total Number of Routes: 4
       Network            NextHop        MED      LocPrf        PrefVal  Path/Ogn
 *>    10.0.107.0/24      10.0.101.1                            0        100 200?
 *>    172.16.1.0/24      0.0.0.0        0                      0        i
 *>    172.16.2.0/24      10.0.101.1                            0        100 200 100 400i
 *>    172.16.3.0/24      10.0.101.1                            0        100 200?
 VPN-Instance xs, Router ID 10.10.10.10:
 Total Number of Routes: 4
       Network            NextHop        MED      LocPrf        PrefVal  Path/Ogn
 *>    10.0.106.0/24      10.0.100.1                            0        100 200?
 *>    192.168.1.0        0.0.0.0        0                      0        i
 *>    192.168.2.0        10.0.100.1              0             100 200 100 400i
 *>    192.168.3.0        10.0.100.1              0             100 200?
```

通过以上输出可以看出，CE 设备能学习到各个站点的路由。

步骤 11：测试。

```
[CE1]tracert -vpn-instance cw -a 172.16.1.1 172.16.2.1
 traceroute to cw 172.16.2.1(172.16.2.1), max hops: 30 ,packet length: 40,press
CTRL_C to break
 1 10.0.101.1 20 ms  30 ms  10 ms
 2 10.0.13.3  40 ms  30 ms  50 ms
 3 10.0.113.1 30 ms  30 ms  30 ms
 4 10.0.113.2 30 ms  50 ms  30 ms
 5 10.0.103.1 40 ms  30 ms  40 ms
 6 10.0.105.1 60 ms  50 ms  40 ms
 7 10.0.105.2 70 ms  40 ms  40 ms
[CE1]tracert -vpn-instance xs -a 192.168.1.1 192.168.2.1
 traceroute to xs 192.168.2.1(192.168.2.1), max hops: 30 ,packet length: 40,press
CTRL_C to break
```

```
 1 10.0.100.1 20 ms  20 ms  10 ms
 2 10.0.13.3  40 ms  30 ms  40 ms
 3 10.0.112.1 40 ms  40 ms  30 ms
 4 10.0.112.2 30 ms  40 ms  30 ms
 5 10.0.102.1 30 ms  40 ms  40 ms
 6 10.0.104.1 60 ms  40 ms  50 ms
 7 10.0.104.2 70 ms  50 ms  60 ms
```

通过以上输出可以看出，分支站点通信经过了总部。

PC1 访问 192.168.1.1，PC1 测试结果如图 6–6 所示。

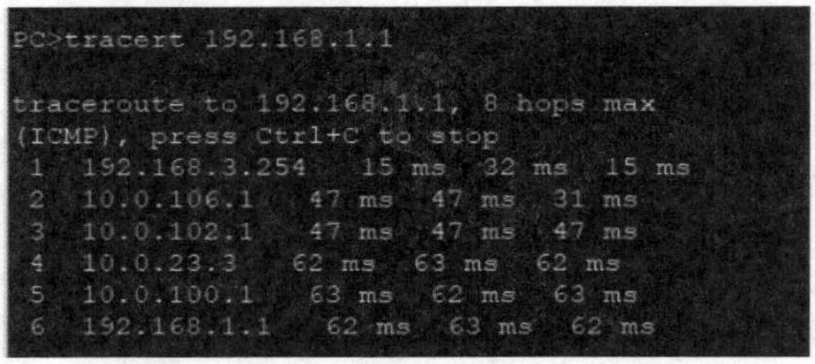

```
PC>tracert 192.168.1.1

traceroute to 192.168.1.1, 8 hops max
(ICMP), press Ctrl+C to stop
 1  192.168.3.254   15 ms  32 ms  15 ms
 2  10.0.106.1   47 ms  47 ms  31 ms
 3  10.0.102.1   47 ms  47 ms  47 ms
 4  10.0.23.3   62 ms  63 ms  62 ms
 5  10.0.100.1   63 ms  62 ms  63 ms
 6  192.168.1.1   62 ms  63 ms  62 ms
```

图 6–6　PC1 测试结果

PC2 访问 172.16.1.1，PC2 测试结果如图 6–7 所示。

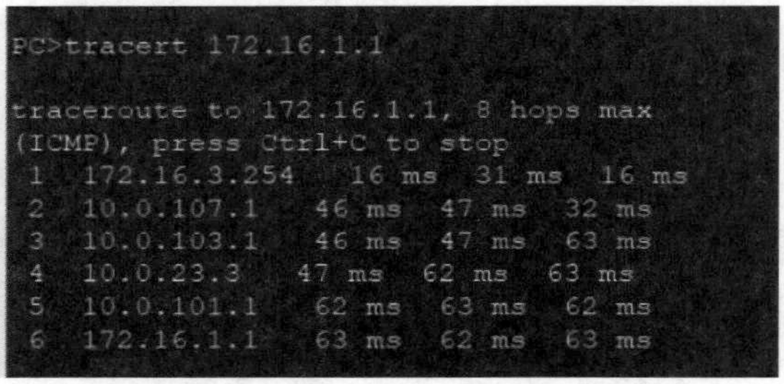

```
PC>tracert 172.16.1.1

traceroute to 172.16.1.1, 8 hops max
(ICMP), press Ctrl+C to stop
 1  172.16.3.254   16 ms  31 ms  16 ms
 2  10.0.107.1   46 ms  47 ms  32 ms
 3  10.0.103.1   46 ms  47 ms  63 ms
 4  10.0.23.3   47 ms  62 ms  63 ms
 5  10.0.101.1   62 ms  63 ms  62 ms
 6  172.16.1.1   63 ms  62 ms  63 ms
```

图 6–7　PC2 测试结果

第 7 章

MPLS VPN 跨域

7.1 MPLS VPN 跨域概述

随着 MPLS VPN 解决方案的广泛应用，服务的终端用户的规格和范围也在增长，一个企业内部的站点数目越来越大，某个地理位置与另外一个服务提供商相连的需求变得非常普遍。例如，国内运营商的不同城域网之间或者相互协作的运营商的骨干网之间，都存在跨越不同自治域的情况。

一般的 MPLS VPN 体系结构都是在一个 AS 内运行的，任何 VPN 的路由信息都只能在一个 AS 内按需扩散，没有提供 AS 内的 VPN 信息向其他 AS 扩散的功能。因此，为了支持运营商之间的 VPN 路由信息交换，就需要扩展现有的协议和修改 MPLS VPN 体系结构，提供一个不同于基本的 MPLS VPN 体系结构所提供的互联模型——跨域（Inter-AS）的 MPLS VPN，以便可以穿过运营商间的链路来发布路由前缀和标签信息。

RFC 中提供了以下 3 种跨域 VPN 解决方案。

（1）跨域 VPN-OptionA（Inter-Provider Backbones OptionA）方式：需要跨域的 VPN 在 ASBR（AS Boundary Router）之间通过专用的接口管理自己的 VPN 路由，也称为 VRF-to-VRF。

（2）跨域 VPN-OptionB（Inter-Provider Backbones Option B）方式：ASBR 之间通过 MP-EBGP 发布标签 VPNv4 路由，也称为 EBGP redistribution of labeled VPNv4 routes。

（3）跨域 VPN-OptionC（Inter-Provider Backbones Option C）方式：PE 之间通过 Multi-Hop MP-EBGP 发布标签 VPNv4 路由，也称为 Multihop EBGP redistribution of labeled VPNv4 routes。

7.1.1 跨域 VPN-OptionA 方式

1. 跨域 VPN-OptionA 方式概述

跨域 VPN-OptionA 方式是基本 BGP/MPLS IP VPN 在跨域环境下的应用，ASBR 之间不需要运行 MPLS，也不需要为跨域进行特殊配置。这种方式下，两个 AS 的边界 ASBR 直接相连，ASBR 同时也是各自所在 AS 的 PE。两个 ASBR 都把对端 ASBR 看作自己的 CE 设备，将会为每一个 VPN 创建 VPN 实例，使用 EBGP 方式向对端发布 VPNv4 路由。

在图 7-1 中，对于 AS 100 的 ASBR1 来说，AS 200 的 ASBR2 只是它的一台 CE 设备；同样，对于 ASBR2 来说，ASBR1 也只是一台接入的 CE 设备。另外，VPN LSP 表示私网隧道；LSP 表示公网隧道。

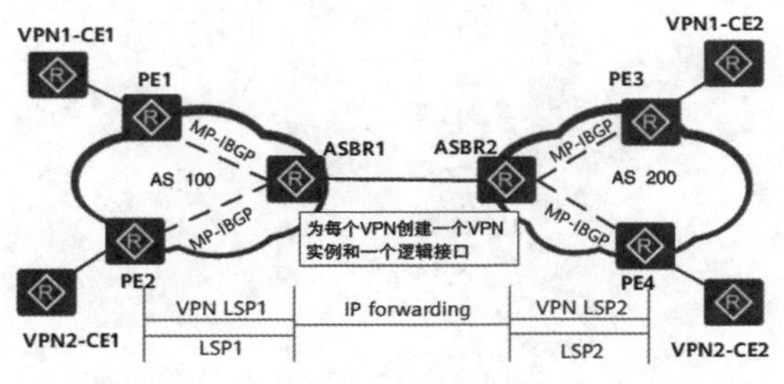

图 7-1 跨域 VPN-OptionA 方式的组网图

（1）跨域 VPN-OptionA 方式的路由信息发布。在 PE 和 ASBR 之间运行 MP-IBGP 协议交换 VPNv4 路由信息。两个 ASBR 之间可以运行普通的 PE-CE 路由协议（BGP 或 IGP 多实例）或静态路由来交互 VPN 信息，但这是不同 AS 之间的交互，建议使用 EBGP。

例如，CE1 将目的地址为 10.1.1.1/24 的路由发布给 CE2，其流程如图 7-2 所示。其中，D 表示目的地址；NH 表示下一跳；L1 和 L2 表示所携带的私网标签。图 7–2 中省略了公网 IGP 路由和标签的分配。

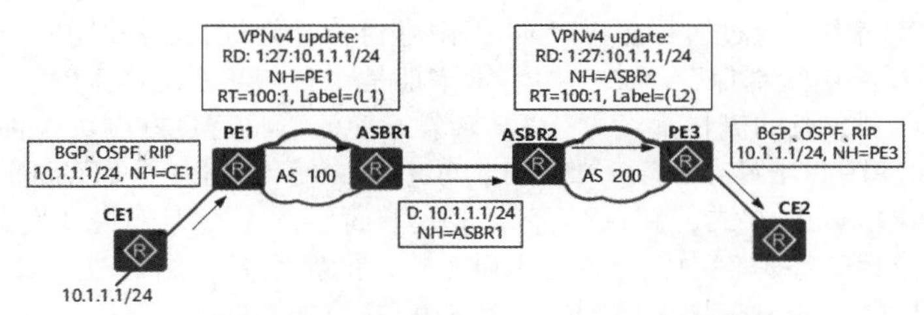

图 7–2　跨域 VPN–Option 方式的路由信息发布流程

（2）跨域 VPN–OptionA 方式的报文转发。以 LSP 为公网隧道，跨域 VPN–OptionA 方式的报文转发流程如图 7–3 所示。其中，L1 和 L2 表示所携带的私网标签；Lx 和 Ly 表示公网外层隧道标签。

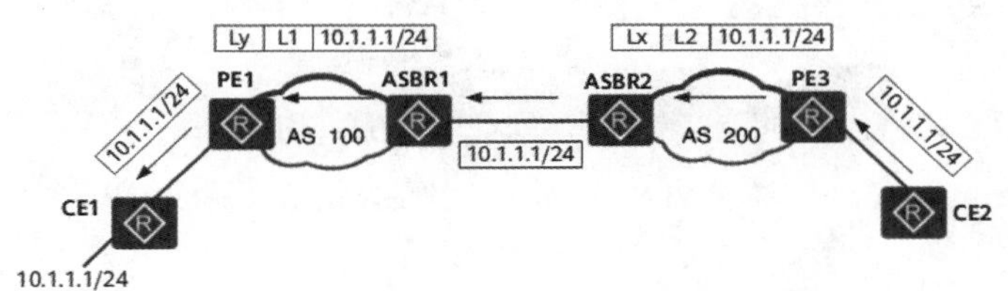

图 7–3　跨域 VPN–OptionA 方式的报文转发流程

2. 跨域 VPN–OptionA 方式的优缺点

（1）优点：配置简单。ASBR 之间不需要运行 MPLS，也不需要为跨域进行特殊配置。

（2）缺点：可扩展性差。ASBR 需要管理所有 VPN 路由，为每个 VPN 创建 VPN 实例。这将导致 ASBR 上的 VPNv4 路由数量过大。另外，由于 ASBR 之间是普通的 IP 转发，要求为每个跨域的 VPN 使用不同的接口，从而提高了对 PE 设备的要求。如果跨越多个自治域，中间域必须支持 VPN 业务，不仅配置量大，而且对中间域影响大。在需要跨域的 VPN 数量比较少的情况下，可以优先考虑使用。

7.1.2　跨域 VPN–OptionB 方式

1. 跨域 VPN–OptionB 方式概述

如图 7–4 所示，在跨域 VPN–OptionB 方式中，两个 ASBR 通过 MP–EBGP 交换它们从各自 AS 的 PE 设备接收的标签 VPNv4 路由。其中，VPN LSP 表示私网隧道；LSP 表示公网隧道。

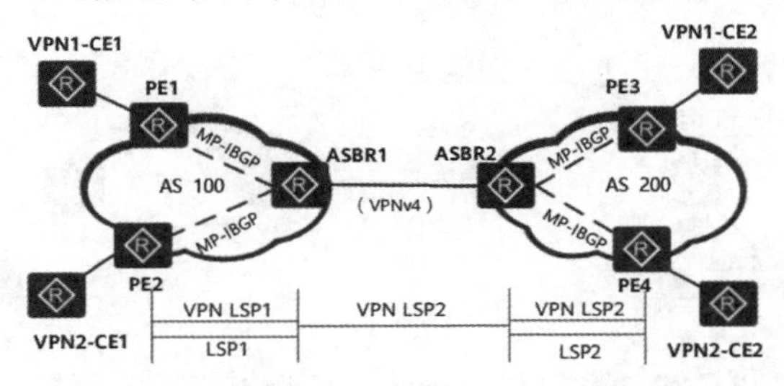

图 7–4　跨域 VPN–OptionB 方式的组网图

在跨域VPN-OptionB方式中，ASBR接收本域内和域外传过来的所有跨域VPNv4路由，再把VPNv4路由发布出去。但在MPLS VPN的基本实现中，PE上只保存与本地VPN实例的VPN Target相匹配的VPN路由。通过对标签VPNv4路由进行特殊处理，让ASBR不进行VPN Target匹配即可把收到的VPN路由全部保存下来，而不管本地是否有与它匹配的VPN实例。

这种方式的优点是所有的流量都经过ASBR转发，使流量具有良好的可控性，但ASBR的负担重。另外，可以同时使用BGP路由策略（如对RT的过滤），使ASBR上只保存部分VPNv4路由。

（1）跨域VPN-OptionB方式的路由信息发布。下面以图7-5为例说明跨域VPN-OptionB方式的路由信息发布流程。在本例中，CE1将10.1.1.1/24的路由发布给CE2，NH表示下一跳，L1、L2和L3表示所携带的私网标签省略了公网IGP路由和标签的分配。

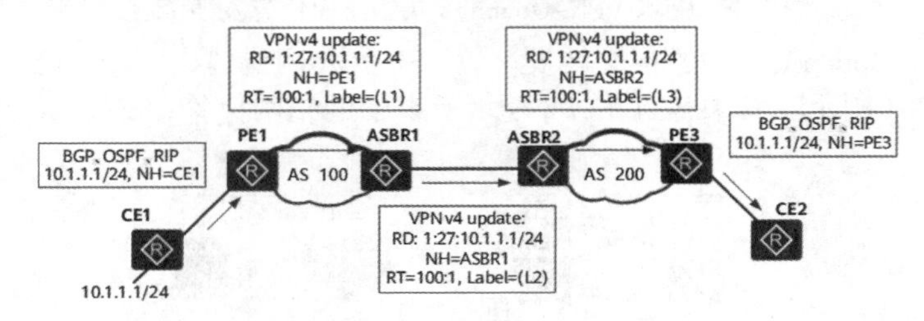

图7-5　跨域VPN-OptionB方式的路由信息发布流程

具体过程如下：

① CE1通过BGP、OSPF或RIP方式将路由发布给AS 100内的PE1。

② AS 100内的PE1先通过MP-IBGP方式把标签VPNv4路由发布给AS 100的ASBR1，或发布给RR，由RR反射给ASBR1。

③ ASBR1通过MP-EBGP方式把标签VPNv4路由发布给ASBR2。由于MP-EBGP在传递路由时需要改变路由的下一跳，因此ASBR1向外发布时给这些VPNv4路由信息分配新标签。

④ ASBR2通过MP-IBGP方式把标签VPNv4路由发布给AS 200内的PE3，或发布给RR，由RR反射给PE3。当ASBR2向域内的MP-IBGP对等体发布路由时，将下一跳改为自己。

⑤ AS 200内的PE3通过BGP、OSPF或RIP方式将路由发布给CE2。

⑥在ASBR1和ASBR2上都对VPNv4路由交换内层标签，域间的标签由BGP携带，因此ASBR之间不需要运行LDP或RSVP（Resource Reservation Protocol，资源预留协议）等协议。

（2）跨域VPN-OptionB方式的报文转发。在跨域VPN-OptionB方式的报文转发中，在两个ASBR上都要对VPN的LSP进行一次交换。以LSP为公网隧道，跨域VPN-OptionB方式的报文转发流程如7-6所示。其中，L1、L2和L3表示所携带的私网标签；Lx和Ly表示公网外层隧道标签。

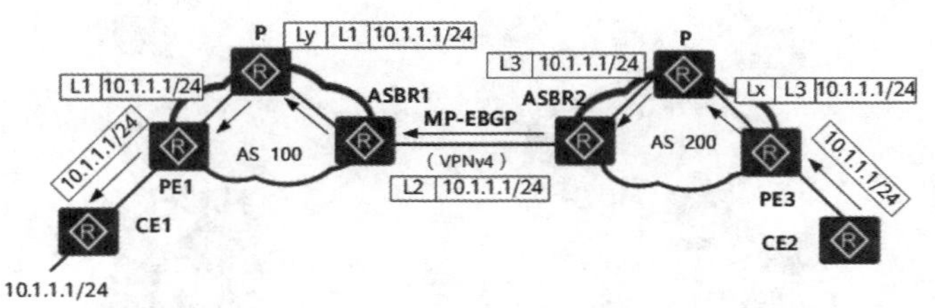

图7-6　跨域VPN-OptionB方式的报文转发流程

2. 跨域 VPN-OptionB 方式的特点

不同于跨域 VPN-OptionA 方式，跨域 VPN-OptionB 方式不受 ASBR 之间互连链路数目的限制。但跨域 VPN-OptionA 方式有一定的局限性，VPN 的路由信息是通过 AS 之间的 ASBR 来保存和扩散的，当 VPN 路由较多时，ASBR 负担重，容易成为故障点。因此，在 MP-EBGP 方案中，需要维护 VPN 路由信息的 ASBR 一般不再负责公网 IP 转发。

7.1.3 跨域 VPN-OptionC 方式

1. 跨域 VPN-OptionC 方式概述

前面小节介绍的两种方式都能够满足跨域 VPN 的组网需求，但这两种方式也都需要 ASBR 参与 VPNv4 路由的维护和发布。当每个 AS 都有大量的 VPN 路由需要交换时，ASBR 就很可能阻碍网络的进一步扩展。

解决上述问题的方案如下：ASBR 不维护或发布 VPNv4 路由，PE 之间直接交换 VPNv4 路由。

- ASBR 通过 MP-IBGP 向各自 AS 内的 PE 设备发布标签 IPv4 路由，并将到达本 AS 内 PE 的标签 IPv4 路由通告给它在对端 AS 内的 ASBR 对等体，过渡 AS 内的 ASBR 也通告标签 VPNv4 路由。这样，在入口 PE 和出口 PE 之间建立一条 LSP。
- 不同 AS 内的 PE 之间通过建立 Multi-Hop 方式的 EBGP 连接，交换标签 VPNv4 路由。
- ASBR 上不保存标签 VPNv4 路由，相互之间也不通告标签 VPNv4 路由。

图 7-7 所示为跨域 VPN-OptionC 方式组网图。其中，VPN LSP 表示私网隧道；LSP 表示公网隧道。BGP LSP 的主要作用是在两个 PE 之间相互交换 LoopBack 信息，由两部分组成。例如，在图 7-7 中，从 PE1 到 PE3 方向建立 BGP LSP1；从 PE3 到 PE1 方向建立 BGP LSP2。

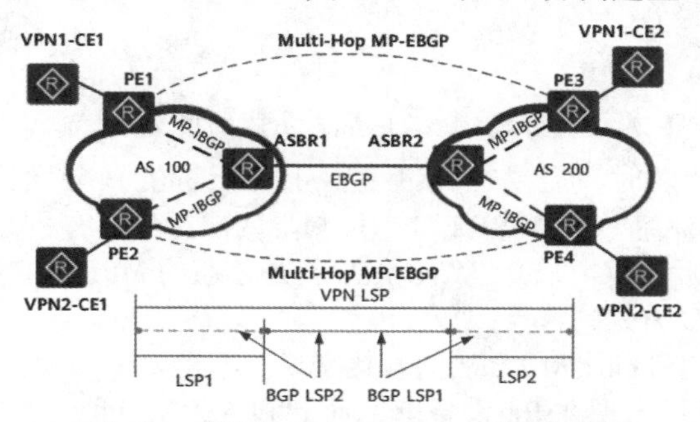

图 7-7 跨域 VPN-OptionC 方式组网图

为了提高可扩展性，可以在每个 AS 内指定一个 RR，由 RR 保存所有 VPNv4 路由，与本 AS 内的 PE 交换 VPNv4 路由信息。两个 AS 内的 RR 之间建立 MP-EBGP 连接，通告 VPNv4 路由，如图 7-8 所示。

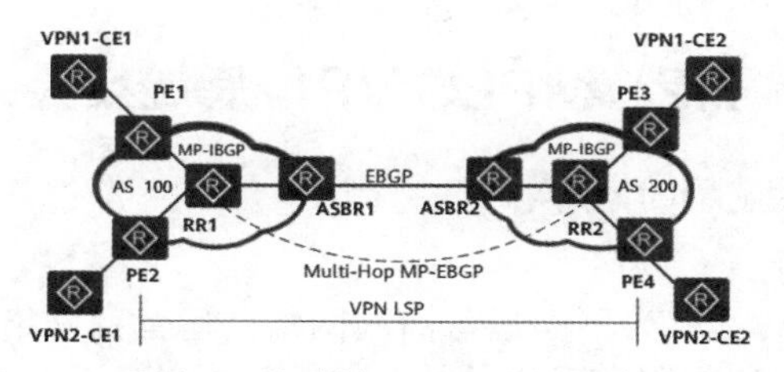

图 7-8 采用 RR 的跨域 VPN-OptionC 方式组网图

（1）跨域 VPN-OptionC 方式的路由发布。跨域 VPN-OptionC 方式的关键实现是公网跨域隧道的建立。例如，在 CE1 中有一条 10.1.1.1/24 的路由信息，其路由信息发布流程如图 7-9 所示。其中，D 表示目的地址；NH 表示下一跳；L3 表示所携带的私网标签；L9 和 L10 表示 BGP LSP 的标签。图 7-9 中省略了公网 IGP 路由和标签的分配。

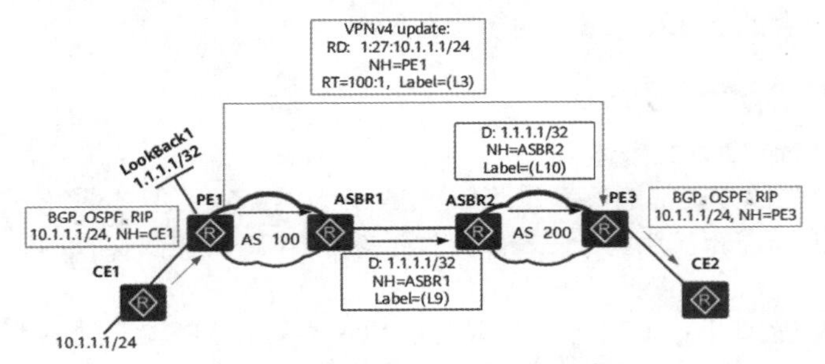

图 7-9　跨域 VPN-OptionC 方式的路由信息发布流程

（2）跨域 VPN-OptionC 方式的报文转发。以 LSP 为公网隧道，跨域 VPN-OptionC 方式的报文转发流程如图 7-10 所示。其中，L3 所携带的表示私网标签；L9 和 L10 表示 BGP LSP 的标签；Lx 和 Ly 表示域内公网外层隧道标签。

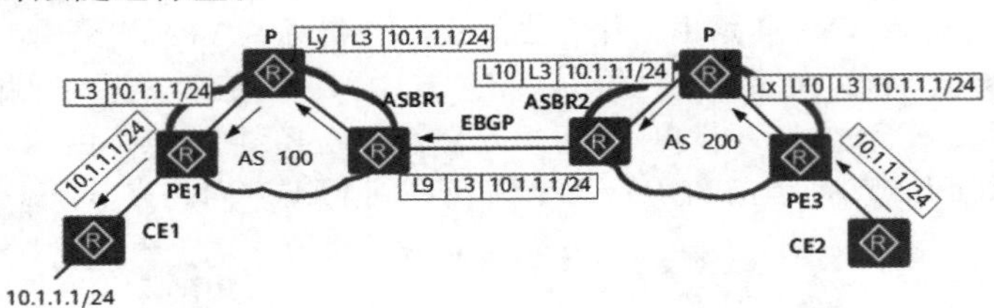

图 7-10　跨域 VPN-OptionC 方式的报文转发流程

报文从 PE3 向 PE1 转发时，需要在 PE3 上打上三层标签，分别为 VPN 的路由标签、BGP LSP 的标签和公网 LSP 的标签。到 ASBR2 时，只剩下两层标签，分别是 VPN 的路由标签和 BGP LSP 的标签；进入 ASBR1 后，BGP LSP 终结，然后就是普通的 MPLS VPN 的转发流程。

2. 跨域 VPN-OptionC 方式的特点

（1）VPN 路由在入口 PE 和出口 PE 之间直接交换，不需要中间设备的保存和转发。

（2）VPN 的路由信息只出现在 PE 设备上，而 P 和 ASBR 只负责报文的转发，使中间域的设备可以不支持 MPLS VPN 业务，只需支持 MPLS 转发，ASBR 设备不再成为性能瓶颈。因此，跨域 VPN-OptionC 方式更适合在跨越多个 AS 时使用。

（3）更适合支持 MPLS VPN 的负载分担。

（4）缺点是维护一条端到端的 PE 连接管理代价较大。

7.2 MPLS VPN 跨域实验

7.2.1 配置跨域 OptionA

扫一扫，看视频

1. 实验目的

AS 100 和 AS 200 为不同运营商的网络，运营商网络内部运行 OSPF 协议。使用 MPLS 跨域 OptionA 组网实现公司 A 互通、公司 B 互通。

2. 实验拓扑

配置跨域 OptionA 的实验拓扑如图 7-11 所示。

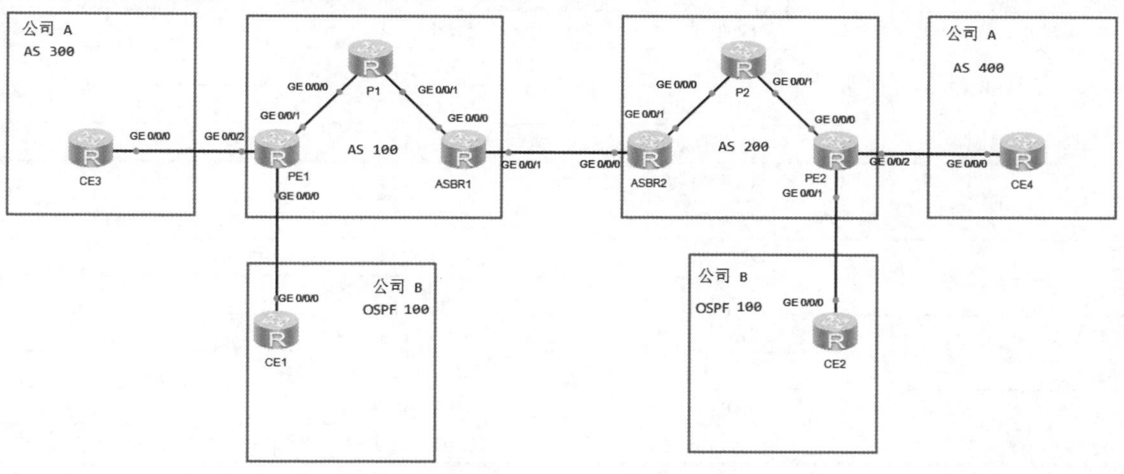

图 7-11　配置跨域 OptionA 的实验拓扑

3. 实验步骤

步骤 1：配置 IP 地址，OptionA IP 地址规划表见表 7-1。

表 7-1　OptionA IP 地址规划表

设备名称	接口编号	IP 地址	所属 VPN 实例
PE1	G0/0/0	17.1.1.1/24	B
	G0/0/1	12.1.1.1/24	
	G0/0/2	19.1.1.1/24	A
	LoopBack 0	1.1.1.1/32	
P1	G0/0/0	12.1.1.2/24	
	G0/0/1	23.1.1.1/24	
	LoopBack 0	2.2.2.2/24	
ASBR1	G0/0/0	23.1.1.2/24	
	G0/0/1.10	100.1.1.1/24	A
	G0/0/1.20	200.1.1.1/24	B
	LoopBack 0	3.3.3.3/32	
PE2	G0/0/0	56.1.1.2/24	
	G0/0/1	28.1.1.2/24	B
	G0/0/2	210.1.1.2/24	A
	LoopBack 0	6.6.6.6/32	
P2	G0/0/0	45.1.1.2/24	
	G0/0/1	56.1.1.1/24	
	LoopBack 0	5.5.5.5/24	

续表

设备名称	接口编号	IP 地址	所属 VPN 实例
ASBR2	G0/0/1	45.1.1.1/24	
	G0/0/0.10	100.1.1.2/24	A
	G0/0/0.20	200.1.1.2/24	B
	LoopBack 0	4.4.4.4/32	
CE1	G0/0/0	17.1.1.7/24	
	LoopBack 0	7.7.7.7/32	
CE2	G0/0/0	28.1.1.8/24	
	LoopBack 0	8.8.8.8/32	
CE3	G0/0/0	19.1.1.9/24	
	LoopBack 0	9.9.9.9/32	
CE4	G0/0/0	210.1.1.0/24	
	LoopBack 0	10.10.10.10/32	

步骤 2：配置运营商网络的 IGP、MPLS 及 MPLS LDP 协议。

（1）配置运营商网络的 IGP。

PE1 的配置：

```
[PE1]ospf 1
[PE1-ospf-1]area 0
[PE1-ospf-1-area-0.0.0.0]network 1.1.1.1 0.0.0.0
[PE1-ospf-1-area-0.0.0.0]network 12.1.1.0 0.0.0.255
```

P1 的配置：

```
[P1]ospf 1
[P1-ospf-1]area 0
[P1-ospf-1-area-0.0.0.0]network 2.2.2.2 0.0.0.0
[P1-ospf-1-area-0.0.0.0]network 12.1.1.0 0.0.0.255
[P1-ospf-1-area-0.0.0.0]network 23.1.1.0 0.0.0.255
```

ASBR1 的配置：

```
[ASBR1]ospf 1
[ASBR1-ospf-1]area 0
[ASBR1-ospf-1-area-0.0.0.0]network 3.3.3.3 0.0.0.0
[ASBR1-ospf-1-area-0.0.0.0]network 23.1.1.0 0.0.0.255
```

ASBR2 的配置：

```
[ASBR2]ospf 1
[ASBR2-ospf-1]area 0
[ASBR2-ospf-1-area-0.0.0.0]network 4.4.4.4 0.0.0.0
[ASBR2-ospf-1-area-0.0.0.0]network 45.1.1.0 0.0.0.255
```

P2 的配置：

```
[P2]ospf 1
[P2-ospf-1]area 0
```

```
[P2-ospf-1-area-0.0.0.0]network 5.5.5.5 0.0.0.0
[P2-ospf-1-area-0.0.0.0]network 45.1.1.0 0.0.0.255
[P2-ospf-1-area-0.0.0.0]network 56.1.1.0 0.0.0.255
```

PE2 的配置：

```
[PE2]ospf 1
[PE2-ospf-1]area 0.0.0.0
[PE2-ospf-1-area-0.0.0.0]network 6.6.6.6 0.0.0.0
[PE2-ospf-1-area-0.0.0.0]network 56.1.1.0 0.0.0.255
```

（2）配置运营商网络的 MPLS 、MPLS LDP 协议，建立公网隧道。

PE1 的配置：

```
[PE1]mpls lsr-id 1.1.1.1
[PE1]mpls
[PE1-mpls]mpls ldp
[PE1]interface G0/0/1
[PE1-GigabitEthernet0/0/1]mpls
[PE1-GigabitEthernet0/0/1]mpls ldp
```

P1 的配置：

```
[P1]mpls lsr-id 2.2.2.2
[P1]mpls
[P1-mpls]mpls ldp
[P1]interface G0/0/0
[P1-GigabitEthernet0/0/0]mpls
[P1-GigabitEthernet0/0/0]mpls ldp
[P1]interface G0/0/1
[P1-GigabitEthernet0/0/1]mpls
[P1-GigabitEthernet0/0/1]mpls ldp
```

ASBR1 的配置：

```
[ASBR1]mpls lsr-id 3.3.3.3
[ASBR1]mpls
[ASBR1-mpls]mpls ldp
[ASBR1]interface G0/0/0
[ASBR1-GigabitEthernet0/0/0]mpls
[ASBR1-GigabitEthernet0/0/0]mpls ldp
```

PE2 的配置：

```
[PE2]mpls lsr-id 6.6.6.6
[PE2]mpls
[PE2-mpls]mpls ldp
[PE2]interface G0/0/0
[PE2-GigabitEthernet0/0/0]mpls
[PE2-GigabitEthernet0/0/0]mpls ldp
```

P2 的配置：

```
[P2]mpls lsr-id 5.5.5.5
[P2]mpls
```

```
[P2-mpls]mpls ldp
[P2]interface G0/0/0
[P2-GigabitEthernet0/0/0]mpls
[P2-GigabitEthernet0/0/0]mpls ldp
[P2]interface G0/0/1
[P2-GigabitEthernet0/0/1]mpls
[P2-GigabitEthernet0/0/1]mpls ldp
```

ASBR2 的配置：

```
[ASBR2]mpls lsr-id 4.4.4.4
[ASBR2]mpls
[ASBR2-mpls]mpls ldp
[ASBR2]interface G0/0/1
[ASBR2-GigabitEthernet0/0/1]mpls
[ASBR2-GigabitEthernet0/0/1]mpls ldp
```

（3）查看 AS 100 和 AS 200 通过 MPLS 建立的 LSP。

查看 PE1 的 LSP：

```
<PE1>display mpls lsp
-------------------------------------------------------------------------------
                 LSP Information: LDP LSP
-------------------------------------------------------------------------------
FEC                In/Out Label        In/Out IF              Vrf Name
2.2.2.2/32         NULL/3              -/GE0/0/1
2.2.2.2/32         1024/3              -/GE0/0/1
3.3.3.3/32         NULL/1024           -/GE0/0/1
3.3.3.3/32         1025/1024           -/GE0/0/1
1.1.1.1/32         3/NULL              -/-
```

查看 PE2 的 LSP：

```
<PE2>display mpls lsp
-------------------------------------------------------------------------------
                 LSP Information: LDP LSP
-------------------------------------------------------------------------------
FEC                In/Out Label        In/Out IF              Vrf Name
4.4.4.4/32         NULL/1024           -/GE0/0/0
4.4.4.4/32         1024/1024           -/GE0/0/0
5.5.5.5/32         NULL/3              -/GE0/0/0
5.5.5.5/32         1025/3              -/GE0/0/0
6.6.6.6/32         3/NULL              -/-
```

通过以上输出可以看出，表示 AS 100 和 AS 200 的公网隧道已经建立完毕。

步骤 3： 配置 PE 和 CE 之间的路由协议。

（1）创建 PE 设备的 VPN 实例，按照题目需求配置好相应的 RT 及 RD 值，并且将接口加入对应的 VPN 实例。

① PE1 的配置。

配置 VPN 实例：

```
[PE1]ip vpn-instance A
```

```
[PE1-vpn-instance-A]ipv4-family
[PE1-vpn-instance-A-af-ipv4]route-distinguisher 100:1
[PE1-vpn-instance-A-af-ipv4]vpn-target 100:1 export-extcommunity
[PE1-vpn-instance-A-af-ipv4]vpn-target 100:1 import-extcommunity
[PE1]ip vpn-instance B
[PE1-vpn-instance-B]ipv4-family
[PE1-vpn-instance-B-af-ipv4]route-distinguisher 200:1
[PE1-vpn-instance-B-af-ipv4]vpn-target 200:1 export-extcommunity
[PE1-vpn-instance-B-af-ipv4]vpn-target 200:1 import-extcommunity
```

将接口加入 VPN 实例：

```
[PE1]interface GigabitEthernet0/0/0
[PE1-GigabitEthernet0/0/0]ip binding vpn-instance B
[PE1-GigabitEthernet0/0/0]ip address 17.1.1.1 255.255.255.0
[PE1-GigabitEthernet0/0/0]
[PE1]interface GigabitEthernet0/0/2
[PE1-GigabitEthernet0/0/2]ip binding vpn-instance A
[PE1-GigabitEthernet0/0/2]ip address 19.1.1.1 255.255.255.0
```

② PE2 的配置。

配置 VPN 实例：

```
[PE2]ip vpn-instance A
[PE2-vpn-instance-A]ipv4-family
[PE2-vpn-instance-A-af-ipv4]route-distinguisher 100:1
[PE2-vpn-instance-A-af-ipv4]vpn-target 100:1 export-extcommunity
[PE2]ip vpn-instance B
[PE2-vpn-instance-B]ipv4-family
[PE2-vpn-instance-B-af-ipv4]route-distinguisher 200:1
[PE2-vpn-instance-B-af-ipv4]vpn-target 200:1 export-extcommunity
[PE2-vpn-instance-B-af-ipv4]vpn-target 200:1 import-extcommunity
```

将接口加入 VPN 实例：

```
[PE2]interface GigabitEthernet0/0/1
[PE2-GigabitEthernet0/0/1]ip binding vpn-instance B
[PE2-GigabitEthernet0/0/1]ip address 28.1.1.2 255.255.255.0
[PE2]interface GigabitEthernet0/0/2
[PE2-GigabitEthernet0/0/2]ip binding vpn-instance A
[PE2-GigabitEthernet0/0/2]ip address 210.1.1.2 255.255.255.0
```

（2）配置 PE 和 CE 之间的路由协议。

① PE1 和 CE1 的 OSPF 协议。

PE1 的配置：

```
[PE1]ospf 100 vpn-instance B
[PE1-ospf-100]area 0
[PE1-ospf-100-area-0.0.0.0]network 28.1.1.0 0.0.0.255
```

CE1 的配置：

```
[CE1]ospf 1
[CE1-ospf-1]area 0
```

```
[CE1-ospf-1-area-0.0.0.0]network 7.7.7.7 0.0.0.0
[CE1-ospf-1-area-0.0.0.0]network 17.1.1.0 0.0.0.255
```

查看 OSPF 邻居关系：

```
[PE1]display ospf 100 peer brief
          OSPF Process 100 with Router ID 17.1.1.1
                    Peer Statistic Information
 ----------------------------------------------------------------------------
 Area Id           Interface                        Neighbor id      State
 0.0.0.0           GigabitEthernet0/0/0             17.1.1.7         Full
 ----------------------------------------------------------------------------
```

查看 PE1 的 VPN 实例 B 的路由：

```
[PE1]display ip routing-table vpn-instance B
Route Flags: R - relay, D - download to fib
------------------------------------------------------------------------------
Routing Tables: B
         Destinations : 5        Routes : 5
Destination/Mask    Proto    Pre   Cost  Flags    NextHop        Interface
       7.7.7.7/32   OSPF     10    1     D        17.1.1.7       GigabitEthernet0/0/0
      17.1.1.0/24   Direct   0     0     D        17.1.1.1       GigabitEthernet0/0/0
      17.1.1.1/32   Direct   0     0     D        127.0.0.1      GigabitEthernet0/0/0
    17.1.1.255/32   Direct   0     0     D        127.0.0.1      GigabitEthernet0/0/0
255.255.255.255/32  Direct   0     0     D        127.0.0.1      InLoopBack0
```

通过以上输出可以看出，PE1 的 VPN 实例 B 学习到了 CE1 的 7.7.7.7/32 的路由。

②配置 PE1 和 CE3 的 BGP。

PE1 的配置：

```
[PE1]bgp 100
[PE1-bgp]ipv4-family vpn-instance A
[PE1-bgp-A]peer 19.1.1.9 as-number 300
```

CE3 的配置：

```
[CE3]bgp 300
[CE3-bgp]peer 19.1.1.1 as-number 100
[CE3-bgp]network 9.9.9.9 255.255.255.255
```

查看 PE1 的 VPNv4 路由表：

```
[PE1]display bgp vpnv4 vpn-instance A routing-table
 BGP Local router ID is 12.1.1.1
 Status codes: * - valid, > - best, d - damped,
               h - history,  i - internal, s - suppressed, S - Stale
               Origin : i - IGP, e - EGP, ? - incomplete
 VPN-Instance A, Router ID 12.1.1.1:
 Total Number of Routes: 1
       Network          NextHop       MED      LocPrf      PrefVal      Path/Ogn
*>     9.9.9.9/32       19.1.1.9      0                    0            300i
```

通过以上输出可以看出，PE1 学习到了 CE3 的 9.9.9.9/32 的路由。

③ PE2 和 CE2 的 OSPF 协议。

PE2 的配置：

```
[PE2]ospf 100 vpn-instance B
[PE2-ospf-100]area 0.0.0.0
[PE2-ospf-100-area-0.0.0.0]network 28.1.1.0 0.0.0.255
```

CE2 的配置：

```
[CE2]ospf 1
[CE2-ospf-1]area 0
[CE2-ospf-1-area-0.0.0.0]network 8.8.8.8 0.0.0.0
[CE2-ospf-1-area-0.0.0.0]network 28.1.1.0 0.0.0.255
```

查看 PE2 的 VPN 实例 B 的路由：

```
[PE2]display ip routing-table vpn-instance B
Route Flags: R - relay, D - download to fib
------------------------------------------------------------------------------
Routing Tables: B
         Destinations : 5        Routes : 5
Destination/Mask     Proto    Pre    Cost   Flags    NextHop        Interface
       8.8.8.8/32    OSPF     10     1      D        28.1.1.8       GigabitEthernet0/0/1
      28.1.1.0/24    Direct   0      0      D        28.1.1.2       GigabitEthernet0/0/1
      28.1.1.2/32    Direct   0      0      D        127.0.0.1      GigabitEthernet0/0/1
    28.1.1.255/32    Direct   0      0      D        127.0.0.1      GigabitEthernet0/0/1
255.255.255.255/32   Direct   0      0      D        127.0.0.1      InLoopBack0
```

通过由以上输出可以看出，PE2 学习到了 CE2 的 8.8.8.8/32 的路由。

④配置 PE2 和 CE4 的 BGP 协议。

PE2 的配置：

```
[PE2]bgp 200
[PE2-bgp]ipv4-family vpn-instance A
[PE2-bgp-A]peer 210.1.1.10 as-number 400
```

CE4 的配置：

```
[CE4]bgp 400
[CE4-bgp]peer 210.1.1.2 as-number 200
[CE4-bgp]network 10.10.10.10 255.255.255.255
```

查看 PE2 的 VPNv4 路由表：

```
[PE2]display bgp vpnv4 vpn-instance A routing-table
 BGP Local router ID is 56.1.1.2
 Status codes: * - valid, > - best, d - damped,
               h - history,  i - internal, s - suppressed, S - Stale
               Origin : i - IGP, e - EGP, ? - incomplete
 VPN-Instance A, Router ID 56.1.1.2:
 Total Number of Routes: 1
        Network          NextHop       MED      LocPrf       PrefVal       Path/Ogn
*>      10.10.10.10/32  210.1.1.10    0                     0             400i
```

通过以上输出可以看出，PE2 学习到了 CE4 的路由。

步骤 4：配置 ASBR 的 VPN 实例。

MPLS 跨域的 OptionA 将本 AS 内的 ASBR 看作一个 PE 设备，将对端 AS 内的 ASBR 看作一个 CE 设备。因此 ASBR 上都需要配置 VPN 实例。

（1）创建 VPN 实例。

ASBR1 的配置：

```
[ASBR1]ip vpn-instance A
[ASBR1-vpn-instance-A]ipv4-family
[ASBR1-vpn-instance-A-af-ipv4]route-distinguisher 100:1
[ASBR1-vpn-instance-A-af-ipv4]vpn-target 100:1 export-extcommunity
[ASBR1-vpn-instance-A-af-ipv4]vpn-target 100:1 import-extcommunity
[ASBR1]ip vpn-instance B
[ASBR1-vpn-instance-B]ipv4-family
[ASBR1-vpn-instance-B-af-ipv4]route-distinguisher 200:1
[ASBR1-vpn-instance-B-af-ipv4]vpn-target 200:1 export-extcommunity
[ASBR1-vpn-instance-B-af-ipv4]vpn-target 200:1 import-extcommunity
```

ASBR2 的配置：

```
[ASBR2]ip vpn-instance A
[ASBR2-vpn-instance-A]ipv4-family
[ASBR2-vpn-instance-A-af-ipv4]route-distinguisher 100:1
[ASBR2-vpn-instance-A-af-ipv4]vpn-target 100:1 export-extcommunity
[ASBR2-vpn-instance-A-af-ipv4]vpn-target 100:1 import-extcommunity
[ASBR2]ip vpn-instance B
[ASBR2-vpn-instance-B]ipv4-family
[ASBR2-vpn-instance-B-af-ipv4]route-distinguisher 200:1
[ASBR2-vpn-instance-B-af-ipv4]vpn-target 200:1 export-extcommunity
[ASBR2-vpn-instance-B-af-ipv4]vpn-target 200:1 import-extcommunity
```

（2）将接口加入到 VPN 实例中。由于 ASBR 之间只有一条物理线路，无法同时加入到两个 VPN 实例中，因此本例使用两个子接口来承担不同 VPN 实例业务的流量。

ASBR1 的配置：

```
[ASBR1]interface GigabitEthernet0/0/1.10
[ASBR1-GigabitEthernet0/0/1.10]dot1q termination vid 10
[ASBR1-GigabitEthernet0/0/1.10]ip binding vpn-instance A
[ASBR1-GigabitEthernet0/0/1.10]ip address 100.1.1.1 255.255.255.0
[ASBR1-GigabitEthernet0/0/1.10]arp broadcast enable
[ASBR1]interface GigabitEthernet0/0/1.20
[ASBR1-GigabitEthernet0/0/1.20]dot1q termination vid 20
[ASBR1-GigabitEthernet0/0/1.20]ip binding vpn-instance B
[ASBR1-GigabitEthernet0/0/1.20]ip address 200.1.1.1 255.255.255.0
[ASBR1-GigabitEthernet0/0/1.20]arp broadcast enable
```

ASBR2 的配置：

```
[ASBR2]interface GigabitEthernet0/0/0.10
[ASBR2-GigabitEthernet0/0/0.10]dot1q termination vid 10
[ASBR2-GigabitEthernet0/0/0.10]ip binding vpn-instance A
[ASBR2-GigabitEthernet0/0/0.10]ip address 100.1.1.2 255.255.255.0
```

```
[ASBR2-GigabitEthernet0/0/0.10]arp broadcast enable
[ASBR2]interface GigabitEthernet0/0/0.20
[ASBR2-GigabitEthernet0/0/0.20]dot1q termination vid 20
[ASBR2-GigabitEthernet0/0/0.20]ip binding vpn-instance B
[ASBR2-GigabitEthernet0/0/0.20]ip address 200.1.1.2 255.255.255.0
[ASBR2-GigabitEthernet0/0/0.20]arp broadcast enable
```

步骤 5: 配置 MP-BGP 传递 CE 的私网路由。

（1）配置 PE 和 ASBR 的 MP-BGP 邻居关系。

PE1 的配置：

```
[PE1]bgp 100
[PE1-bgp]peer 3.3.3.3 as-number 100
[PE1-bgp]peer 3.3.3.3 connect-interface LoopBack0
[PE1-bgp]ipv4-family vpnv4
[PE1-bgp-af-vpnv4]peer 3.3.3.3 enable
```

ASBR1 的配置：

```
[ASBR1]bgp 100
[ASBR1-bgp]peer 1.1.1.1 as-number 100
[ASBR1-bgp]peer 1.1.1.1 connect-interface LoopBack0
[ASBR1-bgp]ipv4-family vpnv4
[ASBR1-bgp-af-vpnv4]peer 1.1.1.1 enable
```

PE2 的配置：

```
[PE2]bgp 200
[PE2-bgp]peer 4.4.4.4 as-number 200
[PE2-bgp]peer 4.4.4.4 connect-interface LoopBack0
[PE2-bgp]ipv4-family vpnv4
[PE2-bgp-af-vpnv4]peer 4.4.4.4 enable
```

ASBR2 的配置：

```
[ASBR2]bgp 200
[ASBR2-bgp]peer 6.6.6.6 as-number 200
[ASBR2-bgp]peer 6.6.6.6 connect-interface LoopBack0
[ASBR2-bgp]ipv4-family vpnv4
[ASBR2-bgp-af-vpnv4]peer 6.6.6.6 enable
```

查看 PE1 的 VPNv4 邻居关系：

```
[PE1]display bgp vpnv4 all  peer
 BGP local router ID : 12.1.1.1
 Local AS number : 100
 Total number of peers : 2                Peers in established state : 2
  Peer         V   AS   MsgRcvd   MsgSent   OutQ   Up/Down    State       PrefRcv
  3.3.3.3      4   100  68        70        0      01:03:16   Established 3
  Peer of IPv4-family for vpn instance :
 VPN-Instance A, Router ID 12.1.1.1:
  19.1.1.9     4   300  133       134       0      02:10:38   Established 1
```

PE1 和 ASBR1 建立了 VPNv4 邻居关系。

查看 PE2 的 VPNv4 邻居关系：

```
[PE2]display bgp vpnv4 all peer
 BGP local router ID : 56.1.1.2
 Local AS number : 200
 Total number of peers : 2                 Peers in established state : 2
  Peer         V   AS    MsgRcvd    MsgSent    OutQ    Up/Down     State         PrefRcv
  4.4.4.4      4   200   65         67         0       01:01:53    Established   3
  Peer of IPv4-family for vpn instance :
 VPN-Instance A, Router ID 56.1.1.2:
  210.1.1.10   4   400   136        137        0       02:13:04    Established   1
```

PE2 和 ASBR2 建立了 VPNv4 邻居关系。

（2）配置 ASBR 之间的 BGP 邻居关系。

ASBR1 的配置：

```
[ASBR1]bgp 100
[ASBR1-bgp]ipv4-family vpn-instance A
[ASBR1-bgp-A]peer 100.1.1.2 as-number 200
[ASBR1-bgp-A]ipv4-family vpn-instance B
[ASBR1-bgp-B]peer 200.1.1.2 as-number 200
```

ASBR2 的配置：

```
[ASBR2]bgp 200
[ASBR2-bgp]ipv4-family vpn-instance A
[ASBR2-bgp-A]peer 100.1.1.1 as-number 100
[ASBR2-bgp-A]ipv4-family vpn-instance B
[ASBR2-bgp-B]peer 200.1.1.1 as-number 100
```

查看 ASBR1 的 BGP 邻居关系建立情况：

```
[ASBR1]display bgp vpnv4 all peer
 BGP local router ID : 23.1.1.2
 Local AS number : 100
 Total number of peers : 3                 Peers in established state : 3
  Peer         V   AS    MsgRcvd    MsgSent    OutQ    Up/Down     State         PrefRcv
  1.1.1.1      4   100   74         74         0       01:09:01    Established   3
  Peer of IPv4-family for vpn instance :
 VPN-Instance A, Router ID 23.1.1.2:
  100.1.1.2    4   200   65         66         0       01:02:39    Established   1
 VPN-Instance B, Router ID 23.1.1.2:
  200.1.1.2    4   200   66         67         0       01:02:27    Established   2
```

通过以上输出可以看出，ASBR 之间的 BGP 邻居关系已经建立完毕。

（3）将 PE 的 VPN 实例 B 的 OSPF 路由和 BGP 路由做双向引入，由于 VPN 实例 B 全部运行在 BGP 中，无须引入。

PE1 的配置：

```
[PE1]ospf 100 vpn-instance B
[PE1-ospf-100]import-route bgp
[PE1]bgp 100
```

```
[PE1-bgp]ipv4-family vpn-instance B
[PE1-bgp-B]import-route ospf 100
```

PE2 的配置：

```
[PE2]ospf 100 vpn-instance B
[PE2-ospf-100]import-route bgp
[PE2]bgp 200
[PE2-bgp]ipv4-family vpn-instance B
[PE2-bgp-B]import-route ospf 100
```

查看 PE1 的 VPNv4 路由表中是否存在用于整个 CE 设备的私网路由信息：

```
[PE1]display bgp vpnv4 all routing-table
 BGP Local router ID is 12.1.1.1
 Status codes: * - valid, > - best, d - damped,
               h - history,  i - internal, s - suppressed, S - Stale
               Origin : i - IGP, e - EGP, ? - incomplete
 Total number of routes from all PE: 6
 Route Distinguisher: 100:1
       Network            NextHop        MED       LocPrf       PrefVal       Path/Ogn
 *>    9.9.9.9/32         19.1.1.9       0                      0             300i
 *>i   10.10.10.10/32     3.3.3.3        100       0            200           400i

 Route Distinguisher: 200:1
       Network            NextHop        MED       LocPrf       PrefVal       Path/Ogn
 *>    7.7.7.7/32         0.0.0.0        2                      0             ?
 *>i   8.8.8.8/32         3.3.3.3                  100          0             200?
 *>    17.1.1.0/24        0.0.0.0        0                      0             ?
 *>i   28.1.1.0/24        3.3.3.3                  100          0             200?

 VPN-Instance A, Router ID 12.1.1.1:
 Total Number of Routes: 2
       Network            NextHop        MED       LocPrf       PrefVal       Path/Ogn
 *>    9.9.9.9/32         19.1.1.9       0                      0             300i
 *>i   10.10.10.10/32     3.3.3.3                  100          0             200 400i

 VPN-Instance B, Router ID 12.1.1.1:
 Total Number of Routes: 4
       Network            NextHop        MED       LocPrf       PrefVal       Path/Ogn
 *>    7.7.7.7/32         0.0.0.0        2                      0             ?
 *>i   8.8.8.8/32         3.3.3.3                  100          0             200?
 *>    17.1.1.0/24        0.0.0.0        0                      0             ?
 *>i   28.1.1.0/24        3.3.3.3                  100          0             200?
```

通过以上输出可以看出，PE1 的 VPNv4 路由表中存在 7.7.7.7/32、8.8.8.8/32、9.9.9.9/32、10.10.10.10/32 的路由信息。

查看 CE1 的路由表：

```
[CE1]display ip routing-table
Route Flags: R - relay, D - download to fib
```

```
------------------------------------------------------------------------------
Routing Tables: Public
         Destinations : 10       Routes : 10
Destination/Mask    Proto   Pre  Cost  Flags  NextHop     Interface
      7.7.7.7/32    Direct  0    0     D      127.0.0.1   LoopBack0
      8.8.8.8/32    O_ASE   150  1     D      17.1.1.1    GigabitEthernet0/0/0
     17.1.1.0/24    Direct  0    0     D      17.1.1.7    GigabitEthernet0/0/0
     17.1.1.7/32    Direct  0    0     D      127.0.0.1   GigabitEthernet0/0/0
   17.1.1.255/32    Direct  0    0     D      127.0.0.1   GigabitEthernet0/0/0
     28.1.1.0/24    O_ASE   150  1     D      17.1.1.1    GigabitEthernet0/0/0
```

通过以上输出可以看出，CE1 学习到了 CE2 的 8.8.8.8/32 的路由信息。

查看 CE3 的路由表：

```
[CE3]display ip routing-table
Route Flags: R - relay, D - download to fib
------------------------------------------------------------------------------
Routing Tables: Public
         Destinations : 9        Routes : 9
Destination/Mask    Proto   Pre  Cost  Flags  NextHop     Interface
      9.9.9.9/32    Direct  0    0     D      127.0.0.1   LoopBack0
  10.10.10.10/32    EBGP    255  0     D      19.1.1.1    GigabitEthernet0/0/0
     19.1.1.0/24    Direct  0    0     D      19.1.1.9    GigabitEthernet0/0/0
     19.1.1.9/32    Direct  0    0     D      127.0.0.1   GigabitEthernet0/0/0
   19.1.1.255/32    Direct  0    0     D      127.0.0.1   GigabitEthernet0/0/0
```

通过以上输出可以看出，CE3 学习到了 CE4 的 9.9.9.9/32 的路由信息。

使用 CE1 ping CE2，并且在 PE1 的 G0/0/1 接口和 ASBR1 的 G0/0/1 接口查看抓包结果。

```
[CE1]ping 8.8.8.8
  PING 8.8.8.8: 56  data bytes, press CTRL_C to break
    Reply from 8.8.8.8: bytes=56 Sequence=1 ttl=249 time=150 ms
    Reply from 8.8.8.8: bytes=56 Sequence=2 ttl=249 time=50 ms
    Reply from 8.8.8.8: bytes=56 Sequence=3 ttl=249 time=50 ms
    Reply from 8.8.8.8: bytes=56 Sequence=4 ttl=249 time=50 ms
    Reply from 8.8.8.8: bytes=56 Sequence=5 ttl=249 time=50 ms
  --- 8.8.8.8 ping statistics ---
    5 packet(s) transmitted
    5 packet(s) received
    0.00% packet loss
round-trip min/avg/max = 50/70/150 ms
```

图 7-12 所示为 PE1 的 G0/0/1 接口的抓包结果，可以看出流量通过 MPLS 隧道转发，最终会发给 ASBR1（此处根据 PE1 的 VPNv4 路由表 8.8.8.8 的下一跳为 3.3.3.3，即可知道下一跳设备是 ASBR1）。

```
> Frame 12: 106 bytes on wire (848 bits), 106 bytes captured (848 bits) on interface -, id 0
> Ethernet II, Src: HuaweiTe_e4:7d:b4 (00:e0:fc:e4:7d:b4), Dst: HuaweiTe_3d:09:1c (00:e0:fc:3d:09:1c)
> MultiProtocol Label Switching Header, Label: 1024, Exp: 0, S: 0, TTL: 254   ① 外层LDP分配的标签
> MultiProtocol Label Switching Header, Label: 1028, Exp: 0, S: 1, TTL: 254   ② 内层MP-BGP分配的标签
> Internet Protocol Version 4, Src: 17.1.1.7, Dst: 8.8.8.8
> Internet Control Message Protocol
```

图 7-12　PE1 的 G0/0/1 接口的抓包结果

图 7-13 所示为 ASBR1 的 G0/0/1 接口的抓包结果，可以看到不携带任何标签信息，也就是说 ASBR 之间的转发是纯 IP 报文转发，不再是隧道转发。

```
> Frame 5: 102 bytes on wire (816 bits), 102 bytes captured (816 bits) on interface -, id 0
> Ethernet II, Src: HuaweiTe_99:36:08 (00:e0:fc:99:36:08), Dst: HuaweiTe_60:0c:fd (00:e0:fc:60:0c:fd)
> 802.1Q Virtual LAN, PRI: 0, DEI: 0, ID: 20
> Internet Protocol Version 4, Src: 17.1.1.7, Dst: 8.8.8.8
> Internet Control Message Protocol
```

图 7-13　ASBR1 的 G0/0/1 接口的抓包结果

在 ASBR2 的 G0/0/1 接口查看抓包结果，如图 7-14 所示。

```
> Frame 9: 106 bytes on wire (848 bits), 106 bytes captured (848 bits) on interface -, id 0
> Ethernet II, Src: HuaweiTe_60:0c:fe (00:e0:fc:60:0c:fe), Dst: HuaweiTe_57:59:ae (00:e0:fc:57:59:ae)
> MultiProtocol Label Switching Header, Label: 1025, Exp: 0, S: 0, TTL: 251
> MultiProtocol Label Switching Header, Label: 1027, Exp: 0, S: 1, TTL: 251
> Internet Protocol Version 4, Src: 17.1.1.7, Dst: 8.8.8.8
> Internet Control Message Protocol
```

图 7-14　ASBR2 的 G0/0/1 接口的抓包结果

由图 7-14 可以看出，流量有进入到 MPLS 隧道，然后发往 PE2。

通过图 7-12 和图 7-14 的对比可以发现，流量的私网标签不一样，由于 PE2 发送 8.8.8.8/32 的私网路由给 ASBR2 时会分配标签 1027，而 ASBR2 和 ASBR1 之间传递的是纯 IP 的路由，并不携带标签，因此 ASBR1 给 PE1 发送 8.8.8.8/32 的路由时会再次分配私网标签 1028。

OptionA 的部署方式在现网中并不常见，原因是 ASBR 需要存储大量的 CE 的私网路由，并且需要为私网路由分配标签，还需要转发跨域的流量。因此，对 ASBR 的设备性能需求较高。但是配置简单，如果跨域的 VPN 数量不多，则可以考虑使用。

7.2.2　配置跨域 OptionB

1. 实验目的

AS 100 和 AS 200 为不同运营商的网络，运营商网络内部运行 OSPF 协议。使用 MPLS 跨域 OptionB 组网实现公司 A 互通、公司 B 互通。

扫一扫，看视频

2. 实验拓扑

配置跨域 OptionB 的实验拓扑如图 7-15 所示。

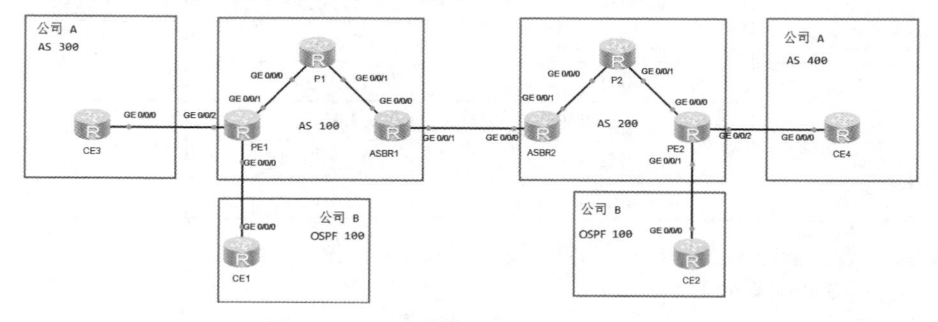

图 7-15　配置跨域 OptionB 的实验拓扑

3. 实验步骤

步骤 1：配置 IP 地址，OptionB IP 地址规划表见表 7-2。

表 7-2　OptionB IP 地址规划表

设备名称	接口编号	IP 地址	所属 VPN 实例
PE1	G0/0/0	17.1.1.1/24	B
	G0/0/1	12.1.1.1/24	
	G0/0/2	19.1.1.1/24	A
	LoopBack 0	1.1.1.1/32	
P1	G0/0/0	12.1.1.2/24	
	G0/0/1	23.1.1.1/24	
	LoopBack 0	2.2.2.2/24	
ASBR1	G0/0/0	23.1.1.2/24	
	G0/0/1	34.1.1.1/24	
	LoopBack 0	3.3.3.3/32	
PE2	G0/0/0	56.1.1.2/24	
	G0/0/1	28.1.1.2/24	B
	G0/0/2	210.1.1.2/24	A
	LoopBack 0	6.6.6.6/32	
P2	G0/0/0	45.1.1.2/24	
	G0/0/1	56.1.1.1/24	
	LoopBack 0	5.5.5.5/24	
ASBR2	G0/0/1	45.1.1.1/24	
	G0/0/0	34.1.1.2/24	
	LoopBack 0	4.4.4.4/32	
CE1	G0/0/0	17.1.1.7/24	
	LoopBack 0	7.7.7.7/32	
CE2	G0/0/0	28.1.1.8/24	
	LoopBack 0	8.8.8.8/32	
CE3	G0/0/0	19.1.1.9/24	
	LoopBack 0	9.9.9.9/32	
CE4	G0/0/0	210.1.1.0/24	
	LoopBack 0	10.10.10.10/32	

步骤 2：配置运营商网络的 IGP、MPLS 及 MPLS LDP 协议。

（1）配置运营商网络的 IGP。

PE1 的配置：

```
[PE1]ospf 1
[PE1-ospf-1]area 0
[PE1-ospf-1-area-0.0.0.0]network 1.1.1.1 0.0.0.0
[PE1-ospf-1-area-0.0.0.0]network 12.1.1.0 0.0.0.255
```

P1 的配置：

```
[P1]ospf 1
[P1-ospf-1]area 0
[P1-ospf-1-area-0.0.0.0]network 2.2.2.2 0.0.0.0
[P1-ospf-1-area-0.0.0.0]network 12.1.1.0 0.0.0.255
[P1-ospf-1-area-0.0.0.0]network 23.1.1.0 0.0.0.255
```

ASBR1 的配置：

```
[ASBR1]ospf 1
[ASBR1-ospf-1]area 0
[ASBR1-ospf-1-area-0.0.0.0]network 3.3.3.3 0.0.0.0
[ASBR1-ospf-1-area-0.0.0.0]network 23.1.1.0 0.0.0.255
```

ASBR2 的配置：

```
[ASBR2]ospf 1
[ASBR2-ospf-1]area 0
[ASBR2-ospf-1-area-0.0.0.0]network 4.4.4.4 0.0.0.0
[ASBR2-ospf-1-area-0.0.0.0]network 45.1.1.0 0.0.0.255
```

P2 的配置：

```
[P2]ospf 1
[P2-ospf-1]area 0
[P2-ospf-1-area-0.0.0.0]network 5.5.5.5 0.0.0.0
[P2-ospf-1-area-0.0.0.0]network 45.1.1.0 0.0.0.255
[P2-ospf-1-area-0.0.0.0]network 56.1.1.0 0.0.0.255
```

PE2 的配置：

```
[PE2]ospf 1
[PE2-ospf-1]area 0.0.0.0
[PE2-ospf-1-area-0.0.0.0]network 6.6.6.6 0.0.0.0
[PE2-ospf-1-area-0.0.0.0]network 56.1.1.0 0.0.0.255
```

（2）配置运营商网络的 MPLS 、MPLS LDP 协议，建立公网隧道。

PE1 的配置：

```
[PE1]mpls lsr-id 1.1.1.1
[PE1]mpls
[PE1-mpls]mpls ldp
[PE1]interface G0/0/1
[PE1-GigabitEthernet0/0/1]mpls
[PE1-GigabitEthernet0/0/1]mpls ldp
```

P1 的配置：

```
[P1]mpls lsr-id 2.2.2.2
[P1]mpls
[P1-mpls]mpls ldp
[P1]interface G0/0/0
[P1-GigabitEthernet0/0/0]mpls
[P1-GigabitEthernet0/0/0]mpls ldp
```

```
[P1]interface G0/0/1
[P1-GigabitEthernet0/0/1]mpls
[P1-GigabitEthernet0/0/1]mpls ldp
```

ASBR1 的配置：

```
[ASBR1]mpls lsr-id 3.3.3.3
[ASBR1]mpls
[ASBR1-mpls]mpls ldp
[ASBR1]interface G0/0/0
[ASBR1-GigabitEthernet0/0/0]mpls
[ASBR1-GigabitEthernet0/0/0]mpls ldp
```

PE2 的配置：

```
[PE2]mpls lsr-id 6.6.6.6
[PE2]mpls
[PE2-mpls]mpls ldp
[PE2]interface G0/0/0
[PE2-GigabitEthernet0/0/0]mpls
[PE2-GigabitEthernet0/0/0]mpls ldp
```

P2 的配置：

```
[P2]mpls lsr-id 5.5.5.5
[P2]mpls
[P2-mpls]mpls ldp
[P2]interface G0/0/0
[P2-GigabitEthernet0/0/0]mpls
[P2-GigabitEthernet0/0/0]mpls ldp
[P2]interface G0/0/1
[P2-GigabitEthernet0/0/1]mpls
[P2-GigabitEthernet0/0/1]mpls ldp
```

ASBR2 的配置：

```
[ASBR2]mpls lsr-id 4.4.4.4
[ASBR2]mpls
[ASBR2-mpls]mpls ldp
[ASBR2]interface G0/0/1
[ASBR2-GigabitEthernet0/0/1]mpls
[ASBR2-GigabitEthernet0/0/1]mpls ldp
```

查看 AS 100 和 AS 200 通过 MPLS 建立的 LSP。

查看 PE1 的 LSP：

```
<PE1>display mpls lsp
-------------------------------------------------------------------------------
                LSP Information: LDP LSP
-------------------------------------------------------------------------------
FEC                In/Out Label          In/Out IF              Vrf Name
2.2.2.2/32         NULL/3                -/GE0/0/1
2.2.2.2/32         1024/3                -/GE0/0/1
3.3.3.3/32         NULL/1024             -/GE0/0/1
```

```
3.3.3.3/32          1025/1024               -/GE0/0/1
1.1.1.1/32          3/NULL                  -/-
```

查看 PE2 的 LSP：

```
<PE2>display mpls lsp
------------------------------------------------------------------------------
                 LSP Information: LDP LSP
------------------------------------------------------------------------------
FEC                 In/Out Label            In/Out IF               Vrf Name
4.4.4.4/32          NULL/1024               -/GE0/0/0
4.4.4.4/32          1024/1024               -/GE0/0/0
5.5.5.5/32          NULL/3                  -/GE0/0/0
5.5.5.5/32          1025/3                  -/GE0/0/0
6.6.6.6/32          3/NULL                  -/-
```

通过以上输出可以看出，AS 100 和 AS 200 的公网隧道已经建立完毕。

步骤 3： 配置 PE 和 CE 之间的路由协议。

（1）创建 PE 设备的 VPN 实例，按照题目需求配置好相应的 RT 及 RD 值，并且将接口加入对应的 VPN 实例。

① PE1 的配置。

配置 VPN 实例：

```
[PE1]ip vpn-instance A
[PE1-vpn-instance-A]ipv4-family
[PE1-vpn-instance-A-af-ipv4]route-distinguisher 100:1
[PE1-vpn-instance-A-af-ipv4]vpn-target 100:1 export-extcommunity
[PE1-vpn-instance-A-af-ipv4]vpn-target 100:1 import-extcommunity
[PE1]ip vpn-instance B
[PE1-vpn-instance-B]ipv4-family
[PE1-vpn-instance-B-af-ipv4]route-distinguisher 200:1
[PE1-vpn-instance-B-af-ipv4]vpn-target 200:1 export-extcommunity
[PE1-vpn-instance-B-af-ipv4]vpn-target 200:1 import-extcommunity
```

将接口加入 VPN 实例：

```
[PE1]interface GigabitEthernet0/0/0
[PE1-GigabitEthernet0/0/0]ip binding vpn-instance B
[PE1-GigabitEthernet0/0/0]ip address 17.1.1.1 255.255.255.0
[PE1-GigabitEthernet0/0/0]
[PE1]interface GigabitEthernet0/0/2
[PE1-GigabitEthernet0/0/2]ip binding vpn-instance A
[PE1-GigabitEthernet0/0/2]ip address 19.1.1.1 255.255.255.0
```

② PE2 的配置。

配置 VPN 实例：

```
[PE2]ip vpn-instance A
[PE2-vpn-instance-A]ipv4-family
[PE2-vpn-instance-A-af-ipv4]route-distinguisher 100:1
[PE2-vpn-instance-A-af-ipv4]vpn-target 100:1 export-extcommunity
```

```
[PE2]ip vpn-instance B
[PE2-vpn-instance-B]ipv4-family
[PE2-vpn-instance-B-af-ipv4]route-distinguisher 200:1
[PE2-vpn-instance-B-af-ipv4]vpn-target 200:1 export-extcommunity
[PE2-vpn-instance-B-af-ipv4]vpn-target 200:1 import-extcommunity
```

将接口加入 VPN 实例：

```
[PE2]interface GigabitEthernet0/0/1
[PE2-GigabitEthernet0/0/1]ip binding vpn-instance B
[PE2-GigabitEthernet0/0/1]ip address 28.1.1.2 255.255.255.0
[PE2]interface GigabitEthernet0/0/2
[PE2-GigabitEthernet0/0/2]ip binding vpn-instance A
[PE2-GigabitEthernet0/0/2]ip address 210.1.1.2 255.255.255.0
```

（2）配置 PE 和 CE 之间的路由协议。

①配置 PE1 和 CE1 的 OSPF 协议。

PE1 的配置：

```
[PE1]ospf 100 vpn-instance B
[PE1-ospf-100]area 0
[PE1-ospf-100-area-0.0.0.0]network 28.1.1.0 0.0.0.255
```

CE1 的配置：

```
[CE1]ospf 1
[CE1-ospf-1]area 0
[CE1-ospf-1-area-0.0.0.0]network 7.7.7.7 0.0.0.0
[CE1-ospf-1-area-0.0.0.0]network 17.1.1.0 0.0.0.255
```

查看 PE1 的 OSPF 邻居关系：

```
[PE1]display ospf 100 peer brief
         OSPF Process 100 with Router ID 17.1.1.1
                  Peer Statistic Information
 ----------------------------------------------------------------------------
 Area Id          Interface                      Neighbor id          State
 0.0.0.0          GigabitEthernet0/0/0           17.1.1.7             Full
 ----------------------------------------------------------------------------
```

查看 PE1 的 VPN 实例 B 的路由：

```
[PE1]display ip routing-table vpn-instance B
Route Flags: R - relay, D - download to fib
------------------------------------------------------------------------------
Routing Tables: B
         Destinations : 5        Routes : 5
Destination/Mask    Proto   Pre  Cost  Flags   NextHop       Interface
       7.7.7.7/32   OSPF    10   1     D       17.1.1.7      GigabitEthernet0/0/0
      17.1.1.0/24   Direct  0    0     D       17.1.1.1      GigabitEthernet0/0/0
      17.1.1.1/32   Direct  0    0     D       127.0.0.1     GigabitEthernet0/0/0
    17.1.1.255/32   Direct  0    0     D       127.0.0.1     GigabitEthernet0/0/0
255.255.255.255/32  Direct  0    0     D       127.0.0.1     InLoopBack0
```

通过以上输出可以看出，在 PE1 的 VPN 实例 B 上学习到了 CE1 的 7.7.7.7/32 的路由。

②配置 PE1 和 CE3 的 BGP。

PE1 的配置：

```
[PE1]bgp 100
[PE1-bgp]ipv4-family vpn-instance A
[PE1-bgp-A]peer 19.1.1.9 as-number 300
```

CE3 的配置：

```
[CE3]bgp 300
[CE3-bgp]peer 19.1.1.1 as-number 100
[CE3-bgp]network 9.9.9.9 255.255.255.255
```

查看 PE1 的 VPNv4 路由表：

```
[PE1]display bgp vpnv4 vpn-instance A routing-table
 BGP Local router ID is 12.1.1.1
 Status codes: * - valid, > - best, d - damped,
               h - history,  i - internal, s - suppressed, S - Stale
               Origin : i - IGP, e - EGP, ? - incomplete
 VPN-Instance A, Router ID 12.1.1.1:
 Total Number of Routes: 1
        Network            NextHop        MED       LocPrf        PrefVal        Path/Ogn
*>      9.9.9.9/32         19.1.1.9       0                       0              300i
```

通过以上输出可以看出，PE1 学习到了 CE3 的 9.9.9.9/32 的路由。

③配置 PE2 和 CE2 的 OSPF 协议。

PE2 的配置：

```
[PE2]ospf 100 vpn-instance B
[PE2-ospf-100]area 0
[PE2-ospf-100-area-0.0.0.0]network 28.1.1.0 0.0.0.255
```

CE2 的配置：

```
[CE2]ospf 1
[CE2-ospf-1]area 0
[CE2-ospf-1-area-0.0.0.0]network 8.8.8.8 0.0.0.0
[CE2-ospf-1-area-0.0.0.0]network 28.1.1.0 0.0.0.255
```

查看 PE2 的 VPN 实例 B 的路由表：

```
[PE2]display  ip routing-table vpn-instance B
Route Flags: R - relay, D - download to fib
------------------------------------------------------------------------------
Routing Tables: B

         Destinations : 5         Routes : 5

Destination/Mask     Proto     Pre    Cost   Flags     NextHop         Interface
       8.8.8.8/32    OSPF      10     1      D         28.1.1.8        GigabitEthernet0/0/1
      28.1.1.0/24    Direct    0      0      D         28.1.1.2        GigabitEthernet0/0/1
      28.1.1.2/32    Direct    0      0      D         127.0.0.1       GigabitEthernet0/0/1
```

```
    28.1.1.255/32    Direct   0     0     D      127.0.0.1     GigabitEthernet0/0/1
255.255.255.255/32   Direct   0     0     D      127.0.0.1     InLoopBack0
```

通过以上输出可以看出，PE2 学习到了 CE2 的 8.8.8.8/32 的路由。

④配置 PE2 和 CE4 的 BGP 协议。

PE2 的配置：

```
[PE2]bgp 200
[PE2-bgp]ipv4-family vpn-instance A
[PE2-bgp-A]peer 210.1.1.10 as-number 400
```

CE4 的配置：

```
[CE4]bgp 400
[CE4-bgp]peer 210.1.1.2 as-number 200
[CE4-bgp]network 10.10.10.10 255.255.255.255
```

查看 PE2 的 VPNv4 路由表：

```
[PE2]display  bgp vpnv4 vpn-instance A routing-table
 BGP Local router ID is 56.1.1.2
 Status codes: * - valid, > - best, d - damped,
               h - history,  i - internal, s - suppressed, S - Stale
               Origin : i - IGP, e - EGP, ? - incomplete
 VPN-Instance A, Router ID 56.1.1.2:
 Total Number of Routes: 1
          Network            NextHop      MED     LocPrf        PrefVal       Path/Ogn

*>        10.10.10.10/32     210.1.1.10   0                     0             400i
```

通过以上输出可以看出，PE2 学习到了 CE4 的路由。

将 PE 的 VPN 实例 B 的 OSPF 路由和 BGP 路由做双向引入，由于 VPN 实例 B 全部运行在 BGP 中，因此无须引入。

PE1 的配置：

```
[PE1]ospf 100 vpn-instance B
[PE1-ospf-100]import-route bgp
[PE1]bgp 100
[PE1-bgp]ipv4-family vpn-instance B
[PE1-bgp-B]import-route ospf 100
```

PE2 的配置：

```
[PE2]ospf 100 vpn-instance B
[PE2-ospf-100]import-route bgp
[PE2]bgp 200
[PE2-bgp]ipv4-family vpn-instance B
[PE2-bgp-B]import-route ospf 100
```

步骤 4： 配置 MP-BGP 传递 CE 的私网路由。

（1）配置 PE1 和 ASBR 的 MP-BGP 邻居关系。

PE1 的配置：

```
[PE1]bgp 100
```

```
[PE1-bgp]peer 3.3.3.3 as-number 100
[PE1-bgp]peer 3.3.3.3 connect-interface LoopBack0
[PE1-bgp]ipv4-family vpnv4
[PE1-bgp-af-vpnv4]peer 3.3.3.3 enable
```

ASBR1 的配置：

```
[ASBR1]bgp 100
[ASBR1-bgp]peer 1.1.1.1 as-number 100
[ASBR1-bgp]peer 1.1.1.1 connect-interface LoopBack0
[ASBR1-bgp]ipv4-family vpnv4
[ASBR1-bgp-af-vpnv4]peer 1.1.1.1 enable
```

PE2 的配置：

```
[PE2]bgp 200
[PE2-bgp]peer 4.4.4.4 as-number 200
[PE2-bgp]peer 4.4.4.4 connect-interface LoopBack0
[PE2-bgp]ipv4-family vpnv4
[PE2-bgp-af-vpnv4]peer 4.4.4.4 enable
```

ASBR2 的配置：

```
[ASBR2]bgp 200
[ASBR2-bgp]peer 6.6.6.6 as-number 200
[ASBR2-bgp]peer 6.6.6.6 connect-interface LoopBack0
[ASBR2-bgp]ipv4-family vpnv4
[ASBR2-bgp-af-vpnv4]peer 6.6.6.6 enable
```

查看 PE1 的 VPNv4 的邻居关系：

```
[PE1]display  bgp vpnv4 all  peer
 BGP local router ID : 12.1.1.1
 Local AS number : 100
 Total number of peers : 2                Peers in established state : 2
  Peer         V   AS    MsgRcvd    MsgSent   OutQ   Up/Down    State        PrefRcv
  3.3.3.3      4   100   68         70        0      01:03:16   Established  3
  Peer of IPv4-family for vpn instance :
 VPN-Instance A, Router ID 12.1.1.1:
  19.1.1.9     4   300   133        134       0      02:10:38   Established  1
```

通过以上输出可以看出，PE1 和 ASBR1 建立了 VPNv4 邻居关系。

```
[PE2]display  bgp vpnv4 all  peer
 BGP local router ID : 56.1.1.2
 Local AS number : 200
 Total number of peers : 2                Peers in established state : 2
  Peer         V   AS    MsgRcvd    MsgSent   OutQ   Up/Down    State        PrefRcv
  4.4.4.4      4   200   65         67        0      01:01:53   Established  3
  Peer of IPv4-family for vpn instance :
 VPN-Instance A, Router ID 56.1.1.2:
  210.1.1.10   4   400   136        137       0      02:13:04   Established  1
```

通过以上输出可以看出，PE2 和 ASBR2 建立了 VPNv4 邻居关系。

（2）配置ASBR之间的MP-EBGP对等体关系。

ASBR1的配置：

```
[ASBR1]bgp 100
[ASBR1-bgp]peer 34.1.1.2 as-number 200
[ASBR1-bgp]ipv4-family vpnv4
//由于ASBR上不配置VPN实例，因此无法识别RT，需要关闭RT检测功能(如果不关闭则无法接收VPNv4路由)
[ASBR1-bgp-af-vpnv4]undo policy vpn-target
[ASBR1-bgp-af-vpnv4]peer 34.1.1.2 enable
```

ASBR2的配置：

```
[ASBR2]bgp 200
[ASBR2-bgp]peer 34.1.1.1 as-number 100
[ASBR2-bgp]ipv4-family unicast
[ASBR2-bgp-af-ipv4]ipv4-family vpnv4
[ASBR2-bgp-af-vpnv4]undo policy vpn-target
[ASBR2-bgp-af-vpnv4]peer 34.1.1.1 enable
```

查看ASBR之间的VPNv4邻居是否建立：

```
[ASBR1]display bgp vpnv4 all peer
 BGP local router ID : 23.1.1.2
 Local AS number : 100
 Total number of peers : 2                    Peers in established state : 2
  Peer          V   AS    MsgRcvd    MsgSent   OutQ   Up/Down    State        PrefRcv

  1.1.1.1       4   100   91         94        0      01:19:49   Established  3
  34.1.1.2      4   200   96         97        0      01:18:51   Established  3
```

通过以上输出可以看出，ASBR1已经和ASBR2建立了VPNv4邻居关系。

（3）将ASBR之间的直连接口开启MPLS功能（开启后当ASBR1收到PE1的VPNv4路由时，会传递给ASBR2并且携带私网标签；如果不开启，则不分配私网标签，ASBR之间的隧道会中断）。

ASBR1的配置：

```
[ASBR1]interface G0/0/1
[ASBR1-GigabitEthernet0/0/1]mpls
```

ASBR2的配置：

```
[ASBR2]interface G0/0/0
[ASBR2-GigabitEthernet0/0/0]mpls
```

在ASBR1上查看LSP隧道的建立情况：

```
[ASBR1]display mpls lsp
-------------------------------------------------------------------------------
                 LSP Information: L3VPN  LSP
-------------------------------------------------------------------------------
FEC                  In/Out Label           In/Out IF              Vrf Name
9.9.9.9/32           1045/1031              -/-                    ASBR LSP
17.1.1.0/24          1049/1033              -/-                    ASBR LSP
10.10.10.10/32       1046/1049              -/-                    ASBR LSP
7.7.7.7/32           1047/1032              -/-                    ASBR LSP
```

```
28.1.1.0/24          1050/1053              -/-                 ASBR LSP
8.8.8.8/32           1048/1051              -/-                 ASBR LSP
-------------------------------------------------------------------------------
                  LSP Information: LDP LSP
-------------------------------------------------------------------------------
FEC                  In/Out Label           In/Out IF           Vrf Name
3.3.3.3/32           3/NULL                 -/-
2.2.2.2/32           NULL/3                 -/GE0/0/0
2.2.2.2/32           1024/3                 -/GE0/0/0
1.1.1.1/32           NULL/1025              -/GE0/0/0
1.1.1.1/32           1025/1025              -/GE0/0/0
```

通过以上输出可以看出，ASBR 会为所有的私网路由分配标签，即在 ASBR 之间使用 MP-EBGP 建立了一条私网 LSP 隧道。

查看 CE1 和 CE3 是否能学习到对端站点的路由信息：

```
[CE1]display ip routing-table
Route Flags: R - relay, D - download to fib
------------------------------------------------------------------------------
Routing Tables: Public
         Destinations : 10        Routes : 10
Destination/Mask    Proto    Pre   Cost   Flags   NextHop      Interface
       7.7.7.7/32   Direct   0     0      D       127.0.0.1    LoopBack0
       8.8.8.8/32   OSPF     10    2      D       17.1.1.1     GigabitEthernet0/0/0
      17.1.1.0/24   Direct   0     0      D       17.1.1.7     GigabitEthernet0/0/0
      17.1.1.7/32   Direct   0     0      D       127.0.0.1    GigabitEthernet0/0/0
    17.1.1.255/32   Direct   0     0      D       127.0.0.1    GigabitEthernet0/0/0
      28.1.1.0/24   O_ASE    150   1      D       17.1.1.1     GigabitEthernet0/0/0
[CE3]display  ip routing-table
Route Flags: R - relay, D - download to fib
------------------------------------------------------------------------------
Routing Tables: Public
         Destinations : 9         Routes : 9
Destination/Mask    Proto    Pre   Cost   Flags   NextHop      Interface
       9.9.9.9/32   Direct   0     0      D       127.0.0.1    LoopBack0
   10.10.10.10/32   EBGP     255   0      D       19.1.1.1     GigabitEthernet0/0/0
      19.1.1.0/24   Direct   0     0      D       19.1.1.9     GigabitEthernet0/0/0
      19.1.1.9/32   Direct   0     0      D       127.0.0.1    GigabitEthernet0/0/0
    19.1.1.255/32   Direct   0     0      D       127.0.0.1    GigabitEthernet0/0/0
```

步骤 5： 测试。

```
[CE1]ping 8.8.8.8
  PING 8.8.8.8: 56  data bytes, press CTRL_C to break
    Reply from 8.8.8.8: bytes=56 Sequence=1 ttl=249 time=60 ms
    Reply from 8.8.8.8: bytes=56 Sequence=2 ttl=249 time=50 ms
    Reply from 8.8.8.8: bytes=56 Sequence=3 ttl=249 time=60 ms
    Reply from 8.8.8.8: bytes=56 Sequence=4 ttl=249 time=50 ms
    Reply from 8.8.8.8: bytes=56 Sequence=5 ttl=249 time=50 ms
  --- 8.8.8.8 ping statistics ---
```

```
    5 packet(s) transmitted
    5 packet(s) received
    0.00% packet loss
round-trip min/avg/max = 50/54/60 ms
[CE3]ping -a 9.9.9.9 10.10.10.10
  PING 10.10.10.10: 56  data bytes, press CTRL_C to break
    Reply from 10.10.10.10: bytes=56 Sequence=1 ttl=249 time=60 ms
    Reply from 10.10.10.10: bytes=56 Sequence=2 ttl=249 time=60 ms
    Reply from 10.10.10.10: bytes=56 Sequence=3 ttl=249 time=50 ms
    Reply from 10.10.10.10: bytes=56 Sequence=4 ttl=249 time=60 ms
    Reply from 10.10.10.10: bytes=56 Sequence=5 ttl=249 time=50 ms
  --- 10.10.10.10 ping statistics ---
    5 packet(s) transmitted
    5 packet(s) received
    0.00% packet loss
round-trip min/avg/max = 50/56/60 ms
```

使用 CE3 ping CE4。PE1 的 G0/0/1 接口、ASBR1 的 G0/0/1 接口、ASBR2 的 G0/0/1 接口的抓包结果如图 7-16~ 图 7-18 所示。

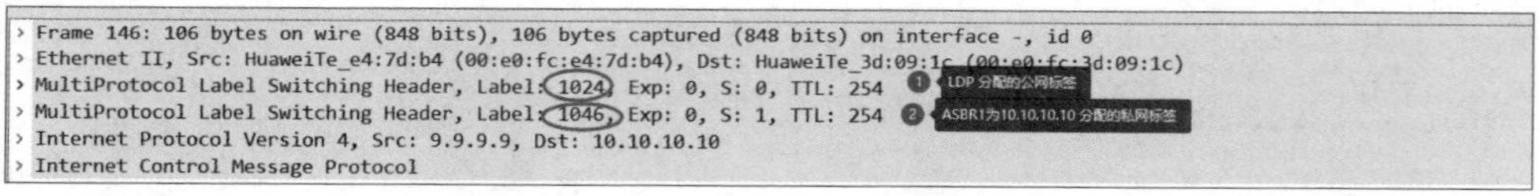

```
> Frame 146: 106 bytes on wire (848 bits), 106 bytes captured (848 bits) on interface -, id 0
> Ethernet II, Src: HuaweiTe_e4:7d:b4 (00:e0:fc:e4:7d:b4), Dst: HuaweiTe_3d:09:1c (00:e0:fc:3d:09:1c)
> MultiProtocol Label Switching Header, Label: 1024, Exp: 0, S: 0, TTL: 254   ❶ LDP 分配的公网标签
> MultiProtocol Label Switching Header, Label: 1046, Exp: 0, S: 1, TTL: 254   ❷ ASBR1为10.10.10.10 分配的私网标签
> Internet Protocol Version 4, Src: 9.9.9.9, Dst: 10.10.10.10
> Internet Control Message Protocol
```

图 7-16　PE1 的 G0/0/1 接口的抓包结果

```
> Frame 5: 102 bytes on wire (816 bits), 102 bytes captured (816 bits) on interface -, id 0
> Ethernet II, Src: HuaweiTe_99:36:08 (00:e0:fc:99:36:08), Dst: HuaweiTe_60:0c:fd (00:e0:fc:60:0c:fd)
> MultiProtocol Label Switching Header, Label: 1049, Exp: 0, S: 1, TTL: 252   ❶ ASBR2为10.10.10.10/32 给ASBR1分配的私网标签，数据转发过程并不携带公网标签。
> Internet Protocol Version 4, Src: 9.9.9.9, Dst: 10.10.10.10
> Internet Control Message Protocol
```

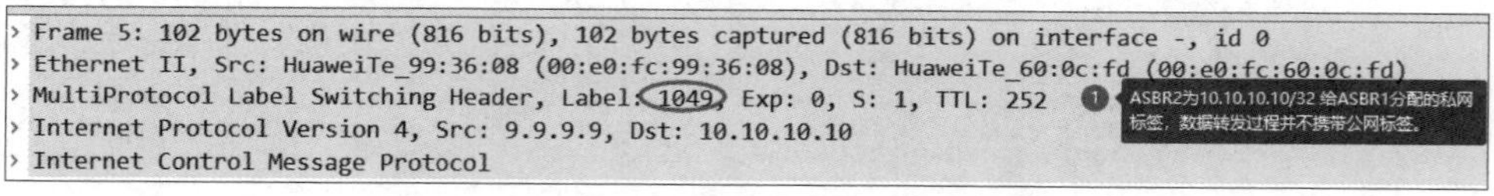

图 7-17　ASBR1 的 G0/0/1 接口的抓包结果

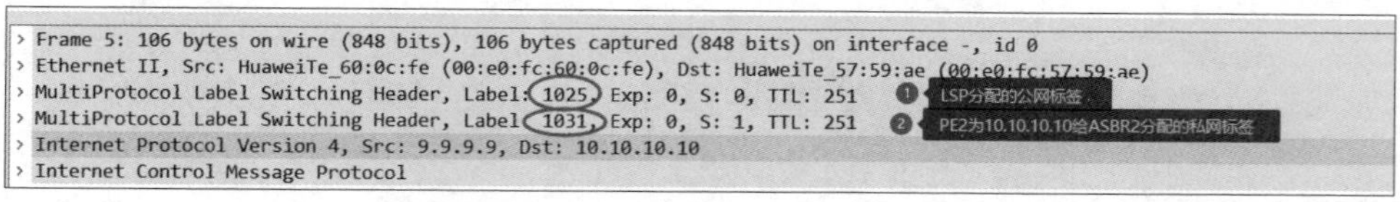

```
> Frame 5: 106 bytes on wire (848 bits), 106 bytes captured (848 bits) on interface -, id 0
> Ethernet II, Src: HuaweiTe_60:0c:fe (00:e0:fc:60:0c:fe), Dst: HuaweiTe_57:59:ae (00:e0:fc:57:59:ae)
> MultiProtocol Label Switching Header, Label: 1025, Exp: 0, S: 0, TTL: 251   ❶ LSP分配的公网标签
> MultiProtocol Label Switching Header, Label 1031, Exp: 0, S: 1, TTL: 251   ❷ PE2为10.10.10.10给ASBR2分配的私网标签
> Internet Protocol Version 4, Src: 9.9.9.9, Dst: 10.10.10.10
> Internet Control Message Protocol
```

图 7-18　ASBR2 的 G0/0/1 接口的抓包结果

通过以上输出结果可以看出，在数据转发的过程中，流量在 AS 内部使用的是由 LDP 建立的 LSP 隧道进行转发，而在 AS 之间依靠的是由 MP-EBGP 建立的私网隧道进行转发。

OptionB 方式的优缺点如下。

（1）优点：ASBR 之间不受物理链路的限制。所有流量都要经过 ASBR，流量具有很好的可控性，但是 ASBR 的负担较重，可在 ASBR 上针对 RT 进行过滤，保留部分路由。

（2）缺点：VPNv4 的路由都是通过 ASBR 来保存及传递的，ASBR 需要存储多个实例的路由以及占用标签空间。

7.2.3　配置跨域 OptionC 方式一（RR 场景）

1. 实验目的

AS 100 和 AS 200 为不同运营商的网络，运营商网络内部运行 OSPF 协议。使用 MPLS 跨域 OptionC 方式一组网（RR 场景）实现公司 A 互通、公司 B 互通。

2. 实验拓扑

配置跨域 OptionC 方式一（RR 场景）的实验拓扑如图 7–19 所示。

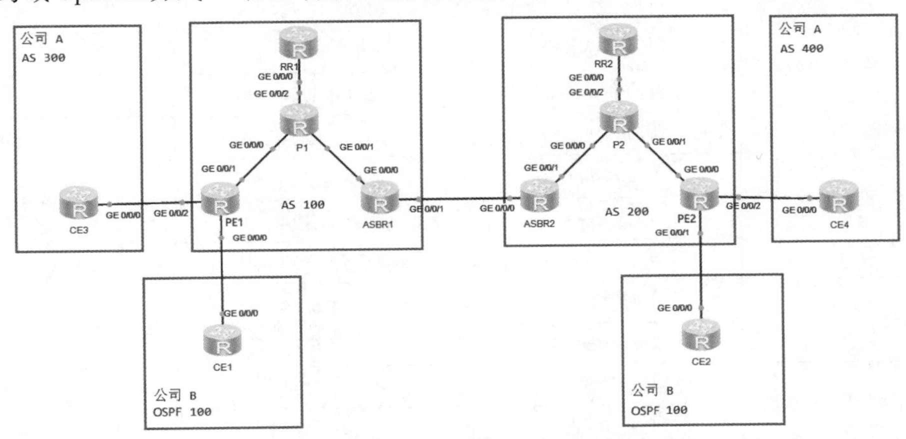

图 7–19　配置跨域 OptionC 方式一（RR 场景）的实验拓扑

3. 实验步骤

步骤 1： 配置 IP 地址，配置跨域 OptionC 方式一（RR 场景）IP 地址规划表见表 7–3。

表 7–3　配置跨域 OptionC 方式一（RR 场景）IP 地址规划表

设备名称	接口编号	IP 地址	所属 VPN 实例
PE1	G0/0/0	17.1.1.1/24	B
	G0/0/1	12.1.1.1/24	
	G0/0/2	19.1.1.1/24	A
	LoopBack 0	1.1.1.1/32	
P1	G0/0/0	12.1.1.2/24	
	G0/0/1	23.1.1.1/24	
	G0/0/2	112.1.1.1/24	
	LoopBack 0	2.2.2.2/24	
ASBR1	G0/0/0	23.1.1.2/24	
	G0/0/1	34.1.1.1/24	
	LoopBack 0	3.3.3.3/32	
PE2	G0/0/0	56.1.1.2/24	
	G0/0/1	28.1.1.2/24	B

设备名称	接口编号	IP 地址	所属 VPN 实例
PE2	G0/0/2	210.1.1.2/24	A
	LoopBack 0	6.6.6.6/32	
P2	G0/0/0	45.1.1.2/24	
	G0/0/1	56.1.1.1/24	
	G0/0/2	112.1.1.1/24	
	LoopBack 0	5.5.5.5/24	
ASBR2	G0/0/1	45.1.1.1/24	
	G0/0/0	34.1.1.2/24	
	LoopBack 0	4.4.4.4/32	
CE1	G0/0/0	17.1.1.7/24	
	LoopBack 0	7.7.7.7/32	
CE2	G0/0/0	28.1.1.8/24	
	LoopBack 0	8.8.8.8/32	
CE3	G0/0/0	19.1.1.9/24	
	LoopBack 0	9.9.9.9/32	
CE4	G0/0/0	210.1.1.0/24	
	LoopBack 0	10.10.10.10/32	
RR1	G0/0/0	112.1.1.12/24	
	LoopBack 0	12.12.12.12/32	
RR2	G0/0/0	213.1.1.13/24	
	LoopBack 0	13.13.13.13/32	

步骤 2: 配置运营商网络的 IGP、MPLS 及 MPLS LDP 协议，RR 无须运行 MPLS 以及 MPLS LDP。

（1）配置运营商网络的 IGP。

PE1 的配置：

```
[PE1]ospf 1
[PE1-ospf-1]area 0
[PE1-ospf-1-area-0.0.0.0]network 1.1.1.1 0.0.0.0
[PE1-ospf-1-area-0.0.0.0]network 12.1.1.0 0.0.0.255
```

P1 的配置：

```
[P1]ospf 1
[P1-ospf-1]area 0
[P1-ospf-1-area-0.0.0.0]network 2.2.2.2 0.0.0.0
[P1-ospf-1-area-0.0.0.0]network 12.1.1.0 0.0.0.255
[P1-ospf-1-area-0.0.0.0]network 23.1.1.0 0.0.0.255
[P1-ospf-1-area-0.0.0.0]network 112.1.1.0 0.0.0.255
```

ASBR1 的配置：

```
[ASBR1]ospf 1
[ASBR1-ospf-1]area 0
[ASBR1-ospf-1-area-0.0.0.0]network 3.3.3.3 0.0.0.0
[ASBR1-ospf-1-area-0.0.0.0]network 23.1.1.0 0.0.0.255
```

RR1 的配置：

```
[RR1]ospf
[RR1-ospf-1]area  0
[RR1-ospf-1-area-0.0.0.0]network 112.1.1.0 0.0.0.255
[RR1-ospf-1-area-0.0.0.0]network 12.12.12.12 0.0.0.0
```

ASBR2 的配置：

```
[ASBR2]ospf 1
[ASBR2-ospf-1]area 0
[ASBR2-ospf-1-area-0.0.0.0]network 4.4.4.4 0.0.0.0
[ASBR2-ospf-1-area-0.0.0.0]network 45.1.1.0 0.0.0.255
```

P2 的配置：

```
[P2]ospf 1
[P2-ospf-1]area 0
[P2-ospf-1-area-0.0.0.0]network 5.5.5.5 0.0.0.0
[P2-ospf-1-area-0.0.0.0]network 45.1.1.0 0.0.0.255
[P2-ospf-1-area-0.0.0.0]network 56.1.1.0 0.0.0.255
[P2-ospf-1-area-0.0.0.0]network 213.1.1.0 0.0.0.255
```

PE2 的配置：

```
[PE2]ospf 1
[PE2-ospf-1]area 0.0.0.0
[PE2-ospf-1-area-0.0.0.0]network 6.6.6.6 0.0.0.0
[PE2-ospf-1-area-0.0.0.0]network 56.1.1.0 0.0.0.255
```

RR2 的配置：

```
[RR2]ospf
[RR2-ospf-1]area  0
[RR2-ospf-1-area-0.0.0.0]network  213.1.1.0 0.0.0.255
[RR2-ospf-1-area-0.0.0.0]network  13.13.13.13 0.0.0.0
```

（2）配置运营商网络的 MPLS 、MPLS LDP 协议，建立公网隧道。

PE1 的配置：

```
[PE1]mpls lsr-id 1.1.1.1
[PE1]mpls
[PE1-mpls]mpls ldp
[PE1]interface G0/0/1
[PE1-GigabitEthernet0/0/1]mpls
[PE1-GigabitEthernet0/0/1]mpls ldp
```

P1 的配置：

```
[P1]mpls lsr-id 2.2.2.2
```

```
[P1]mpls
[P1-mpls]mpls ldp
[P1]interface G0/0/0
[P1-GigabitEthernet0/0/0]mpls
[P1-GigabitEthernet0/0/0]mpls ldp
[P1]interface G0/0/1
[P1-GigabitEthernet0/0/1]mpls
[P1-GigabitEthernet0/0/1]mpls ldp
```

ASBR1 的配置：

```
[ASBR1]mpls lsr-id 3.3.3.3
[ASBR1]mpls
[ASBR1-mpls]mpls ldp
[ASBR1]interface G0/0/0
[ASBR1-GigabitEthernet0/0/0]mpls
[ASBR1-GigabitEthernet0/0/0]mpls ldp
```

PE2 的配置：

```
[PE2]mpls lsr-id 6.6.6.6
[PE2]mpls
[PE2-mpls]mpls ldp
[PE2]interface G0/0/0
[PE2-GigabitEthernet0/0/0]mpls
[PE2-GigabitEthernet0/0/0]mpls ldp
```

P2 的配置：

```
[P2]mpls lsr-id 5.5.5.5
[P2]mpls
[P2-mpls]mpls ldp
[P2]interface G0/0/0
[P2-GigabitEthernet0/0/0]mpls
[P2-GigabitEthernet0/0/0]mpls ldp
[P2]interface G0/0/1
[P2-GigabitEthernet0/0/1]mpls
[P2-GigabitEthernet0/0/1]mpls ldp
```

ASBR2 的配置：

```
[ASBR2]mpls lsr-id 4.4.4.4
[ASBR2]mpls
[ASBR2-mpls]mpls ldp
[ASBR2]interface G0/0/1
[ASBR2-GigabitEthernet0/0/1]mpls
[ASBR2-GigabitEthernet0/0/1]mpls ldp
```

（3）查看 AS 100 和 AS 200 通过 MPLS 建立的 LSP。

查看 PE1 的 LSP：

```
<PE1>display mpls lsp
-------------------------------------------------------------------------------
```

```
                 LSP Information: LDP LSP
-------------------------------------------------------------------------------
FEC                 In/Out Label            In/Out IF               Vrf Name
2.2.2.2/32          NULL/3                  -/GE0/0/1
2.2.2.2/32          1024/3                  -/GE0/0/1
3.3.3.3/32          NULL/1024               -/GE0/0/1
3.3.3.3/32          1025/1024               -/GE0/0/1
1.1.1.1/32          3/NULL                  -/-
```

查看 PE2 的 LSP：

```
<PE2>display mpls lsp
-------------------------------------------------------------------------------
                 LSP Information: LDP LSP
-------------------------------------------------------------------------------
FEC                 In/Out Label            In/Out IF               Vrf Name
4.4.4.4/32          NULL/1024               -/GE0/0/0
4.4.4.4/32          1024/1024               -/GE0/0/0
5.5.5.5/32          NULL/3                  -/GE0/0/0
5.5.5.5/32          1025/3                  -/GE0/0/0
6.6.6.6/32          3/NULL                  -/-
```

通过以上输出可以看出，AS 100 和 AS 200 的公网隧道已经建立完毕。

步骤 3： 配置 PE 和 CE 之间的路由协议。

（1）配置 PE 设备的 VPN 实例。

PE1 的配置：

```
[PE1]ip vpn-instance A
[PE1-vpn-instance-A]ipv4-family
[PE1-vpn-instance-A-af-ipv4]route-distinguisher 100:1
[PE1-vpn-instance-A-af-ipv4]vpn-target 100:1 export-extcommunity
[PE1-vpn-instance-A-af-ipv4]vpn-target 100:1 import-extcommunity
[PE1]ip vpn-instance B
[PE1-vpn-instance-B]ipv4-family
[PE1-vpn-instance-B-af-ipv4]route-distinguisher 200:1
[PE1-vpn-instance-B-af-ipv4]vpn-target 200:1 export-extcommunity
[PE1-vpn-instance-B-af-ipv4]vpn-target 200:1 import-extcommunity
```

PE2 的配置：

```
[PE2]ip vpn-instance A
[PE2-vpn-instance-A]ipv4-family
[PE2-vpn-instance-A-af-ipv4]route-distinguisher 100:1
[PE2-vpn-instance-A-af-ipv4]vpn-target 100:1 export-extcommunity
[PE2-vpn-instance-A-af-ipv4]vpn-target 100:1 import-extcommunity
[PE2]ip vpn-instance B
[PE2-vpn-instance-B]ipv4-family
[PE2-vpn-instance-B-af-ipv4]route-distinguisher 200:1
[PE2-vpn-instance-B-af-ipv4]vpn-target 200:1 export-extcommunity
[PE2-vpn-instance-B-af-ipv4]vpn-target 200:1 import-extcommunity
```

（2）将对应的接口加入到 VPN 实例。

PE1 的配置：

```
[PE1]interface GigabitEthernet0/0/0
[PE1-GigabitEthernet0/0/0]ip binding vpn-instance B
[PE1-GigabitEthernet0/0/0]ip address 17.1.1.1 255.255.255.0
[PE1]interface GigabitEthernet0/0/2
[PE1-GigabitEthernet0/0/2]ip binding vpn-instance A
[PE1-GigabitEthernet0/0/2]ip address 19.1.1.1 255.255.255.0
```

PE2 的配置：

```
[PE2]interface GigabitEthernet0/0/1
[PE2-GigabitEthernet0/0/1]ip binding vpn-instance B
[PE2-GigabitEthernet0/0/1]ip address 28.1.1.2 255.255.255.0
[PE2]interface GigabitEthernet0/0/2
[PE2-GigabitEthernet0/0/2]ip binding vpn-instance A
[PE2-GigabitEthernet0/0/2]ip address 210.1.1.2 255.255.255.0
```

（3）配置 PE1 和 CE1 之间的路由协议。

①配置 PE1 和 CE1 的 OSPF 协议。

PE1 的配置：

```
[PE1]ospf 100 vpn-instance B
[PE1-ospf-100]area 0
[PE1-ospf-100-area-0.0.0.0]network 28.1.1.0 0.0.0.255
```

CE1 的配置：

```
[CE1]ospf 1
[CE1-ospf-1]area 0
[CE1-ospf-1-area-0.0.0.0]network 7.7.7.7 0.0.0.0
[CE1-ospf-1-area-0.0.0.0]network 17.1.1.0 0.0.0.255
```

查看 PE1 的 OSPF 邻居关系：

```
[PE1]display ospf 100 peer brief
        OSPF Process 100 with Router ID 17.1.1.1
                 Peer Statistic Information
 ----------------------------------------------------------------------------
 Area Id          Interface                       Neighbor id       State
 0.0.0.0          GigabitEthernet0/0/0            17.1.1.7          Full
 ----------------------------------------------------------------------------
```

查看 PE1 的 VPN 实例 B 的路由表：

```
[PE1]display ip routing-table vpn-instance B
Route Flags: R - relay, D - download to fib
------------------------------------------------------------------------------
Routing Tables: B
        Destinations : 5        Routes : 5
Destination/Mask    Proto   Pre  Cost  Flags   NextHop      Interface
       7.7.7.7/32   OSPF    10   1     D       17.1.1.7     GigabitEthernet0/0/0
      17.1.1.0/24   Direct  0    0     D       17.1.1.1     GigabitEthernet0/0/0
```

```
        17.1.1.1/32    Direct  0    0    D       127.0.0.1      GigabitEthernet0/0/0
      17.1.1.255/32    Direct  0    0    D       127.0.0.1      GigabitEthernet0/0/0
255.255.255.255/32     Direct  0    0    D       127.0.0.1      InLoopBack0
```

通过以上输出可以看出，在 PE1 的 VPN 实例 B 上学习到了 CE1 的 7.7.7.7/32 的路由。

②配置 PE1 和 CE3 的 BGP。

PE1 的配置：

```
[PE1]bgp 100
[PE1-bgp]ipv4-family vpn-instance A
[PE1-bgp-A]peer 19.1.1.9 as-number 300
```

CE3 的配置：

```
[CE3]bgp 300
[CE3-bgp]peer 19.1.1.1 as-number 100
[CE3-bgp]network 9.9.9.9 255.255.255.255
```

查看 PE1 的 VPNv4 路由表：

```
[PE1]display bgp vpnv4 vpn-instance A routing-table
 BGP Local router ID is 12.1.1.1
 Status codes: * - valid, > - best, d - damped,
               h - history,  i - internal, s - suppressed, S - Stale
               Origin : i - IGP, e - EGP, ? - incomplete
 VPN-Instance A, Router ID 12.1.1.1:
 Total Number of Routes: 1
        Network            NextHop        MED      LocPrf      PrefVal      Path/Ogn
*>      9.9.9.9/32         19.1.1.9       0                    0            300i
```

通过以上输出可以看出，PE1 学习到了 CE3 的 9.9.9.9/32 的路由。

③配置 PE2 和 CE2 的 OSPF 协议。

PE2 的配置：

```
[PE2]ospf 100 vpn-instance B
[PE2-ospf-100]area 0
[PE2-ospf-100-area-0.0.0.0]network 28.1.1.0 0.0.0.255
```

CE2 的配置：

```
[CE2]ospf 1
[CE2-ospf-1]area 0
[CE2-ospf-1-area-0.0.0.0]network 8.8.8.8 0.0.0.0
[CE2-ospf-1-area-0.0.0.0]network 28.1.1.0 0.0.0.255
```

查看 PE2 的 VPN 实例 B 的路由表：

```
[PE2]display  ip routing-table vpn-instance B
Route Flags: R - relay, D - download to fib
------------------------------------------------------------------------------
Routing Tables: B
         Destinations : 5        Routes : 5
Destination/Mask    Proto   Pre  Cost  Flags    NextHop        Interface
      8.8.8.8/32    OSPF    10   1     D        28.1.1.8       GigabitEthernet0/0/1
     28.1.1.0/24    Direct  0    0     D        28.1.1.2       GigabitEthernet0/0/1
```

```
      28.1.1.2/32    Direct   0     0     D        127.0.0.1     GigabitEthernet0/0/1
     28.1.1.255/32   Direct   0     0     D        127.0.0.1     GigabitEthernet0/0/1
255.255.255.255/32   Direct   0     0     D        127.0.0.1     InLoopBack0
```

通过以上输出可以看出，PE2 学习到了 CE2 的 8.8.8.8/32 的路由。

④ PE2 和 CE4 的 BGP 协议。

PE2 的配置：

```
[PE2]bgp 200
[PE2-bgp]ipv4-family vpn-instance A
[PE2-bgp-A]peer 210.1.1.10 as-number 400
```

CE4 的配置：

```
[CE4]bgp 400
[CE4-bgp]peer 210.1.1.2 as-number 200
[CE4-bgp]network 10.10.10.10 255.255.255.255
```

查看 PE2 的 VPNv4 路由表：

```
[PE2]display  bgp vpnv4 vpn-instance A routing-table
 BGP Local router ID is 56.1.1.2
 Status codes: * - valid, > - best, d - damped,
               h - history,  i - internal, s - suppressed, S - Stale
               Origin : i - IGP, e - EGP, ? - incomplete
 VPN-Instance A, Router ID 56.1.1.2:
 Total Number of Routes: 1
        Network           NextHop      MED     LocPrf        PrefVal      Path/Ogn
*>      10.10.10.10/32    210.1.1.10   0                     0            400i
```

通过以上输出可以看出，PE2 学习到了 CE4 的路由。

将 PE1 的 VPN 实例 B 的 OSPF 路由和 BGP 路由做双向引入，由于 VPN 实例 B 全部运行在 BGP 中，因此无须引入。

PE1 的配置：

```
[PE1]ospf 100 vpn-instance B
[PE1-ospf-100]import-route bgp
[PE1]bgp 100
[PE1-bgp]ipv4-family vpn-instance B
[PE1-bgp-B]import-route ospf 100
```

PE2 的配置：

```
[PE2]ospf 100 vpn-instance B
[PE2-ospf-100]import-route bgp
[PE2]bgp 200
[PE2-bgp]ipv4-family vpn-instance B
[PE2-bgp-B]import-route ospf 100
```

步骤 4：配置 AS 100 和 AS 200 之间的 BGP 邻居关系。此步骤的目的是让 PE1 和 PE2 学习到对端的环回口路由，PE1 和 PE2 的环回口路由作为 VPNv4 的下一跳，防止下一跳不可达。RR1 和 RR2 学习到对端的环回口路由（RR1 和 RR2 后续需要建立 MP-BGP 的邻居关系，需要环回口地址可达才能建立 TCP 连接）。

AS 100 内的 BGP 邻居关系：RR1 和 PE1、P1、ASBR1 建立 IBGP 邻居关系，RR1 作为反射器。

AS 200 内的 BGP 邻居关系：RR2 和 PE2、P2、ASBR2 建立 IBGP 邻居关系，RR1 作为反射器。

AS 100 和 AS 200 之间的邻居关系：ASBR1 和 ASBR2 建立 EBGP 邻居关系。

（1）按需求配置 BGP 邻居关系。

PE1 的配置：

```
[PE1]bgp 100
[PE1-bgp]peer 12.12.12.12 as-number 100
[PE1-bgp]peer 12.12.12.12 connect-interface LoopBack0
```

P1 的配置：

```
[P1]bgp 100
[P1-bgp]peer 12.12.12.12 as-number 100
[P1-bgp]peer 12.12.12.12 connect-interface LoopBack0
```

ASBR1 的配置：

```
[ASBR1]bgp 100
[ASBR1-bgp]peer 12.12.12.12 as-number 100
[ASBR1-bgp]peer 12.12.12.12 connect-interface LoopBack0
[ASBR1-bgp]peer  12.12.12.12 next-hop-local
[ASBR1-bgp]peer  34.1.1.2 as-number 200
```

RR1 的配置：

```
[RR1]bgp 100
[RR1-bgp]peer 1.1.1.1 as-number 100
[RR1-bgp]peer 1.1.1.1 connect-interface LoopBack0
[RR1-bgp]peer 1.1.1.1 reflect-client
[RR1-bgp]peer 2.2.2.2 as-number 100
[RR1-bgp]peer 2.2.2.2 connect-interface LoopBack0
[RR1-bgp]peer 2.2.2.2 reflect-client
[RR1-bgp]peer 3.3.3.3 as-number 100
[RR1-bgp]peer 3.3.3.3 connect-interface LoopBack0
[RR1-bgp]peer 3.3.3.3 reflect-client
```

PE2 的配置：

```
[PE2]bgp 200
[PE2-bgp]peer 13.13.13.13 as-number 200
[PE2-bgp]peer 13.13.13.13 connect-interface LoopBack0
```

P2 的配置：

```
[P2]bgp 200
[P2-bgp]peer 13.13.13.13 as-number 200
[P2-bgp]peer 13.13.13.13 connect-interface LoopBack0
```

ASBR2 的配置：

```
[ASBR2]bgp 200
[ASBR2-bgp]peer 13.13.13.13 as-number 200
[ASBR2-bgp]peer 13.13.13.13 connect-interface LoopBack0
[ASBR2-bgp]peer  13.13.13.13 next-hop-local
```

```
[ASBR2-bgp]peer 34.1.1.1 as-number 100
```

RR2 的配置：

```
[RR2]bgp 200
[RR2-bgp]peer 4.4.4.4 as-number 200
[RR2-bgp]peer 4.4.4.4 connect-interface LoopBack0
[RR2-bgp]peer 4.4.4.4 reflect-client
[RR2-bgp]peer 5.5.5.5 as-number 200
[RR2-bgp]peer 5.5.5.5 connect-interface LoopBack0
[RR2-bgp]peer 5.5.5.5 reflect-client
[RR2-bgp]peer 6.6.6.6 as-number 200
[RR2-bgp]peer 6.6.6.6 connect-interface LoopBack0
[RR2-bgp]peer 6.6.6.6 reflect-client
```

查看 RR1 和 RR2 的邻居关系的建立情况：

```
[RR1]display bgp peer
 BGP local router ID : 112.1.1.12
 Local AS number : 100
 Total number of peers : 3                Peers in established state : 3
  Peer         V    AS    MsgRcvd    MsgSent   OutQ    Up/Down     State        PrefRcv
  1.1.1.1      4    100   5          5         0       00:03:43    Established  0
  2.2.2.2      4    100   5          5         0       00:03:38    Established  0
  3.3.3.3      4    100   5          5         0       00:03:32    Established  0
[RR2]display bgp peer
 BGP local router ID : 213.1.1.13
 Local AS number : 200
 Total number of peers : 3                Peers in established state : 3

  Peer         V    AS    MsgRcvd    MsgSent   OutQ    Up/Down     State        PrefRcv

  4.4.4.4      4    200   2          2         0       00:00:38    Established  0
  5.5.5.5      4    200   2          2         0       00:00:29    Established  0
  6.6.6.6      4    200   2          2         0       00:00:15    Established  0
[ASBR1]display bgp peer
 BGP local router ID : 23.1.1.2
 Local AS number : 100
 Total number of peers : 2                Peers in established state : 2
  Peer         V    AS    MsgRcvd    MsgSent   OutQ    Up/Down     State        PrefRcv
  12.12.12.12 4    100   6          7         0       00:04:19    Established  0
  34.1.1.2     4    200   4          6         0       00:02:22    Established  0
[ASBR2]display bgp peer
 BGP local router ID : 34.1.1.2
 Local AS number : 200
 Total number of peers : 2                Peers in established state : 2
  Peer         V    AS    MsgRcvd    MsgSent   OutQ    Up/Down     State        PrefRcv
  13.13.13.13 4    200   4          5         0       00:02:07    Established  0
  34.1.1.1     4    100   4          4         0       00:02:38    Established  0
```

通过以上输出可以看出，邻居关系已经按照需求建立好了。

（2）分别在 ASBR 设备宣告本 AS 的 RR 和 PE 设备的环回口路由。

ASBR1 的配置：

```
[ASBR1]bgp 100
[ASBR1-bgp]network 1.1.1.1 255.255.255.255
[ASBR1-bgp]network 12.12.12.12 255.255.255.255
```

ASBR2 的配置：

```
[ASBR2]bgp 200
[ASBR2-bgp]network 6.6.6.6 255.255.255.255
[ASBR2-bgp]network 13.13.13.13 255.255.255.255
```

查看 PE1、PE2 和 ASBR 设备是否能学习到 BGP 路由表：

```
[PE1]display bgp routing-table
 BGP Local router ID is 12.1.1.1
 Status codes: * - valid, > - best, d - damped,
               h - history,  i - internal, s - suppressed, S - Stale
               Origin : i - IGP, e - EGP, ? - incomplete
 Total Number of Routes: 4
        Network          NextHop      MED      LocPrf       PrefVal      Path/Ogn
*>i     1.1.1.1/32       3.3.3.3      2        100          0            i
*>i     6.6.6.6/32       3.3.3.3      2        100          0            200i
*>i     12.12.12.12/32   3.3.3.3      2        100          0            i
*>i     13.13.13.13/32   3.3.3.3      2        100          0            200i
[PE2]display  bgp routing-table
 BGP Local router ID is 56.1.1.2
 Status codes: * - valid, > - best, d - damped,
               h - history,  i - internal, s - suppressed, S - Stale
               Origin : i - IGP, e - EGP, ? - incomplete
 Total Number of Routes: 4
        Network          NextHop      MED      LocPrf       PrefVal      Path/Ogn
*>i     1.1.1.1/32       4.4.4.4      2        100          0            100i
*>i     6.6.6.6/32       4.4.4.4      2        100          0            i
*>i     12.12.12.12/32   4.4.4.4      2        100          0            100i
*>i     13.13.13.13/32   4.4.4.4      2        100          0            i
```

通过以上输出可以看出，AS 100 和 AS 200 内的设备已经可以学习到对应的 BGP 路由。

步骤 5：建立 AS 100 和 AS 200 的 MP-BGP 邻居关系。

AS 100 内的 PE1 和 RR1 建立 MP-IBGP 邻居关系。

AS 200 内的 PE2 和 RR2 建立 MP-IBGP 邻居关系。

RR1 和 RR2 建立 MP-EBGP 邻居关系。

PE1 的配置：

```
[PE1]bgp 100
[PE1-bgp]ipv4-family vpnv4
[PE1-bgp-af-vpnv4]peer 12.12.12.12 enable  //使能与 RR1 的 VPNv4 邻居关系
```

RR1 的配置：

```
[RR1]bgp 100
```

```
[RR1-bgp]peer 13.13.13.13 as-number 200
Error: The peer already exists in AS 200.
[RR1-bgp]peer 13.13.13.13 ebgp-max-hop 10   //配置EBGP邻居的多跳
[RR1-bgp]peer 13.13.13.13 connect-interface LoopBack0
[RR1-bgp]ipv4-family vpnv4
[RR1-bgp-af-vpnv4]undo policy vpn-target    //关闭RT的检测
[RR1-bgp-af-vpnv4]peer 1.1.1.1 enable
[RR1-bgp-af-vpnv4]peer 1.1.1.1 next-hop-invariable //传递VPNv4路由时下一跳保持不变
[RR1-bgp-af-vpnv4]peer 13.13.13.13 enable
[RR1-bgp-af-vpnv4]peer 13.13.13.13 next-hop-invariable
```

PE2的配置：

```
[PE2]bgp 200
[PE2-bgp]ipv4-family vpnv4
[PE2-bgp-af-vpnv4]peer 13.13.13.13 enable
```

RR2的配置：

```
[RR2]bgp 200
[RR2-bgp]peer 12.12.12.12 as-number 100
[RR2-bgp]peer 12.12.12.12 ebgp-max-hop 10
[RR2-bgp]peer 12.12.12.12 connect-interface LoopBack0
[RR2-bgp]ipv4-family vpnv4
[RR2-bgp-af-vpnv4]undo policy vpn-target
[RR2-bgp-af-vpnv4]peer 6.6.6.6 enable
[RR2-bgp-af-vpnv4]peer 6.6.6.6 next-hop-invariable
[RR2-bgp-af-vpnv4]peer 12.12.12.12 enable
[RR2-bgp-af-vpnv4]peer 12.12.12.12 next-hop-invariable
```

注意：

（1）由于RR之间非直连，因此RR之间建立MP-EBGP时需要配置EBGP邻居的多跳。

（2）RR与PE或RR建立邻居关系时，必须配置传递路由下一跳不变，由于后续隧道的建立是基于VPNv4路由的下一跳建立的，因此需要保证对端PE学习到的VPNv4路由的下一跳为本端PE的环回口地址。

（3）RR设备无须配置VPN实例，因此需要关闭RT检测。

查看MP-BGP的邻居建立情况：

```
[RR1]display bgp vpnv4 all peer
 BGP local router ID : 112.1.1.12
 Local AS number : 100
 Total number of peers : 2                 Peers in established state : 2
  Peer          V    AS   MsgRcvd    MsgSent    OutQ   Up/Down     State        PrefRcv
  1.1.1.1       4    100  24         28         0      00:18:55    Established  3
  13.13.13.13   4    200  24         25         0      00:14:58    Established  3

[RR2]display bgp vpnv4 all peer
 BGP local router ID : 213.1.1.13
 Local AS number : 200
```

```
 Total number of peers : 2                 Peers in established state : 2
  Peer          V    AS    MsgRcvd    MsgSent   OutQ    Up/Down     State          PrefRcv
  6.6.6.6       4    200   21         24        0       00:15:12    Established    3
  12.12.12.12 4      100   24         25        0       00:15:20    Established    3
```

通过以上输出可以看出，RR设备已经跟对端RR和本端PE建立好了MP-BGP邻居关系。

查看PE设备是否能学习到对端站点的VPNv4路由：

```
[PE1]display  bgp vpnv4 all  routing-table
 BGP Local router ID is 12.1.1.1
 Status codes: * - valid, > - best, d - damped,
               h - history,  i - internal, s - suppressed, S - Stale
               Origin : i - IGP, e - EGP, ? - incomplete
 Total number of routes from all PE: 6
 Route Distinguisher: 100:1
        Network          NextHop       MED     LocPrf       PrefVal      Path/Ogn
 *>     9.9.9.9/32       19.1.1.9      0                    0            300i
 *>i    10.10.10.10/32   6.6.6.6               100          0            200 400i

 Route Distinguisher: 200:1
        Network          NextHop       MED     LocPrf       PrefVal      Path/Ogn
 *>     7.7.7.7/32       0.0.0.0       2                    0            ?
 *>i    8.8.8.8/32       6.6.6.6               100          0            200?
 *>     17.1.1.0/24      0.0.0.0       0                    0            ?
 *>i    28.1.1.0/24      6.6.6.6               100          0            200?

 VPN-Instance A, Router ID 12.1.1.1:
Total Number of Routes: 2
        Network          NextHop       MED     LocPrf       PrefVal      Path/Ogn
 *>     9.9.9.9/32       19.1.1.9      0                    0            300i
 i      10.10.10.10/32   6.6.6.6               100          0            200 400i

 VPN-Instance B, Router ID 12.1.1.1:
 Total Number of Routes: 4
        Network          NextHop       MED     LocPrf       PrefVal      Path/Ogn
 *>     7.7.7.7/32       0.0.0.0       2                    0            ?
 i      8.8.8.8/32       6.6.6.6               100          0            200?
 *>     17.1.1.0/24      0.0.0.0       0                    0            ?
 i      28.1.1.0/24      6.6.6.6               100          0            200?
```

以上输出为PE1的VPNv4路由表，可以看到VPNv4路由表中已经接收到了对端站点的私网路由，但是VPN实例的路由表中并不认为对端的私网路由是有效路由，因此也不会将这些路由发送给CE设备。其原因是隧道还未建立。

步骤6：建立BGP隧道。

（1）为ASBR之间的物理链路开启MPLS功能。

ASBR1的配置：

```
[ASBR1]interface GigabitEthernet0/0/1
```

```
[ASBR1-GigabitEthernet0/0/1]mpls
```

ASBR2 的配置：

```
[ASBR2]interface GigabitEthernet0/0/0
[ASBR2-GigabitEthernet0/0/0]mpls
```

（2）创建标签分配的策略。创建两个策略，分别用于与 ASBR 建立邻居关系以及与 RR 建立邻居关系时使用。

ASBR1 的配置：

```
[ASBR1]route-policy asbr-asbr permit node 10
[ASBR1-route-policy]apply mpls-label    //策略 asbr-asbr 表示当发布 BGP 路由给 ASBR
                                        //设备时，给此路由分配 MPLS 标签
[ASBR1]route-policy asbr-RR permit node 10
[ASBR1-route-policy]if-match mpls-label
[ASBR1-route-policy]apply mpls-label    //策略 asbr-RR 表示当发布 BGP 路由给 RR 设备时，
                                        //如果此路由携带标签，那么给此路由分配 MPLS 标签
```

ASBR2 的配置：

```
[ASBR2]route-policy asbr-asbr permit node 10
[ASBR2-route-policy]apply mpls-label
[ASBR2-route-policy]route-policy asbr-RR permit node 10
[ASBR2-route-policy]if-match mpls-label
[ASBR2-route-policy]apply mpls-label
```

（3）ASBR 与 ASBR、RR 建立 BGP 邻居关系时应用策略，并且开启 ASBR 与 ASBR、RR 以及 RR 和 PE 之间的 BGP 标签交互能力。

ASBR1 的配置：

```
[ASBR1]bgp 100
[ASBR1-bgp]peer 12.12.12.12 route-policy asbr-RR export
[ASBR1-bgp]peer 12.12.12.12 label-route-capability //开启标签交互能力
[ASBR1-bgp]peer 34.1.1.2 route-policy asbr-asbr export
[ASBR1-bgp]peer 34.1.1.2 label-route-capability
```

ASBR2 的配置：

```
[ASBR2]bgp 200
[ASBR2-bgp]peer 13.13.13.13 route-policy asbr-RR export
[ASBR2-bgp]peer 13.13.13.13 label-route-capability
[ASBR2-bgp]peer 34.1.1.1 route-policy asbr-asbr export
[ASBR2-bgp]peer 34.1.1.1 label-route-capability
```

RR1 的配置：

```
[RR1]bgp 100
[RR1-bgp]peer 1.1.1.1 label-route-capability
[RR1-bgp]peer 3.3.3.3 label-route-capability
```

RR2 的配置：

```
[RR2]bgp 200
[RR2-bgp]peer 4.4.4.4 label-route-capability
[RR2-bgp]peer 6.6.6.6 label-route-capability
```

PE1 的配置：

```
[PE1]bgp 100
[PE1-bgp]peer 12.12.12.12 label-route-capability
```

PE2 的配置：

```
[PE2]bgp 200
[PE2-bgp]peer 13.13.13.13 label-route-capability
```

查看 LSP 的建立情况，以 6.6.6.6/32 为例。

```
[ASBR1]display mpls lsp
-------------------------------------------------------------------------------
                 LSP Information: BGP  LSP
-------------------------------------------------------------------------------
FEC                In/Out Label          In/Out IF             Vrf Name
12.12.12.12/32     1025/NULL             -/-
1.1.1.1/32         1027/NULL             -/-
13.13.13.13/32     NULL/1025             -/-
6.6.6.6/32         NULL/1027             -/-
6.6.6.6/32         1029/1027             -/-
13.13.13.13/32     1030/1025             -/-
-------------------------------------------------------------------------------
                 LSP Information: LDP LSP
-------------------------------------------------------------------------------
FEC                In/Out Label          In/Out IF             Vrf Name
3.3.3.3/32         3/NULL                -/-
2.2.2.2/32         NULL/3                -/GE0/0/0
2.2.2.2/32         1024/3                -/GE0/0/0
12.12.12.12/32     NULL/1025             -/GE0/0/0
12.12.12.12/32     1026/1025             -/GE0/0/0
1.1.1.1/32         NULL/1026             -/GE0/0/0
1.1.1.1/32         1028/1026             -/GE0/0/0

[PE1]display mpls lsp
-------------------------------------------------------------------------------
                 LSP Information: BGP  LSP
-------------------------------------------------------------------------------
FEC                In/Out Label          In/Out IF             Vrf Name
9.9.9.9/32         1027/NULL             -/-                       A
17.1.1.0/24        1028/NULL             -/-            B
7.7.7.7/32         1029/NULL             -/-            B
13.13.13.13/32     NULL/1030             -/-
6.6.6.6/32         NULL/1029             -/-
-------------------------------------------------------------------------------
                 LSP Information: LDP LSP
-------------------------------------------------------------------------------
FEC                In/Out Label          In/Out IF             Vrf Name
2.2.2.2/32         NULL/3                -/GE0/0/1
```

```
2.2.2.2/32          1024/3                  -/GE0/0/1
3.3.3.3/32          NULL/1024               -/GE0/0/1
3.3.3.3/32          1025/1024               -/GE0/0/1
12.12.12.12/32      NULL/1025               -/GE0/0/1
12.12.12.12/32      1026/1025               -/GE0/0/1
1.1.1.1/32          3/NULL                  -/-
```

通过以上输出可以看出，去往对端 VPNv4 路由的下一跳地址有对应的隧道，此时再次查看 PE1 的路由表，观察对端的私网 VPNv4 路由是否是有效路由。

```
[PE1]display bgp vpnv4 all routing-table
 BGP Local router ID is 12.1.1.1
 Status codes: * - valid, > - best, d - damped,
               h - history,  i - internal, s - suppressed, S - Stale
               Origin : i - IGP, e - EGP, ? - incomplete
 Total number of routes from all PE: 6
 Route Distinguisher: 100:1
        Network           NextHop       MED     LocPrf       PrefVal      Path/Ogn
*>      9.9.9.9/32        19.1.1.9      0                    0            300i
*>i     10.10.10.10/32    6.6.6.6               100          0            200 400i

 Route Distinguisher: 200:1
        Network           NextHop       MED     LocPrf       PrefVal      Path/Ogn
*>      7.7.7.7/32        0.0.0.0       2                    0            ?
*>i     8.8.8.8/32        6.6.6.6               100          0            200?
*>      17.1.1.0/24       0.0.0.0       0                    0            ?
*>i     28.1.1.0/24       6.6.6.6               100          0            200?

 VPN-Instance A, Router ID 12.1.1.1:
 Total Number of Routes: 2
        Network           NextHop       MED     LocPrf       PrefVal      Path/Ogn
*>      9.9.9.9/32        19.1.1.9      0                    0            300i
*>i     10.10.10.10/32    6.6.6.6               100          0            200 400i

 VPN-Instance B, Router ID 12.1.1.1:
 Total Number of Routes: 4
        Network           NextHop       MED     LocPrf       PrefVal      Path/Ogn
*>      7.7.7.7/32        0.0.0.0       2                    0            ?
*>i     8.8.8.8/32        6.6.6.6               100          0            200?
*>      17.1.1.0/24       0.0.0.0       0                    0            ?
*>i     28.1.1.0/24       6.6.6.6               100          0            200?
```

通过以上输出可以看出，下一跳为 6.6.6.6 的 VPNv4 路由被 VPN 实例所优选，并且会更新给对应的 CE 设备。

查看 CE1 和 CE3 的路由表：

```
[CE1]display ip routing-table
Route Flags: R - relay, D - download to fib
------------------------------------------------------------------------------
```

```
Routing Tables: Public
         Destinations : 10        Routes : 10
 Destination/Mask    Proto   Pre   Cost  Flags     NextHop        Interface
       7.7.7.7/32    Direct  0     0     D         127.0.0.1      LoopBack0
       8.8.8.8/32    OSPF    10    2     D         17.1.1.1       GigabitEthernet0/0/0
      17.1.1.0/24    Direct  0     0     D         17.1.1.7       GigabitEthernet0/0/0
      17.1.1.7/32    Direct  0     0     D         127.0.0.1      GigabitEthernet0/0/0
    17.1.1.255/32    Direct  0     0     D         127.0.0.1      GigabitEthernet0/0/0
      28.1.1.0/24    O_ASE   150   1     D         17.1.1.1       GigabitEthernet0/0/0
      127.0.0.0/8    Direct  0     0     D         127.0.0.1      InLoopBack0
     127.0.0.1/32    Direct  0     0     D         127.0.0.1      InLoopBack0
127.255.255.255/32   Direct  0     0     D         127.0.0.1      InLoopBack0
255.255.255.255/32   Direct  0     0     D         127.0.0.1      InLoopBack0

[CE3]display ip routing-table
Route Flags: R - relay, D - download to fib
------------------------------------------------------------------------------
Routing Tables: Public
         Destinations : 9         Routes : 9
Destination/MaskProto        Pre   Cost  Flags     NextHop        Interface
       9.9.9.9/32    Direct  0     0     D         127.0.0.1      LoopBack0
   10.10.10.10/32    EBGP    255   0     D         19.1.1.1       GigabitEthernet0/0/0
      19.1.1.0/24    Direct  0     0     D         19.1.1.9       GigabitEthernet0/0/0
      19.1.1.9/32    Direct  0     0     D         127.0.0.1      GigabitEthernet0/0/0
    19.1.1.255/32    Direct  0     0     D         127.0.0.1      GigabitEthernet0/0/0
      127.0.0.0/8    Direct  0     0     D         127.0.0.1      InLoopBack0
     127.0.0.1/32    Direct  0     0     D         127.0.0.1      InLoopBack0
127.255.255.255/32   Direct  0     0     D         127.0.0.1      InLoopBack0
255.255.255.255/32   Direct  0     0     D         127.0.0.1      InLoopBack0
```

通过以上输出可以看出，CE1 和 CE3 能够学习到对端站点的私网路由。

步骤 7：测试网络连通性，并且在 PE1 的 G0/0/1 接口抓包。

```
[CE1]ping 8.8.8.8
  PING 8.8.8.8: 56  data bytes, press CTRL_C to break
    Reply from 8.8.8.8: bytes=56 Sequence=1 ttl=249 time=60 ms
    Reply from 8.8.8.8: bytes=56 Sequence=2 ttl=249 time=70 ms
    Reply from 8.8.8.8: bytes=56 Sequence=3 ttl=249 time=60 ms
    Reply from 8.8.8.8: bytes=56 Sequence=4 ttl=249 time=50 ms
    Reply from 8.8.8.8: bytes=56 Sequence=5 ttl=249 time=50 ms
  --- 8.8.8.8 ping statistics ---
    5 packet(s) transmitted
    5 packet(s) received
    0.00% packet loss
round-trip min/avg/max = 50/58/70 ms

[CE3]ping -a 9.9.9.9 10.10.10.10
  PING 10.10.10.10: 56  data bytes, press CTRL_C to break
```

```
    Reply from 10.10.10.10: bytes=56 Sequence=1 ttl=249 time=50 ms
    Reply from 10.10.10.10: bytes=56 Sequence=2 ttl=249 time=50 ms
    Reply from 10.10.10.10: bytes=56 Sequence=3 ttl=249 time=60 ms
    Reply from 10.10.10.10: bytes=56 Sequence=4 ttl=249 time=60 ms
    Reply from 10.10.10.10: bytes=56 Sequence=5 ttl=249 time=50 ms
  --- 10.10.10.10 ping statistics ---
    5 packet(s) transmitted
    5 packet(s) received
    0.00% packet loss
  round-trip min/avg/max = 50/54/60 ms
```

如图 7-20 所示，有三层标签，至于这些标签在何时使用，如何使用，需要了解以下整个过程。

```
Frame 3032: 110 bytes on wire (880 bits), 110 bytes captured (880 bits) on interface -, id 0
Ethernet II, Src: HuaweiTe_e4:7d:b4 (00:e0:fc:e4:7d:b4), Dst: HuaweiTe_3d:09:1c (00:e0:fc:3d:09:1c)
MultiProtocol Label Switching Header, Label: 1024, Exp: 0, S: 0, TTL: 254   ① LDP 分配的公网标签
MultiProtocol Label Switching Header, Label: 1028, Exp: 0, S: 0, TTL: 254   ② BGP 分配的BGP标签
MultiProtocol Label Switching Header, Label: 1031, Exp: 0, S: 1, TTL: 254   ③ MP-BGP分配的私网标签
Internet Protocol Version 4, Src: 17.1.1.7, Dst: 8.8.8.8
Internet Control Message Protocol
```

图 7-20　PE1 的 G0/0/1 接口的抓包结果

（1）CE1 将流量发给 PE1（此时是纯 IP 流量）。

（2）PE1 从 G0/0/0 接口收到流量后，将查看对应的 VPN 实例的路由表，可以看到分配的私网标签是 1031，迭代的下一跳地址为 6.6.6.6。此时该报文将打上私网标签 1031。

```
 <PE1>display bgp vpnv4 all routing-table 8.8.8.8
 BGP local router ID : 12.1.1.1
 Local AS number : 100
 Total routes of Route Distinguisher(200:1): 1
 BGP routing table entry information of 8.8.8.8/32:
 Label information (Received/Applied): 1031/NULL //PE2 给其分配的私网标签是 1031
 From: 6.6.6.6 (56.1.1.2)
 Route Duration: 01h01m35s
 Relay IP Nexthop: 12.1.1.2
 Relay IP Out-Interface: GigabitEthernet0/0/1
 Relay Tunnel Out-Interface: GigabitEthernet0/0/1
 Relay token: 0xa
Original nexthop: 6.6.6.6// 迭代下一跳为 6.6.6.6
-------------------------------------------
```

（3）查看去往 6.6.6.6 的 BGP 路由信息。此时将流量打上第二层标签 1028。

```
 <PE1>display bgp routing-table 6.6.6.6

 BGP local router ID : 12.1.1.1
 Local AS number : 100
 Paths:   1 available, 1 best, 1 select
 BGP routing table entry information of 6.6.6.6/32:
 Label information (Received/Applied): 1028/NULL//ASBR1 给其分配的 BGP 隧道标签是 1028
 From: 3.3.3.3 (23.1.1.2)
 Route Duration: 00h49m55s
```

```
 Relay IP Nexthop: 12.1.1.2
 Relay IP Out-Interface: GigabitEthernet0/0/1
 Relay Tunnel Out-Interface: GigabitEthernet0/0/1
 Relay token: 0x3
 Original nexthop: 3.3.3.3 //迭代下一跳为3.3.3.3
------------------------
```

（4）查看去往3.3.3.3的MPLS LSP隧道。此时将流量打上第三层标签1024。

```
<PE1>display mpls lsp
                  LSP Information: LDP LSP
-------------------------------------------------------------------------------
FEC                 In/Out Label        In/Out IF                   Vrf Name
2.2.2.2/32          NULL/3              -/GE0/0/1
2.2.2.2/32          1024/3              -/GE0/0/1
3.3.3.3/32          NULL/1024           -/GE0/0/1       //去往3.3.3.3的流量迭代进入该
                                                        //隧道，并打上标签1024
3.3.3.3/32          1025/1024           -/GE0/0/1
1.1.1.1/32          3/NULL              -/-
```

（5）此流量沿着AS 100内部LDP建立的LSP隧道由P1设备将流量发给ASBR1。P1设备是3.3.3.3的次末跳，此时将直接弹出外层标签1024。ASBR1收到的报文只有两层标签，如图7-21所示。

```
Frame 19: 106 bytes on wire (848 bits), 106 bytes captured (848 bits) on interface -, id 0
Ethernet II, Src: HuaweiTe_3d:09:1d (00:e0:fc:3d:09:1d), Dst: HuaweiTe_99:36:07 (00:e0:fc:99:36:07)
MultiProtocol Label Switching Header, Label: 1028, Exp: 0, S: 0, TTL: 253
MultiProtocol Label Switching Header, Label: 1031, Exp: 0, S: 1, TTL: 254
Internet Protocol Version 4, Src: 17.1.1.7, Dst: 8.8.8.8
Internet Control Message Protocol
```

图7-21 ASBR1的G0/0/1接口的抓包结果

（6）ASBR1收到此报文后，查看MPLS LSP标签，并且会将1028交换为1026转发给ASBR2。注意此时用的是BGP的LSP。

```
[ASBR1]display mpls lsp
-------------------------------------------------------------------------------
                  LSP Information: BGP  LSP
-------------------------------------------------------------------------------
FEC                 In/Out Label        In/Out IF                Vrf Name
1.1.1.1/32          1026/NULL           -/-
6.6.6.6/32          NULL/1026           -/-
6.6.6.6/32          1028/1026           -/-
```

（7）ASBR2收到此报文后，再次查看MPLS LSP标签表项。

```
[ASBR2]display mpls lsp in-label 1026 verbose
-------------------------------------------------------------------------------
                  LSP Information: BGP  LSP
-------------------------------------------------------------------------------
  No                    :  1
  VrfIndex              :
  RD Value              :  0:0
```

```
Fec                    :  6.6.6.6/32
Nexthop                :  -------
In-Label               :  1026
Out-Label              :  NULL
In-Interface           :  ----------
Out-Interface          :  ----------
LspIndex               :  4096
Token                  :  0x0
LsrType                :  Egress
Outgoing token         :  0x3  // 迭代进入 0x3 隧道
Label Operation        :  POPGO// 执行 POPGO 的动作，表示将标签弹出并加上另外一个公网标签
Mpls-Mtu               :  ------
TimeStamp              :  4658sec
FrrToken               :  0x0
FrrOutgoingToken       :  0x0
BGPKey                 :  -------
BackupBGPKey           :  -------
FrrOutLabel            :  -------
```

查看 0x3 隧道，出标签为 1025，此时 ASBR2 发出去的流量将存在两层标签。外层为 1025，由 LDP 分配。内层标签还是 1031，并且沿着 LSP 隧道发给 PE2。PE2 收到后将查看内层标签 1031 和对应的 VPN 实例路由表，把流量发给 CE2。

```
[ASBR2]display  tunnel-info tunnel-id  3
Tunnel ID:                    0x3
Tunnel Token:                 3
Type:                         lsp
Destination:                  6.6.6.6
Out Slot:                     0
Instance ID:                  0
Out Interface:                GigabitEthernet0/0/1
Out Label:                    1025
Next Hop:                     45.1.1.2
Lsp Index:                    6147
```

7.2.4 配置跨域 OptionC 方式二（RR 场景）

1. 实验目的

AS 100 和 AS 200 为不同运营商的网络，运营商网络内部运行 OSPF 协议。使用 MPLS 跨域 OptionC 方式二组网实现公司 A 互通、公司 B 互通。

2. 实验拓扑

配置跨域 OptionC 方式二（RR 场景）的实验拓扑如图 7-22 所示。

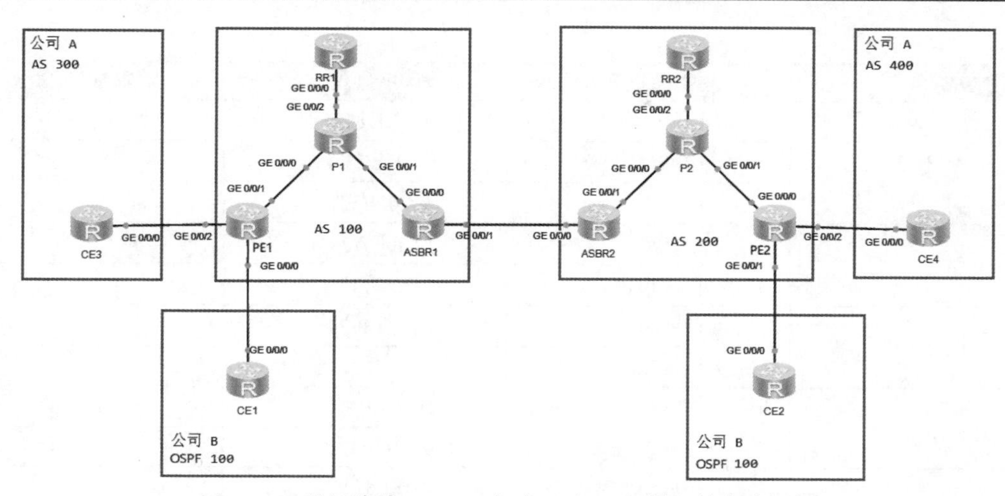

图 7-22　配置跨域 OptionC 方式二（RR 场景）的实验拓扑

3. 实验步骤

步骤 1： 配置 IP 地址，配置跨域 OptionC 方式二（RR 场景）IP 地址规划表见表 7-4。

表 7-4　配置跨域 OptionC 方式二（RR 场景）IP 地址规划表

设备名称	接口编号	IP 地址	所属 VPN 实例
PE1	G0/0/0	17.1.1.1/24	B
	G0/0/1	12.1.1.1/24	
	G0/0/2	19.1.1.1/24	A
	LoopBack 0	1.1.1.1/32	
P1	G0/0/0	12.1.1.2/24	
	G0/0/1	23.1.1.1/24	
	G0/0/2	112.1.1.1/24	
	LoopBack 0	2.2.2.2/24	
ASBR1	G0/0/0	23.1.1.2/24	
	G0/0/1	34.1.1.1/24	
	LoopBack 0	3.3.3.3/32	
PE2	G0/0/0	56.1.1.2/24	
	G0/0/1	28.1.1.2/24	B
	G0/0/2	210.1.1.2/24	A
	LoopBack 0	6.6.6.6/32	
P2	G0/0/0	45.1.1.2/24	
	G0/0/1	56.1.1.1/24	
	G0/0/2	112.1.1.1/24	
	LoopBack 0	5.5.5.5/24	

续表

设备名称	接口编号	IP 地址	所属 VPN 实例
ASBR2	G0/0/1	45.1.1.1/24	
	G0/0/0	34.1.1.2/24	
	LoopBack 0	4.4.4.4/32	
CE1	G0/0/0	17.1.1.7/24	
	LoopBack 0	7.7.7.7/32	
CE2	G0/0/0	28.1.1.8/24	
	LoopBack 0	8.8.8.8/32	
CE3	G0/0/0	19.1.1.9/24	
	LoopBack 0	9.9.9.9/32	
CE4	G0/0/0	210.1.1.0/24	
	LoopBack 0	10.10.10.10/32	
RR1	G0/0/0	112.1.1.12/24	
	LoopBack 0	12.12.12.12/32	
RR2	G0/0/0	213.1.1.13/24	
	LoopBack 0	13.13.13.13/32	

步骤 2: 配置运营商网络的 IGP、MPLS 及 MPLS LDP 协议，RR 无须运行 MPLS 以及 MPLS LDP。

（1）配置运营商网络的 IGP。

PE1 的配置：

```
[PE1]ospf 1
[PE1-ospf-1]area 0
[PE1-ospf-1-area-0.0.0.0]network 1.1.1.1 0.0.0.0
[PE1-ospf-1-area-0.0.0.0]network 12.1.1.0 0.0.0.255
```

P1 的配置：

```
[P1]ospf 1
[P1-ospf-1]area 0
[P1-ospf-1-area-0.0.0.0]network 2.2.2.2 0.0.0.0
[P1-ospf-1-area-0.0.0.0]network 12.1.1.0 0.0.0.255
[P1-ospf-1-area-0.0.0.0]network 23.1.1.0 0.0.0.255
[P1-ospf-1-area-0.0.0.0]network 112.1.1.0 0.0.0.255
```

ASBR1 的配置：

```
[ASBR1]ospf 1
[ASBR1-ospf-1]area 0
[ASBR1-ospf-1-area-0.0.0.0]network 3.3.3.3 0.0.0.0
[ASBR1-ospf-1-area-0.0.0.0]network 23.1.1.0 0.0.0.255
```

RR1 的配置：

```
[RR1]ospf
```

```
[RR1-ospf-1]area 0
[RR1-ospf-1-area-0.0.0.0]network 112.1.1.0 0.0.0.255
[RR1-ospf-1-area-0.0.0.0]network 12.12.12.12 0.0.0.0
```

ASBR2 的配置：

```
[ASBR2]ospf 1
[ASBR2-ospf-1]area 0
[ASBR2-ospf-1-area-0.0.0.0]network 4.4.4.4 0.0.0.0
[ASBR2-ospf-1-area-0.0.0.0]network 45.1.1.0 0.0.0.255
```

P2 的配置：

```
[P2]ospf 1
[P2-ospf-1]area 0
[P2-ospf-1-area-0.0.0.0]network 5.5.5.5 0.0.0.0
[P2-ospf-1-area-0.0.0.0]network 45.1.1.0 0.0.0.255
[P2-ospf-1-area-0.0.0.0]network 56.1.1.0 0.0.0.255
[P2-ospf-1-area-0.0.0.0]network 213.1.1.0 0.0.0.255
```

PE2 的配置：

```
[PE2]ospf 1
[PE2-ospf-1]area 0.0.0.0
[PE2-ospf-1-area-0.0.0.0]network 6.6.6.6 0.0.0.0
[PE2-ospf-1-area-0.0.0.0]network 56.1.1.0 0.0.0.255
```

RR2 的配置：

```
[RR2]ospf
[RR2-ospf-1]area  0
[RR2-ospf-1-area-0.0.0.0]network 213.1.1.0 0.0.0.255
[RR2-ospf-1-area-0.0.0.0]network 13.13.13.13 0.0.0.0
```

（2）配置运营商网络的 MPLS 、MPLS LDP 协议，建立公网隧道。

PE1 的配置：

```
[PE1]mpls lsr-id 1.1.1.1
[PE1]mpls
[PE1-mpls]mpls ldp
[PE1]interface G0/0/1
[PE1-GigabitEthernet0/0/1]mpls
[PE1-GigabitEthernet0/0/1]mpls ldp
```

P1 的配置：

```
[P1]mpls lsr-id 2.2.2.2
[P1]mpls
[P1-mpls]mpls ldp
[P1]interface G0/0/0
[P1-GigabitEthernet0/0/0]mpls
[P1-GigabitEthernet0/0/0]mpls ldp
[P1]interface G0/0/1
[P1-GigabitEthernet0/0/1]mpls
[P1-GigabitEthernet0/0/1]mpls ldp
```

ASBR1 的配置：

```
[ASBR1]mpls lsr-id 3.3.3.3
[ASBR1]mpls
[ASBR1-mpls]mpls ldp
[ASBR1]interface G0/0/0
[ASBR1-GigabitEthernet0/0/0]mpls
[ASBR1-GigabitEthernet0/0/0]mpls ldp
```

PE2 的配置：

```
[PE2]mpls lsr-id 6.6.6.6
[PE2]mpls
[PE2-mpls]mpls ldp
[PE2]interface G0/0/0
[PE2-GigabitEthernet0/0/0]mpls
[PE2-GigabitEthernet0/0/0]mpls ldp
```

P2 的配置：

```
[P2]mpls lsr-id 5.5.5.5
[P2]mpls
[P2-mpls]mpls ldp
[P2]interface G0/0/0
[P2-GigabitEthernet0/0/0]mpls
[P2-GigabitEthernet0/0/0]mpls ldp
[P2]interface G0/0/1
[P2-GigabitEthernet0/0/1]mpls
[P2-GigabitEthernet0/0/1]mpls ldp
```

ASBR2 的配置：

```
[ASBR2]mpls lsr-id 4.4.4.4
[ASBR2]mpls
[ASBR2-mpls]mpls ldp
[ASBR2]interface G0/0/1
[ASBR2-GigabitEthernet0/0/1]mpls
[ASBR2-GigabitEthernet0/0/1]mpls ldp
```

（3）查看 AS 100 和 AS 200 通过 MPLS 建立的 LSP。

查看 PE1 的 LSP：

```
<PE1>display mpls lsp
-------------------------------------------------------------------------------
                 LSP Information: LDP LSP
-------------------------------------------------------------------------------
FEC                 In/Out Label          In/Out IF              Vrf Name
2.2.2.2/32          NULL/3                -/GE0/0/1
2.2.2.2/32          1024/3                -/GE0/0/1
3.3.3.3/32          NULL/1024             -/GE0/0/1
3.3.3.3/32          1025/1024             -/GE0/0/1
1.1.1.1/32          3/NULL                -/-
```

查看 PE2 的 LSP：

```
<PE2>display mpls lsp
-------------------------------------------------------------------------------
                 LSP Information: LDP LSP
-------------------------------------------------------------------------------
FEC                  In/Out Label         In/Out IF              Vrf Name
4.4.4.4/32           NULL/1024            -/GE0/0/0
4.4.4.4/32           1024/1024            -/GE0/0/0
5.5.5.5/32           NULL/3               -/GE0/0/0
5.5.5.5/32           1025/3               -/GE0/0/0
6.6.6.6/32           3/NULL               -/-
```

通过以上输出可以看出，AS 100 和 AS 200 的公网隧道已经建立完毕。

步骤 3: 配置 PE 和 CE 之间的路由协议。

（1）配置 PE 设备的 VPN 实例。

PE1 的配置：

```
[PE1]ip vpn-instance A
[PE1-vpn-instance-A]ipv4-family
[PE1-vpn-instance-A-af-ipv4]route-distinguisher 100:1
[PE1-vpn-instance-A-af-ipv4]vpn-target 100:1 export-extcommunity
[PE1-vpn-instance-A-af-ipv4]vpn-target 100:1 import-extcommunity
[PE1]ip vpn-instance B
[PE1-vpn-instance-B]ipv4-family
[PE1-vpn-instance-B-af-ipv4]route-distinguisher 200:1
[PE1-vpn-instance-B-af-ipv4]vpn-target 200:1 export-extcommunity
[PE1-vpn-instance-B-af-ipv4]vpn-target 200:1 import-extcommunity
```

PE2 的配置：

```
[PE2]ip vpn-instance A
[PE2-vpn-instance-A]ipv4-family
[PE2-vpn-instance-A-af-ipv4]route-distinguisher 100:1
[PE2-vpn-instance-A-af-ipv4]vpn-target 100:1 export-extcommunity
[PE2-vpn-instance-A-af-ipv4]vpn-target 100:1 import-extcommunity
[PE2]ip vpn-instance B
[PE2-vpn-instance-B]ipv4-family
[PE2-vpn-instance-B-af-ipv4]route-distinguisher 200:1
[PE2-vpn-instance-B-af-ipv4]vpn-target 200:1 export-extcommunity
[PE2-vpn-instance-B-af-ipv4]vpn-target 200:1 import-extcommunity
```

（2）将对应的接口加入到 VPN 实例。

PE1 的配置：

```
[PE1]interface GigabitEthernet0/0/0
[PE1-GigabitEthernet0/0/0]ip binding vpn-instance B
[PE1-GigabitEthernet0/0/0]ip address 17.1.1.1 255.255.255.0
[PE1]interface GigabitEthernet0/0/2
[PE1-GigabitEthernet0/0/2]ip binding vpn-instance A
[PE1-GigabitEthernet0/0/2]ip address 19.1.1.1 255.255.255.0
```

PE2 的配置：

```
[PE2]interface GigabitEthernet0/0/1
[PE2-GigabitEthernet0/0/1]ip binding vpn-instance B
[PE2-GigabitEthernet0/0/1]ip address 28.1.1.2 255.255.255.0
[PE2]interface GigabitEthernet0/0/2
[PE2-GigabitEthernet0/0/2]ip binding vpn-instance A
[PE2-GigabitEthernet0/0/2]ip address 210.1.1.2 255.255.255.0
```

（3）配置 PE 和 CE 之间的路由协议。

①配置 PE1 和 CE1 的 OSPF 协议。

PE1 的配置：

```
[PE1]ospf 100 vpn-instance B
[PE1-ospf-100]area 0
[PE1-ospf-100-area-0.0.0.0]network 28.1.1.0 0.0.0.255
```

CE1 的配置：

```
[CE1]ospf 1
[CE1-ospf-1]area 0
[CE1-ospf-1-area-0.0.0.0]network 7.7.7.7 0.0.0.0
[CE1-ospf-1-area-0.0.0.0]network 17.1.1.0 0.0.0.255
```

查看 PE1 的 OSPF 邻居关系：

```
[PE1]display ospf 100 peer brief
         OSPF Process 100 with Router ID 17.1.1.1
                  Peer Statistic Information
 ----------------------------------------------------------------------------
 Area Id          Interface                         Neighbor id      State
 0.0.0.0          GigabitEthernet0/0/0              17.1.1.7         Full
 ----------------------------------------------------------------------------
```

查看 PE1 的 VPN 实例 B 的路由表：

```
[PE1]display ip routing-table vpn-instance B
Route Flags: R - relay, D - download to fib
------------------------------------------------------------------------------
Routing Tables: B
         Destinations : 5        Routes : 5
Destination/Mask    Proto   Pre  Cost  Flags    NextHop         Interface
       7.7.7.7/32   OSPF    10   1       D      17.1.1.7        GigabitEthernet0/0/0
      17.1.1.0/24   Direct  0    0       D      17.1.1.1        GigabitEthernet0/0/0
      17.1.1.1/32   Direct  0    0       D      127.0.0.1       GigabitEthernet0/0/0
    17.1.1.255/32   Direct  0    0       D      127.0.0.1       GigabitEthernet0/0/0
255.255.255.255/32  Direct  0    0       D      127.0.0.1       InLoopBack0
```

通过以上输出可以看出，在 PE1 的 VPN 实例 B 上学习到了 CE1 的 7.7.7.7/32 的路由。

②配置 PE1 和 CE3 的 BGP。

PE1 的配置：

```
[PE1]bgp 100
[PE1-bgp]ipv4-family vpn-instance A
```

```
[PE1-bgp-A]peer 19.1.1.9 as-number 300
```

CE3 的配置：

```
[CE3]bgp 300
[CE3-bgp]peer 19.1.1.1 as-number 100
[CE3-bgp]network 9.9.9.9 255.255.255.255
```

查看 PE1 的 VPNv4 路由表：

```
[PE1]display bgp vpnv4 vpn-instance A routing-table
 BGP Local router ID is 12.1.1.1
 Status codes: * - valid, > - best, d - damped,
               h - history,  i - internal, s - suppressed, S - Stale
               Origin : i - IGP, e - EGP, ? - incomplete
 VPN-Instance A, Router ID 12.1.1.1:
 Total Number of Routes: 1
       Network            NextHop        MED     LocPrf      PrefVal      Path/Ogn
*>     9.9.9.9/32         19.1.1.9       0                   0            300i
```

通过以上输出可以看出，PE1 学习到了 CE3 的 9.9.9.9/32 的路由。

③配置 PE2 和 CE2 的 OSPF 协议。

PE2 的配置：

```
[PE2]ospf 100 vpn-instance B
[PE2-ospf-100]area 0
[PE2-ospf-100-area-0.0.0.0]network 28.1.1.0 0.0.0.255
```

CE2 的配置：

```
[CE2]ospf 1
[CE2-ospf-1]area 0
[CE2-ospf-1-area-0.0.0.0]network 8.8.8.8 0.0.0.0
[CE2-ospf-1-area-0.0.0.0]network 28.1.1.0 0.0.0.255
```

查看 PE2 的 VPN 实例 B 的路由表：

```
[PE2]display  ip routing-table vpn-instance B
Route Flags: R - relay, D - download to fib
------------------------------------------------------------------------------
Routing Tables: B
         Destinations : 5        Routes : 5
 Destination/Mask    Proto    Pre    Cost   Flags    NextHop       Interface
       8.8.8.8/32    OSPF     10     1      D        28.1.1.8      GigabitEthernet0/0/1
      28.1.1.0/24    Direct   0      0      D        28.1.1.2      GigabitEthernet0/0/1
      28.1.1.2/32    Direct   0      0      D        127.0.0.1     GigabitEthernet0/0/1
    28.1.1.255/32    Direct   0      0      D        127.0.0.1     GigabitEthernet0/0/1
255.255.255.255/32   Direct   0      0      D        127.0.0.1     InLoopBack0
```

通过以上输出可以看出，PE2 学习到了 CE2 的 8.8.8.8/32 的路由。

④配置 PE2 和 CE4 的 BGP 协议。

PE2 的配置：

```
[PE2]bgp 200
[PE2-bgp]ipv4-family vpn-instance A
```

```
[PE2-bgp-A]peer 210.1.1.10 as-number 400
```

CE4 的配置：

```
[CE4]bgp 400
[CE4-bgp]peer 210.1.1.2 as-number 200
[CE4-bgp]network 10.10.10.10 255.255.255.255
```

查看 PE2 的 VPNv4 路由表：

```
[PE2]display  bgp vpnv4 vpn-instance A routing-table
 BGP Local router ID is 56.1.1.2
 Status codes: * - valid, > - best, d - damped,
               h - history,  i - internal, s - suppressed, S - Stale
               Origin : i - IGP, e - EGP, ? - incomplete
 VPN-Instance A, Router ID 56.1.1.2:

 Total Number of Routes: 1
        Network            NextHop        MED      LocPrf        PrefVal      Path/Ogn
*>      10.10.10.10/32     210.1.1.10     0                      0            400i
```

通过以上输出可以看出，PE2 学习到了 CE4 的路由。

将 PE 的 VPN 实例 B 的 OSPF 路由和 BGP 路由做双向引入，由于 VPN 实例 B 全部运行在 BGP 中，无须引入。

PE1 的配置：

```
[PE1]ospf 100 vpn-instance B
[PE1-ospf-100]import-route bgp
[PE1]bgp 100
[PE1-bgp]ipv4-family vpn-instance B
[PE1-bgp-B]import-route ospf 100
```

PE2 的配置：

```
[PE2]ospf 100 vpn-instance B
[PE2-ospf-100]import-route bgp
[PE2]bgp 200
[PE2-bgp]ipv4-family vpn-instance B
[PE2-bgp-B]import-route ospf 100
```

步骤 4： 配置 ASBR 之间的 EBGP 邻居关系，并宣告 PE 和 RR 的环回口路由。

（1）配置 ASBR 之间的 EBGP 邻居关系。

ASBR1 的配置：

```
[ASBR1]bgp 100
[ASBR1-bgp]peer  34.1.1.2 as-number 200
```

ASBR2 的配置：

```
[ASBR2]bgp 200
[ASBR2-bgp]peer  34.1.1.1 as-number 100
```

（2）宣告 PE 和 RR 的环回口路由。宣告 RR 的环回口路由是为了让 RR 后续能建立 MP-EBGP 邻居关系，宣告 PE 的环回口路由是为了让 VPNv4 路由的下一跳可达。

ASBR1 的配置：

```
[ASBR1]bgp 100
[ASBR1-bgp]network  1.1.1.1 32
[ASBR1-bgp]network  12.12.12.12 32
```

ASBR2 的配置：

```
[ASBR2]bgp 200
[ASBR2-bgp]network 6.6.6.6 32
[ASBR2-bgp]network 13.13.13.13 32
```

（3）在 ASBR 设备上引入 BGP 路由，目的是让 PE 设备能学习到对端 PE 的路由。

ASBR1 的配置：

```
[ASBR1]ospf
[ASBR1-ospf-1]import-route bgp
```

ASBR2 的配置：

```
[ASBR2]ospf
[ASBR2-ospf-1]import-route  bgp
```

查看 PE 和 RR 设备是否能学习到对端的 BGP 路由表：

```
<PE1>display  ip routing-table  protocol ospf
Route Flags: R - relay, D - download to fib
------------------------------------------------------------------------------
Public routing table : OSPF
         Destinations : 7           Routes : 7
OSPF routing table status : <Active>
         Destinations : 7           Routes : 7
Destination/Mask    Proto    Pre   Cost  Flags    NextHop      Interface
       2.2.2.2/32   OSPF     10    1     D        12.1.1.2     GigabitEthernet0/0/1
       3.3.3.3/32   OSPF     10    2     D        12.1.1.2     GigabitEthernet0/0/1
       6.6.6.6/32   O_ASE    150   1     D        12.1.1.2     GigabitEthernet0/0/1
   12.12.12.12/32   OSPF     10    2     D        12.1.1.2     GigabitEthernet0/0/1
   13.13.13.13/32   O_ASE    150   1     D        12.1.1.2     GigabitEthernet0/0/1
      23.1.1.0/24   OSPF     10    2     D        12.1.1.2     GigabitEthernet0/0/1
     112.1.1.0/24   OSPF     10    2     D        12.1.1.2     GigabitEthernet0/0/1

<PE2>display  ip routing-table protocol ospf
Route Flags: R - relay, D - download to fib
------------------------------------------------------------------------------
Public routing table : OSPF
         Destinations : 7           Routes : 7
OSPF routing table status : <Active>
         Destinations : 7           Routes : 7
Destination/Mask    Proto    Pre   Cost  Flags    NextHop      Interface
       1.1.1.1/32   O_ASE    150   1     D        56.1.1.1     GigabitEthernet0/0/0
       4.4.4.4/32   OSPF     10    2     D        56.1.1.1     GigabitEthernet0/0/0
       5.5.5.5/32   OSPF     10    1     D        56.1.1.1     GigabitEthernet0/0/0
   12.12.12.12/32   O_ASE    150   1     D        56.1.1.1     GigabitEthernet0/0/0
```

```
   13.13.13.13/32    OSPF    10    2    D    56.1.1.1    GigabitEthernet0/0/0
     45.1.1.0/24    OSPF    10    2    D    56.1.1.1    GigabitEthernet0/0/0
    213.1.1.0/24    OSPF    10    2    D    56.1.1.1    GigabitEthernet0/0/0

<RR1>display  ip routing-table  protocol ospf
Route Flags: R - relay, D - download to fib
------------------------------------------------------------------------------
Public routing table : OSPF
         Destinations : 7         Routes : 7
OSPF routing table status : <Active>
         Destinations : 7         Routes : 7
Destination/Mask    Proto    Pre   Cost   Flags    NextHop        Interface
       1.1.1.1/32   OSPF     10    2      D        112.1.1.1      GigabitEthernet0/0/0
       2.2.2.2/32   OSPF     10    1      D        112.1.1.1      GigabitEthernet0/0/0
       3.3.3.3/32   OSPF     10    2      D        112.1.1.1      GigabitEthernet0/0/0
       6.6.6.6/32   O_ASE    150   1      D        112.1.1.1      GigabitEthernet0/0/0
      12.1.1.0/24   OSPF     10    2      D        112.1.1.1      GigabitEthernet0/0/0
   13.13.13.13/32   O_ASE    150   1      D        112.1.1.1      GigabitEthernet0/0/0
      23.1.1.0/24   OSPF     10    2      D        112.1.1.1      GigabitEthernet0/0/0

<RR2>display  ip routing-table protocol  ospf
Route Flags: R - relay, D - download to fib
------------------------------------------------------------------------------
Public routing table : OSPF
         Destinations : 7         Routes : 7
OSPF routing table status : <Active>
         Destinations : 7         Routes : 7
Destination/Mask    Proto    Pre   Cost   Flags    NextHop        Interface
       1.1.1.1/32   O_ASE    150   1      D        213.1.1.2      GigabitEthernet0/0/0
       4.4.4.4/32   OSPF     10    2      D        213.1.1.2      GigabitEthernet0/0/0
       5.5.5.5/32   OSPF     10    1      D        213.1.1.2      GigabitEthernet0/0/0
       6.6.6.6/32   OSPF     10    2      D        213.1.1.2      GigabitEthernet0/0/0
   12.12.12.12/32   O_ASE    150   1      D        213.1.1.2      GigabitEthernet0/0/0
      45.1.1.0/24   OSPF     10    2      D        213.1.1.2      GigabitEthernet0/0/0
      56.1.1.0/24   OSPF     10    2      D        213.1.1.2      GigabitEthernet0/0/0
```

通过以上输出可以看出，PE 和 RR 设备通过 OSPF 学习到了对端 PE 和 RR 的环回口路由。

步骤 5: 配置 PE1 和 RR1、PE2 和 RR2 的 MP-IBGP 邻居关系，以及 RR1 和 RR2 的 MP-EBGP 邻居关系。

PE1 的配置：

```
[PE1]bgp 100
[PE1-bgp]peer 12.12.12.12 as-number 100
[PE1-bgp]peer 12.12.12.12 connect-interface LoopBack0
[PE1-bgp]ipv4-family vpnv4
[PE1-bgp-af-vpnv4]peer 12.12.12.12 enable
```

PE2 的配置：

```
[PE2]bgp 200
```

```
[PE2-bgp]peer 13.13.13.13 as-number 200
[PE2-bgp]peer 13.13.13.13 connect-interface LoopBack0
[PE2-bgp]ipv4-family vpnv4
[PE2-bgp-af-vpnv4]peer 13.13.13.13 enable
```

RR1 的配置：

```
[RR1]bgp 100
[RR1-bgp]peer 1.1.1.1 as-number 100
[RR1-bgp]peer 1.1.1.1 connect-interface LoopBack0
[RR1-bgp]peer 13.13.13.13 as-number 200
[RR1-bgp]peer 13.13.13.13 ebgp-max-hop 10          //配置EBGP多跳
[RR1-bgp]peer 13.13.13.13 connect-interface LoopBack0
[RR1-bgp]ipv4-family vpnv4
[RR1-bgp-af-vpnv4]undo policy vpn-target           //关闭RT检测
[RR1-bgp-af-vpnv4]peer 1.1.1.1 enable
[RR1-bgp-af-vpnv4]peer 1.1.1.1 next-hop-invariable //配置传递路由下一跳不变
[RR1-bgp-af-vpnv4]peer 13.13.13.13 enable
[RR1-bgp-af-vpnv4]peer 13.13.13.13 next-hop-invariable
```

RR2 的配置：

```
[RR2]bgp 200
[RR2-bgp]peer 6.6.6.6 as-number 200
[RR2-bgp]peer 6.6.6.6 connect-interface LoopBack0
[RR2-bgp]peer 12.12.12.12 as-number 100
[RR2-bgp]peer 12.12.12.12 ebgp-max-hop 10
[RR2-bgp]peer 12.12.12.12 connect-interface LoopBack0
[RR2-bgp]ipv4-family vpnv4
[RR2-bgp-af-vpnv4]undo policy vpn-target
[RR2-bgp-af-vpnv4]peer 6.6.6.6 enable
[RR2-bgp-af-vpnv4]peer 6.6.6.6 next-hop-invariable
[RR2-bgp-af-vpnv4]peer 12.12.12.12 enable
[RR2-bgp-af-vpnv4]peer 12.12.12.12 next-hop-invariable
```

查看 PE1 和 PE2 是否能学习到对端的 VPNv4 路由：

```
[PE1]display bgp vpnv4 all routing-table
 BGP Local router ID is 12.1.1.1
 Status codes: * - valid, > - best, d - damped,
               h - history,  i - internal, s - suppressed, S - Stale
               Origin : i - IGP, e - EGP, ? - incomplete
 Total number of routes from all PE: 6
 Route Distinguisher: 100:1
        Network           NextHop       MED       LocPrf       PrefVal       Path/Ogn
 *>     9.9.9.9/32        19.1.1.9      0                      0             300i
 *>i    10.10.10.10/32    6.6.6.6                 100          0             200 400i

 Route Distinguisher: 200:1
        Network           NextHop       MED       LocPrf       PrefVal       Path/Ogn
 *>     7.7.7.7/32        0.0.0.0       2                      0             ?
```

```
 *>i     8.8.8.8/32         6.6.6.6               100        0          200?
 *>      17.1.1.0/24        0.0.0.0       0                  0          ?
 *>i     28.1.1.0/24        6.6.6.6               100        0          200?

 VPN-Instance A, Router ID 12.1.1.1:
 Total Number of Routes: 2
         Network            NextHop       MED     LocPrf     PrefVal    Path/Ogn
 *>      9.9.9.9/32         19.1.1.9      0                  0          300i
   i     10.10.10.10/32     6.6.6.6               100        0          200 400i

 VPN-Instance B, Router ID 12.1.1.1:
 Total Number of Routes: 4
         Network            NextHop       MED     LocPrf     PrefVal    Path/Ogn
 *>      7.7.7.7/32         0.0.0.0       2                  0          ?
   i     8.8.8.8/32         6.6.6.6               100        0          200?
 *>      17.1.1.0/24        0.0.0.0       0                  0          ?
   i     28.1.1.0/24        6.6.6.6               100        0          200?

[PE2]display bgp vpnv4 all routing-table
 BGP Local router ID is 56.1.1.2
 Status codes: * - valid, > - best, d - damped,
               h - history,  i - internal, s - suppressed, S - Stale
               Origin : i - IGP, e - EGP, ? - incomplete
 Total number of routes from all PE: 6
 Route Distinguisher: 100:1
         Network            NextHop       MED     LocPrf     PrefVal    Path/Ogn
 *>i     9.9.9.9/32         1.1.1.1               100        0          100 300i
 *>      10.10.10.10/32     210.1.1.10    0                  0          400i

 Route Distinguisher: 200:1
         Network            NextHop       MED     LocPrf     PrefVal    Path/Ogn
 *>i     7.7.7.7/32         1.1.1.1               100        0          100?
 *>      8.8.8.8/32         0.0.0.0       2                  0          ?
 *>i     17.1.1.0/24        1.1.1.1               100        0          100?
 *>      28.1.1.0/24        0.0.0.0       0                  0          ?
 VPN-Instance A, Router ID 56.1.1.2:
 Total Number of Routes: 2
         Network            NextHop       MED     LocPrf     PrefVal    Path/Ogn
   i     9.9.9.9/32         1.1.1.1               100        0          100 300i
 *>      10.10.10.10/32     210.1.1.10    0                  0          400i

 VPN-Instance B, Router ID 56.1.1.2:
 Total Number of Routes: 4
         Network            NextHop       MED     LocPrf     PrefVal    Path/Ogn
   i     7.7.7.7/32         1.1.1.1               100        0          100?
 *>      8.8.8.8/32         0.0.0.0       2                  0          ?
   i     17.1.1.0/24        1.1.1.1               100        0          100?
```

```
*>      28.1.1.0/24      0.0.0.0      0                    0            ?
```

通过以上输出可以看出，PE 设备的 VPNv4 路由表中已经接收到了对端站点的私网路由，但是 VPN 实例的路由表中并不认为对端的私网路由是有效路由，因此也不会将这些路由发送给 CE 设备。其原因是隧道还未建立。

步骤 6：配置 BGP 隧道。

需要注意的是，在 MPLS VPN 跨域 OptionC 方式中，VPNv4 路由的下一跳地址（1.1.1.1/32、6.6.6.6/32）的路由是通过 OSPF 学习到的，LDP 可以为其分配标签，因此只需要在 ASBR 之间建立 BGP 隧道即可。

（1）为 ASBR 之间的物理链路开启 MPLS 功能。

ASBR1 的配置：

```
[ASBR1]interface GigabitEthernet0/0/1
[ASBR1-GigabitEthernet0/0/1]mpls
```

ASBR2 的配置：

```
[ASBR2]interface GigabitEthernet0/0/0
[ASBR2-GigabitEthernet0/0/0]mpls
```

（2）创建标签分配的策略。

ASBR1 的配置：

```
[ASBR1]route-policy ASBR-ASBR permit node 10
[ASBR1-route-policy]apply mpls-label
```

ASBR2 的配置：

```
[ASBR2]route-policy ASBR-ASBR permit node 10
[ASBR2-route-policy]apply mpls-label
```

（3）调用标签分配策略并开启 ASBR 之间的标签交互能力。

ASBR1 的配置：

```
[ASBR1]bgp 100
[ASBR1-bgp]peer 34.1.1.2 route-policy ASBR-ASBR export
[ASBR1-bgp]peer 34.1.1.2 label-route-capability
```

ASBR2 的配置：

```
[ASBR2]bgp 200
[ASBR2-bgp]peer  34.1.1.1 route-policy ASBR-ASBR export
[ASBR2-bgp]peer  34.1.1.1 label-route-capability
```

以 ASBR1 为例，查看 ASBR1 的 MPLS LSP。

```
[ASBR1]display mpls lsp
-------------------------------------------------------------------------------
                 LSP Information: BGP  LSP
-------------------------------------------------------------------------------
FEC                 In/Out Label        In/Out IF          Vrf Name
12.12.12.12/32      1026/NULL           -/-
1.1.1.1/32          1027/NULL           -/-
6.6.6.6/32          NULL/1026           -/-
```

```
13.13.13.13/32      NULL/1027           -/-
------------------------------------------------------------------------------
                    LSP Information: LDP LSP
------------------------------------------------------------------------------
FEC                 In/Out Label        In/Out IF            Vrf Name
1.1.1.1/32          NULL/1024           -/GE0/0/0
1.1.1.1/32          1024/1024           -/GE0/0/0
2.2.2.2/32          NULL/3              -/GE0/0/0
2.2.2.2/32          1025/3              -/GE0/0/0
3.3.3.3/32          3/NULL              -/-
```

通过以上输出可以看出，ASBR1 为 6.6.6.6 分配了 BGP 的 LSP，说明配置生效。但是并没有针对 6.6.6.6 的 LDP LSP，因此需要在 ASBR 设备上进行配置，为公网 BGP 路由建立 LDP LSP，保证隧道的连续。

（4）配置公网 BGP 路由 LDP LSP。

ASBR1 的配置：

```
[ASBR1]mpls
[ASBR1-mpls]lsp-trigger bgp-label-route
```

ASBR2 的配置：

```
[ASBR2]mpls
[ASBR2-mpls]lsp-trigger bgp-label-route
```

查看 ASBR 的 MPLS LSP：

```
[ASBR1]display mpls lsp
------------------------------------------------------------------------------
                    LSP Information: BGP LSP
------------------------------------------------------------------------------
FEC                 In/Out Label        In/Out IF            Vrf Name
12.12.12.12/32      1026/NULL           -/-
1.1.1.1/32          1027/NULL           -/-
13.13.13.13/32      NULL/1026           -/-
6.6.6.6/32          NULL/1027           -/-
------------------------------------------------------------------------------
                    LSP Information: LDP LSP
------------------------------------------------------------------------------
FEC                 In/Out Label        In/Out IF            Vrf Name
2.2.2.2/32          NULL/3              -/GE0/0/0
2.2.2.2/32          1024/3              -/GE0/0/0
3.3.3.3/32          3/NULL              -/-
1.1.1.1/32          NULL/1025           -/GE0/0/0
1.1.1.1/32          1025/1025           -/GE0/0/0
6.6.6.6/32          1028/1027           -/-
13.13.13.13/32      1029/1026           -/-

[ASBR2]display mpls  lsp
------------------------------------------------------------------------------
```

```
                    LSP Information: BGP LSP
-------------------------------------------------------------------------------
FEC                 In/Out Label       In/Out IF          Vrf Name
13.13.13.13/32      1026/NULL          -/-
6.6.6.6/32          1027/NULL          -/-
12.12.12.12/32      NULL/1026          -/-
1.1.1.1/32          NULL/1027          -/-
-------------------------------------------------------------------------------
                    LSP Information: LDP LSP
-------------------------------------------------------------------------------
FEC                 In/Out Label       In/Out IF          Vrf Name
4.4.4.4/32          3/NULL             -/-
5.5.5.5/32          NULL/3             -/GE0/0/1
5.5.5.5/32          1024/3             -/GE0/0/1
6.6.6.6/32          NULL/1025          -/GE0/0/1
6.6.6.6/32          1025/1025          -/GE0/0/1
1.1.1.1/32          1028/1027          -/-
12.12.12.12/32      1029/1026          -/-
```

通过以上输出可以看出，ASBR 设备为对端 PE 的环回口路由建立了 LDP LSP。

查看 PE1 和 PE2 的 MPLS LSP，查看是否为 VPNv4 路由的下一跳建立了对应的 MPLS LSP。

```
<PE1>display mpls lsp
-------------------------------------------------------------------------------
                    LSP Information: BGP LSP
-------------------------------------------------------------------------------
FEC                 In/Out Label       In/Out IF          Vrf Name
9.9.9.9/32          1026/NULL          -/-                A
17.1.1.0/24         1027/NULL          -/-                B
7.7.7.7/32          1028/NULL          -/-                B
-------------------------------------------------------------------------------
                    LSP Information: LDP LSP
-------------------------------------------------------------------------------
FEC                 In/Out Label       In/Out IF          Vrf Name
1.1.1.1/32          3/NULL             -/-
2.2.2.2/32          NULL/3             -/GE0/0/1
2.2.2.2/32          1024/3             -/GE0/0/1
3.3.3.3/32          NULL/1025          -/GE0/0/1
3.3.3.3/32          1025/1025          -/GE0/0/1
13.13.13.13/32      NULL/1026          -/GE0/0/1
13.13.13.13/32      1029/1026          -/GE0/0/1
6.6.6.6/32          NULL/1027          -/GE0/0/1
6.6.6.6/32          1030/1027          -/GE0/0/1
```

```
<PE2>display mpls lsp
-------------------------------------------------------------------------------
                    LSP Information: BGP LSP
```

```
---------------------------------------------------------------------------
FEC                 In/Out Label        In/Out IF             Vrf Name
10.10.10.10/32      1026/NULL           -/-                   A
28.1.1.0/24         1027/NULL           -/-                   B
8.8.8.8/32          1028/NULL           -/-                   B
---------------------------------------------------------------------------
                LSP Information: LDP LSP
---------------------------------------------------------------------------
FEC                 In/Out Label        In/Out IF             Vrf Name
6.6.6.6/32          3/NULL              -/-
4.4.4.4/32          NULL/1025           -/GE0/0/0
4.4.4.4/32          1024/1025           -/GE0/0/0
5.5.5.5/32          NULL/3              -/GE0/0/0
5.5.5.5/32          1025/3              -/GE0/0/0
1.1.1.1/32          NULL/1026           -/GE0/0/0
1.1.1.1/32          1029/1026           -/GE0/0/0
12.12.12.12/32      NULL/1027           -/GE0/0/0
12.12.12.12/32      1030/1027           -/GE0/0/0
```

通过以上输出可以看出，PE1 为 6.6.6.6/32 这个 FEC 建立了 MPLS LSP，PE2 为 1.1.1.1/32 这个 FEC 建立了 MPLS LSP。

注意：

在MPLS VPN OptionB方式二中，PE和ASBR之间的隧道是通过LDP建立的，而ASBR和ASBR之间的隧道是通过BGP建立的。

再次查看 PE 设备的路由表，观察远端 PE 发送过来的 VPNv4 路由在 VPN 实例中是否是有效路由：

```
<PE1>display bgp vpnv4 all routing-table
BGP Local router ID is 12.1.1.1
Status codes: * - valid, > - best, d - damped,
              h - history,  i - internal, s - suppressed, S - Stale
              Origin : i - IGP, e - EGP, ? - incomplete
Total number of routes from all PE: 6
Route Distinguisher: 100:1
      Network           NextHop       MED     LocPrf      PrefVal      Path/Ogn
*>    9.9.9.9/32        19.1.1.9      0                   0            300i
*>i   10.10.10.10/32    6.6.6.6               100         0            200 400i

Route Distinguisher: 200:1
      Network           NextHop       MED     LocPrf      PrefVal      Path/Ogn
*>    7.7.7.7/32        0.0.0.0       2                   0            ?
*>i   8.8.8.8/32        6.6.6.6               100         0            200?
*>    17.1.1.0/24       0.0.0.0       0                   0            ?
*>i   28.1.1.0/24       6.6.6.6               100         0            200?

VPN-Instance A, Router ID 12.1.1.1:
```

```
Total Number of Routes: 2
        Network          NextHop     MED      LocPrf      PrefVal     Path/Ogn
*>      9.9.9.9/32       19.1.1.9    0                    0           300i
*>i     10.10.10.10/32   6.6.6.6              100         0           200 400i

VPN-Instance B, Router ID 12.1.1.1:
Total Number of Routes: 4
        Network          NextHop     MED      LocPrf      PrefVal     Path/Ogn
*>      7.7.7.7/32       0.0.0.0     2                    0           ?
*>i     8.8.8.8/32       6.6.6.6              100         0           200?
*>      17.1.1.0/24      0.0.0.0     0                    0           ?
*>i     28.1.1.0/24      6.6.6.6              100         0           200?

<PE2>display bgp vpnv4 all routing-table
BGP Local router ID is 56.1.1.2
Status codes: * - valid, > - best, d - damped,
              h - history,  i - internal, s - suppressed, S - Stale
              Origin : i - IGP, e - EGP, ? - incomplete
Total number of routes from all PE: 6
Route Distinguisher: 100:1
        Network          NextHop     MED      LocPrf      PrefVal     Path/Ogn
*>i     9.9.9.9/32       1.1.1.1              100         0           100 300i
*>      10.10.10.10/32   210.1.1.10  0                    0           400i

Route Distinguisher: 200:1
        Network          NextHop     MED      LocPrf      PrefVal     Path/Ogn
*>i     7.7.7.7/32       1.1.1.1              100         0           100?
*>      8.8.8.8/32       0.0.0.0     2                    0           ?
*>i     17.1.1.0/24      1.1.1.1              100         0           100?
*>      28.1.1.0/24      0.0.0.0     0                    0           ?

VPN-Instance A, Router ID 56.1.1.2:
Total Number of Routes: 2
        Network          NextHop     MED      LocPrf      PrefVal     Path/Ogn
*>i     9.9.9.9/32       1.1.1.1              100         0           100 300i
*>      10.10.10.10/32   210.1.1.10  0                    0           400i

VPN-Instance B, Router ID 56.1.1.2:
Total Number of Routes: 4
        Network          NextHop     MED      LocPrf      PrefVal     Path/Ogn
*>i     7.7.7.7/32       1.1.1.1              100         0           100?
*>      8.8.8.8/32       0.0.0.0     2                    0           ?
*>i     17.1.1.0/24      1.1.1.1              100         0           100?
*>      28.1.1.0/24      0.0.0.0     0                    0           ?
```

通过以上输出可以看出，远端PE发送过来的路由在本端都为有效路由，并且会更新给CE设备。

查看 CE1 和 CE3 的路由表：

```
[CE1]display ip routing-table
Route Flags: R - relay, D - download to fib
------------------------------------------------------------------------------
Routing Tables: Public
         Destinations : 10        Routes : 10
 Destination/Mask    Proto    Pre   Cost  Flags     NextHop        Interface
       7.7.7.7/32    Direct   0     0     D         127.0.0.1      LoopBack0
       8.8.8.8/32    OSPF     10    2     D         17.1.1.1       GigabitEthernet0/0/0
      17.1.1.0/24    Direct   0     0     D         17.1.1.7       GigabitEthernet0/0/0
      17.1.1.7/32    Direct   0     0     D         127.0.0.1      GigabitEthernet0/0/0
    17.1.1.255/32    Direct   0     0     D         127.0.0.1      GigabitEthernet0/0/0
      28.1.1.0/24    O_ASE    150   1     D         17.1.1.1       GigabitEthernet0/0/0
     127.0.0.0/8     Direct   0     0     D         127.0.0.1      InLoopBack0
     127.0.0.1/32    Direct   0     0     D         127.0.0.1      InLoopBack0
127.255.255.255/32   Direct   0     0     D         127.0.0.1      InLoopBack0
255.255.255.255/32   Direct   0     0     D         127.0.0.1      InLoopBack0

[CE3]display ip routing-table
Route Flags: R - relay, D - download to fib
------------------------------------------------------------------------------
Routing Tables: Public
         Destinations : 9         Routes : 9
 Destination/Mask    Proto    Pre   Cost  Flags     NextHop        Interface
       9.9.9.9/32    Direct   0     0     D         127.0.0.1      LoopBack0
   10.10.10.10/32    EBGP     255   0     D         19.1.1.1       GigabitEthernet0/0/0
      19.1.1.0/24    Direct   0     0     D         19.1.1.9       GigabitEthernet0/0/0
      19.1.1.9/32    Direct   0     0     D         127.0.0.1      GigabitEthernet0/0/0
    19.1.1.255/32    Direct   0     0     D         127.0.0.1      GigabitEthernet0/0/0
     127.0.0.0/8     Direct   0     0     D         127.0.0.1      InLoopBack0
     127.0.0.1/32    Direct   0     0     D         127.0.0.1      InLoopBack0
127.255.255.255/32   Direct   0     0     D         127.0.0.1      InLoopBack0
255.255.255.255/32    Direct  0     0     D         127.0.0.1      InLoopBack0
```

通过以上输出可以看出，CE1 和 CE3 能够学习到对端站点的私网路由。

步骤 7：测试网络连通性。

使用 CE1 ping CE2，并且在 PE1 的 G0/0/1 接口抓包，如图 7-23 所示。

```
[CE1]ping 8.8.8.8
  PING 8.8.8.8: 56  data bytes, press CTRL_C to break
    Reply from 8.8.8.8: bytes=56 Sequence=1 ttl=249 time=60 ms
    Reply from 8.8.8.8: bytes=56 Sequence=2 ttl=249 time=50 ms
    Reply from 8.8.8.8: bytes=56 Sequence=3 ttl=249 time=40 ms
    Reply from 8.8.8.8: bytes=56 Sequence=4 ttl=249 time=40 ms
    Reply from 8.8.8.8: bytes=56 Sequence=5 ttl=249 time=50 ms
  --- 8.8.8.8 ping statistics ---
    5 packet(s) transmitted
    5 packet(s) received
```

```
    0.00% packet loss
    round-trip min/avg/max = 40/48/60 ms
```

```
> Frame 14: 106 bytes on wire (848 bits), 106 bytes captured (848 bits) on interface -, id 0
> Ethernet II, Src: HuaweiTe_e4:7d:b4 (00:e0:fc:e4:7d:b4), Dst: HuaweiTe_3d:09:1c (00:e0:fc:3d:09:1c)
> MultiProtocol Label Switching Header, Label: 1027, Exp: 0, S: 0, TTL: 254 ① MPLS LDP为6.6.6.6分配的公网标签
> MultiProtocol Label Switching Header, Label: 1028, Exp: 0, S: 1, TTL: 254 ② MP-BGP为私网路由8.8.8.8分配的私网标签
> Internet Protocol Version 4, Src: 17.1.1.7, Dst: 8.8.8.8
> Internet Control Message Protocol
```

图 7-23　PE1 的 G0/0/1 接口抓包结果

从图 7-23 中可以看到有两层标签，关于具体的数据如何传递，可以了解以下整个过程。

（1）CE1 将流量发给 PE1（此时是纯 IP 流量）。

（2）PE1 从 G0/0/0 接口收到流量后，将查看对应的 VPN 实例的路由表，可以看到分配的私网标签是 1028，迭代的下一跳地址为 6.6.6.6。此时该报文将打上私网标签 1028。

```
[PE1]display bgp vpnv4 all routing-table  8.8.8.8
 BGP local router ID : 12.1.1.1
 Local AS number : 100
 Total routes of Route Distinguisher(200:1): 1
 BGP routing table entry information of 8.8.8.8/32:
 Label information (Received/Applied): 1028/NULL      //PE2 给其分配的私网标签为 1028
 From: 12.12.12.12 (112.1.1.12)
 Route Duration: 00h26m19s
 Relay IP Nexthop: 12.1.1.2
 Relay IP Out-Interface: GigabitEthernet0/0/1
 Relay Tunnel Out-Interface: GigabitEthernet0/0/1
 Relay token: 0x7
 Original nexthop: 6.6.6.6// 迭代下一跳为 6.6.6.6
```

（3）去往 6.6.6.6。此时将流量迭代进入 LDP LSP，打上公网标签 1027。

```
[PE1]display mpls lsp
-------------------------------------------------------------------------------
                 LSP Information: BGP LSP
-------------------------------------------------------------------------------
FEC                In/Out Label            In/Out IF                Vrf Name
9.9.9.9/32         1026/NULL               -/-                      A
17.1.1.0/24        1027/NULL               -/-                      B
7.7.7.7/32         1028/NULL               -/-                      B
-------------------------------------------------------------------------------
                 LSP Information: LDP LSP
-------------------------------------------------------------------------------
FEC                In/Out Label            In/Out IF                Vrf Name
1.1.1.1/32         3/NULL                  -/-
2.2.2.2/32         NULL/3                  -/GE0/0/1
2.2.2.2/32         1024/3                  -/GE0/0/1
3.3.3.3/32         NULL/1025               -/GE0/0/1
3.3.3.3/32         1025/1025               -/GE0/0/1
13.13.13.13/32     NULL/1026               -/GE0/0/1
```

```
13.13.13.13/32      1029/1026              -/GE0/0/1
6.6.6.6/32          NULL/1027              -/GE0/0/1
6.6.6.6/32          1030/1027              -/GE0/0/1
```

（4）此流量由 P1 设备沿着 AS 100 内部 LDP 建立的 LSP 隧道将流量发给 ASBR1，P1 设备收到如标签为 1027 的流量，将把标签交换为 1029，并转发给 ASBR1。

```
[P1]display mpls lsp in-label 1027
-------------------------------------------------------------------------------
                  LSP Information: LDP LSP
-------------------------------------------------------------------------------
FEC                 In/Out Label           In/Out IF              Vrf Name
6.6.6.6/32          1027/1029              -/GE0/0/1
```

（5）ASBR1 收到 1029 的标签，将执行 SWAPPUSH 操作，将标签 1029 交换为 1026 且迭代到 BGP 的 LSP 隧道。

```
[ASBR1]display mpls lsp in-label 1029 verbose
-------------------------------------------------------------------------------
                  LSP Information: LDP LSP
-------------------------------------------------------------------------------
  No                      :  1
  VrfIndex                :
  Fec                     :  6.6.6.6/32
  Nexthop                 :  34.1.1.2
  In-Label                :  1029
  Out-Label               :  1026
  In-Interface            :  ----------
  Out-Interface           :  ----------
  LspIndex                :  6150
  Token                   :  0x0
  FrrToken                :  0x0
  LsrType                 :  Egress
  Outgoing token          :  0x0
  Label Operation         :  SWAPPUSH
  Mpls-Mtu                :  ------
  TimeStamp               :  1890sec
  Bfd-State               :  ---
  BGPKey                  :  0x4
```

（6）ASBR2 收到如标签为 1026 的标签，将执行 POPGO 操作，将标签弹出，并且根据 LDP LSP 为其打上 1024 的新标签。

```
[ASBR2]display mpls  lsp  in-label 1026 verbose
-------------------------------------------------------------------------------
                  LSP Information: BGP LSP
-------------------------------------------------------------------------------
  No                      :  1
  VrfIndex                :
  RD Value                :  0:0
```

```
   Fec                     :  6.6.6.6/32
   Nexthop                 :  -------
   In-Label                :  1026
   Out-Label               :  NULL
   In-Interface            :  ----------
   Out-Interface           :  ----------
   LspIndex                :  4096
   Token                   :  0x0
   LsrType                 :  Egress
   Outgoing token          :  0x1        //迭代进入 0x1 隧道
   Label Operation         :  POPGO      //弹出 BGP 标签，进入 LDP 隧道
   Mpls-Mtu                :  ------
   TimeStamp               :  2169sec
   FrrToken                :  0x0
   FrrOutgoingToken        :  0x0
   BGPKey                  :  -------
   BackupBGPKey            :  -------
   FrrOutLabel             :  -------

 [ASBR2]display  tunnel-info tunnel-id  1
 Tunnel ID:                     0x1
 Tunnel Token:                  1
 Type:                          lsp
 Destination:                   6.6.6.6
 Out Slot:                      0
 Instance ID:                   0
 Out Interface:                 GigabitEthernet0/0/1
 Out Label:                     1024
 Next Hop:                      45.1.1.2
 Lsp Index:                     6144
```

（7）此时数据的外层标签为1024，内层标签为1028，将沿着LSP隧道发给PE2。PE2收到后将查看内层标签1028及对应的VPN实例路由表，把流量发给CE2。

第8章

EVPN技术

8.1　EVPN概述

EVPN（Ethernet Virtual Private Network，以太网虚拟专用网）是一种用于二层网络互联的VPN技术。EVPN技术采用类似于BGP/MPLS IP VPN的机制，通过扩展BGP协议，使用扩展后的可达性信息，使不同站点的二层网络之间的MAC地址学习和发布过程从数据平面转移到控制平面。

EVPN颠覆了传统L2VPN数据平面学习MAC地址的方式，引入控制平面学习MAC和IP指导数据转发，实现了转控分离。

EVPN解决了传统L2VPN的典型问题，实现双活、快速收敛、简化运维等价值。

EVPN的控制平面采用MP-BGP，数据平面支持多种类型的隧道，如MPLS、GRE隧道和SRv6。

8.1.1　EVPN基本术语

（1）ES（Ethernet Segment，以太网段）代表用户站点（设备或网络）连接到PE的一组以太链路，使用ESI（Ethernet Segment Identifier，以太网段标识符）来表示。

（2）EVI（EVPN Instance，EVPN实例）代表一个EVPN实例，用于标识一个EVPN客户。

（3）MAC-VRF是PE上属于EVI的MAC地址表。

（4）RD（Route Distinguisher，路由标识）是EVPN的唯一标识，用于区分EVI。

（5）RT（Route Target，路由目标）用于控制EVPN路由的引入。

（6）DF（Designated Forwarder，指定转发器）用于在CE多归属场景下只转发一份BUM流量至CE。

（7）ESI Label是EVPN Type1路由所携带的扩展团体属性。在多归属场景下，用于实现快速收敛和水平分割。

（8）BUM（Broadcast、Unknown Unicast、Multicast）Label是由Type3路由携带的，用于转发BUM流量。

（9）单播Label由Type2路由携带，用于转发单播流量。

8.1.2　EVPN路由

EVPN定义了一种新的BGP NLRI（Network Layer Reachable Information，网络层可达信息）来承载所有的EVPN路由，称为EVPN NLRI。

EVPN NLRI被MP-BGP携带。MP-BGP支持多协议扩展，定义EVPN的AFI（Address Family Identifier，地址族标识）为25，SAFI（Subsequent Address Family Identifier，子地址族标识）为70。EVPN路由见表8-1。

表8-1　EVPN路由

路由类型	作　用	受　益
(Type1) Ethernet A-D Route	别名、MAC地址批量撤销、多活指示、通告ESI标签	环路避免、快速收敛、负载分担
(Type2)MAC/IP Advertisement Route	MAC地址学习通告、MAC/IP绑定、MAC地址移动性	ARP抑制、主机迁移
(Type3) Inclusive Multicast Route	组播隧道端点自动发现、组播类型自动发现	支持BUM流量转发
(Type4) Ethernet Segment Route	ES成员自动发现、DF选举	多活、单活支持

8.2 EVPN 实验

8.2.1 配置二层 EVPN

1. 实验目的

某公司两个不同的站点出口设备为 CE1 和 CE2，需要在两个站点之间使用 EVPN 实现二层流量互访。

2. 实验拓扑

配置二层 EVPN 的实验拓扑如图 8-1 所示。

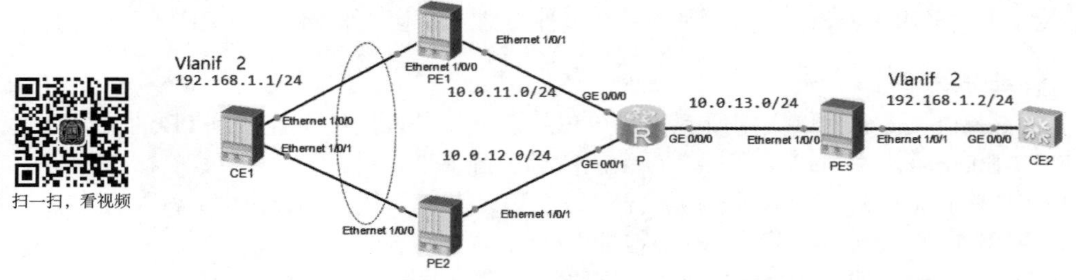

图 8-1 配置二层 EVPN 的实验拓扑

3. 实验步骤

步骤 1： 配置 IP 地址，配置二层 EVPN IP 地址规划表见表 8-2。

表 8-2 配置二层 EVPN IP 地址规划表

设备名称	接口编号	IP 地址
PE1	Ethernet1/0/1	10.0.11.1/24
	LoopBack 0	1.1.1.1/32
PE2	Ethernet1/0/1	10.0.12.1/24
	LoopBack 0	2.2.2.2/32
PE3	Ethernet1/0/1	10.0.13.2/24
	LoopBack 0	3.3.3.3/32
P	G0/0/0	10.0.11.2/24
	G0/0/1	10.0.12.2/24
	G0/0/2	10.0.13.1/24
	LoopBack 0	4.4.4.4/32
CE1	Vlanif1	192.168.1.1/24
CE2	Vlanif2	192.168.1.2/24

步骤 2： 配置 PE 和 P 设备的 IGP、MPLS 以及 MPLS LDP。配置期间在设备的系统中输入 undo dcn 命令，以关闭设备的 DCN 功能。

（1）配置设备的 OSPF 协议。

PE1 的配置：

```
[PE1]ospf 1
[PE1-ospf-1]area 0
[PE1-ospf-1-area-0.0.0.0]network 10.0.11.1 0.0.0.0
[PE1-ospf-1-area-0.0.0.0]network 1.1.1.1 0.0.0.0
```

PE2 的配置：

```
[PE2]ospf
[PE2-ospf-1]area 0
[PE2-ospf-1-area-0.0.0.0]network 10.0.12.1 0.0.0.0
[PE2-ospf-1-area-0.0.0.0]network 2.2.2.2 0.0.0.0
```

PE3 的配置：

```
[PE3]ospf
[PE3-ospf-1]area 0
[PE3-ospf-1-area-0.0.0.0]network 10.0.13.2 0.0.0.0
[PE3-ospf-1-area-0.0.0.0]network 3.3.3.3 0.0.0.0
```

P 的配置：

```
[P]ospf
[P-ospf-1]area 0
[P-ospf-1-area-0.0.0.0]network 10.0.11.2 0.0.0.0
[P-ospf-1-area-0.0.0.0]network 10.0.12.2 0.0.0.0
[P-ospf-1-area-0.0.0.0]network 10.0.13.1 0.0.0.0
[P-ospf-1-area-0.0.0.0]network 4.4.4.4 0.0.0.0
```

（2）配置设备的 MPLS、MPLS LDP，建立公网隧道。

PE1 的配置：

```
[PE1]mpls lsr-id 1.1.1.1
[PE1]mpls
[PE1-mpls]mpls ldp
[PE1-mpls-ldp]q
[PE1]interface  Ethernet 1/0/1
[PE1-Ethernet1/0/1]mpls
[PE1-Ethernet1/0/1]mpls ldp
```

PE2 的配置：

```
[PE2]mpls lsr-id 2.2.2.2
[PE2]mpls
[PE2-mpls]mpls ldp
[PE2-mpls-ldp]q
[PE2]interface  Ethernet 1/0/1
[PE2-Ethernet1/0/1]mpls
[PE2-Ethernet1/0/1]mpls ldp
```

PE3 的配置：

```
[PE3]mpls lsr-id 3.3.3.3
[PE3]mpls
[PE3-mpls]mpls ldp
```

```
[PE3-mpls-ldp]q
[PE3]interface  Ethernet 1/0/0
[PE3-Ethernet1/0/0]mpls
[PE3-Ethernet1/0/0]mpls ldp
```

P 的配置：

```
[P]mpls lsr-id 4.4.4.4
[P]mpls
[P-mpls]mpls ldp
[P-mpls-ldp]q
[P]interface GigabitEthernet0/0/0
[P-GigabitEthernet0/0/0]mpls
[P-GigabitEthernet0/0/0]mpls ldp
[P]interface GigabitEthernet0/0/1
[P-GigabitEthernet0/0/1]mpls
[P-GigabitEthernet0/0/1]mpls ldp
[P]interface GigabitEthernet0/0/2
[P-GigabitEthernet0/0/2]mpls
[P-GigabitEthernet0/0/2]mpls ldp
```

查看设备的 LSP 建立情况，以 PE1 为例：

```
[PE1]display  mpls lsp
Flag after Out IF: (I) - RLFA Iterated LSP, (I*) - Normal and RLFA Iterated LSP
Flag after LDP FRR: (L) - Logic FRR LSP
-------------------------------------------------------------------------------
                  LSP Information: LDP LSP
-------------------------------------------------------------------------------
FEC                 In/Out Label           In/Out IF               Vrf Name
1.1.1.1/32          3/NULL                 -/-
2.2.2.2/32          NULL/1024              -/Eth1/0/1
2.2.2.2/32          48120/1024             -/Eth1/0/1
3.3.3.3/32          NULL/1026              -/Eth1/0/1
3.3.3.3/32          48122/1026             -/Eth1/0/1
4.4.4.4/32          NULL/3                 -/Eth1/0/1
4.4.4.4/32          48121/3                -/Eth1/0/1
```

通过以上输出可以看出，公网的 LSP 隧道已经建立完成。

步骤 3：配置 EVPN 实例，并且绑定到 BD 模式中，配置 EVPN 的源 IP 地址。

PE1 的配置：

```
[PE1]evpn vpn-instance 1 bd-mode //指定创建EVPN的BD模式实例
[PE1-evpn-instance-1]route-distinguisher 100:1
[PE1-evpn-instance-1]vpn-target 1:1 export-extcommunity
[PE1-evpn-instance-1]vpn-target 1:1 import-extcommunity
[PE1-evpn-instance-1]q
[PE1]bridge-domain 10              //将BD域10绑定到EVPN的BD模式实例中
[PE1-bd10]evpn binding vpn-instance 1
[PE1-bd10]q
[PE1]evpn source-address 1.1.1.1 //指定EVPN的源IP地址
```

PE2 的配置：

```
[PE2]evpn vpn-instance 1 bd-mode
[PE2-evpn-instance-1]route-distinguisher 100:2
[PE2-evpn-instance-1]vpn-target 1:1 export-extcommunity
[PE2-evpn-instance-1]vpn-target 1:1 import-extcommunity
[PE2-evpn-instance-1]q
[PE2]bridge-domain 10
[PE2-bd10]evpn binding vpn-instance 1
[PE2-bd10]q
[PE2]evpn source-address 2.2.2.2
```

PE3 的配置：

```
[PE3]evpn vpn-instance 1 bd-mode
[PE3-evpn-instance-1]route-distinguisher 100:3
[PE3-evpn-instance-1]vpn-target 1:1 export-extcommunity
[PE3-evpn-instance-1]vpn-target 1:1 import-extcommunity
[PE3-evpn-instance-1]q
[PE3]bridge-domain 10
[PE3-bd10]evpn binding vpn-instance 1
[PE3-bd10]q
[PE3]evpn source-address 3.3.3.3
```

步骤 4：配置 PE1 和 PE2 的 e–trunk 来实现双活。

（1）CE 双规侧配置 e–trunk，绑定到 eth–trunk 中，做跨设备链路聚合。

PE1 的配置：

```
[PE1]lacp e-trunk system-id 1111-1111-1111 //配置e-trunk的System ID，两端需要一致
[PE1]lacp e-trunk priority 1                //配置e-trunk的优先级
[PE1]e-trunk 1
[PE1-e-trunk-1]peer-address 2.2.2.2 source-address 1.1.1.1 //配置e-trunk的对端设
                                            //备地址为2.2.2.2,本端的IP地址为1.1.1.1
[PE1-e-trunk-1]q
[PE1]interface Eth-Trunk10                  //创建聚合口
[PE1-Eth-Trunk10]mode lacp-static           //配置为LACP模式
[PE1-Eth-Trunk10]e-trunk 1                  //绑定e-trunk
[PE1-Eth-Trunk10]e-trunk mode force-master  //设置前置为主动端
[PE1-Eth-Trunk10]esi 0000.1111.1111.1111.1111  //配置连接CE链路的ESI，两端必须一致
```

PE2 的配置：

```
[PE2]lacp e-trunk system-id 1111-1111-1111
[PE2]lacp e-trunk priority 1
[PE2]e-trunk 1
[PE2-e-trunk-1]peer-address 1.1.1.1 source-address 2.2.2.2
[PE2]interface Eth-Trunk10
[PE2-Eth-Trunk10]mode lacp-static
[PE2-Eth-Trunk10]e-trunk 1
[PE2-Eth-Trunk10]esi 0000.1111.1111.1111.1111  //配置连接CE链路的ESI，两端必须一致
```

（2）创建 eth–trunk 子接口，模式改为 L2，绑定到 BD 中，并且将物理接口绑定到 eth–trunk 中。

PE1 的配置：

```
[PE1]interface Eth-Trunk10.1 mode l2              //进入聚合口的子接口，模式改为二层模式
[PE1-Eth-Trunk10.1]encapsulation dot1q vid 2     //指定二层子接口接收带指定802.1q Tag
                                                  //封装的报文
[PE1-Eth-Trunk10.1]rewrite pop single    //配置EVC二层子接口的流动作是pop，对接收的报
                                         //文进行剥除VLAN Tag操作
[PE1-Eth-Trunk10.1]bridge-domain 10      //绑定BD域10
[PE1-Eth-Trunk10.1]q
[PE1]interface Ethernet1/0/0
[PE1-Ethernet1/0/0]eth-trunk 10          //将物理接口绑定到聚合口中
```

PE2 的配置：

```
[PE2]interface Eth-Trunk10.1 mode l2
[PE2-Eth-Trunk10.1]encapsulation dot1q vid 2
[PE2-Eth-Trunk10.1]rewrite pop single
[PE2-Eth-Trunk10.1]bridge-domain 10
[PE2-Eth-Trunk10.1]q
[PE2]interface Ethernet1/0/0
[PE2-Ethernet1/0/0]eth-trunk 10
```

（3）配置 CE1 的链路聚合。

```
[CE1]Vlan 2
[CE1-vlan2]q
[CE1]Int vlanif 2
[CE1-Vlanif2]Ip address 192.168.1.1 24
[CE1-Vlanif2]q
[CE1]interface Eth-Trunk10
[CE1-Eth-Trunk10]portswitch
[CE1-Eth-Trunk10]port link-type trunk
[CE1-Eth-Trunk10]port trunk allow-pass vlan 2
[CE1-Eth-Trunk10]mode lacp-static
[CE1-Eth-Trunk10]trunkport Ethernet 1/0/0
[CE1-Eth-Trunk10]trunkport Ethernet 1/0/1
```

查看 PE 端的 e-trunk 的状态，可以看到 PE1 和 PE2 之间的 e-trunk 建立成功：

```
[PE1]display  e-trunk 1
                        The E-Trunk information
E-TRUNK-ID : 1                          Revert-Delay-Time (s) : 120
Priority : 100                          System-ID : 707b-e8d0-7ab4
Peer-IP : 2.2.2.2                       Source-IP : 1.1.1.1
State : Backup                          Causation : PRI
Send-Period (100ms) : 10                Fail-Time (100ms) : 200
Receive : 618                           Send : 634
RecDrop : 0                             SndDrop : 0
Peer-Priority : 100                     Peer-System-ID : 707b-e800-7a2e
Peer-Fail-Time (100ms) : 200            BFD-Session : -
Description : -
Sequence : Disable
```

```
Dynamic-BFD : Disabled                    BFD-State : -
TX (ms) : -                               RX (ms) : -
Multiplier : -
---------------------------------------------------------------------------------
                            The Member information
Type       ID    LocalPhyState  Work-Mode     State   Causation         Remote-ID
Eth-Trunk 10     Up             force-master  Master  FORCE_MASTER      10
```

查看 CE1 的链路聚合状态：

```
[CE1]display  eth-trunk 10
Eth-Trunk10's state information is:
Local:
LAG ID: 10                        WorkingMode: STATIC
Preempt Delay: Disabled           Hash arithmetic: According to flow
System Priority: 32768            System ID: 707b-e864-7581
Least Active-linknumber: 1        Max Active-linknumber: 32
Operate status: up                Number Of Up Ports In Trunk: 2
Timeout Period: Slow
---------------------------------------------------------------------------------
ActorPortName    Status     PortType    PortPri   PortNo   PortKey   PortState   Weight
Ethernet1/0/0    Selected   100M        32768     1        2593      10111100    1
Ethernet1/0/1    Selected   100M        32768     2        2593      10111100    1

Partner:
---------------------------------------------------------------------------------
ActorPortName    SysPri   SystemID         PortPri    PortNo    PortKey    PortState
Ethernet1/0/0    1        1111-1111-1111   32768      1         2593       10111100
Ethernet1/0/1    1        1111-1111-1111   32768      32769     2593       10111100
```

通过以上输出可以看出，CE1 和 PE1、PE2 之间的链路聚合状态为 UP。

步骤 5：配置 PE3 和 CE2 的互联接口。

PE3 的配置：

```
[PE3]interface Ethernet1/0/1.1 mode l2
[PE3-Ethernet1/0/1.1]encapsulation dot1q vid 2
[PE3-Ethernet1/0/1.1]rewrite pop single
[PE3-Ethernet1/0/1.1]bridge-domain 10
[PE3-Ethernet1/0/1.1]q
[PE3]interface Ethernet1/0/1
[PE3-Ethernet1/0/1]esi 0000.2222.2222.2222.2222
```

CE2 的配置：

```
[CE2]Vlan 2
[CE2-vlan2]q
[CE2]interface Vlanif2
[CE2-Vlanif2]ip address 192.168.1.2 255.255.255.0
[CE2-Vlanif2]q
[CE2]interface GigabitEthernet0/0/1
[CE2-GigabitEthernet0/0/1]port link-type trunk
```

```
[CE2-GigabitEthernet0/0/1]port trunk allow-pass vlan 2
```

步骤 6: 配置本端、远端 MAC 路由快速重路由功能。

PE1 的配置:

```
[PE1]evpn
[PE1-evpn]vlan-extend private enable
[PE1-evpn]vlan-extend redirect enable
[PE1-evpn]local-remote frr enable
```

PE2 的配置:

```
[PE2]evpn
[PE2-evpn]vlan-extend private enable
[PE2-evpn]vlan-extend redirect enable
[PE2-evpn]local-remote frr enable
```

PE3 的配置:

```
[PE3]evpn
[PE3-evpn]vlan-extend private enable
[PE3-evpn]vlan-extend redirect enable
[PE3-evpn]local-remote frr enable
```

步骤 7: 配置 PE 之间的 EVPN 邻居关系。

PE1 的配置:

```
[PE1]bgp 100
[PE1-bgp]peer 2.2.2.2 as-number 100
[PE1-bgp]peer 2.2.2.2 connect-interface LoopBack0
[PE1-bgp]peer 3.3.3.3 as-number 100
[PE1-bgp]peer 3.3.3.3 connect-interface LoopBack0
[PE1-bgp]
[PE1-bgp]l2vpn-family evpn
[PE1-bgp-af-evpn]peer 2.2.2.2 enable
Warning: This operation will reset the peer session. Continue? [Y/N]:y
[PE1-bgp-af-evpn]peer 3.3.3.3 enable
Warning: This operation will reset the peer session. Continue? [Y/N]:y
```

PE2 的配置:

```
[PE2]bgp 100
[PE2-bgp]peer 1.1.1.1 as-number 100
[PE2-bgp]peer 1.1.1.1 connect-interface LoopBack0
[PE2-bgp]peer 3.3.3.3 as-number 100
[PE2-bgp]peer 3.3.3.3 connect-interface LoopBack0
[PE2-bgp]l2vpn-family evpn
[PE2-bgp-af-evpn]peer 1.1.1.1 enable
Warning: This operation will reset the peer session. Continue? [Y/N]:y
[PE2-bgp-af-evpn]peer 3.3.3.3 enable
Warning: This operation will reset the peer session. Continue? [Y/N]:y
```

PE3 的配置:

```
[PE3]bgp 100
```

```
[PE3-bgp]peer 1.1.1.1 as-number 100
[PE3-bgp]peer 1.1.1.1 connect-interface LoopBack0
[PE3-bgp]peer 2.2.2.2 as-number 100
[PE3-bgp]peer 2.2.2.2 connect-interface LoopBack0
[PE3-bgp]l2vpn-family evpn
[PE3-bgp-af-evpn]peer 1.1.1.1 enable
Warning: This operation will reset the peer session. Continue? [Y/N]:y
[PE3-bgp-af-evpn]peer 2.2.2.2 enable
Warning: This operation will reset the peer session. Continue? [Y/N]:y
```

查看 BGP 的 EVPN 邻居关系：

```
[PE1]display bgp evpn peer

 BGP local router ID : 10.0.11.1
 Local AS number : 100
 Total number of peers : 2                  Peers in established state : 1

  Peer          V    AS    MsgRcvd    MsgSent    OutQ    Up/Down     State         PrefRcv
  2.2.2.2       4    100   23         20         0       00:11:30    Established   4
  3.3.3.3       4    100   0          0          0       00:00:03    Established   2
```

测试 CE 之间的连通性：

```
[CE2]ping 192.168.1.1
  PING 192.168.1.1: 56  data bytes, press CTRL_C to break
    Reply from 192.168.1.1: bytes=56 Sequence=1 ttl=255 time=5 ms
    Reply from 192.168.1.1: bytes=56 Sequence=2 ttl=255 time=2 ms
    Reply from 192.168.1.1: bytes=56 Sequence=3 ttl=255 time=1 ms
    Reply from 192.168.1.1: bytes=56 Sequence=4 ttl=255 time=1 ms
    Reply from 192.168.1.1: bytes=56 Sequence=5 ttl=255 time=2 ms

  --- 192.168.1.1 ping statistics ---
    5 packet(s) transmitted
    5 packet(s) received
    0.00% packet loss
    round-trip min/avg/max = 1/2/5 ms
```

查看 CE2 的 MAC 地址表，可以看到 CE2 学习到了 CE1 的 MAC 地址：

```
<CE2>display mac-address
MAC address table of slot 0:
-------------------------------------------------------------------------------
MAC Address     VLAN/      PEVLAN     CEVLAN     Port     Type      LSP/LSR-ID
                VSI/SI                                              MAC-Tunnel
-------------------------------------------------------------------------------
707b-e864-7581  2          -          -          GE0/0/1  dynamic   0/-
-------------------------------------------------------------------------------
Total matching items on slot 0 displayed = 1
```

下面通过几张 EVPN 路由表来看一下 L2 EVPN 的工作过程。

（1）PE之间建立MP-BGP的邻居关系，使用Type3路由分配BUM流量标签，产生BUM流量转发表。

```
[PE1]display bgp evpn all routing-table inclusive-route
 Local AS number : 100
 BGP Local router ID is 10.0.11.1
 Status codes: * - valid, > - best, d - damped, x - best external, a - add path,
               h - history,  i - internal, s - suppressed, S - Stale
               Origin : i - IGP, e - EGP, ? - incomplete
 EVPN address family:
 Number of Inclusive Multicast Routes: 3
 Route Distinguisher: 100:1
       Network(EthTagId/IpAddrLen/OriginalIp)                  NextHop
 *>    0:32:1.1.1.1                                            127.0.0.1
 Route Distinguisher: 100:2
       Network(EthTagId/IpAddrLen/OriginalIp)                  NextHop
 *>i   0:32:2.2.2.2                                            2.2.2.2
 Route Distinguisher: 100:3
       Network(EthTagId/IpAddrLen/OriginalIp)                  NextHop
 *>i   0:32:3.3.3.3                                            3.3.3.3

 EVPN-Instance 1:
 Number of Inclusive Multicast Routes: 3
       Network(EthTagId/IpAddrLen/OriginalIp)                  NextHop
 *>    0:32:1.1.1.1                                            127.0.0.1
 *>i   0:32:2.2.2.2                                            2.2.2.2
 *>i   0:32:3.3.3.3                                            3.3.3.3
```

查看其中一条的详细信息：

```
[PE1]display bgp evpn all routing-table inclusive-route  0:32:2.2.2.2
 BGP local router ID : 10.0.11.1
 Local AS number : 100
 Total routes of Route Distinguisher(100:2): 1
 BGP routing table entry information of 0:32:2.2.2.2:
 Label information (Received/Applied): 48126/NULL //发送BUM流量给2.2.2.2时，将打上
                                                  //标签48126
 From: 2.2.2.2 (10.0.12.1)
 Route Duration: 0d00h08m23s
 Relay IP Nexthop: 10.0.11.2
 Relay Tunnel Out-Interface: LDP LSP //流量将迭代到公网的LSP隧道并转发到对端
 Original nexthop: 2.2.2.2
 Qos information : 0x0
 Ext-Community: RT <1 : 1>
  AS-path Nil, origin incomplete, localpref 100, pref-val 0, valid, internal,
best, select, pre 255, IGP cost 2
 PMSI: Flags 0, Ingress Replication, Label 0:0:0(48126), Tunnel Identifier:2.2.2.2
 Route Type: 3 (Inclusive Multicast Route)
 Ethernet Tag ID: 0, Originator IP:2.2.2.2/32
 Not advertised to any peer yet
```

（2）PE 设备交互 Type4 路由，传递 ESI 信息，并且进行 DF 的选举（多归场景下面选举）。

```
[PE1]display bgp evpn all routing-table es-route
 Local AS number : 100
 BGP Local router ID is 10.0.11.1
 Status codes: * - valid, > - best, d - damped, x - best external, a - add path,
               h - history,  i - internal, s - suppressed, S - Stale
               Origin : i - IGP, e - EGP, ? - incomplete
 EVPN address family:
 Number of ES Routes: 3 //三台 PE 设备 ESI
 Route Distinguisher: 1.1.1.1:0
       Network(ESI)                                            NextHop
 *>    0000.1111.1111.1111.1111                                127.0.0.1
 Route Distinguisher: 2.2.2.2:0
       Network(ESI)                                            NextHop
 *>i   0000.1111.1111.1111.1111                                2.2.2.2
 Route Distinguisher: 3.3.3.3:0
       Network(ESI)                                            NextHop
 *>i   0000.2222.2222.2222.2222                                3.3.3.3

 EVPN-Instance 1:
 Number of ES Routes: 2 //仅将 ESI 相同的路由加入 EVPN 实例路由表中
       Network(ESI)                                            NextHop
 *>    0000.1111.1111.1111.1111                                127.0.0.1
 * i                                                           2.2.2.2
```

查看 DF 的选举结果。在 PE1 和 PE2 双归场景下，PE1 被选举为 DF，转发 BUM 流量。

```
[PE1]display evpn vpn-instance name 1 df result
ESI Count: 1
ESI: 0000.1111.1111.1111.1111
 IFName Eth-Trunk10.1:
  DF Result    : Primary
```

（3）PE 通过 Type1 类路由分发 ESI 标签。ESI 标签用于水平分割，以防止来自同一 ES 的流量又绕回到该 ES。以 PE1 为例，PE1 为将 Type1 类路由发送给所有的邻居，携带 ESI 和所分配标签的对应关系。如果发送数据给相同的 ESI 设备时，会携带 ESI 标签，对端设备收到数据后，将不转发，从而实现水平分割。

```
[PE1]display bgp evpn all routing-table ad-route
 Local AS number : 100
 BGP Local router ID is 10.0.11.1
 Status codes: * - valid, > - best, d - damped, x - best external, a - add path,
               h - history,  i - internal, s - suppressed, S - Stale
               Origin : i - IGP, e - EGP, ? - incomplete
 EVPN address family:
 Number of A-D Routes: 6
 Route Distinguisher: 100:1
       Network(ESI/EthTagId)                                   NextHop
 *>    0000.1111.1111.1111.1111:0                              127.0.0.1
```

```
Route Distinguisher: 100:2
      Network(ESI/EthTagId)                                    NextHop
*>i   0000.1111.1111.1111.1111:0                               2.2.2.2
Route Distinguisher: 100:3
      Network(ESI/EthTagId)                                    NextHop
*>i   0000.2222.2222.2222.2222:0                               3.3.3.3
Route Distinguisher: 1.1.1.1:0
      Network(ESI/EthTagId)                                    NextHop
*>    0000.1111.1111.1111.1111:4294967295                      127.0.0.1
Route Distinguisher: 2.2.2.2:0
      Network(ESI/EthTagId)                                    NextHop
*>i   0000.1111.1111.1111.1111:4294967295                      2.2.2.2
Route Distinguisher: 3.3.3.3:0
      Network(ESI/EthTagId)                                    NextHop
*>i   0000.2222.2222.2222.2222:4294967295                      3.3.3.3

EVPN-Instance 1:
Number of A-D Routes: 5
      Network(ESI/EthTagId)                                    NextHop
*>    0000.1111.1111.1111.1111:0                               127.0.0.1
* i                                                            2.2.2.2
*>i   0000.1111.1111.1111.1111:4294967295                      2.2.2.2
*>i   0000.2222.2222.2222.2222:0                               3.3.3.3
*>i   0000.2222.2222.2222.2222:4294967295                      3.3.3.3
```

设备收到CE发送的ARP请求报文后，首先生成MAC-VRF表项，然后向所有的邻居发送Type2路由，为这个MAC地址分配公网标签。

```
[PE1]display bgp evpn all routing-table mac-route
 Local AS number : 100
 BGP Local router ID is 10.0.11.1
 Status codes: * - valid, > - best, d - damped, x - best external, a - add path,
               h - history,  i - internal, s - suppressed, S - Stale
               Origin : i - IGP, e - EGP, ? - incomplete
 EVPN address family:
 Number of Mac Routes: 3
 Route Distinguisher: 100:1
       Network(EthTagId/MacAddrLen/MacAddr/IpAddrLen/IpAddr)   NextHop
 *>    0:48:707b-e864-7581:0:0.0.0.0                           0.0.0.0
 Route Distinguisher: 100:2
       Network(EthTagId/MacAddrLen/MacAddr/IpAddrLen/IpAddr)   NextHop
 *>i   0:48:707b-e864-7581:0:0.0.0.0                           2.2.2.2
 Route Distinguisher: 100:3
       Network(EthTagId/MacAddrLen/MacAddr/IpAddrLen/IpAddr)   NextHop
 *>i   0:48:4c1f-ccf6-0e60:0:0.0.0.0                           3.3.3.3

 EVPN-Instance 1:
 Number of Mac Routes: 3
```

```
    Network(EthTagId/MacAddrLen/MacAddr/IpAddrLen/IpAddr)  NextHop
*>i 0:48:4c1f-ccf6-0e60:0:0.0.0.0                          3.3.3.3
*>  0:48:707b-e864-7581:0:0.0.0.0                          0.0.0.0
* i                                                        2.2.2.2
```

8.2.2　配置三层 EVPN

1. 实验目的

在 PE1 和 PE2 之间建立 EVPN 邻居关系，使用 EVPN 邻居关系传递 CE 之间的路由以实现三层互通。

2. 实验拓扑

配置三层 EVPN 的实验拓扑如图 8-2 所示。

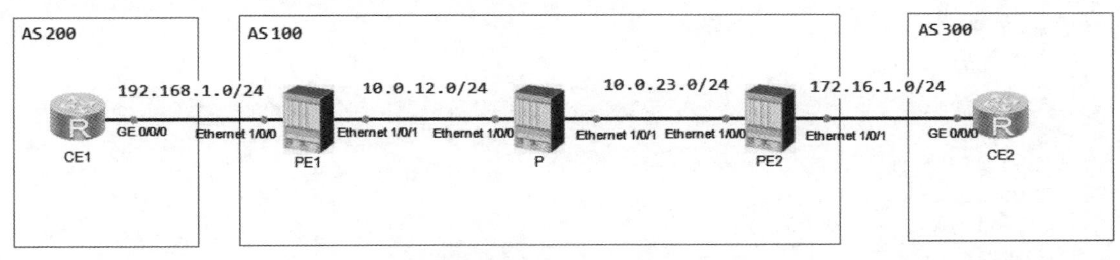

图 8-2　配置三层 EVPN 的实验拓扑

3. 实验步骤

步骤 1：配置 IP 地址，配置三层 EVPN IP 地址规划表见表 8-3。

表 8-3　配置三层 EVPN IP 地址规划表

设备名称	接口编号	IP 地址
PE1	Ethernet1/0/0	192.168.1.1/24
	LoopBack 0	1.1.1.1/32
	Ethernet1/0/1	10.0.12.1/24
PE2	Ethernet1/0/0	10.0.23.3/24
	LoopBack 0	3.3.3.3/32
	Ethernet1/0/1	172.16.1.1/24
P	Ethernet1/0/0	10.0.12.2/24
	LoopBack 0	2.2.2.2/32
	Ethernet1/0/1	10.0.23.2/24
CE1	G0/0/0	192.168.1.2/24
	LoopBack 0	10.10.10.10/32
CE2	G0/0/0	172.16.1.2/24
	LoopBack 0	20.20.20.20/32

步骤 2：配置 AS 100 的 IGP、MPLS 及 MPLS LDP 协议。

PE1 的配置：

```
[PE1]ospf
[PE1-ospf-1]area 0
[PE1-ospf-1-area-0.0.0.0]network 10.0.12.1 0.0.0.0
[PE1-ospf-1-area-0.0.0.0]network 1.1.1.1 0.0.0.0
[PE1]mpls lsr-id 1.1.1.1
[PE1]mpls
[PE1-mpls]mpls ldp
[PE1-mpls-ldp]q
[PE1]interface ethernet 1/0/1
[PE1-Ethernet1/0/1]mpls
[PE1-Ethernet1/0/1]mpls ldp
```

PE2 的配置：

```
[PE2]ospf
[PE2-ospf-1]area 0
[PE2-ospf-1-area-0.0.0.0]network 10.0.23.3 0.0.0.0
[PE2-ospf-1-area-0.0.0.0]network 3.3.3.3 0.0.0.0
[PE2]mpls lsr-id 3.3.3.3
[PE2]mpls
[PE2-mpls]mpls ldp
[PE2-mpls-ldp]q
[PE2]interface ethernet 1/0/0
[PE2-Ethernet1/0/0]mpls
[PE2-Ethernet1/0/0]mpls ldp
```

P 的配置：

```
[P]ospf
[P-ospf-1]area 0
[P-ospf-1-area-0.0.0.0]network 10.0.12.2 0.0.0.0
[P-ospf-1-area-0.0.0.0]network 2.2.2.2 0.0.0.0
[P-ospf-1-area-0.0.0.0]network 10.0.23.2 0.0.0.0
[P]mpls lsr-id 2.2.2.2
[P]mpls
[P-mpls]mpls ldp
[P-mpls-ldp]q
[P]interface ethernet 1/0/0
[P-Ethernet1/0/0]mpls
[P-Ethernet1/0/0]mpls ldp
[P-Ethernet1/0/0]q
[P]interface ethernet 1/0/1
[P-Ethernet1/0/1]mpls
[P-Ethernet1/0/1]mpls ldp
```

步骤 3：配置 PE 设备的 VPN 实例。

PE1 的配置：

```
[PE1]ip vpn-instance 1
```

```
[PE1-vpn-instance-1]ipv4-family
[PE1-vpn-instance-1-af-ipv4]route-distinguisher 100:1
[PE1-vpn-instance-1-af-ipv4]vpn-target 1:1 export-extcommunity evpn
[PE1-vpn-instance-1-af-ipv4]vpn-target 1:1 import-extcommunity evpn
[PE1-vpn-instance-1-af-ipv4]evpn mpls routing-enable  //使能EVPN生成和发布IP
                                                      //前缀路由和IRB路由的功能
[PE1-vpn-instance-1-af-ipv4]interface Ethernet1/0/0
[PE1-Ethernet1/0/0]undo shutdown
[PE1-Ethernet1/0/0]ip binding vpn-instance 1
Info: All IPv4 and IPv6 related configurations on this interface are removed.
[PE1-Ethernet1/0/0]ip address 192.168.1.1 255.255.255.0
```

PE2 的配置：

```
[PE2]ip vpn-instance 1
[PE2-vpn-instance-1]ipv4-family
[PE2-vpn-instance-1-af-ipv4]route-distinguisher 100:2
[PE2-vpn-instance-1-af-ipv4]vpn-target 1:1 export-extcommunity evpn
[PE2-vpn-instance-1-af-ipv4]vpn-target 1:1 import-extcommunity evpn
[PE2-vpn-instance-1-af-ipv4]evpn mpls routing-enable
[PE2-vpn-instance-1-af-ipv4]q
[PE2-vpn-instance-1]q
[PE2]interface Ethernet1/0/1
[PE2-Ethernet1/0/1]undo shutdown
[PE2-Ethernet1/0/1]ip binding vpn-instance 1
[PE2-Ethernet1/0/1]ip address 172.16.1.1 255.255.255.0
```

步骤 4：建立 BGP ENPN 的对等体关系。

PE1 的配置：

```
[PE1]bgp 100
[PE1-bgp]peer 3.3.3.3 as-number 100
[PE1-bgp]peer 3.3.3.3 connect-interface LoopBack0
[PE1-bgp]l2vpn-family evpn
[PE1-bgp-af-evpn]peer 3.3.3.3 enable
Warning: This operation will reset the peer session. Continue? [Y/N]:y
```

PE2 的配置：

```
[PE2]bgp 100
[PE2-bgp]peer 1.1.1.1 as-number 100
[PE2-bgp]peer 1.1.1.1 connect-interface LoopBack0
[PE2-bgp]l2vpn-family evpn
[PE2-bgp-af-evpn]peer 1.1.1.1 enable
Warning: This operation will reset the peer session. Continue? [Y/N]:y
```

查看 EVPN 的邻居关系的建立情况：

```
[PE1]display  bgp evpn peer
 BGP local router ID : 10.0.11.1
 Local AS number : 100
 Total number of peers : 2                Peers in established state : 2
```

```
  Peer            V    AS    MsgRcvd    MsgSent    OutQ    Up/Down     State          PrefRcv
  3.3.3.3         4    100   216        216        0       02:46:11    Established    1
```

步骤 5：在 PE 和 CE 之间建立 BGP 邻居关系。

PE1 的配置：

```
[PE1]bgp 100
[PE1-bgp]ipv4-family vpn-instance 1
[PE1-bgp-1]advertise l2vpn evpn
[PE1-bgp-1]peer 192.168.1.2 as-number 200
```

PE2 的配置：

```
[PE2]bgp 100
[PE2-bgp]ipv4-family vpn-instance 1
[PE2-bgp-1]advertise l2vpn evpn
[PE2-bgp-1]peer 172.16.1.2 as-number 300
```

CE1 的配置：

```
[CE1]bgp 200
[CE1-bgp]peer 192.168.1.1 as-number 100
[CE1-bgp]network 10.10.10.10 255.255.255.255
```

CE2 的配置：

```
[CE2]bgp 300
[CE2-bgp]peer 172.16.1.1 as-number 100
[CE2-bgp]network 20.20.20.20 32
```

查看 EVPN 的路由表，可以看到产生了 5 类的路由，描述 IP 前缀信息：

```
[PE1]display bgp evpn all routing-table
 Local AS number : 100
 BGP Local router ID is 10.0.12.1
 Status codes: * - valid, > - best, d - damped, x - best external, a - add path,
               h - history,  i - internal, s - suppressed, S - Stale
               Origin : i - IGP, e - EGP, ? - incomplete
 EVPN address family:
 Number of Ip Prefix Routes: 2                        //Prefix Routes 表示 5 类路由
 Route Distinguisher: 100:1
       Network(EthTagId/IpPrefix/IpPrefixLen)                         NextHop
 *>    0:10.10.10.10:32                                                 0.0.0.0
 Route Distinguisher: 100:2
       Network(EthTagId/IpPrefix/IpPrefixLen)                         NextHop
 *>i   0:20.20.20.20:32                                                 3.3.3.3

 EVPN-Instance __RD_1_100_1__:
 Number of Ip Prefix Routes: 2
       Network(EthTagId/IpPrefix/IpPrefixLen)                         NextHop
 *>    0:10.10.10.10:32                                                 0.0.0.0
 *>i   0:20.20.20.20:32                                                 3.3.3.3
[PE1]display bgp evpn all routing-table prefix-route 0:20.20.20.20:32
 BGP local router ID : 10.0.12.1
```

```
 Local AS number : 100
 Total routes of Route Distinguisher(100:2): 1
 BGP routing table entry information of 0:20.20.20.20:32:
 Label information (Received/Applied): 48122/NULL //48122 为 20.20.20.20 路由的私网标签
 From: 3.3.3.3 (10.0.23.3)
 Route Duration: 0d00h01m32s
 Relay IP Nexthop: 10.0.12.2
 Relay Tunnel Out-Interface: LDP LSP
 Original nexthop: 3.3.3.3
 Qos information : 0x0
 Ext-Community: RT <1 : 1>
  AS-path 300, origin igp, MED 0, localpref 100, pref-val 0, valid, internal,
best, select, pre 255, IGP cost 2
 Route Type: 5 (Ip Prefix Route)
 Ethernet Tag ID: 0, IP Prefix/Len: 20.20.20.20/32, ESI: 0000.0000.0000.0000.0000,
GW IP Address: 0.0.0.0
 Not advertised to any peer yet
```

在 PE1 的 Ethernet1/0/1 接口抓包查看结果，如图 8-3 所示。

```
[PE1]ping -a 10.10.10.10 20.20.20.20
  PING 20.20.20.20: 56  data bytes, press CTRL_C to break
    Reply from 20.20.20.20: bytes=56 Sequence=1 ttl=252 time=30 ms
    Reply from 20.20.20.20: bytes=56 Sequence=2 ttl=252 time=40 ms
    Reply from 20.20.20.20: bytes=56 Sequence=3 ttl=252 time=60 ms
    Reply from 20.20.20.20: bytes=56 Sequence=4 ttl=252 time=40 ms
    Reply from 20.20.20.20: bytes=56 Sequence=5 ttl=252 time=40 ms

  --- 20.20.20.20 ping statistics ---
    5 packet(s) transmitted
    5 packet(s) received
    0.00% packet loss
    round-trip min/avg/max = 30/42/60 ms
```

```
> Frame 28: 106 bytes on wire (848 bits), 106 bytes captured (848 bits) on interface -, id 0
> Ethernet II, Src: 38:4e:75:01:01:01 (38:4e:75:01:01:01), Dst: 38:4e:75:02:01:00 (38:4e:75:02:01:00)
> MultiProtocol Label Switching Header, Label: 48121, Exp: 0, S: 0, TTL: 254   ① LDP 分配的公网标签
> MultiProtocol Label Switching Header, Label: 48122, Exp: 0, S: 1, TTL: 254   ② BGP EVPN 分配的私网标签
> Internet Protocol Version 4, Src: 10.10.10.10, Dst: 20.20.20.20
> Internet Control Message Protocol
```

图 8-3　PE1 的 Ethernet1/0/1 接口的抓包结果

第 9 章

IPv6 路由

9.1 IPv6 路由概述

随着万物互联时代的到来，IPv4 地址空间不足，IPv6 取代 IPv4 势在必行。IPv6 的网络环境中通常会用到一些路由协议，如 OSPFv3、IS-IS IPv6、BGP4+。

9.1.1 OSPFv3

OSPF 是 IETF 定义的一种基于链路状态的内部网关路由协议。目前针对 IPv4 协议使用的是 OSPF Version 2（OSPFv2），针对 IPv6 协议使用的是 OSPF Version 3（OSPFv3）。

OSPFv3 与 OSPFv2 的区别与相同点如下：

（1）网络类型和接口类型。

（2）接口状态机和邻居状态机。

（3）链路状态数据库（LSDB）。

（4）泛洪机制（Flooding Mechanism）。

（5）相同类型的报文：Hello 报文、DD 报文、LSR 报文、LSU 报文和 LSAck 报文。

（6）路由计算基本相同。

OSPFv3 的特点如下：

（1）OSPFv3 基于链路运行以及拓扑计算，而不再是网段。

（2）OSPFv3 支持一个链路上存在多个实例。

（3）OSPFv3 报文和 LSA 中去掉了 IP 地址的意义，并且重构了报文格式和 LSA 格式。

（4）OSPFv3 报文和 Router LSA/Network LSA 中不包含 IP 地址。

（5）OSPFv3 的 LSA 中定义了 LSA 的泛洪范围。

（6）OSPFv3 中创建了新的 LSA 承载 IPv6 地址和前缀。

（7）OSPFv3 邻居不再由 IP 地址标识，只由 Router ID 标识。

OSPFv3 的 LSA 类型：OSPFv3 新增了两类 LSA，包括 Link LSA 和 Intra Area Prefix LSA。

（1）Link LSA：用于向路由器宣告各个链路上对应的链路本地地址及其所配置的 IPv6 全局地址，仅在链路内泛洪。

（2）Intra Area Prefix LSA：用于向其他路由器宣告本路由器或本网络（广播网及 NBMA）的 IPv6 全局地址信息，在区域内泛洪。

9.1.2 IS-IS IPv6

IS-IS 最初是为 OSI 网络设计的一种基于链路状态算法的动态路由协议。之后为了提供对 IPv4 的路由支持，扩展应用到 IPv4 网络，称为集成 IS-IS。

扫一扫，看视频

随着 IPv6 网络的建设，同样需要动态路由协议为 IPv6 报文的转发提供准确有效的路由信息。IS-IS 路由协议结合自身具有良好的扩展性的特点，实现了对 IPv6 网络层协议的支持，可以发现和生成 IPv6 路由。

IETF 的 draft-ietf-isis-ipv6-05 中规定了 IS-IS 为支持 IPv6 所新增的内容。为了支持 IPv6 路由的处理和计算，IS-IS 新增了两个 TLV（Type-Length-Value，类型 - 长度 - 值）和一个新的 NLPID（Network Layer Protocol Identifier，网络层协议标识）。

新增的两个TLV分别为236号TLV和232号TLV。

（1）236号TLV（IPv6 Reachability，IPv6可达性）：通过定义路由信息前缀、度量值等信息来说明网络的可达性。

（2）232号TLV（IPv6 Interface Address，IPv6接口地址）：相当于IPv4中的IP Interface Address TLV，只不过把原来的32比特的IPv4地址改为128比特的IPv6地址。

NLPID是标识网络层协议报文的一个8比特字段，IPv6的NLPID值为142（0x8E）。如果IS-IS支持IPv6，那么向外发布IPv6路由时必须携带NLPID值。

9.1.3 BGP4+

传统的BGP-4只能管理IPv4单播路由信息，对于使用其他网络层协议（如IPv6、组播等）的应用，在跨AS传播时就受到一定限制。BGP多协议扩展（BGP4+）MP-BGP就是为了提供对多种网络层协议的支持，对BGP-4进行的扩展。目前的MP-BGP使用扩展属性和地址族来实现对IPv6、组播和VPN相关内容的支持，BGP协议原有的报文机制和路由机制并没有改变。

BGP使用的报文中，与IPv4相关的三处信息都由Update报文携带，这三处信息分别是NLRI字段、Next_Hop属性和Aggregator属性。

为实现对多种网络层协议的支持，BGP需要将网络层协议的信息反映到NLRI及Next_Hop。因此，MP-BGP引入了以下两个新的可选非过渡路径属性。

（1）MP_REACH_NLRI：Multiprotocol Reachable NLRI，多协议可达NLRI。用于发布可达路由及下一跳信息。MP_REACH_NLRI格式如图9-1所示。

Address Family Identifier
Subsequent Address Family Identifier
Length of Next Hop Network Address
Network Address of Next Hop
Reserved
Network Layer Reachability Information

图9-1　MP_REACH_NLRI格式

当传递IPv6路由时，AFI=2、SAFI=1（单播）、SAFI=2（组播）。

下一跳地址长度字段的通常值为16，表示下一跳地址为下一跳路由器的全球单播地址。

保留字段恒等于0。

NLRI字段是可变长字段，表示路由前缀和掩码信息。

（2）MP_UNREACH_NLRI：Multiprotocol Unreachable NLRI，多协议不可达NLRI。用于撤销不可达路由。MP_UNREACH_NLRI格式如图9-2所示。

Address Family Identifier
Subsequent Address Family Identifier
Withdrawn Routes

图9-2　MP_UNREACH_NLRI格式

当撤销IPv6路由时，AFI=2、SAFI=1（单播）、SAFI=2（组播）。

Withdrawn Routes字段表示需要撤销的路由前缀及掩码。

9.2　IPv6 路由实验

9.2.1　配置 OSPFv3

1. 实验目的

每台设备都运行，AR1、AR2 和 AR3 属于 Area 0，AR3 和 AR4 属于 Area 1。每台设备上运行 LoopBack 0，设备的 IP 地址配置为 x::x（x 为设备编号。例如，AR1 的 LoopBack 0 的地址为 1::1/128），最终实现全网互通。

2. 实验拓扑

配置 OSPFv3 的实验拓扑如图 9-3 所示。

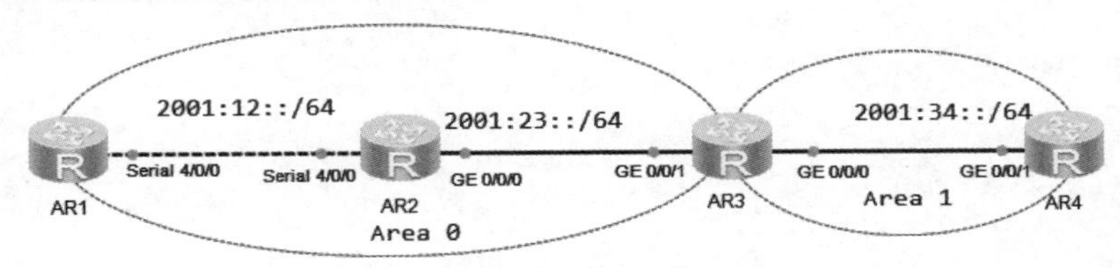

图 9-3　配置 OSPFv3 的实验拓扑

3. 实验步骤

步骤 1： 配置 IP 地址，OSPFv3 实验 IP 规划表见表 9-1。

表 9-1　OSPFv3 实验 IP 规划表

设备名称	接口编号	IP 地址
AR1	S4/0/0	2001:12::1/64
	LoopBack 0	1::1/128
AR2	G0/0/0	2001:23::2/64
	S4/0/0	2001:12::2/64
	LoopBack 0	2::2/128
AR3	G0/0/0	2001:34::3/64
	G0/0/1	2001:23::3/64
	LoopBack 0	3::3/128
AR4	G0/0/1	2001:32::4/64
	LoopBack 0	4::4/128

步骤 2： 配置 OSPFv3。

AR1 的配置：

```
[AR1]ospfv3
[AR1-ospfv3-1]router-id 1.1.1.1
[AR1]interface  LoopBack 0
[AR1-LoopBack0]ospfv3 1 area 0
[AR1]interface s4/0/0
[AR1-Serial4/0/0]ospfv3 1 area 0
```

AR2 的配置：

```
[AR2]ospfv3
[AR2-ospfv3-1]router-id 2.2.2.2
[AR2]interface  g0/0/0
[AR2]int s4/0/0
[AR2-Serial4/0/0]ospfv3 1 area 0
[AR2-Serial4/0/0]quit
[AR2]int g0/0/0
[AR2-GigabitEthernet0/0/0]ospfv3 1 area 0
[AR2-GigabitEthernet0/0/0]quit
[AR2]int loopback0
[AR2-LoopBack0]ospfv3 1 area 0
[AR2-LoopBack0]quit
```

AR3 的配置：

```
[AR3]ospfv3
[AR3-ospfv3-1]router-id 3.3.3.3
[AR3]int g0/0/1
[AR3-GigabitEthernet0/0/1]ospfv3 1 area 0
[AR3-GigabitEthernet0/0/1]quit
[AR3]int g0/0/0
[AR3-GigabitEthernet0/0/0]ospfv3 1 area 1
[AR3-GigabitEthernet0/0/0]int loopback0
[AR3-LoopBack0]ospfv3 1 area 0
[AR3-LoopBack0]quit
```

AR4 的配置：

```
[AR4]ospfv3
[AR4-ospfv3-1]router-id 4.4.4.4
[AR4]int g0/0/1
[AR4-GigabitEthernet0/0/1]ospfv3 1 area 1
[AR4-GigabitEthernet0/0/1]quit
[AR4]interface LoopBack 0
[AR4-LoopBack0]ospfv3 1 area 1
[AR4-LoopBack0]quit
```

查看邻居关系是否建立成功：

```
 [AR3]display ospfv3 peer
OSPFv3 Process (1)
OSPFv3 Area (0.0.0.0)
Neighbor ID      Pri      State             Dead Time     Interface      Instance ID
2.2.2.2          1        Full/DR           00:00:38      GE0/0/1        0
OSPFv3 Area (0.0.0.1)
Neighbor ID      Pri      State             Dead Time     Interface      Instance ID
4.4.4.4          1        Full/Backup       00:00:38      GE0/0/0        0
```

通过以上输出可以看出，AR3 分别与 AR2 及 AR4 建立了 OSPFv3 的邻居关系。

查看 AR4 的 IPv6 路由表：

```
[AR4]display ipv6 routing-table protocol ospfv3
Public Routing Table : OSPFv3
Summary Count : 7

OSPFv3 Routing Table's Status : < Active >
Summary Count : 5

 Destination  : 1::1                      PrefixLength : 128
 NextHop      : FE80::2E0:FCFF:FE98:3BF7  Preference   : 10
 Cost         : 50                        Protocol     : OSPFv3
 RelayNextHop : ::                        TunnelID     : 0x0
 Interface    : GigabitEthernet0/0/1      Flags        : D

 Destination  : 2::2                      PrefixLength : 128
 NextHop      : FE80::2E0:FCFF:FE98:3BF7  Preference   : 10
 Cost         : 2                         Protocol     : OSPFv3
 RelayNextHop : ::                        TunnelID     : 0x0
 Interface    : GigabitEthernet0/0/1      Flags        : D

 Destination  : 3::3                      PrefixLength : 128
 NextHop      : FE80::2E0:FCFF:FE98:3BF7  Preference   : 10
 Cost         : 1                         Protocol     : OSPFv3
 RelayNextHop : ::                        TunnelID     : 0x0
 Interface    : GigabitEthernet0/0/1      Flags        : D

 Destination  : 2001:12::                 PrefixLength : 64
 NextHop      : FE80::2E0:FCFF:FE98:3BF7  Preference   : 10
 Cost         : 50                        Protocol     : OSPFv3
 RelayNextHop : ::                        TunnelID     : 0x0
 Interface    : GigabitEthernet0/0/1      Flags        : D

 Destination  : 2001:23::                 PrefixLength : 64
 NextHop      : FE80::2E0:FCFF:FE98:3BF7  Preference   : 10
 Cost         : 2                         Protocol     : OSPFv3
 RelayNextHop : ::                        TunnelID     : 0x0
 Interface    : GigabitEthernet0/0/1      Flags        : D

OSPFv3 Routing Table's Status : < Inactive >
Summary Count : 2

 Destination  : 4::4                      PrefixLength : 128
 NextHop      : ::                        Preference   : 10
 Cost         : 0                         Protocol     : OSPFv3
 RelayNextHop : ::                        TunnelID     : 0x0
 Interface    : LoopBack0                 Flags        :
```

```
Destination  : 2001:34::                    PrefixLength  : 64
NextHop      : ::                           Preference    : 10
Cost         : 1                            Protocol      : OSPFv3
RelayNextHop : ::                           TunnelID      : 0x0
Interface    : GigabitEthernet0/0/1         Flags         :
```

通过以上输出可以看出，AR4 通过 OSPFv3 学习到了其他 3 台路由器的环回口路由。

测试网络连通性：

```
[AR4]ping ipv6 1::1
  PING 1::1 : 56  data bytes, press CTRL_C to break
    Reply from 1::1
    bytes=56 Sequence=1 hop limit=62  time = 30 ms
    Reply from 1::1
    bytes=56 Sequence=2 hop limit=62  time = 40 ms
    Reply from 1::1
    bytes=56 Sequence=3 hop limit=62  time = 40 ms
    Reply from 1::1
    bytes=56 Sequence=4 hop limit=62  time = 30 ms
    Reply from 1::1
    bytes=56 Sequence=5 hop limit=62  time = 40 ms
  --- 1::1 ping statistics ---
    5 packet(s) transmitted
    5 packet(s) received
    0.00% packet loss
    round-trip min/avg/max = 30/36/40 ms
```

查看 AR4 产生的 1 类 LSA：

```
[AR4]display ospfv3 lsdb router
           OSPFv3 Router with ID (4.4.4.4) (Process 1)
               Router-LSA (Area 0.0.0.1)
  LS Age: 429
  LS Type: Router-LSA
  Link State ID: 0.0.0.0
  Originating Router: 3.3.3.3
  LS Seq Number: 0x80000005
  Retransmit Count: 0
  Checksum: 0x5E9D
  Length: 40
  Flags: 0x01 (-|-|-|-|B)
  Options: 0x000013 (-|R|-|-|E|V6)
    Link connected to: a Transit Network
      Metric: 1
      Interface ID: 0x3
      Neighbor Interface ID: 0x3
      Neighbor Router ID: 3.3.3.3
  LS Age: 405
  LS Type: Router-LSA
  Link State ID: 0.0.0.0
```

```
    Originating Router: 4.4.4.4
    LS Seq Number: 0x80000005
    Retransmit Count: 0
    Checksum: 0x4BAC
    Length: 40
    Flags: 0x00 (-|-|-|-|-)
    Options: 0x000013 (-|R|-|-|E|V6)
      Link connected to: a Transit Network
        Metric: 1
        Interface ID: 0x4
        Neighbor Interface ID: 0x3
        Neighbor Router ID: 3.3.3.3
```

通过以上输出可以看出，AR4 通过 Transit 的链路使用自己的 ID 为 3 的接口与 Router ID 为 3.3.3.3 的路由建立了 OSPFv3 的邻居关系。另外，此处并没有路由信息，只是描述了拓扑信息。那么区域间的路由信息怎么传递呢？这时就需要 OSPFv3 新增加的 9 类 LSA 来描述区域间的路由信息了。

查看 AR4 产生的 9 类 LSA：

```
  [AR4]display ospfv3 lsdb intra-prefix
             OSPFv3 Router with ID (4.4.4.4) (Process 1)
                 Intra-Area-Prefix-LSA (Area 0.0.0.1)
    LS Age: 586
    LS Type: Intra-Area-Prefix-LSA
    Link State ID: 0.0.0.1
    Originating Router: 3.3.3.3
    LS Seq Number: 0x80000004
    Retransmit Count: 0
    Checksum: 0x252C
    Length: 44
    Number of Prefixes: 1
    Referenced LS Type: 0x2002
    Referenced Link State ID: 0.0.0.3
    Referenced Originating Router: 3.3.3.3
    Prefix: 2001:34::/64
    Prefix Options: 0 (-|-|-|-|-)
    Metric: 0
    LS Age: 562
    LS Type: Intra-Area-Prefix-LSA
    Link State ID: 0.0.0.1
    Originating Router: 4.4.4.4
    LS Seq Number: 0x80000001
    Retransmit Count: 0
    Checksum: 0xF45E
    Length: 52
    Number of Prefixes: 1
    Referenced LS Type: 0x2001
    Referenced Link State ID: 0.0.0.0
```

```
  Referenced Originating Router: 4.4.4.4
  Prefix: 4::4/128
  Prefix Options: 2 (-|-|-|LA|-)
  Metric: 0
```

通过以上输出可以看出，AR4通过产生的9类LSA描述了本设备的接口的路由信息分别为2001:34::/64、4::4/128。9类LSA泛洪范围为本区域，因此本区域的所有设备都能够学习到AR4的路由信息。了解了9类LSA，那么8类LSA的作用呢？

查看OSPFv3的路由表：

```
[AR4]display ipv6 routing-table protocol ospfv3
Public Routing Table : OSPFv3
Summary Count : 7
OSPFv3 Routing Table's Status : < Active >
Summary Count : 5
 Destination  : 1::1                          PrefixLength : 128
 NextHop      : FE80::2E0:FCFF:FE98:3BF7      Preference   : 10
 Cost         : 50                            Protocol     : OSPFv3
 RelayNextHop : ::                            TunnelID     : 0x0
 Interface    : GigabitEthernet0/0/1          Flags        : D
--------------------- 此处只截取了一部分路由
```

通过以上输出可以看出，去往1::1/128的下一跳地址为FE80::2E0:FCFF:FE98:3BF7，这是一个链路本地地址，实际上是AR3的S4/0/0接口的链路本地地址，那么AR4是如何知道AR3的接口链路本地地址的呢？这个就是通过8类LSA来描述的。

查看AR4产生的8类LSA：

```
[AR4]display ospfv3 lsdb link
           OSPFv3 Router with ID (4.4.4.4) (Process 1)
               Link-LSA (Interface GigabitEthernet0/0/1)
  LS Age: 896
  LS Type: Link-LSA
  Link State ID: 0.0.0.3
  Originating Router: 3.3.3.3
  LS Seq Number: 0x80000001
  Retransmit Count: 0
  Checksum: 0x4614
  Length: 56
  Priority: 1
  Options: 0x000013 (-|R|-|-|E|V6)
  Link-Local Address: FE80::2E0:FCFF:FE98:3BF7
  Number of Prefixes: 1
  Prefix: 2001:34::/64
  Prefix Options: 0 (-|-|-|-|-)
  LS Age: 748
  LS Type: Link-LSA
  Link State ID: 0.0.0.4
  Originating Router: 4.4.4.4
  LS Seq Number: 0x80000001
```

```
    Retransmit Count: 0
    Checksum: 0xADA7
    Length: 56
    Priority: 1
    Options: 0x000013 (-|R|-|-|E|V6)
    Link-Local Address: FE80::2E0:FCFF:FED0:3BBF
    Number of Prefixes: 1
    Prefix: 2001:34::/64
    Prefix Options: 0 (-|-|-|-|-)
```

通过以上输出可以看出，AR3 产生了 8 类 LSA 中存在接口的链路本地地址，这样 AR4 就可以使用这个地址作为 IPv6 路由的下一跳地址了。

9.2.2　配置 OSPFv3 多实例

1. 实验目的

配置接口 IP 地址，在设备上配置 OSPFv3 多实例，实现 AR2 的环回口能够访问 AR6 的环回口以及 AR3 的环回口能够访问 AR5 的环回口。

2. 实验拓扑

配置 OSPFv3 多实例的实验拓扑如图 9-4 所示。

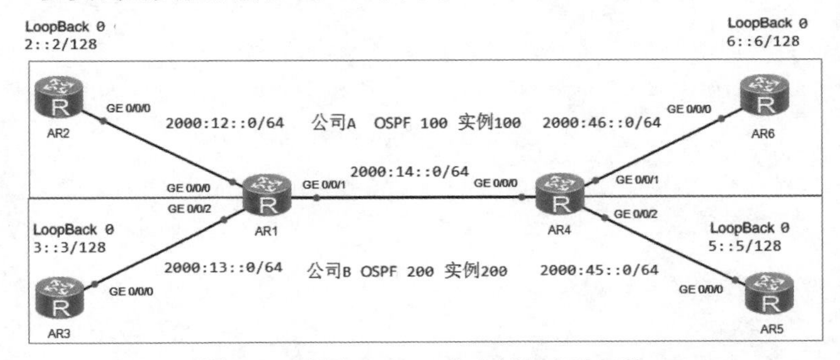

图 9-4　配置 OSPFv3 多实例的实验拓扑

3. 实验步骤

步骤 1：配置 IP 地址，OSPFv3 多实例实验 IP 规划表见表 9-2。

表 9-2　OSPFv3 多实例实验 IP 规划表

设备名称	接口编号	IP 地址
AR1	G0/0/0	2000:12::1/64
	G0/0/1	2000:14::1/64
	G0/0/2	2000:13::1/64
	LoopBack 0	1::1/128
AR2	G0/0/0	2000:12::2/64
	LoopBack 0	2::2/128
AR3	G0/0/0	2000:13::3/64
	LoopBack 0	3::3/128

续表

设备名称	接口编号	IP 地址
AR4	G0/0/0	2000:14::4/64
	G0/0/1	2000:46:4/64
	G0/0/2	2000:45:4/64
	LoopBack 0	4::4/128
AR5	G0/0/0	2000:45::5/64
	LoopBack 0	5::5/128
AR6	G0/0/0	2000:46::6/64
	LoopBack 0	6::6/128

步骤 2：配置 OSPFv3。

AR1 的配置：

```
[AR1]ospfv3 100                              //配置 OSPFv3 100 给公司 A
[AR1-ospfv3-100]router-id 1.1.1.1
[AR1]ospfv 200                               //配置 OSPFv3 200 给公司 B
[AR1-ospfv3-200]router-id 1.1.1.1
[AR1]interface g0/0/1
[AR1-GigabitEthernet0/0/1]ospfv3 100 area 0 instance 100
[AR1-GigabitEthernet0/0/1]ospfv3 200 area 0 instance 200
[AR1]int g0/0/0
[AR1-GigabitEthernet0/0/0]ospfv3 100 area 0
[AR1]int g0/0/2
[AR1-GigabitEthernet0/0/0]ospfv3 200 area 0
```

AR4 的配置：

```
[AR4]ospfv3 100
[AR4-ospfv3-100]router-id 4.4.4.4
[AR4]ospfv3 200
[AR4-ospfv3-200]router-id 4.4.4.4
[AR4]interface g0/0/0
[AR4-GigabitEthernet0/0/0]ospfv3 100 area 0 instance 100
[AR4-GigabitEthernet0/0/0]ospfv3 200 area 0 instance 200
[AR4]interface g0/0/1
[AR4-GigabitEthernet0/0/1]ospfv3 100 area 0
[AR4]int g0/0/2
[AR4-GigabitEthernet0/0/2]ospfv3 200 area 0
```

AR2 的配置：

```
[AR2]ospfv3
[AR2-ospfv3-1]router-id 2.2.2.2
[AR2]int g0/0/0
[AR2-GigabitEthernet0/0/0]ospfv3 1 area 0
[AR2]int lo0
[AR2-LoopBack0]ospfv3 1 area 0
```

AR3 的配置：

```
[AR3]ospfv3
[AR3-ospfv3-1]router-id 3.3.3.3
[AR3]int g0/0/0
[AR3-GigabitEthernet0/0/0]ospfv3 1 area 0
[AR3]int lo0
[AR3-LoopBack0]ospfv3 1 area 0
```

AR5 的配置：

```
[AR5]ospfv3
[AR5-ospfv3-1]router-id 5.5.5.5
[AR5]int g0/0/0
[AR5-GigabitEthernet0/0/0]ospfv3 1 area 0
[AR5]int lo0
[AR5-LoopBack0]ospfv3 1 area 0
```

AR6 的配置：

```
[AR6]ospfv3
[AR6-ospfv3-1]router-id 6.6.6.6
[AR6]int g0/0/0
[AR6-GigabitEthernet0/0/0]ospfv3 1 area 0
[AR6]int lo0
[AR6-LoopBack0]ospfv3 1 area 0
```

查看 AR5 的邻居关系：

```
[AR1]display ospfv3 peer
OSPFv3 Process (100)
OSPFv3 Area (0.0.0.0)
Neighbor ID     Pri    State          Dead Time      Interface      Instance ID
2.2.2.2         1      Full/Backup    00:00:37       GE0/0/0        0
4.4.4.4         1      Full/Backup    00:00:33       GE0/0/1        100

OSPFv3 Process (200)
OSPFv3 Area (0.0.0.0)
Neighbor ID     Pri    State          Dead Time      Interface      Instance ID
4.4.4.4         1      Full/Backup    00:00:31       GE0/0/1        200
3.3.3.3         1      Full/Backup    00:00:33       GE0/0/2        0
```

通过以上输出可以看出，通过 G0/0/1 接口建立了两个 OSPFv3 的邻居关系，从而实现可以在一条链路上建立多个邻居关系。在 OSPFv2 上，在一条链路上只能建立一个邻居关系，OSPFv3 通过实例实现了链路的复用。

查看 AR2 的路由表：

```
[AR2]display ipv6 routing-table protocol ospfv3
Public Routing Table : OSPFv3
Summary Count : 5

OSPFv3 Routing Table's Status : < Active >
```

```
Summary Count : 3

 Destination  : 6::6                          PrefixLength  : 128
 NextHop      : FE80::2E0:FCFF:FED6:5041      Preference    : 10
 Cost         : 3                             Protocol      : OSPFv3
 RelayNextHop : ::                            TunnelID      : 0x0
 Interface    : GigabitEthernet0/0/0          Flags         : D

 Destination  : 2000:14::                     PrefixLength  : 64
 NextHop      : FE80::2E0:FCFF:FED6:5041      Preference    : 10
 Cost         : 2                             Protocol      : OSPFv3
 RelayNextHop : ::                            TunnelID      : 0x0
 Interface    : GigabitEthernet0/0/0          Flags         : D

 Destination  : 2000:46::                     PrefixLength  : 64
 NextHop      : FE80::2E0:FCFF:FED6:5041      Preference    : 10
 Cost         : 3                             Protocol      : OSPFv3
 RelayNextHop : ::                            TunnelID      : 0x0
 Interface    : GigabitEthernet0/0/0          Flags         : D
```

通过以上输出可以看出，只有 6::6/128 的路由，没有公司 B 的路由。

查看 AR3 的路由表：

```
[AR3]display ipv6 routing-table protocol ospfv3
Public Routing Table : OSPFv3
Summary Count : 5

OSPFv3 Routing Table's Status : < Active >
Summary Count : 3

 Destination  : 5::5                          PrefixLength  : 128
 NextHop      : FE80::2E0:FCFF:FED6:5043      Preference    : 10
 Cost         : 3                             Protocol      : OSPFv3
 RelayNextHop : ::                            TunnelID      : 0x0
 Interface    : GigabitEthernet0/0/0          Flags         : D

 Destination  : 2000:14::                     PrefixLength  : 64
 NextHop      : FE80::2E0:FCFF:FED6:5043      Preference    : 10
 Cost         : 2                             Protocol      : OSPFv3
 RelayNextHop : ::                            TunnelID      : 0x0
 Interface    : GigabitEthernet0/0/0          Flags         : D

 Destination  : 2000:45::                     PrefixLength  : 64
 NextHop      : FE80::2E0:FCFF:FED6:5043      Preference    : 10
 Cost         : 3                             Protocol      : OSPFv3
 RelayNextHop : ::                            TunnelID      : 0x0
 Interface    : GigabitEthernet0/0/0          Flags         : D
```

通过以上输出可以看出，只有 5::5/128 的路由，没有公司 A 的路由。

9.2.3 配置 BGP4+

1. 实验环境

配置接口 IP 地址，AS 100 内部运行 IS-IS IPv6，AR2 作为 AS 100 的 RR，AR1、AR3、AR4 作为 AR2 的反射器客户端。配置相应的 BGP4+，在 AR1 上创建环回口 100，IPv6 地址为 2002::1/128，通告给 BGP 进程，使其他几台路由器能够学习到 BGP4+ 的路由信息。

2. 实验拓扑

配置 BGP4+ 的实验拓扑如图 9-5 所示。

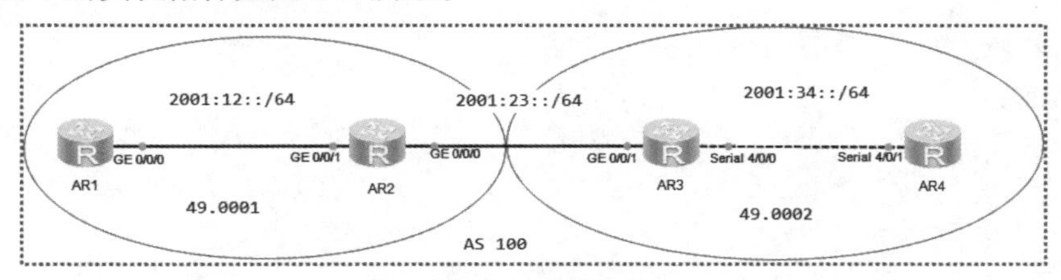

图 9-5　配置 BGP4+ 的实验拓扑

3. 实验步骤

步骤 1： 配置 BGP4+IP 地址规划表见表 9-3。

表 9-3　配置 BGP4+ IP 地址规划表

设备名称	接口编号	IP 地址
AR1	G0/0/0	2001:12::1/64
	LoopBack 0	1::1/128
AR2	G0/0/0	2001:23::2/64
	G0/0/1	2001:12::2/64
	LoopBack 0	2::2/128
AR3	G0/0/1	2001:23::3/64
	S4/0/0	2001:34::3/64
	LoopBack 0	3::3/128
AR4	S4/0/1	2001:34::4/64
	LoopBack 0	4::4/128

步骤 2： 配置 IGP 协议。

AR1 的配置：

```
[AR1]isis
[AR1-isis-1]network-entity 49.0001.0000.0000.0001.00 //配置NET地址
[AR1-isis-1]is-level level-1                 //配置路由器类型为 level-1 设备
[AR1-isis-1]ipv6 enable topology ipv6        //使能 IS-IS 的 IPv6 功能并且开启 IS-IS 的多
                                             //拓扑能力
[AR1]interface g0/0/0
[AR1-GigabitEthernet0/0/0]isis ipv6 enable  //接口使能 IS-IS IPv6 功能
[AR1]interface LoopBack 0
[AR1-LoopBack0]isis ipv6 enable
```

AR2 的配置：

```
[AR2]isis
[AR2-isis-1]network-entity 49.0001.0000.0000.0002.00
[AR2-isis-1]ipv6 enable topology ipv6
[AR2]interface  g0/0/0
[AR2-GigabitEthernet0/0/0]isis ipv6 enable
[AR2]interface  g0/0/1
[AR2-GigabitEthernet0/0/1]isis ipv6 enable
[AR2]interface  LoopBack 0
[AR2-LoopBack0]isis ipv6 enable
```

AR3 的配置：

```
[AR3]isis
[AR3-isis-1]network-entity 49.0002.0000.0000.0003.00
[AR3-isis-1]is-level level-2
[AR3-isis-1]ipv6 enable topology ipv6
[AR3]interface  g0/0/1
[AR3-GigabitEthernet0/0/1]isis  ipv6 enable
[AR3]interface  s4/0/0
[AR3-Serial4/0/0]isis ipv6 enable
[AR3]interface  LoopBack 0
[AR3-LoopBack0]isis ipv6 enable
```

AR4 的配置：

```
[AR4]isis
[AR4-isis-1]network-entity 49.0002.0000.0000.0004.00
[AR4-isis-1]is-level level-2
[AR4-isis-1]ipv6 enable topology ipv6
[AR4]interface s4/0/1
[AR4-Serial4/0/1]isis ipv6 enable
[AR4]interface LoopBack 0
[AR4-LoopBack0]isis ipv6 enable
```

步骤 3：配置 BGP4+。

AR2 的配置：

```
[AR2]bgp 100
[AR2-bgp]router-id 2.2.2.2                          //手动配置 Router ID
[AR2-bgp]peer 1::1 as-number 100                    //配置 IDGP 对等体 1::1
[AR2-bgp]peer 1::1 connect-interface LoopBack 0     //配置 TCP 连接接口为 LoopBack 0
[AR2-bgp]ipv6-family unicast                        //进入 IPv6 单播地址族
[AR2-bgp-af-ipv6]peer 1::1 enable                   //使能 1::1 邻居关系
[AR2-bgp-af-ipv6]peer 1::1 reflect-client
[AR2-bgp]peer 3::3 as-number 100
[AR2-bgp]peer 3::3 connect-interface LoopBack 0
[AR2-bgp]peer 4::4 as-number 100
[AR2-bgp]peer 4::4 connect-interface LoopBack 0
[AR2-bgp]ipv6-family unicast
[AR2-bgp-af-ipv6]peer 3::3 enable
```

```
[AR2-bgp-af-ipv6]peer 4::4 enable
[AR2-bgp-af-ipv6]peer 3::3 reflect-client
[AR2-bgp-af-ipv6]peer 4::4 reflect-client
```

AR1 的配置：

```
[AR1]bgp 100
[AR1-bgp]router-id 1.1.1.1
[AR1-bgp]peer 2::2 as-number 100
[AR1-bgp]peer 2::2 connect-interface LoopBack 0
[AR1-bgp]ipv6-family
[AR1-bgp-af-ipv6]peer 2::2 enable
```

AR3 的配置：

```
[AR3]bgp 100
[AR3-bgp]router-id 3.3.3.3
[AR3-bgp]peer 2::2 as-number 100
[AR3-bgp]peer 2::2 connect-interface LoopBack0
[AR3-bgp]ipv6-family unicast
[AR3-bgp-af-ipv6]peer 2::2 enable
```

AR4 的配置：

```
[AR4]bgp 100
[AR4-bgp]router-id 4.4.4.4
[AR4-bgp]peer 2::2 as-number 100
[AR4-bgp]peer 2::2 connect-interface LoopBack0
[AR4-bgp]ipv6-family unicast
[AR4-bgp-af-ipv6]peer 2::2 enable
```

在 AR2 上查看 BGP 的邻居关系：

```
[AR2]display bgp ipv6 peer
 BGP local router ID : 2.2.2.2
 Local AS number : 100
 Total number of peers : 3                 Peers in established state : 3
  Peer    V    AS     MsgRcvd     MsgSent    OutQ   Up/Down      State        PrefRcv
  1::1    4    100    4           6          0      00:02:56     Established  0
  3::3    4    100    2           7          0      00:00:03     Established  0
  4::4    4    100    2           3          0      00:00:49     Established
```

通过以上输出可以看出，AR2 分别与 AR1、AR3、AR4 建立了 IBGP 的邻居关系。

在 AR1 上创建环回口 100，并且将路由注入到 BGP4+ 中。

AR1 的配置：

```
[AR1]interface LoopBack 100
[AR1-LoopBack100]ipv6 enable
[AR1-LoopBack100]ipv6 address 2002::1 128
[AR1]bgp 100
[AR1-bgp]ipv6-family unicast
[AR1-bgp-af-ipv6]network  2002::1 128 // 注入 BGP 路由信息
```

配置前在 AR4 的 S4/0/0 接口抓包查看现象，抓取的报文为 UPDATE 报文，如图 9-6 所示。

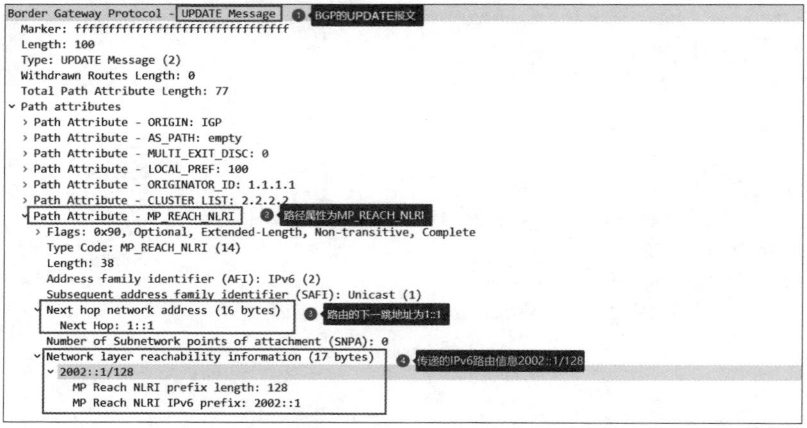

```
Border Gateway Protocol - UPDATE Message        ❶ BGP的UPDATE报文
  Marker: ffffffffffffffffffffffffffffffff
  Length: 100
  Type: UPDATE Message (2)
  Withdrawn Routes Length: 0
  Total Path Attribute Length: 77
 ˅ Path attributes
   › Path Attribute - ORIGIN: IGP
   › Path Attribute - AS_PATH: empty
   › Path Attribute - MULTI_EXIT_DISC: 0
   › Path Attribute - LOCAL_PREF: 100
   › Path Attribute - ORIGINATOR_ID: 1.1.1.1
   › Path Attribute - CLUSTER_LIST: 2.2.2.2
   ˅ Path Attribute - MP_REACH_NLRI        ❷ 路径属性为MP_REACH_NLRI
     › Flags: 0x90, Optional, Extended-Length, Non-transitive, Complete
       Type Code: MP_REACH_NLRI (14)
       Length: 38
       Address family identifier (AFI): IPv6 (2)
       Subsequent address family identifier (SAFI): Unicast (1)
     ˅ Next hop network address (16 bytes)        ❸ 路由的下一跳地址为1::1
         Next Hop: 1::1
       Number of Subnetwork points of attachment (SNPA): 0
     ˅ Network layer reachability information (17 bytes)        ❹ 传递的IPv6路由信息2002::1/128
       ˅ 2002::1/128
           MP Reach NLRI prefix length: 128
           MP Reach NLRI IPv6 prefix: 2002::1
```

图 9-6　AR4 的 S4/0/0 接口的抓包结果

从图 9-6 中可以看出，BGP4+ 不再与传统的 BGP 一样，即不再使用 NLRI 来传递路由信息，而是通过 MP_REACH_NLRI 来传递 IPv6 的路由信息。

在 AR4 中查看 BGP4+ 的路由表：

```
[AR4]display bgp ipv6 routing-table
 BGP Local router ID is 4.4.4.4
 Status codes: * - valid, > - best, d - damped,
               h - history,  i - internal, s - suppressed, S - Stale
               Origin : i - IGP, e - EGP, ? - incomplete
 Total Number of Routes: 1
 *>i Network  : 2002::1                                  PrefixLen    : 128
     NextHop  : 1::1                                     LocPrf       : 100
     MED      : 0                                        PrefVal      : 0
     Label    :
     Path/Ogn : i
```

通过以上输出可以看出，在 AR4 学习到了 AR1 通告的 2002::1/128 的路由信息。

在 AR4 中测试 2002::1/128 的连通性：

```
[AR4]ping ipv6 2002::1
  PING 2002::1 : 56  data bytes, press CTRL_C to break
    Reply from 2002::1
    bytes=56 Sequence=1 hop limit=62  time = 30 ms
    Reply from 2002::1
    bytes=56 Sequence=2 hop limit=62  time = 30 ms
    Reply from 2002::1
    bytes=56 Sequence=3 hop limit=62  time = 30 ms
    Reply from 2002::1
    bytes=56 Sequence=4 hop limit=62  time = 40 ms
    Reply from 2002::1
    bytes=56 Sequence=5 hop limit=62  time = 30 ms
```

```
  --- 2002::1 ping statistics ---
    5 packet(s) transmitted
    5 packet(s) received
    0.00% packet loss
    round-trip min/avg/max = 30/32/40 ms
```

测试结果为可以通信。

接下来在 AR1 上将 LoopBack 100 接口的 IPv6 地址删除，从而来模拟 BGP4+ 撤销路由信息的过程。

AR1 上的配置：

```
[AR1]interface LoopBack 100
[AR1-LoopBack100]undo ipv6 address  2002::1 128 //删除接口的 IPv6 地址 2002::1/128
```

再次查看 AR4 的抓包结果，抓取的报文为 UPDATE 报文，如图 9-7 所示。

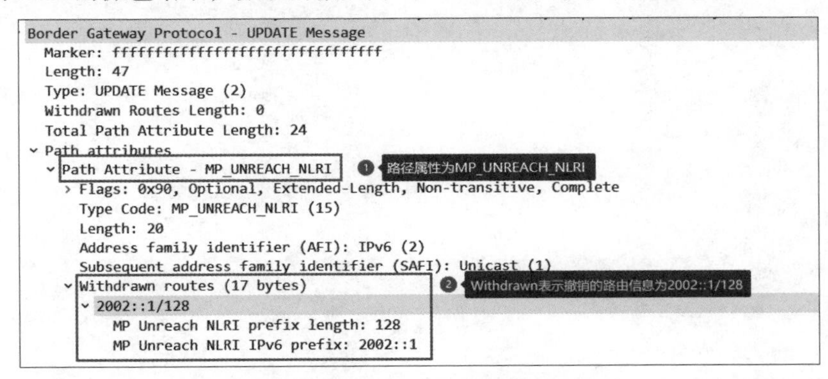

图 9-7　AR4 的 S4/0/0 接口的抓包结果

结果表明，BGP4+ 使用 MP_UNREACH_NLRI 撤销 IPv6 的路由信息。

查看 AR4 的路由表此时为空，代表实验成功。

第 10 章

IPv6 过渡技术

10.1　IPv6 过渡技术概述

当前世界上不同地区对部署 IPv6 的需求程度不一样，并且当前 IPv4 网络仍然占主流地位，因此短时间内 IPv6 和 IPv4 将会共存，可以借助双栈技术、隧道技术、转换技术，实现 IPv4/IPv6 共存业务的互通。

10.1.1　IPv4/IPv6 双栈

设备支持 IPv4/IPv6，IPv4/IPv6 在网络中独立部署，在一段时间内并存，并对现有 IPv4 业务影响较小。IPv4/IPv6 双栈示意图如图 10-1 所示。

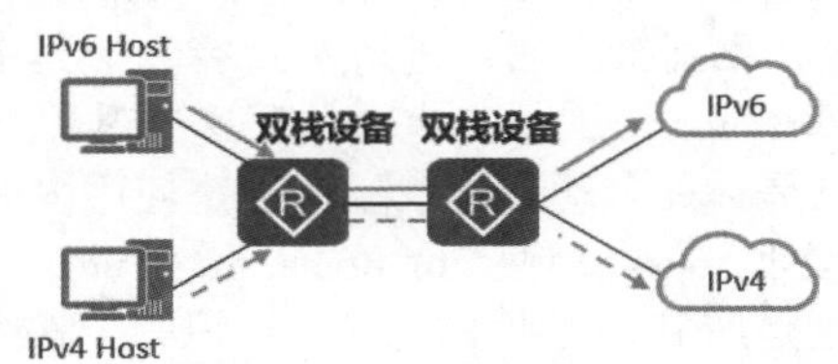

图 10-1　IPv4/IPv6 双栈示意图

演进方案相对简单、易理解。网络规划设计工作量相对更少。

10.1.2　隧道技术 IPv6 over IPv4

IPv6 over IPv4 是通过隧道技术，使 IPv6 报文在 IPv4 网络中传输，实现 IPv6 网络之间的孤岛互联。

1. 手动隧道

手动隧道即边界设备不能自动获得隧道目的 IPv4 地址，需要手动配置，报文才能正确发送至隧道终点。手动隧道分为 IPv6 over IPv4 手动隧道和 IPv6 over IPv4 GRE 隧道。

（1）IPv6 over IPv4 手动隧道。IPv6 over IPv4 手动隧道直接把 IPv6 报文封装到 IPv4 报文中，IPv6 报文作为 IPv4 报文的净载荷。手动隧道的源地址和目的地址也是手动指定的，它提供了一个点到点的连接。

（2）IPv6 over IPv4 GRE 隧道。IPv6 over IPv4 GRE 隧道使用标准的 GRE 隧道技术提供了点到点连接服务，需要手动指定隧道的端点地址。

2. 自动隧道

在自动隧道中，用户仅需要配置设备隧道的起点，隧道的终点由设备自动生成。为了使设备能够自动产生终点，隧道接口的 IPv6 地址采用内嵌 IPv4 地址的特殊 IPv6 地址形式。设备从 IPv6 报文中的目的 IPv6 地址中解析出 IPv4 地址，然后以这个 IPv4 地址代表的节点作为隧道的终点。

根据 IPv6 报文封装的不同，自动隧道又可以分为 IPv4 兼容 IPv6 自动隧道、6to4 隧道和 ISATAP 隧道。

（1）IPv4 兼容 IPv6 自动隧道。IPv4 兼容 IPv6 自动隧道承载的 IPv6 报文的目的地址（自动隧道所使用的特殊地址）是 IPv4 兼容 IPv6 地址。IPv4 兼容 IPv6 地址的前 96 位全部为 0，后 32 位为 IPv4 地址。其地址格式如图 10-2 所示。

下面以图 10-3 为例说明 IPv4 兼容 IPv6 自动隧道的转发机制。

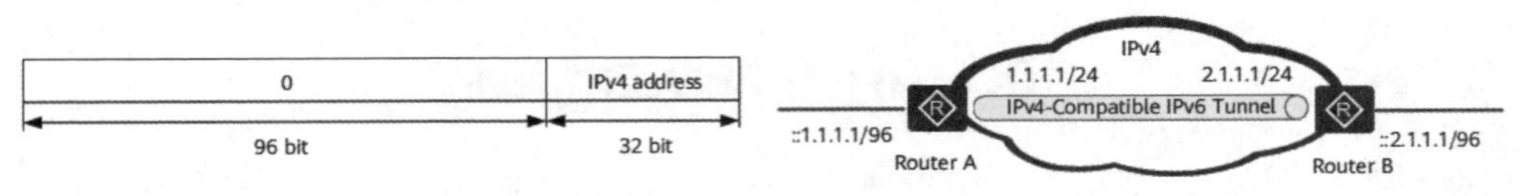

图 10-2　IPv4 兼容 IPv6 地址格式　　　　图 10-3　IPv4 兼容 IPv6 自动隧道的转发机制

需要经过 Router A 发给 Router B 的 IPv6 报文到达 Router A 后，以目的地址 ::2.1.1.1 查找 IPv6 路由，发现路由的下一跳为虚拟的 Tunnel 口。由于 Router A 上配置的隧道类型是 IPv4 兼容 IPv6 自动隧道，因此 Router A 对 IPv6 报文进行了封装。IPv6 报文被封装为 IPv4 报文，IPv4 报文中的源地址为隧道的起始点地址 1.1.1.1，而目的 IP 地址直接从 IPv4 兼容 IPv6 地址 ::2.1.1.1 的后 32 位复制过来，即 2.1.1.1。这个报文被路由器从隧道口发出后，在 IPv4 的网络中被路由转发到目的地址 2.1.1.1，即 Router B。Router B 收到报文后，进行解封装，把其中的 IPv6 报文取出，送给 IPv6 协议栈进行处理。Router B 返回 Router A 的报文也是按照这个过程进行的。

（2）6to4 隧道。6to4 隧道也是一种自动隧道，隧道也是使用内嵌在 IPv6 地址中的 IPv4 地址建立的。与 IPv4 兼容自动隧道不同，6to4 隧道支持 Router 到 Router、Host 到 Router、Router 到 Host、Host 到 Host。这是因为 6to4 地址是用 IPv4 地址作为网络标识的。其地址格式如图 10-4 所示。

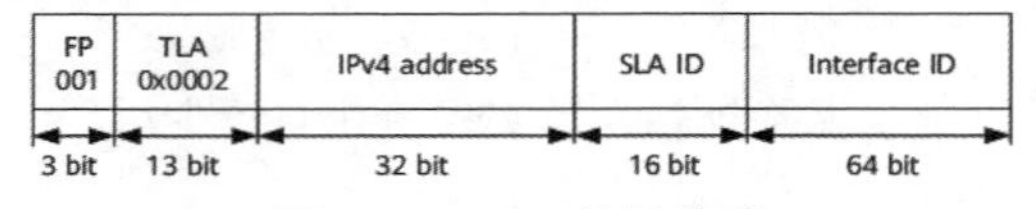

图 10-4　6to4 地址格式

① FP：可聚合全球单播地址的格式前缀（Format Prefix），其值为 001。

② TLA：顶级聚合标识符（Top Level Aggregator），其值为 0x0002。

③ SLA：站点级聚合标识符（Site Level Aggregator）。

6to4 地址可以表示为 2002::/16，而一个 6to4 网络可以表示为 2002:IPv4 地址 ::/48。6to4 地址的网络前缀长度为 64 位，其中前 48 位（2002: a.b.c.d）被分配给路由器上的 IPv4 地址决定，用户不能改变，而后 16 位（SLA）是由用户自己定义的。6to4 隧道的封装和转发过程如图 10-5 所示。

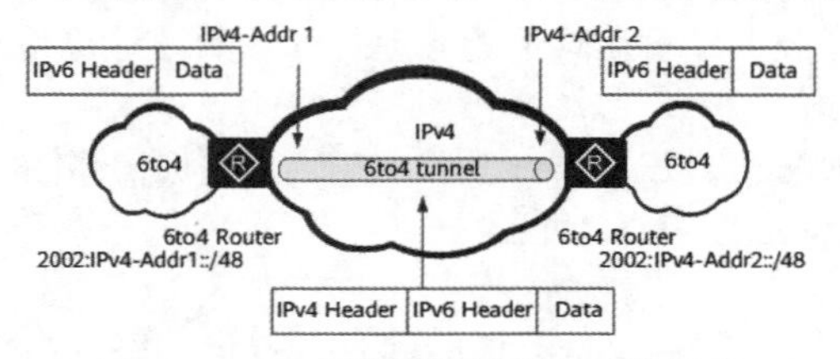

图 10-5　6to4 隧道的封装和转发过程

（3）ISATAP 隧道。ISATAP（Intra-Site Automatic Tunnel Addressing Protocol，站内自动隧道寻址协议）是另外一种自动隧道技术。ISATAP 隧道同样使用了内嵌 IPv4 地址的特殊 IPv6 地址形式，只是与 6to4 地址不同的是，6to4 地址是使用 IPv4 地址作为网络前缀，而 ISATAP 用 IPv4 地址作为接口标识。其地址接口标识符格式如图 10-6 所示。

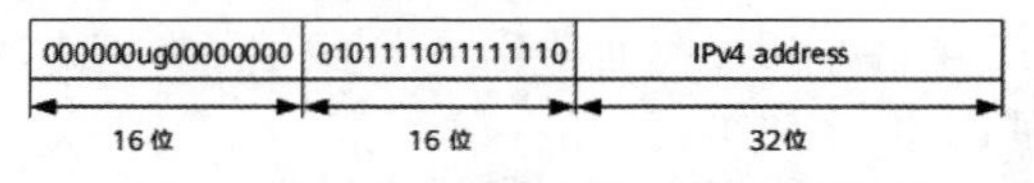

图 10-6　ISATAP 地址接口标识符格式

如果 IPv4 地址是全局唯一的，则 u 位为 1；否则 u 位为 0。g 位是 IEEE 群体 / 个体标志。由于 ISATAP 是通过接口标识来表现的，因此，ISATAP 地址有全局单播地址、链路本地地址、ULA 地址、组播地址等形式。ISATAP 地址的前 64 位是通过向 ISATAP 路由器发送请求得到的，它可以进行地址自动配置。在 ISATAP 隧道的两端设备之间可以运行 ND 协议。ISATAP 隧道将 IPv4 网络看作一个非广播多路访问（Non-Broadcast Muitiple Access，NBMA）链路。

ISATAP 过渡机制允许在现有的 IPv4 网络内部署 IPv6，该技术简单且扩展性很好，可以用于本地站点的过渡。ISATAP 支持 IPv6 站点本地路由和全局 IPv6 路由域，以及自动 IPv6 隧道。ISATAP 同时还可以与 NAT 结合，从而可以使用站点内部非全局唯一的 IPv4 地址。典型的 ISATAP 隧道应用是在站点内部，因此其内嵌的 IPv4 地址不需要是全局唯一的。

10.1.3 隧道技术 6VPE

对于网络中的运营商来说，无须另外新建 IPv6 骨干网络，可以利用现有 IPv4 网络为用户提供 IPv6 服务来连接网络中的 IPv6 孤岛，6PE（IPv6 Provider Edge，IPv6 供应商边缘）正是基于这个理念来设计的。

6PE 是一种在目前的 IPv4 网络中利用 MPLS 隧道技术为不同地区的被分割的 IPv6 网络提供连通服务的解决方案，它是在 ISP 的 PE 上实现 IPv4/IPv6 双协议栈，利用 MP-BGP 为其分配的标签标识 IPv6 路由，从而通过 PE 之间的 IPv4 隧道实现 IPv6 孤岛之间的互通。

6PE 技术本质上相当于将所有通过 6PE 连接的 IPv6 业务都放在一个 VPN 内，无法实现逻辑隔离，因此只能用于开放的、无保护的 IPv6 网络互联，如果需要对所连接的 IPv6 业务进行逻辑隔离，即实现 IPv6 VPN，则需要借助于 6VPE（IPv6 VPN Provider Edge，IPv6 VPN 供应商边缘）技术。

10.2 配置 IPv6 实验

10.2.1 配置 IPv6 over IPv4 手工隧道

1. 实验目的

R1 → R2 → R3 为 IPv4 网络，现在需要在 R1 → R3 上配置 IPv6 over IPv4 手动隧道实现 PC1 能访问 PC2 的 IPv6 网络。

扫一扫，看视频

2. 实验拓扑

配置 IPv6 over IPv4 手工隧道的实验拓扑如图 10-7 所示。

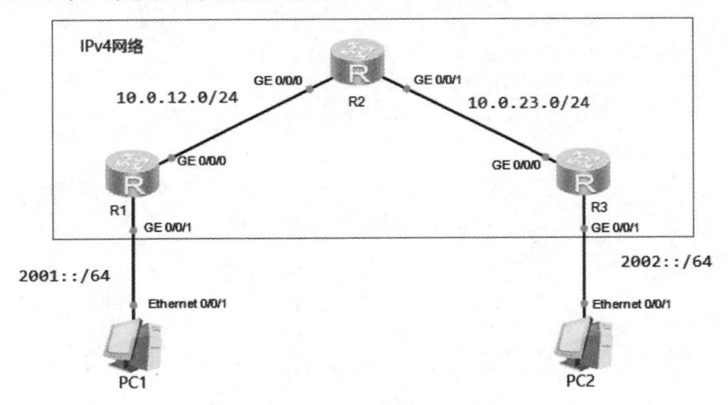

图 10-7 配置 IPv6 over IPv4 手工隧道的实验拓扑

3. 实验步骤

步骤 1: 配置接口的 IPv4 以及 IPv6 地址，配置 IPv6 over IPv4 手工隧道 IP 地址规划表见表 10-1。

表 10-1　配置 IPv6 over IPv4 手工隧道 IP 地址规划表

设备名称	接口编号	IP 地址
R1	G0/0/0	10.0.12.1/24
	G0/0/1	2001::254/64
R2	G0/0/0	10.0.12.2/24
	G0/0/1	10.0.23.2/24
R3	G0/0/0	10.0.23.3/24
	G0/0/1	2002::254/64

步骤 2: 在 R1 和 R3 上配置静态路由，让 R1 和 R3 的 IPv4 网络能够互通。

R1 的配置：

```
[R1]ip route-static 10.0.23.0 255.255.255.0 10.0.12.2
```

R3 的配置：

```
[R3]ip route-static 10.0.12.0 255.255.255.0 10.0.23.2
```

步骤 3: 配置 6to4 的手工隧道。

R1 的配置：

```
[R1]interface Tunnel0/0/0
[R1-Tunnel0/0/0]ipv6 enable
[R1-Tunnel0/0/0]ipv6 address 3001::1/64         //配置隧道的 IPv6 地址
[R1-Tunnel0/0/0]tunnel-protocol ipv6-ipv4       //配置隧道协议为 6to4
[R1-Tunnel0/0/0]source GigabitEthernet0/0/0     //配置隧道的源为 G0/0/0 接口
[R1-Tunnel0/0/0]destination 10.0.23.3           //隧道的目的为 R3 接口的 IPv4 地址
```

R3 的配置：

```
[R3]interface Tunnel0/0/0
[R3-Tunnel0/0/0]ipv6 enable
[R3-Tunnel0/0/0]ipv6 address 3001::2/64
[R3-Tunnel0/0/0]tunnel-protocol ipv6-ipv4
[R3-Tunnel0/0/0]source GigabitEthernet0/0/0
[R3-Tunnel0/0/0]destination 10.0.12.1
```

查看隧道状态（为 up 即可）：

```
[R1]display ipv6 interface brief
*down: administratively down
(l): loopback
(s): spoofing
Interface                        Physical              Protocol
GigabitEthernet0/0/1             up                    up
[IPv6 Address]2001::254
Tunnel0/0/0                      up                    up
[IPv6 Address]3001::1
```

步骤 4: 在 R1 和 R3 上配置去往对端 PC 的 IPv6 路由，本实验使用 OSPFv3。

（1）设备开启 OSPFv3。

R1 的配置：

```
[R1]ospfv3 1
[R1-ospfv3-1]router-id 1.1.1.1
```

R3 的配置：

```
[R3]ospfv3 1
[R3-ospfv3-1]router-id 3.3.3.3
```

（2）设备接口使能 OSPFv3。

R1 的配置：

```
[R1]interface Tunnel0/0/0
[R1-Tunnel0/0/0]ospfv3 1 area 0
[R1-Tunnel0/0/0]interface GigabitEthernet0/0/1
[R1-GigabitEthernet0/0/1]ospfv3 1 area 0
```

R2 的配置：

```
[R3]interface Tunnel0/0/0
[R3-Tunnel0/0/0]ospfv3 1 area 0
[R3-Tunnel0/0/0]interface GigabitEthernet0/0/1
[R3-GigabitEthernet0/0/1]ospfv3 1 area 0
```

查看 OSPFv3 的邻居关系（状态为 Full 即可）：

```
[R1]display ospfv3 peer
OSPFv3 Process (1)
OSPFv3 Area (0.0.0.0)
Neighbor ID        Pri     State           Dead Time      Interface      Instance ID
3.3.3.3            1       Full/-          00:00:40       Tun0/0/0       0
```

查看 IPv6 的路由表，R1 可以通过 OSPFv3 学习到对端的 IPv6 路由，并且下一跳为隧道口：

```
[R1]display ipv6 routing-table protocol ospfv3
Public Routing Table : OSPFv3
Summary Count : 3
OSPFv3 Routing Table's Status : < Active >
Summary Count : 1
 Destination   : 2002:: //对端的目的网段         PrefixLength     : 64
 NextHop       : FE80::A00:1703                Preference       : 10
 Cost          : 1563                          Protocol         : OSPFv3
 RelayNextHop  : ::                            TunnelID         : 0x0
 Interface     : Tunnel0/0/0                   Flags            : D

OSPFv3 Routing Table's Status : < Inactive >
Summary Count : 2
 Destination   : 2001::                        PrefixLength     : 64
 NextHop       : ::                            Preference       : 10
 Cost          : 1                             Protocol         : OSPFv3
 RelayNextHop  : ::                            TunnelID         : 0x0
```

```
Interface     : GigabitEthernet0/0/1        Flags          :

Destination   : 3001::                      PrefixLength   : 64
NextHop       : ::                          Preference     : 10
Cost          : 1562                        Protocol       : OSPFv3
RelayNextHop  : ::                          TunnelID       : 0x0
Interface     : Tunnel0/0/0                 Flags          :
```

配置 PC1 和 PC2 的 IPv6 地址，如图 10–8 和图 10–9 所示。

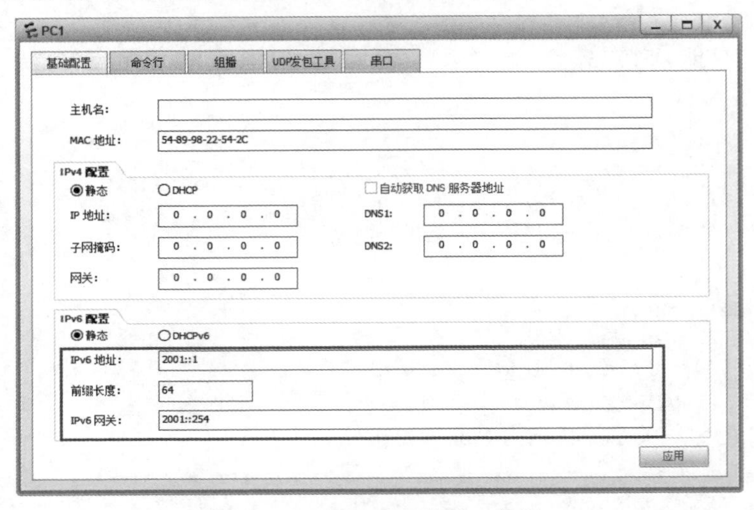

图 10–8　配置 PC1 的 IPv6 地址

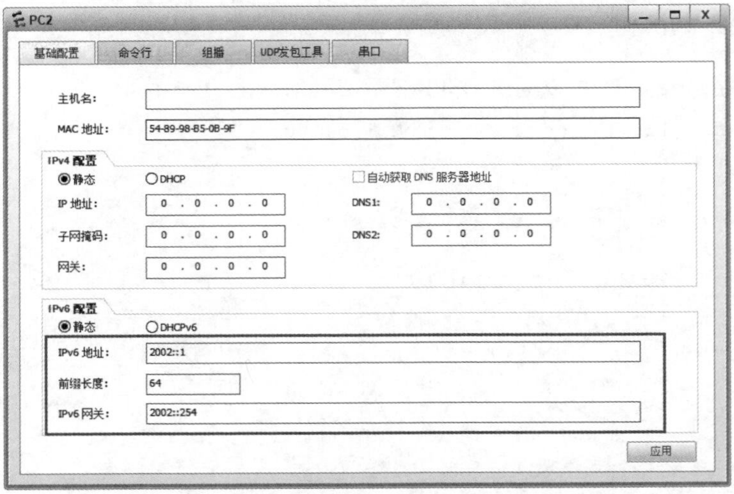

图 10–9　配置 PC2 的 IPv6 地址

测试：使用 PC1 访问 PC2。

```
PC>ping 2002::1
```

```
Ping 2002::1: 32 data bytes, Press Ctrl_C to break
From 2002::1: bytes=32 seq=1 hop limit=0 time<1 ms
From 2002::1: bytes=32 seq=2 hop limit=0 time<1 ms
From 2002::1: bytes=32 seq=3 hop limit=0 time<1 ms
From 2002::1: bytes=32 seq=4 hop limit=0 time<1 ms
From 2002::1: bytes=32 seq=5 hop limit=0 time<1 ms

--- 2002::1 ping statistics ---
  5 packet(s) transmitted
  5 packet(s) received
  0.00% packet loss
  round-trip min/avg/max = 0/0/0 ms
```

10.2.2　配置 6to4 隧道

1. 实验目的

在 R1 和 R3 之间使用 Tunnel 接口创建 6to4 隧道，实现 PC1 和 PC2 互访。

2. 实验拓扑

配置 6to4 隧道的实验拓扑如图 10–10 所示。

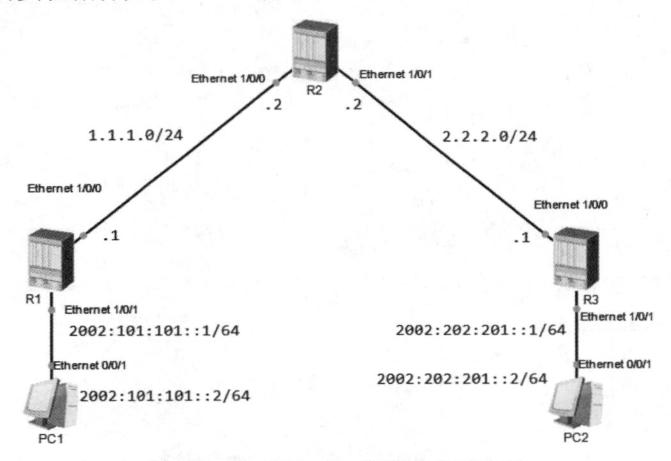

图 10–10　配置 6to4 隧道的实验拓扑

3. 实验步骤

步骤 1：配置 IP 地址，配置 6to4 隧道 IP 地址规划表见表 10–2。

表 10–2　配置 6to4 隧道 IP 地址规划表

设备名称	接口编号	IP 地址
R1	Ethernet 1/0/0	1.1.1.1/24
	Ethernet 1/0/1	2002:101:101::1/64
R2	Ethernet 1/0/0	1.1.1.2/24
	Ethernet 1/0/1	2.2.2.2/24

续表

设备名称	接口编号	IP 地址
R3	Ethernet 1/0/0	2.2.2.1/24
	Ethernet 1/0/1	2002:202:201::1/64

步骤 2：配置 R1 和 R3 的 IPv4 路由，使两端的 IPv4 网络能够互访。

R1 的配置：

```
[R1]system-view immediately
[R1]ip route-static 0.0.0.0 0.0.0.0 1.1.1.2
```

R3 的配置：

```
[R3]system-view immediately
[R3]ip route-static 0.0.0.0 0.0.0.0 2.2.2.2
```

测试 R1 和 R3 的接口 IP 是否能够通信：

```
[R1]ping 2.2.2.1
  PING 2.2.2.1: 56 data bytes, press CTRL_C to break
    Reply from 2.2.2.1: bytes=56 Sequence=1 ttl=254 time=10 ms
    Reply from 2.2.2.1: bytes=56 Sequence=2 ttl=254 time=17 ms
    Reply from 2.2.2.1: bytes=56 Sequence=3 ttl=254 time=11 ms
    Reply from 2.2.2.1: bytes=56 Sequence=4 ttl=254 time=11 ms
    Reply from 2.2.2.1: bytes=56 Sequence=5 ttl=254 time=12 ms

  --- 2.2.2.1 ping statistics ---
    5 packet(s) transmitted
    5 packet(s) received
    0.00% packet loss
round-trip min/avg/max = 10/12/17 ms
```

步骤 3：配置 6to4 隧道。

R1 的配置：

```
[R1]interface Tunnel0
[R1-Tunnel0]ipv6 enable
[R1-Tunnel0]ipv6 address auto link-local   //配置 6to4 隧道的 IPv6 地址
[R1-Tunnel0]tunnel-protocol ipv6-ipv4 6to4 //配置隧道为 6to4 隧道
[R1-Tunnel0]source 1.1.1.1                 //配置隧道的源 IP 地址
```

查看隧道的地址是否配置成功：

```
[R1]display ipv6 int brief
*down: administratively down
!down: FIB overload down
(l): loopback
(s): spoofing
Interface                    Physical                 Protocol                VPN
Tunnel0                      up                       up                      --
[IPv6 Address]FE80::727B:E800:1AC2:6376
```

R3 的配置：

```
[R3]interface Tunnel0
[R3-Tunnel0]ipv6 enable
[R3-Tunnel0]ipv6 address auto link-local
[R3-Tunnel0]tunnel-protocol ipv6-ipv4 6to4
[R3-Tunnel0]source 2.2.2.1
```

步骤 4：配置连接 PC 的接口 IPv6 地址，这个 IPv6 地址的组成为 2002:( ) ::/64，( ) 内为隧道的源 IP 地址转换过来的 IPv6 地址前缀。例如，R1 的 Tunnel 0 口的 IPv4 源地址为 1.1.1.1，那么 ( ) 内的数值为 0101: 0101，即此接口的 IPv6 地址为 2002:101:101::1/64。

R1 的配置：

```
[R1]interface Ethernet1/0/1
[R1-Ethernet1/0/1]ipv6 enable
[R1-Ethernet1/0/1]ipv6 address 2002:101:101::1/64
```

R3 的配置：

```
[R3]interface Ethernet1/0/1
[R3-Ethernet1/0/1]ipv6 enable
[R3-Ethernet1/0/1]ipv6 address 2002:202:201::1/64
```

步骤 5：配置去往对端 IPv6 的静态路由，使用 Tunnel 口作为出接口。

R1 的配置：

```
[R1]ipv6 route-static 2002:: 16 Tunnel 0
```

R3 的配置：

```
[R3]ipv6 route-static 2002:: 16 Tunnel 0
```

步骤 6：使用 PC1 访问 PC2, 并且在 R1 的 E 1/0/0 接口抓包，配置 PC1 的 IPv6 地址，如图 10-11 所示。

图 10-11　配置 PC1 的 IPv6 地址

PC1 测试结果如图 10-12 所示。

图 10-12　PC1 测试结果

R1 的 E1/0/0 接口的抓包结果如图 10-13 所示。

```
Wireshark · 分组 24 · ·
> Frame 24: 114 bytes on wire (912 bits), 114 bytes captured (912 bits) on interface -, id 0
> Ethernet II, Src: 38:4e:75:02:01:00 (38:4e:75:02:01:00), Dst: 38:4e:75:01:01:00 (38:4e:75:01:01:00)
> Internet Protocol Version 4, Src: 1.1.1.1, Dst: 2.2.2.1  ❶ 目的IP为对端隧道的源地址，通过IPv6的目的地址中的202:201获取
> Internet Protocol Version 6, Src: 2002:101:101::2, Dst: 2002:202:201::2
> Internet Control Message Protocol v6
```

图 10-13　R1 的 E1/0/0 接口的抓包结果

10.2.3　配置 6VPE

1. 实验目的

AS 100 为 MPLS 域，在 PE1 和 PE2 上允许 6VPE 实现 CE1 和 CE2 的 IPv6 网络能够通过 AS 100 的 IPv4 网络互访。

2. 实验拓扑

配置 6VPE 的实验拓扑如图 10-14 所示。

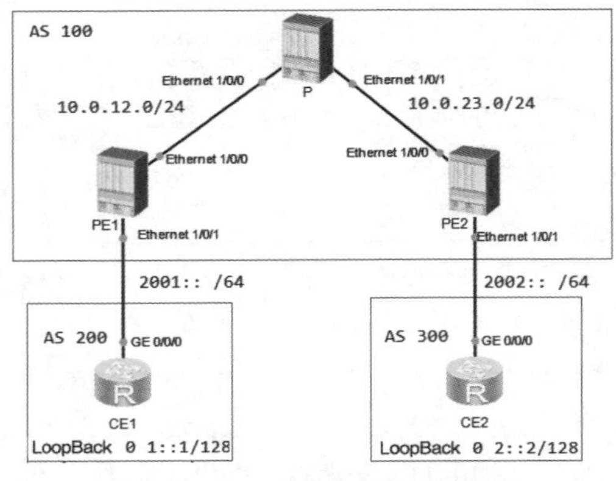

图 10-14　配置 6VPE 的实验拓扑

3. 实验步骤

步骤 1：配置 AS 100 的 IPv4 地址及 CE 设备的 IPv6 地址，配置 6VPE IP 地址规划表见表 10-3。

表 10-3　配置 6VPE IP 地址规划表

设备名称	接口标号	IP 地址
P	Ethernet1/0/0	10.0.12.2/24
	Ethernet1/0/1	10.0.23.2/24
	LoopBack 0	2.2.2.2/32
PE1	Ethernet1/0/0	10.0.12.1/24
	Ethernet1/0/1	2001::1/64
	LoopBack 0	1.1.1.1/32
PE2	Ethernet1/0/0	10.0.23.3/24
	Ethernet1/0/1	2002::1/64
	LoopBack 0	3.3.3.3/32

步骤 2：配置 AS 100 的 IGP 协议。

PE1 的配置：

```
[PE1]ospf 1
[PE1-ospf-1]area 0
[PE1-ospf-1-area-0.0.0.0]network 1.1.1.1 0.0.0.0
[PE1-ospf-1-area-0.0.0.0]network 10.0.12.0 0.0.0.255
```

PE2 的配置：

```
[PE2]ospf 1
[PE2-ospf-1]area 0
[PE2-ospf-1-area-0.0.0.0]network 3.3.3.3 0.0.0.0
[PE2-ospf-1-area-0.0.0.0]network 10.0.23.0 0.0.0.255
```

P 的配置：

```
[P]ospf 1
[P-ospf-1]area 0
[P-ospf-1-area-0.0.0.0]network 2.2.2.2 0.0.0.0
[P-ospf-1-area-0.0.0.0]network 10.0.12.0 0.0.0.255
[P-ospf-1-area-0.0.0.0]network 10.0.23.0 0.0.0.255
```

查看 OSPF 邻居建立情况：

```
[P]dis ospf peer brief
(M) Indicates MADJ neighbor
         OSPF Process 1 with Router ID 10.0.12.2
                  Peer Statistic Information
Total number of peer(s): 2
 Peer(s) in full state: 2
-----------------------------------------------------------------------------
 Area Id          Interface                    Neighbor id          State
```

```
 0.0.0.0           Eth1/0/0        10.0.12.1            Full
 0.0.0.0           Eth1/0/1        10.0.23.3            Full
-----------------------------------------------------------------------------
```

查看路由学习情况，确保PE设备和P设备能学习到各自的环回口路由。

```
[P]display ip routing-table protocol ospf
Route Flags: R - relay, D - download to fib, T - to vpn-instance, B - black hole
route
------------------------------------------------------------------------------
_public_ Routing Table : OSPF
         Destinations : 5        Routes : 5
OSPF routing table status : <Active>
         Destinations : 2        Routes : 2
Destination/Mask    Proto    Pre   Cost        Flags    NextHop        Interface
1.1.1.1/32          OSPF     10    1           D        10.0.12.1      Ethernet1/0/0
3.3.3.3/32          OSPF     10    1           D        10.0.23.3      Ethernet1/0/1
```

步骤3：配置AS 100的MPLS LDP。

PE1的配置：

```
[PE1]mpls lsr-id 1.1.1.1
[PE1]mpls
Info: Mpls starting, please wait... OK!
[PE1-mpls]mpls ldp
[PE1-mpls-ldp]interface Ethernet1/0/0
[PE1-Ethernet1/0/0]mpls
[PE1-Ethernet1/0/0]mpls ldp
```

PE2的配置：

```
[PE2]mpls lsr-id 3.3.3.3
[PE2]mpls
Info: Mpls starting, please wait... OK!
[PE2-mpls]mpls ldp
[PE2-mpls-ldp]interface Ethernet1/0/0
[PE2-Ethernet1/0/0]mpls
[PE2-Ethernet1/0/0]mpls ldp
```

P的配置：

```
[P]mpls lsr-id 2.2.2.2
[P]mpls
Info: Mpls starting, please wait... OK!
[P-mpls]mpls ldp
[P-mpls-ldp]interface Ethernet1/0/0
[P-Ethernet1/0/0]mpls
[P-Ethernet1/0/0]mpls ldp
[P-Ethernet1/0/0]q
[P]interface Ethernet1/0/1
[P-Ethernet1/0/1]mpls
[P-Ethernet1/0/1]mpls ldp
```

查看 MPLS LSP 的建立情况，确保为每一条主机路由建立好 LSP。

```
[PE1]display mpls lsp
Flag after Out IF: (I) - RLFA Iterated LSP, (I*) - Normal and RLFA Iterated LSP
Flag after LDP FRR: (L) - Logic FRR LSP
-------------------------------------------------------------------------------
                  LSP Information: LDP LSP
-------------------------------------------------------------------------------
FEC                In/Out Label           In/Out IF              Vrf Name
1.1.1.1/32         3/NULL                 -/-
2.2.2.2/32         NULL/3                 -/Eth1/0/0
2.2.2.2/32         48120/3                -/Eth1/0/0
3.3.3.3/32         NULL/48120             -/Eth1/0/0
3.3.3.3/32         48121/48120            -/Eth1/0/0
```

步骤 4: 配置 VPN 实例，并且将连接 CE 的接口加入到对应的 VPN 实例中。

PE1 的配置:

```
[PE1]ip vpn-instance 1
[PE1-vpn-instance-1]ipv6-family //配置IPv6地址族的VPN实例
[PE1-vpn-instance-1-af-ipv6]route-distinguisher 1:1             //RD为1::1
[PE1-vpn-instance-1-af-ipv6]vpn-target 1:1 export-extcommunity //RT为1::1 both
[PE1-vpn-instance-1-af-ipv6]vpn-target 1:1 import-extcommunity
[PE1-vpn-instance-1-af-ipv6]q
[PE1-vpn-instance-1]interface Ethernet1/0/1
[PE1-Ethernet1/0/1]ip binding vpn-instance 1
[PE1-Ethernet1/0/1]ipv6 enable
[PE1-Ethernet1/0/1]ipv6 address 2001::1/64
```

PE2 的配置:

```
[PE2]ip vpn-instance 1
[PE2-vpn-instance-1]ipv6-family
[PE2-vpn-instance-1-af-ipv6]route-distinguisher 2:2
[PE2-vpn-instance-1-af-ipv6]vpn-target 1:1 export-extcommunity
[PE2-vpn-instance-1-af-ipv6]vpn-target 1:1 import-extcommunity
[PE2]interface Ethernet1/0/1
[PE2-Ethernet1/0/1]ip binding vpn-instance 1
[PE2-Ethernet1/0/1]ipv6 enable
[PE2-Ethernet1/0/1]ipv6 address 2002::1/64
```

步骤 5: 配置 PE1 和 PE2 的 VPNv6 的邻居关系。

PE1 的配置:

```
[PE1]bgp 100
[PE1-bgp]peer 3.3.3.3 as-number 100
[PE1-bgp]peer 3.3.3.3 connect-interface LoopBack0
[PE1-bgp]ipv6-family vpnv6
[PE1-bgp-af-vpnv6]peer 3.3.3.3 enable  //使能VPNv6邻居关系
Warning: This operation will reset the peer session. Continue? [Y/N]:y
```

PE2 的配置：

```
[PE2]bgp 100
[PE2-bgp]peer 1.1.1.1 as-number 100
[PE2-bgp]peer 1.1.1.1 connect-interface LoopBack0
[PE2-bgp]ipv6-family vpnv6
[PE2-bgp-af-vpnv6]peer 1.1.1.1 enable
Warning: This operation will reset the peer session. Continue? [Y/N]:y
```

步骤 6: 配置 PE 和 CE 设备的 BGP4+ 邻居关系。

PE1 的配置：

```
[PE1]bgp 100
[PE1-bgp]ipv6-family vpn-instance 1
[PE1-bgp-6-1]peer 2001::2 as-number 200
```

PE2 的配置：

```
[PE2]bgp 100
[PE2-bgp]ipv6-family vpn-instance 1
[PE2-bgp-6-1]peer 2002::2 as-number 300
```

CE1 的配置：

```
[CE1]bgp 200
[CE1-bgp]router-id 10.10.10.10
[CE1-bgp]peer 2001::1 as-number 100
[CE1-bgp]ipv6-family unicast
[CE1-bgp-af-ipv6]network 1::1 128
[CE1-bgp-af-ipv6]peer 2001::1 enable
```

CE2 的配置：

```
[CE2]bgp 300
[CE2-bgp]router-id 20.20.20.20
[CE2-bgp]peer 2002::1 as-number 100
[CE2-bgp]ipv6-family unicast
[CE2-bgp-af-ipv6]network 2::2 128
[CE2-bgp-af-ipv6]peer 2002::1 enable
```

查看 PE 设备的 VPNv6 的邻居建立情况：

```
[PE1]display bgp vpnv6 all peer
 BGP local router ID : 10.0.12.1
 Local AS number : 100
 Total number of peers : 2                 Peers in established state : 2
  Peer        V   AS    MsgRcvd    MsgSent    OutQ    Up/Down     State        PrefRcv
  3.3.3.3     4   100   10         9          0       00:04:24    Established  1

  Peer of IPv6-family for vpn instance :
  VPN-Instance 1, Router ID 10.0.12.1:
  Peer        V   AS    MsgRcvd    MsgSent    OutQ    Up/Down     State        PrefRcv
  2001::2     4   200   4          6          0       00:01:16    Established  1
[PE2-bgp]display bgp vpnv6 all peer
```

```
BGP local router ID : 10.0.23.3
Local AS number : 100
Total number of peers : 2        Peers in established state : 2
 Peer          V   AS    MsgRcvd    MsgSent    OutQ    Up/Down     State        PrefRcv
 1.1.1.1       4   100   8          9          0       00:03:28    Established  1

 Peer of IPv6-family for vpn instance :
 VPN-Instance 1, Router ID 10.0.23.3:
 Peer          V   AS    MsgRcvd    MsgSent    OutQ    Up/Down     State        PrefRcv
 2002::2       4   300   3          5          0       00:00:05    Established  1
```

查看 PE 设备的 VPNv6 路由表（以 PE1 为例），可以看到能够学习到对端 CE2 的 IPv6 路由 2::2/128。

```
[PE1]display bgp vpnv6 all routing-table
 Total number of routes from all PE: 2
 Route Distinguisher: 1:1
 *>      Network  : 1::1                                        PrefixLen : 128
         NextHop  : 2001::2                                     LocPrf    :
         MED      : 0                                           PrefVal   : 0
         Label    : NULL/48122
         Path/Ogn : 200i
 Route Distinguisher: 2:2
 *>i     Network  : 2::2                                        PrefixLen : 128
         NextHop  : ::FFFF:3.3.3.3                              LocPrf    : 100
         MED      : 0                                           PrefVal   : 0
         Label    : 48122/NULL
         Path/Ogn : 300i
 VPN-Instance 1, Router ID 10.0.12.1:
 Total Number of Routes: 2
 *>      Network  : 1::1                                        PrefixLen : 128
         NextHop  : 2001::2                                     LocPrf    :
         MED      : 0                                           PrefVal   : 0
         Label    :
         Path/Ogn : 200i
 *>i     Network  : 2::2                                        PrefixLen : 128
         NextHop  : ::FFFF:3.3.3.3                              LocPrf    : 100
         MED      : 0                                           PrefVal   : 0
         Label    : 48122/NULL
         Path/Ogn : 300i
```

查看 CE1 的 IPv6 路由表，可以学习到对端的 2::2/128 的路由：

```
[CE1]display ipv6 routing-table protocol bgp
Public Routing Table : BGP
Summary Count : 1
BGP Routing Table's Status : < Active >
Summary Count : 1
 Destination  : 2::2                              PrefixLength  : 128
```

```
 NextHop       : 2001::1                        Preference    : 255
 Cost          : 0                              Protocol      : EBGP
 RelayNextHop  : ::                             TunnelID      : 0x0
 Interface     : GigabitEthernet0/0/0           Flags         : D
BGP Routing Table's Status : < Inactive >
Summary Count : 0
```

测试 CE 之间的连通性：

```
[CE1]ping ipv6 -a 1::1 2::2
  PING 2::2 : 56  data bytes, press CTRL_C to break
    Reply from 2::2
    bytes=56 Sequence=1 hop limit=61  time = 40 ms
    Reply from 2::2
    bytes=56 Sequence=2 hop limit=61  time = 40 ms
    Reply from 2::2
    bytes=56 Sequence=3 hop limit=61  time = 50 ms
    Reply from 2::2
    bytes=56 Sequence=4 hop limit=61  time = 30 ms
    Reply from 2::2
    bytes=56 Sequence=5 hop limit=61  time = 40 ms
  --- 2::2 ping statistics ---
    5 packet(s) transmitted
    5 packet(s) received
    0.00% packet loss
    round-trip min/avg/max = 30/40/50 ms
```

第 11 章

QoS 技术

11.1 QoS 技术概述

在带宽有限的情况下，QoS（Quality of Service，服务质量）技术应用一个“有保证”的策略对网络流量进行管理，确保不同的流量可以获得不同级别的优先服务。

11.1.1 服务模型

- 尽力而为：默认模型，使用先进先出的方式转发流量。
- 综合服务：资源预留，资源空闲时其他业务流量也不能使用。
- 区分服务：常用的 QoS 模型，使用方式遵循“标记→分类→差分服务”的流程。

11.1.2 QoS 的流量分类

（1）简单流分类：根据报文头部中的优先级字段，把外部优先级映射成内部优先级。

简单流分类是指采用简单的规则，如只根据 IP 报文的 IP 优先级或 DSCP（DS Code Point，差异化服务代码点）值、IPv6 报文的 TC 值、MPLS 报文的 EXP 域值、VLAN 报文的 802.1p 值，对报文进行粗略的分类，以识别出具有不同优先级或服务等级特征的流量，实现外部优先级和内部优先级之间的映射。

① Precedence 字段。IP 报文头的 ToS（Type of Service，服务类型）域由 8 比特组成，其中 3 比特的 Precedence 字段标识了 IP 报文的优先级。IP Precedence/DSCP 字段如图 11-1 所示。

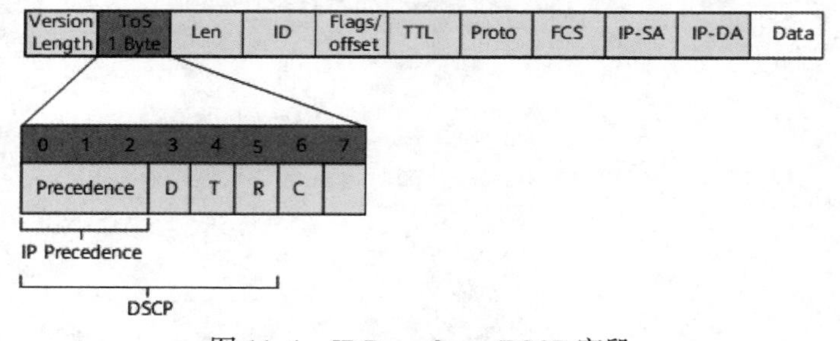

图 11-1 IP Precedence/DSCP 字段

比特 0 ~ 2 表示 Precedence 字段，代表报文传输的 8 个优先级，按照优先级从高到低的顺序取值为 7、6、5、4、3、2、1 和 0。最高优先级是 7，通常是为路由选择或更新网络控制通信保留的，用户级应用仅能使用 0 ~ 5。

除了 Precedence 字段，ToS 域中还包括 D、T 和 R 三个比特。

D 比特表示延迟要求（Delay，0 代表正常延迟；1 代表低延迟）。

T 比特表示吞吐量（Throughput，0 代表正常吞吐量；1 代表高吞吐量）。

R 比特表示可靠性（Reliability，0 代表正常可靠性；1 代表高可靠性）。

② DSCP 字段。RFC 重新定义了 IP 报文中的 ToS 域，增加了 C 比特，表示传输开销（Monetary Cost）。然后，IETF DiffServ 工作组在 RFC 中将 IPv4 报文头 ToS 域中的比特 0 ~ 5 重新定义为 DSCP，并将 ToS 域改名为 DS（Differentiated Service，差异化服务）字段。DSCP 在报文中的位置见图 11-1。

DS 字段的前 6 位（0 ~ 5）用作区分 DSCP，后 2 位（6 位、7 位）是保留位。DS 字段的前 3

位（0 ~ 2）是 CSCP（Class Selector Code Point，类选择代码点），相同的 CSCP 值代表一类 DSCP。DS 节点根据 DSCP 的值选择相应的 PHB（Per-Hop Behavior，逐跳行为）。

通常二层设备之间要交互 VLAN 帧。根据 IEEE 802.1q 定义，VLAN 帧头中的 PRI 字段（802.1p 优先级）或称 CoS（Class of Service，服务等级）字段，标识了服务质量需求。VLAN 帧中的 PRI 字段位置如图 11-2 所示。

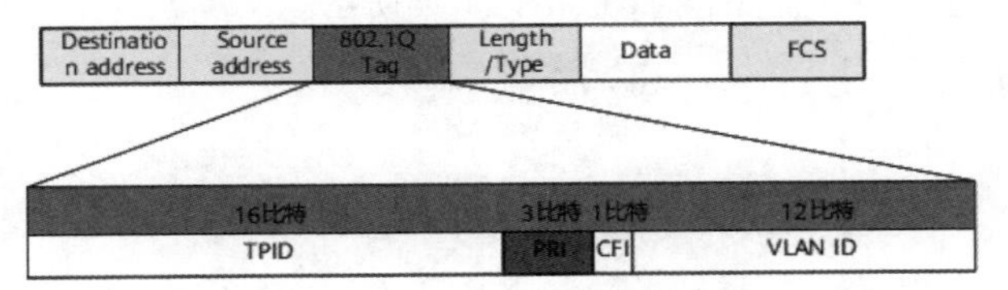

图 11-2　VLAN 帧中的 PRI 字段位置

在 802.1q 头部中包含 3 比特长的 PRI 字段。PRI 字段定义了 8 种业务优先级 CoS，按照优先级从高到低的顺序取值为 7、6、5、4、3、2、1 和 0。

③ MPLS EXP 字段。MPLS 报文与普通的 IP 报文相比增加了标签信息。标签的长度为 4 字节，MPLS 标签的封装格式如图 11-3 所示。

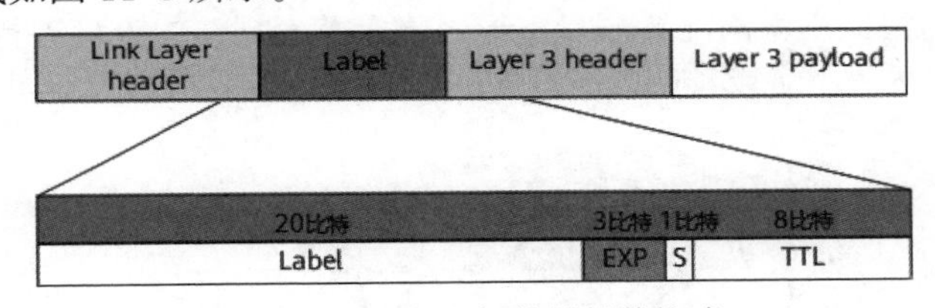

图 11-3　MPLS 标签的封装格式

MPLS 标签共有以下 4 个域。

Label：20 比特，标签值字段，用于转发的指针。

EXP：3 比特，保留字段，用于扩展，现在通常用作 CoS。

S：1 比特，栈底标识。MPLS 支持标签的分层结构，即多重标签，S 值为 1 时表示最底层标签。

TTL：8 比特，与 IP 分组中的 TTL 意义相同。

对于 MPLS 报文，通常将标签信息中的 EXP 域作为 MPLS 报文的 CoS 域，与 IP 网络的 ToS 域等效，用于区分数据流量的服务等级，以支持 MPLS 网络的 DiffServ。EXP 域表示 8 个传输优先级，按照优先级从高到低的顺序取值为 7、6、5、4、3、2、1 和 0。

在 IP 网络中，由 IP 报文的 IP 优先级或 DSCP 标识服务等级。但是对于 MPLS 网络，由于报文的 IP 头对 LSR 设备是不可见的，因此需要在 MPLS 网络的边缘对 MPLS 报文的 EXP 域进行标记。

默认情况下，在 MPLS 网络的边缘，将 IP 报文的 IP 优先级直接复制到 MPLS 报文的 EXP 域；但是在某些情况下，如 ISP 不信任用户网络，或者 ISP 定义的差别服务类别不同于用户网络，则可以根据一定的分类策略，依据内部的服务等级重新设置 MPLS 报文的 EXP 域，而在 MPLS 网络转发的过程中保持 IP 报文的 ToS 域不变。

在 MPLS 网络的中间节点，根据 MPLS 报文的 EXP 域对报文进行分类，并实现拥塞管理、流量监管或流量整形等 PHB 行为。

（2）复杂流分类：根据报文中的优先级字段或五元组信息对流量进行分类，然后标记上对应的优先级（使用 MQC 实现）。

11.1.3 流量限速技术

1. 流量监管

通过监控进入网络的某一流量的规格，限制它在一个允许的范围之内，若某个连接的报文流量过大，就丢弃报文，或者重新设置该报文的优先级，以保护网络资源不受损害。流量监管可配置在设备的接口进出方向。流量监管示意图如图 11-4 所示。

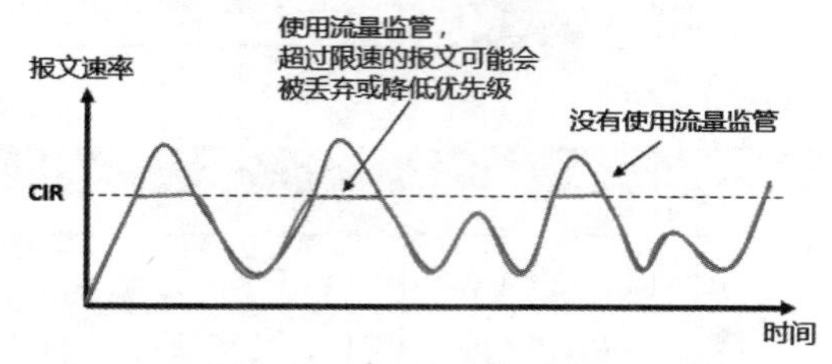

图 11-4 流量监管示意图

2. 流量整形

流量整形是一种主动调整流量输出速率的措施。

通过在上游设备的接口出方向配置流量整形，将上游不规整的流量进行削峰填谷，输出一条比较平整的流量，从而解决下游设备的瞬时拥塞问题。流量整形示意图如图 11-5 所示。

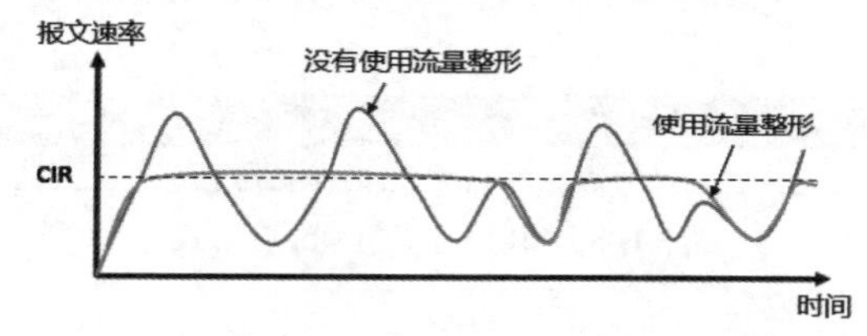

图 11-5 流量整形示意图

流量监管和流量整形有什么区别？流量监管在数据超过最大速率时对报文进行丢弃。流量整形在数据超过最大速率时，先对报文进行缓存，等到带宽足够时，再通过相应的队列技术进行调度并且转发。

11.1.4 拥塞避免技术

当某一队列已经被装满时，传统的处理方法会将后续向该队列发送的报文全部丢弃，直至拥塞解除，这种处理方式称为尾丢弃（Tail Drop）。

尾丢弃会导致以下问题。

（1）TCP 全局同步。

① TCP 启动过程。

②流量过大，导致队列被装满，发生尾丢弃行为。

③由于服务器回复的 TCP 确认包拥塞被丢掉，因此发送方未收到 TCP 确认包，则系统认为网络发生了拥塞，于是同时将 TCP 滑动窗口 Size 减小，整体流量同时也减小。

④网络拥塞消除，发送方又都能收到 TCP 确认包，因此系统认为网络不再拥塞，于是都进入 TCP 慢启动过程，周而复始。

TCP 全局同步示意图如图 11-6 所示。

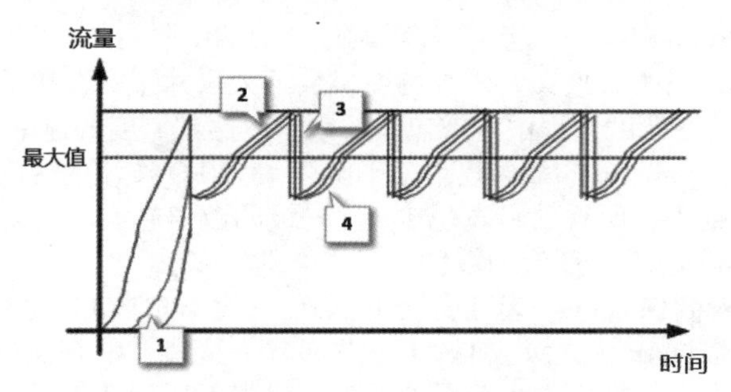

图 11-6 TCP 全局同步示意图

（2）无差别的丢弃：当队列的长度达到最大值后，所有新入队列的报文（缓存在队列尾部）都将被丢弃。

WRED（Weighted Random Early Detection，加权随机先期检测）通过对不同优先级的数据包或队列设置相应的丢弃策略，以实现对不同流量进行区分丢弃。

以图 11-7 为例，IP Precedence=0 的流量低门限为 20、高门限为 40，IP Precedence=2 的流量低门限为 35、高门限为 40，比 IPP=0 的流量晚丢弃。

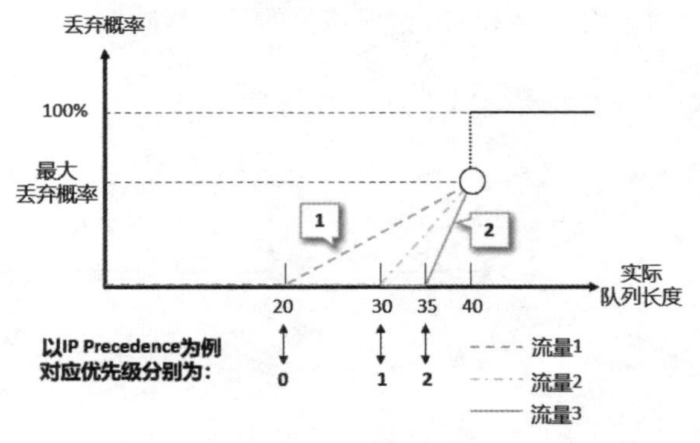

图 11-7 WRED 示意图

WRED 的优点如下：打乱 TCP 滑动窗口的调整时间，避免 TCP 全局同步现象。基于权重实现了不同流量的区分丢弃。

11.1.5 拥塞管理技术

下面介绍拥塞管理技术。

（1）FIFO（First In First Out，先进先出）：FIFO 不对报文进行分类，而是按报文到达接口的先后顺序让报文进入队列，在队列的出口让报文按进队的顺序出队，先进的报文将先出队，后进的报文将后出队。

（2）PQ（Priority Queue，优先级队列）：对高优先级报文进行优先转发，高优先级报文转发完后再转发低优先级报文，可能出现饿死现象。（紧急业务流量。）

（3）WRR（Weighted Round Robin，权重轮巡）：避免了 PQ 调度的“饿死”现象。但是基于报文个数来调度，容易导致包长尺寸不同的报文出现不平等调度；低时延业务得不到及时调度。（简单的逐包数据。）

（4）DRR（Deficit Round Robin，差额循环调度）：DRR 调度同样也是 RR 的扩展，相对于 WRR 而言，解决了 WRR 只关心报文，同等调度机会下大尺寸报文获得的实际带宽要大于小尺寸报文获得的带宽的问题，在调度过程中考虑包长的因素以达到调度的速率公平性。

（5）WFQ（Weighted Fair Queuing，权重公平队列）：WFQ 按流的优先级来分配每个流应占有的带宽。优先级的数值越小，所得的带宽越少；优先级的数值越大，所得的带宽越多。这种方式只有 CBQ 的 default-class 支持。（逐流的数据，报文类别较多）

（6）CBQ（Class Based Queuing，基于分类的队列）：对 WFQ 功能的扩展，为用户提供了自定义类的支持。CBQ 首先根据 IP 优先级或 DSCP 优先级、入接口、IP 报文的五元组等规则来对报文进行分类，然后让不同类别的报文进入不同的队列。对于不匹配任何类别的报文，会送入系统定义的默认类。

11.2 QoS 实验

11.2.1 配置简单流分类

扫一扫，看视频

1. 实验目的

10.1.1.0/24 网段的 PC1 访问 PC3，流量进入 AR1 时，DSCP 字段的优先级为 0，在 AR1 上进行配置，将 10.1.1.0/24 网段流量的优先级映射为内部优先级 46。

2. 实验拓扑

配置简单流分类的实验拓扑如图 11-8 所示。

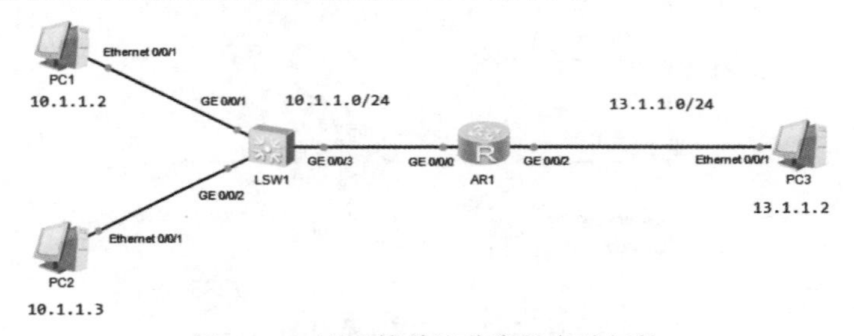

图 11-8　配置简单流分类的实验拓扑

3. 实验步骤

步骤 1：配置 IP 地址。

```
[AR1]interface g0/0/0
[AR1-GigabitEthernet0/0/0]ip address 10.1.1.1 24
[AR1]interface  g0/0/2
[AR1-GigabitEthernet0/0/2]ip address 13.1.1.1 24
```

测试 PC1 或 PC2 与 PC3 的连通性，并在 AR1 的 G0/0/0、G0/0/2 接口抓包，观察 IP 头部的 DSCP 字段数值。

PC1 ping PC3：

```
PC>ping 13.1.1.2
Ping 13.1.1.2: 32 data bytes, Press Ctrl_C to break
From 13.1.1.2: bytes=32 seq=1 ttl=127 time=47 ms
From 13.1.1.2: bytes=32 seq=2 ttl=127 time=31 ms
```

```
From 13.1.1.2: bytes=32 seq=3 ttl=127 time=47 ms
From 13.1.1.2: bytes=32 seq=4 ttl=127 time=31 ms
From 13.1.1.2: bytes=32 seq=5 ttl=127 time=32 ms
--- 13.1.1.2 ping statistics ---
  5 packet(s) transmitted
  5 packet(s) received
  0.00% packet loss
  round-trip min/avg/max = 31/37/47 ms
```

G0/0/0 接口的抓包结果如图 11-9 所示。

```
> Frame 37: 74 bytes on wire (592 bits), 74 bytes captured (592 bits) on interface -, id 0
> Ethernet II, Src: HuaweiTe_e0:12:f6 (54:89:98:e0:12:f6), Dst: HuaweiTe_c9:22:aa (00:e0:fc:c9:22:aa)
v Internet Protocol Version 4, Src: 10.1.1.2, Dst: 13.1.1.2
    0100 .... = Version: 4
    .... 0101 = Header Length: 20 bytes (5)
  > Differentiated Services Field: 0x00 (DSCP: CS0, ECN: Not-ECT)
    Total Length: 60                        此处为DSCP字段，CS0代表流量的IP优先级为0
    Identification: 0xe7e6 (59366)
  > Flags: 0x4000, Don't fragment
    Fragment offset: 0
    Time to live: 128
    Protocol: ICMP (1)
    Header checksum: 0xf9d4 [validation disabled]
    [Header checksum status: Unverified]
    Source: 10.1.1.2
    Destination: 13.1.1.2
> Internet Control Message Protocol
```

图 11-9 G0/0/0 接口的抓包结果

G0/0/2 接口的抓包结果如图 11-10 所示。

```
> Frame 3: 74 bytes on wire (592 bits), 74 bytes captured (592 bits) on interface -, id 0
> Ethernet II, Src: HuaweiTe_c9:22:ac (00:e0:fc:c9:22:ac), Dst: HuaweiTe_6e:29:62 (54:89:98:6e:29:62)
v Internet Protocol Version 4, Src: 10.1.1.2, Dst: 13.1.1.2
    0100 .... = Version: 4
    .... 0101 = Header Length: 20 bytes (5)
  > Differentiated Services Field: 0x00 (DSCP: CS0, ECN: Not-ECT)
    Total Length: 60                      流量发往PC3，优先级字段默认不发生改变
    Identification: 0xe7d1 (59345)
  > Flags: 0x4000, Don't fragment
    Fragment offset: 0
    Time to live: 127
    Protocol: ICMP (1)
    Header checksum: 0xfae9 [validation disabled]
    [Header checksum status: Unverified]
    Source: 10.1.1.2
    Destination: 13.1.1.2
> Internet Control Message Protocol
```

图 11-10 G0/0/2 接口的抓包结果

具体原因是设备上存在 qos map-table，即外部优先级与内部优先级的映射关系：

```
<AR1>display qos map-table dscp-dscp //查看外部 DSCP 到内部 DSCP 的映射关系
Input DSCP       DSCP
-------------------
 0                0
 1                1
 2                2
 3                3
 4                4
```

```
 5            5
 6            6
 7            7
--------------------
此处只截取一部分回显信息！
```

通过以上输出可以看出，当外部优先级为0时，映射的内部优先级也为0，因此想要达到实验目的，仅修改此映射表项即可。

步骤2：配置QoS的映射表项。

```
[AR1]qos map-table dscp-dscp //指定进入DSCP-DSCP视图，即从DSCP到DSCP的映射视图
[AR1-maptbl-dscp-dscp]input 0 output 46        //将外部优先级0映射为内部优先级46
[AR1]interface  g0/0/0
[AR1-GigabitEthernet0/0/0]trust dscp override  //DSCP报文安装映射表的关系进行映射
```

查看QoS的映射表：

```
[AR1]display qos map-table dscp-dscp
Input DSCP      DSCP
-------------------
 0              46
 1              1
 2              2
 3              3
 4              4
 5              5
 6              6
 7              7
```

通过以上输出可以看出，外部优先级0已被映射为内部优先级46。

再次测试PC1或PC2与PC3的连通性，并在AR1的G0/0/0、G0/0/2接口抓包，观察IP头部的DSCP字段数值。

PC1 ping PC3：

```
PC>ping 13.1.1.2
Ping 13.1.1.2: 32 data bytes, Press Ctrl_C to break
From 13.1.1.2: bytes=32 seq=1 ttl=127 time=47 ms
From 13.1.1.2: bytes=32 seq=2 ttl=127 time=31 ms
From 13.1.1.2: bytes=32 seq=3 ttl=127 time=47 ms
From 13.1.1.2: bytes=32 seq=4 ttl=127 time=31 ms
From 13.1.1.2: bytes=32 seq=5 ttl=127 time=32 ms
--- 13.1.1.2 ping statistics ---
  5 packet(s) transmitted
  5 packet(s) received
  0.00% packet loss
  round-trip min/avg/max = 31/37/47 ms
```

AR1的G0/0/0接口的抓包结果如图11-11所示。

```
> Frame 407: 74 bytes on wire (592 bits), 74 bytes captured (592 bits) on interface -, id 0
> Ethernet II, Src: HuaweiTe_e0:12:f6 (54:89:98:e0:12:f6), Dst: HuaweiTe_c9:22:aa (00:e0:fc:c9:22:aa)
v Internet Protocol Version 4, Src: 10.1.1.2, Dst: 13.1.1.2
    0100 .... = Version: 4
    .... 0101 = Header Length: 20 bytes (5)
  > Differentiated Services Field: 0x00 (DSCP: CS0, ECN: Not-ECT)
    Total Length: 60
    Identification: 0xeb0f (60175)
  > Flags: 0x4000, Don't fragment
    Fragment offset: 0
    Time to live: 128
    Protocol: ICMP (1)
    Header checksum: 0xf6ab [validation disabled]
    [Header checksum status: Unverified]
    Source: 10.1.1.2
    Destination: 13.1.1.2
> Internet Control Message Protocol
```

图 11-11　AR1 的 G0/0/0 接口的抓包结果

AR1 的 G0/0/2 接口的抓包结果如图 11-12 所示。

```
> Frame 24: 74 bytes on wire (592 bits), 74 bytes captured (592 bits) on interface -, id 0
> Ethernet II, Src: HuaweiTe_c9:22:ac (00:e0:fc:c9:22:ac), Dst: HuaweiTe_6e:29:62 (54:89:98:6e:29:62)
v Internet Protocol Version 4, Src: 10.1.1.2, Dst: 13.1.1.2
    0100 .... = Version: 4
    .... 0101 = Header Length: 20 bytes (5)
  > Differentiated Services Field: 0xb8 (DSCP: EF PHB, ECN: Not-ECT)
    Total Length: 60                          DSCP值为EF，即优先级为46
    Identification: 0xeb0d (60173)
  > Flags: 0x4000, Don't fragment
    Fragment offset: 0
    Time to live: 127
    Protocol: ICMP (1)
    Header checksum: 0xf6f5 [validation disabled]
    [Header checksum status: Unverified]
    Source: 10.1.1.2
    Destination: 13.1.1.2
> Internet Control Message Protocol
```

图 11-12　AR1 的 G0/0/2 接口的抓包结果

通过以上输出可以看出，10.1.1.0/24 网段的流量进入 G0/0/0 接口时的优先级为 0，在 G0/0/2 接口被转发出去时，优先级为 EF（46），与实验预计效果相符。

11.2.2　配置复杂流分类

1. 实验目的

PC1 和 PC2 属于 10.1.1.0/24 网段，要求在 AR1 上使用复杂流分类，将 PC1 的流量的 DSCP 值映射为 AF41（34），PC2 的流量的 DSCP 值映射为 EF（46）。

扫一扫，看视频

2. 实验拓扑

配置复杂流分类的实验拓扑如图 11-13 所示。

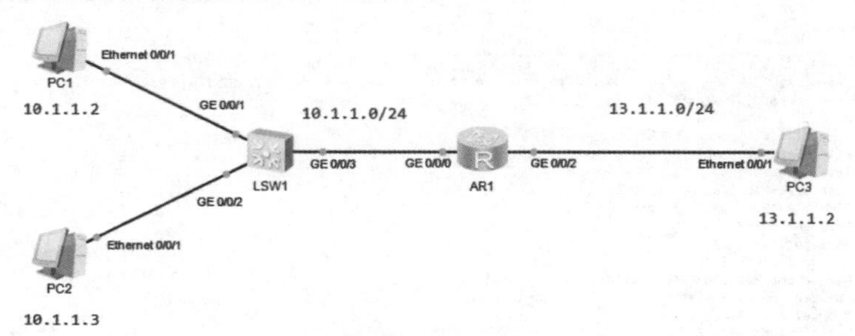

图 11-13　配置复杂流分类的实验拓扑

3. 实验步骤

步骤 1： 配置 IP 地址。

```
[AR1]interface g0/0/0
[AR1-GigabitEthernet0/0/0]ip address 10.1.1.1 24
[AR1]interface g0/0/2
[AR1-GigabitEthernet0/0/2]ip address 13.1.1.1 24
```

步骤 2： 在 AR1 上配置 MQC 实现复杂流分类。

（1）配置 ACL 匹配 PC1 和 PC2 的流量：

```
[AR1]acl 2000
[AR1-acl-basic-2000]rule permit source 10.1.1.2 0
[AR1]acl 2001
[AR1-acl-basic-2001]rule permit source 10.1.1.3 0
```

（2）创建 PC1 和 PC2 的流分类，并调用 ACL：

```
[AR1]traffic classifier pc1                    //创建流分类，命名为pc1
[AR1-classifier-pc1]if-match acl 2000          //配置条件语句，调用ACL 2000
[AR1]traffic classifier pc2
[AR1-classifier-pc2]if-match acl 2001
```

（3）创建 PC1 和 PC2 的流行为，并重标 DSCP 的优先级：

```
[AR1]traffic behavior pc1
[AR1-behavior-pc1]remark dscp af41             //将DSCP值重标记为AF41
[AR1]traffic behavior pc2
[AR1-behavior-pc2]remark dscp ef
```

（4）创建流策略，管理相应的流分类和流行为：

```
[AR1]traffic policy RM                         //创建流策略命名为RM
[AR1-trafficpolicy-RM]classifier pc1 behavior pc1 //将流分类pc1关联到流策略pc1
[AR1-trafficpolicy-RM]classifier pc2 behavior pc2
```

最终能实现的效果是，当 10.1.1.2 的流量被匹配到后，其 DSCP 的数值被重标记为 AF41，当 10.1.1.3 的流量被匹配到后，其 DSCP 的数值被重标记为 EF。

步骤 3： 在接口调用流策略。

```
[AR1]interface g0/0/0
[AR1-GigabitEthernet0/0/0]traffic-policy RM inbound
```

步骤 4： 测试并在 AR1 的 G0/0/2 接口抓包并查看结果，如图 11-14 和图 11-15 所示。

PC1 ping PC3：

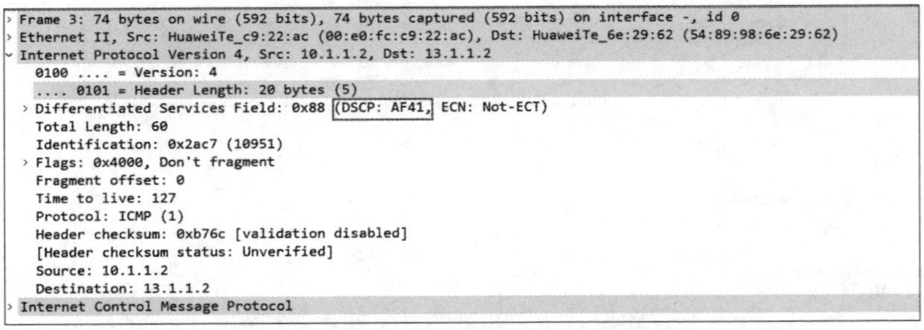

```
> Frame 3: 74 bytes on wire (592 bits), 74 bytes captured (592 bits) on interface -, id 0
> Ethernet II, Src: HuaweiTe_c9:22:ac (00:e0:fc:c9:22:ac), Dst: HuaweiTe_6e:29:62 (54:89:98:6e:29:62)
v Internet Protocol Version 4, Src: 10.1.1.2, Dst: 13.1.1.2
    0100 .... = Version: 4
    .... 0101 = Header Length: 20 bytes (5)
  > Differentiated Services Field: 0x88 (DSCP: AF41, ECN: Not-ECT)
    Total Length: 60
    Identification: 0x2ac7 (10951)
  > Flags: 0x4000, Don't fragment
    Fragment offset: 0
    Time to live: 127
    Protocol: ICMP (1)
    Header checksum: 0xb76c [validation disabled]
    [Header checksum status: Unverified]
    Source: 10.1.1.2
    Destination: 13.1.1.2
> Internet Control Message Protocol
```

图 11-14　AR1 的 G0/0/2 接口的抓包结果（1）

PC2 ping PC3：

```
> Frame 11: 74 bytes on wire (592 bits), 74 bytes captured (592 bits) on interface -, id 0
> Ethernet II, Src: HuaweiTe_c9:22:ac (00:e0:fc:c9:22:ac), Dst: HuaweiTe_6e:29:62 (54:89:98:6e:29:62)
v Internet Protocol Version 4, Src: 10.1.1.3, Dst: 13.1.1.2
    0100 .... = Version: 4
    .... 0101 = Header Length: 20 bytes (5)
  > Differentiated Services Field: 0xb8 (DSCP: EF PHB, ECN: Not-ECT)
    Total Length: 60
    Identification: 0x2af2 (10994)
  > Flags: 0x4000, Don't fragment
    Fragment offset: 0
    Time to live: 127
    Protocol: ICMP (1)
    Header checksum: 0xb710 [validation disabled]
    [Header checksum status: Unverified]
    Source: 10.1.1.3
    Destination: 13.1.1.2
> Internet Control Message Protocol
```

图 11-15　AR1 的 G0/0/2 接口的抓包结果（2）

通过以上输出可以看出，PC1 的流量 DSCP 值被重标记为 AF41，PC2 的流量 DSCP 值被重标记为 EF。

11.2.3　QoS 的基本配置

1. 实验目的

（1）使用 MQC 将 PC1 和 PC2 的流量 DSCP 优先级设置为 EF=46 和 AF41=34。

（2）将 EF 的流量在 AR1 上重标记为 AF21=010010 =18、AF41 的流量重标记为 AF12=001100=12。

（3）在 AR1 的 G0/0/0 接口使用流量监管，配置接口带宽为 800m。

（4）在 AR1 配置拥塞管理和拥塞避免（WRED），队列 1 使用 WFQ 调度流量，低门限为 40、高门限为 80、丢包率为 90；队列 2 使用 WFQ 调度流量，低门限为 60、高门限为 95、丢包率为 80。

2. 实验拓扑

QoS 的基本配置实验拓扑如图 11-16 所示。

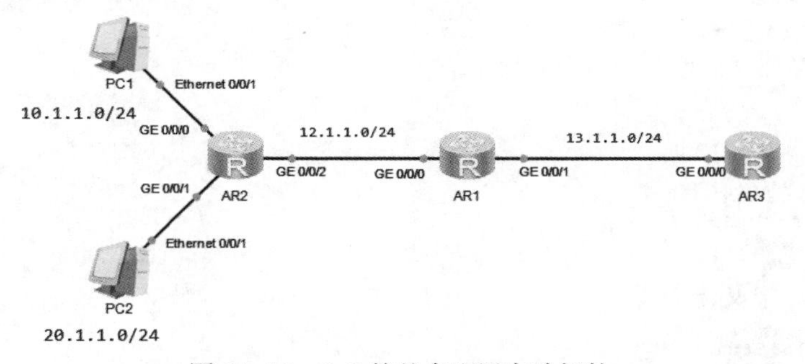

图 11-16　QoS 的基本配置实验拓扑

3. 实验步骤

步骤 1：配置 IP 地址及路由，实现全网互通。

（1）配置设备的 IP 地址。

AR1 的配置：

```
[AR1]interface g0/0/0
[AR1-GigabitEthernet0/0/0]ip address 12.1.1.1 24
```

```
[AR1-GigabitEthernet0/0/0]q
[AR1]int g0/0/1
[AR1-GigabitEthernet0/0/1]ip address 13.1.1.1 24
```

AR2 的配置：

```
[AR2]interface g0/0/0
[AR2-GigabitEthernet0/0/0]ip address 10.1.1.254 24
[AR2-GigabitEthernet0/0/0]q
[AR2]interface g0/0/1
[AR2-GigabitEthernet0/0/1]ip address 20.1.1.254 24
[AR2-GigabitEthernet0/0/1]q
[AR2]interface g0/0/2
[AR2-GigabitEthernet0/0/2]ip address 12.1.1.2 24
```

AR3 的配置：

```
[AR3]interface g0/0/0
[AR3-GigabitEthernet0/0/0]ip address 13.1.1.3 24
```

（2）配置路由条目。

AR1 的配置：

```
[AR1]ip route-static 10.1.1.0 24 12.1.1.2
[AR1]ip route-static 20.1.1.0 24 12.1.1.2
```

AR2 的配置：

```
[AR2]ip route-static 13.1.1.0 24 12.1.1.1
```

AR3 的配置：

```
[AR3]ip route-static 10.1.1.0 24 13.1.1.1
[AR3]ip route-static 20.1.1.0 24 13.1.1.1
```

步骤 2：在 AR2 上对流量进行分类，使用 MQC 将 PC1 和 PC2 的流量 DSCP 优先级设置为 EF=46 和 AF41=34。（复杂流分类，一般在 DS 域边缘配置。）

（1）使用 ACL 分别匹配 PC1 和 PC2 的流量：

```
[AR2]acl number 3000
[AR2-acl-adv-3000]rule 5 permit ip source 10.1.1.0 0.0.0.255
[AR2-acl-adv-3000]q
[AR2]acl number 3001
[AR2-acl-adv-3001]rule 5 permit ip source 20.1.1.0 0.0.0.255
```

（2）使用 MQC 定义流分类：

```
[AR2]traffic classifier pc2
[AR2-classifier-pc2]if-match acl 3001
[AR2-classifier-pc2]q
[AR2]traffic classifier pc1
[AR2-classifier-pc1]if-match acl 3000
```

（3）定义流行为，将 PC1 的流量 DSCP 设置为 EF，将 PC2 的流量 DSCP 设置为 AF41：

```
[AR2]traffic behavior pc2
[AR2-behavior-pc2]remark dscp af41
```

```
[AR2-behavior-pc2]q
[AR2]traffic behavior pc1
[AR2-behavior-pc1]remark dscp ef
```

（4）将流分类和流行为关联到流策略：

```
[AR2]traffic policy remark
[AR2-trafficpolicy-remark]classifier pc1 behavior pc1
[AR2-trafficpolicy-remark]classifier pc2 behavior pc2
```

（5）AR2 的 G0/0/2 接口的出方向调用：

```
[AR2]interface GigabitEthernet0/0/2
[AR2-GigabitEthernet0/0/2]traffic-policy remark outbound
```

步骤 3： 在 AR1 的 G0/0/0 接口对流量进行简单流分类，将 EF 的流量在 AR1 上重标记为 AF21=010010 =18、将 AF41 的流量重标记为 AF12=001 100=12。

（1）配置 QoS 的映射模板：

```
[AR1]qos map-table dscp-dscp
[AR1-maptbl-dscp-dscp]input 34 output 12 //将AF41映射为AF12
[AR1-maptbl-dscp-dscp]input 46 output 18 //将EF映射为AR21
```

（2）信任外部优先级，并且覆盖为内部映射的优先级：

```
[AR1]interface GigabitEthernet0/0/0
[AR1-GigabitEthernet0/0/0]ip address 11.1.1.2 255.255.255.0
[AR1-GigabitEthernet0/0/0]trust dscp override
```

在 AR1 上用命令 display qos map-table dscp-dscp 查看是否映射成功：

```
[AR1]display qos map-table dscp-dscp
Input DSCP        DSCP
--------------------
34                12
35                35
36                36
37                37
38                38
39                39
40                40
41                41
42                42
43                43
44                44
45                45
46                18
47                47
```

步骤 4： 在 AR1 的 G0/0/0 接口使用流量监管，配置接口带宽为 800 m。

```
[AR1]interface GigabitEthernet0/0/0
[AR1-GigabitEthernet0/0/0]qos car inbound cir 819200
```

步骤 5： 在 AR1 配置拥塞管理和拥塞避免，队列 1 使用 WFQ 调度流量，低门限为 40、高门限为 80、丢包率为 90；队列 2 使用 WFQ 调度流量，低门限为 60、高门限为 95、丢包率为 80。队列 1 使用 GTS（流量整形）进行限速，速率为 400m。

（1）设备可以根据收到报文的 DSCP 优先级，自动将流量分配到对应的软件队列中，DSCP 优先级为 0~7，分配的队列为 0；DSCP 的优先级为 8~15，分配的队列为 1。PC1 的流量 DSCP 优先级为 12，流量将分配到队列 1；PC2 的流量 DSCP 优先级为 21，流量将分配到队列 2：

```
[R1]display qos map-table dscp-lp
Input DSCP      LP
--------------------
 0               0
 1               0
 2               0
 3               0
 4               0
 5               0
 6               0
 7               0
 8               1
 9               1
10               1
11               1
12               1
13               1
14               1
15               1
16               2
17               2
```

（2）配置丢弃模板：

```
[AR1]drop-profile af21                //模板名字为 af21
[AR1-drop-profile-af21]wred dscp     //指定当前 WRED 丢弃模板基于 DSCP 优先级进行丢弃
[AR1-drop-profile-af21]dscp af21 low-limit 60 high-limit 95 discard-percentage 80
                                      //DSCP 值为 af21，低门限为 60、高门限为 95、丢包率为 80
[AR1-drop-profile-af21]drop-profile af12
[AR1-drop-profile-af12]wred dscp
[AR1-drop-profile-af12]dscp af12 low-limit 40 high-limit 80 discard-percentage 90
```

（3）配置 QoS 队列模板：

```
[AR1]qos queue-profile 1                            //创建队列模板，名字为 1
[AR1-qos-queue-profile-1]queue 2 weight 20          //配置队列权重为 20，默认为 10
[AR1-qos-queue-profile-1]schedule wfq 1 to 2 //配置队列 1、2 以 WFQ 方式进行流量调度
                                                     //（拥塞管理）
[AR1-qos-queue-profile-1]queue 1 drop-profile af12 //队列 1 应用丢弃模板 af12
[AR1-qos-queue-profile-1]queue 2 drop-profile af21 //队列 2 应用丢弃模板 af21
[AR1-qos-queue-profile-1]queue 1 gts cir 409600    //配置队列 1 的流量整形
```

（4）查看队列模板：

```
[R1-qos-queue-profile-1]display qos queue-profile 1
```

```
Queue-profile: 1
Queue Schedule Weight Length(Bytes/Packets) GTS(CIR/CBS)
-----------------------------------------------------------------
1      WFQ       10           -/-              409600/10240000
2      WFQ       20           -/-              -/-
```

11.2.4　配置 HQoS

1. 实验目的

在企业网络中有 VLAN 10 和 VLAN 20 两个部门，部门中都包含视频和其他的业务流量。

要求使用 HQoS（Hierarchical Quality of Service，层次化服务质量），其中视频报文优先级为 DSCP EF，业务报文优先级为 DSCP AF21。

（1）配置 VLAN 10 和 VLAN 20 的子流策略，基于 DSCP 优先级进行流分类，视频报文进入 LLQ 队列、数据报文进入 AF 队列，对 AF 队列设置丢弃模板。

（2）配置父流策略，基于 VLAN ID 进行流分类，对来自不同 VLAN 的报文进行流量整形，并为其绑定相应的子流策略。

（3）出方向上应用父流策略，实现对不同用户的不同业务流量的区分，提供更为精细化的服务质量。

2. 实验拓扑

配置 HQoS 的实验拓扑如图 11-17 所示。

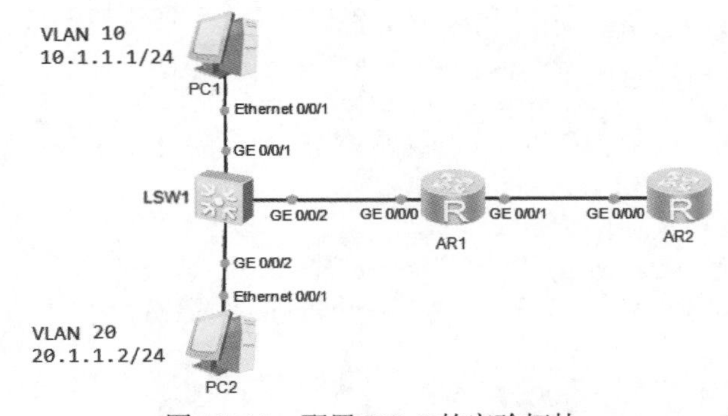

图 11-17　配置 HQoS 的实验拓扑

3. 实验步骤

步骤 1： 配置交换机和路由器的基础配置。

S1 的配置：

```
[S1]vlan batch 10 20
[S1]interface GigabitEthernet0/0/1
[S1-GigabitEthernet0/0/1]port link-type access
[S1-GigabitEthernet0/0/1]port default vlan 10
[S1-GigabitEthernet0/0/1]interface GigabitEthernet0/0/2
[S1-GigabitEthernet0/0/2]port link-type access
[S1-GigabitEthernet0/0/2]port default vlan 20
[S1-GigabitEthernet0/0/2]interface GigabitEthernet0/0/3
```

```
[S1-GigabitEthernet0/0/3]port link-type trunk
[S1-GigabitEthernet0/0/3]port trunk allow-pass vlan 10 20
```

R1 的配置：

```
[R1]interface GigabitEthernet0/0/0.10
[R1-GigabitEthernet0/0/0.10]dot1q termination vid 10
[R1-GigabitEthernet0/0/0.10]ip address 10.1.1.254 255.255.255.0
[R1-GigabitEthernet0/0/0.10]arp broadcast enable
[R1]interface GigabitEthernet0/0/0.20
[R1-GigabitEthernet0/0/0.20]dot1q termination vid 20
[R1-GigabitEthernet0/0/0.20]ip address 20.1.1.254 255.255.255.0
[R1-GigabitEthernet0/0/0.20]arp broadcast enable
[R1]interface GigabitEthernet0/0/1
[R1-GigabitEthernet0/0/1]ip address 12.1.1.1 255.255.255.0
```

R2 的配置：

```
[R2]interface GigabitEthernet0/0/0
[R2-GigabitEthernet0/0/0]ip address 12.1.1.2 255.255.255.0
[R2]ip route-static 10.1.1.0 24 12.1.1.1
[R2]ip route-static 20.1.1.0 24 12.1.1.1
```

步骤 2： 配置子策略，分别用于匹配不同的业务流量，视频报文进入 LLQ 队列、数据报文进入 AF 队列，对 AF 队列设置丢弃模板。

（1）配置流分类：

```
[R1]traffic classifier video
[R1-classifier-video]if-match dscp ef
[R1]traffic classifier data
[R1-classifier-data]if-match dscp af21
```

（2）配置 AF 队列的丢弃模板：

```
[R1]drop-profile AF
[R1-drop-profile-AF]wred dscp
[R1-drop-profile-AF]dscp af21 low-limit 50 high-limit 80 discard-percentage 60
```

（3）配置流行为：

```
[R1]traffic behavior video
[R1-behavior-video]queue llq bandwidth pct 25
[R1]traffic behavior data
[R1-behavior-data]queue af bandwidth pct 74
[R1-behavior-data]drop-profile AF
```

（4）创建子策略，调用流分类和流行为：

```
[R1]traffic policy 1
[R1-trafficpolicy-1]classifier video behavior video
[R1-trafficpolicy-1]classifier data behavior data
```

步骤 3： 创建父策略，用于区分不同的 VLAN 流量。

（1）创建流分类：

```
[R1]traffic classifier vlan10
```

```
[R1-classifier-vlan10]if-match inbound-interface g0/0/0.10
[R1]traffic classifier vlan20
[R1-classifier-vlan20]if-match  inbound-interface g0/0/0.20
```

（2）创建流行为：

```
[R1]traffic behavior vlan10
[R1-behavior-vlan10]gts cir 81920
[R1-behavior-vlan10]traffic-policy 1
[R1]traffic behavior  vlan20
[R1-behavior-vlan20]gts cir 20480
[R1-behavior-vlan20]traffic-policy 1
```

（3）创建父策略，调用 VLAN 10 和 VLAN 20 的流分类和流行为：

```
[R1]traffic policy 2
[R1-trafficpolicy-2]classifier vlan10 behavior vlan10
[R1-trafficpolicy-2]classifier vlan20 behavior vlan20
```

（4）将父策略调用在 AR1 的流量出接口：

```
[R1]interface g0/0/1
[R1-GigabitEthernet0/0/1]traffic-policy 2 outbound
```

查看策略名为 2 的应用情况，如下所示：

```
[R1]display traffic-policy applied-record 2
-------------------------------------------------
  Policy Name:   2
  Policy Index:  1
     Classifier:vlan10     Behavior:vlan10
     Classifier:vlan20     Behavior:vlan20
-------------------------------------------------
 *interface GigabitEthernet0/0/1                        //G0/0/1 接口调用了此策略
    traffic-policy 2 outbound
      slot 0     :  success
      nest Policy :  1
      slot 0     :  success
      nest Policy :  1
      slot 0     :  success
   Classifier: vlan10                                   // 流分类为 VLAN 10
    Operator: OR
    Rule(s)  :
     if-match inbound-interface GigabitEthernet0/0/0.10 // 匹配 G0/0/0.10 接口
     Behavior: vlan10
      General Traffic Shape:
        CIR 81920 (Kbps), CBS 2048000 (byte)            //GTS 限速为 81920 Kbps
        Queue length 1 (Packets)
      Nest Policy :  1
       Classifier: video                                // 子策略下的流分类为 video
        Operator: OR
        Rule(s)  :
         if-match dscp ef                               // 匹配 DSCP EF
```

```
        Behavior: video
          Low-latency:
            Bandwidth 25 (%)                                  //占用总带宽的25%
            Bandwidth 20480 (Kbps) CBS 512000 (Bytes) //实际带宽为81920×0.25=20480
      Classifier: data
       Operator: OR
       Rule(s) :
        if-match dscp af21
        Behavior: data
          Assured Forwarding:
            Bandwidth 74 (%)
            Bandwidth 60620 (Kbps)
            Drop Method: WRED

  Classifier: vlan20
   Operator: OR
   Rule(s) :
    if-match inbound-interface GigabitEthernet0/0/0.20
    Behavior: vlan20
     General Traffic Shape:
       CIR 20480 (Kbps), CBS 512000 (byte)
       Queue length 1 (Packets)
     Nest Policy :  1
     Nest Policy :  1
      Classifier: video
       Operator: OR
       Rule(s) :
        if-match dscp ef
        Behavior: video
          Low-latency:
            Bandwidth 25 (%)
            Bandwidth 5120 (Kbps) CBS 128000 (Bytes)
      Classifier: data
       Operator: OR
       Rule(s) :
        if-match dscp af21
        Behavior: data
          Assured Forwarding:
            Bandwidth 74 (%)
            Bandwidth 15155 (Kbps)
            Drop Method: WRED

-------------------------------------------------
  Policy total applied times: 1.
```

第 12 章

VXLAN 技术

12.1 VXLAN技术概述

为了解决数据中心网络服务器虚拟化以及虚拟机不受限迁移问题，VXLAN（Virtual eXtensible Local Area Network，虚拟扩展局域网）技术应运而生。由于VXLAN技术在本质上属于一种VPN技术，因此，其同样能够应用在园区网络中，以实现分散物理站点之间的二层互联以及站点间的三层互联。

在当前的园区网络中，租户站点与站点之间为了实现二、三层互联，需要部署相关设备以及多种二、三层网络技术。而基于Overlay的VXLAN技术，不感知当前的物理网络，能够在任意路由可达的网络上叠加二层虚拟网络，实现站点与站点之间的二层互联。同时，基于VXLAN三层网关，也能够实现站点与站点之间的三层互联。因此，通过VXLAN技术实现租户不同站点之间的互联更加快速、灵活。

12.1.1 VXLAN的基本概念

下面介绍VXLAN的基本概念。

（1）VNI（VXLAN Network Identifier，VXLAN网络标识）：类似于传统网络中的VXLAN ID，用于区分VXLAN段，不同VXLAN段的租户不能直接进行二层通信。一个租户可以有一个或多个VNI，VNI由24比特组成，支持多达16MB的租户。

（2）BD（Broadcast Domain，广播域）：类似于传统网络中采用VXLAN划分广播域方法，在VXLAN网络中通过BD划分广播域。在VXLAN网络中，将VNI以1∶1方式映射到BD，一个BD就表示一个广播域，同一个BD内的主机就可以进行二层互通。

（3）VTEP（VXLAN Tunnel Endpoints，VXLAN隧道端点）：VTEP可以对VXLAN报文进行封装和解封装。VXLAN报文中源IP地址为源端VTEP的IP地址，目的IP地址为目的端VTEP的IP地址。一对VTEP地址对应着一条VXLAN隧道。在源端封装报文后，通过隧道向目的端VTEP发送封装报文，目的端VTEP对接收到的封装报文进行解封装。

（4）VAP（Virtual Access Point，虚拟接入点）：VXLAN业务接入点，可以基于VXLAN或报文流封装类型接入业务。

①基于VXLAN接入业务：在VTEP上建立VXLAN与BD的一对一或多对一的映射。这样，当VTEP收到业务侧报文后，根据VXLAN与BD的映射关系，实现报文在BD内进行转发。

②基于报文流封装类型接入业务：在VTEP连接下行业务的物理接口上创建二层子接口，并配置不同的流封装类型，使不同的接口接入不同的数据报文。同时，将二层子接口与BD进行一一映射。这样业务侧报文到达VTEP后，便会进入指定的二层子接口，即根据二层子接口与BD的映射关系实现报文在BD内进行转发。

（5）NVE（Network Virtualization Edge，网络虚拟边缘）：NVE是实现网络虚拟化功能的网络实体。报文经过NVE封装转换后，NVE之间就可以基于三层基础网络建立二层虚拟化网络。图12-1中的路由器设备即为NVE。

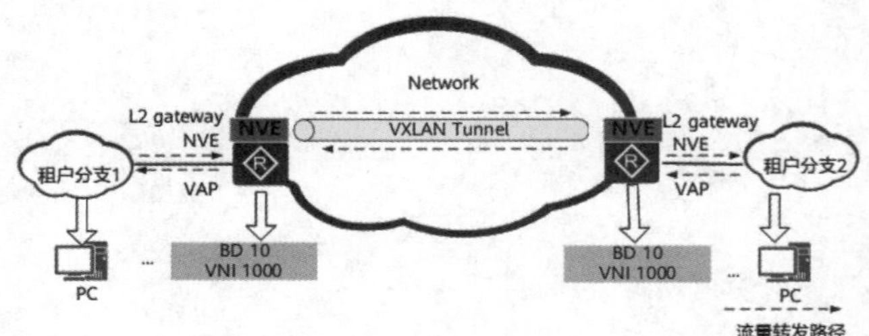

图12-1　VXLAN二层网关示意图

（6）二层网关：类似于传统网络的二层接入设备，在VXLAN网络中通过二层网关解决租户接入 VXLAN 虚拟网络问题，也可以用于同一 VXLAN 虚拟网络的子网通信。

（7）三层网关：类似于传统网络中不同 VXLAN 的用户间不能直接进行二层互访，不同 VNI 之间的 VXLAN 及 VXLAN 和非 VXLAN 之间也不能直接相互通信。为了使 VXLAN 之间，以及 VXLAN 和非 VXLAN 之间能够进行通信，引入了 VXLAN 三层网关的概念。

三层网关用于 VXLAN 虚拟网络的跨子网通信以及外部网络的访问。VXLAN 三层网关示意图如图 12-2 所示。

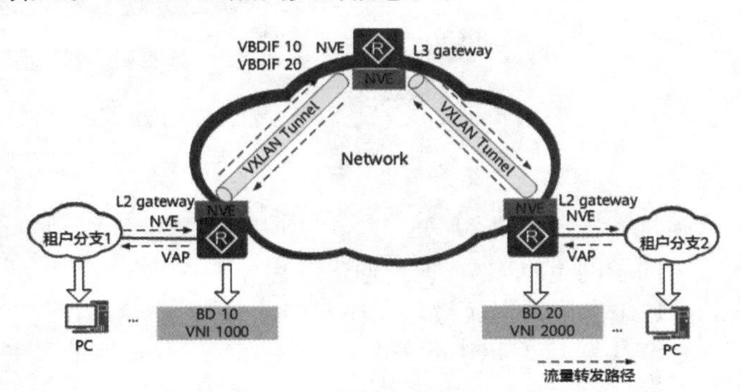

图 12-2　VXLAN 三层网关示意图

（8）VBDIF 接口：类似于传统网络中采用 VXLAN 解决不同广播域互通的方法，在 VXLAN 中引入了 VBDIF 的概念。VBDIF 接口在 VXLAN 三层网关上配置，是基于 BD 创建的三层逻辑接口。通过 VBDIF 接口配置 IP 地址可以实现不同网段的 VXLAN 之间，以及 VXLAN 和非 VXLAN 之间的通信，也可以实现二层网络接入三层网络。

12.1.2　VXLAN 隧道建立方式

VXLAN 隧道建立方式有以下两种。

（1）静态建立：用户通过手工配置本端和远端的 VNI、VTEP IP 地址和头端复制列表（head-end peer-list）来完成。

（2）动态建立：在 VTEP 之间建立 BGP EVPN 对等体，然后对等体之间利用 BGP EVPN 路由来互相传递 VNI 和 VTEP IP 地址信息，从而实现动态地建立 VXLAN 隧道。

12.1.3　BGP EVPN 与 VXLAN

VXLAN 可以看作一个数据平面的协议。因此 VXLAN 隧道的自动建立需要依托 BGP EVPN，BGP EVPN 除了可以自动建立 VXLAN 隧道，还可以完成 VXLAN 控制平面的相关工作。

BGP EVPN 路由分为 Type2 路由、Type3 路由和 Type5 路由，下面分别对其进行简要介绍。

（1）Type2 路由（MAC/IP 路由）：用于主机 MAC 地址 /ARP/IP 路由通告。

在不同场景下，Type2 路由携带的内容不一致，如图 12-3 所示。

主机MAC地址通告	主机ARP通告	主机IP路由通告
Route Distinguisher	Route Distinguisher	Route Distinguisher
Ethernet Segment Identifier	Ethernet Segment Identifier	Ethernet Segment Identifier
Ethernet Tag ID	Ethernet Tag ID	Ethernet Tag ID
MAC Address Length = MAC地址长度	MAC Address Length = MAC地址长度	MAC Address Length = MAC地址长度
MAC Address = MAC地址	MAC Address = MAC地址	MAC Address = MAC地址
IP Address Length	IP Address Length = IP地址长度	IP Address Length = IP地址长度
IP Address	IP Address = IP地址	IP Address = IP地址
MPLS Label1 = VNI（二层）	MPLS Label1 = VNI（二层）	MPLS Label1 = VNI（二层）
MPLS Label2	MPLS Label2	MPLS Label2 = VNI（三层）
相同子网主机互访场景下进行主机MAC地址通告，包含主机MAC信息和二层VNI。	集中式网关部署场景下通告ARP类型路由，包含主机IP信息、MAC信息和二层VNI。	分布式网关部署不同子网互访场景下，进行IRB类型路由通告。它包括主机MAC信息、IP信息、二层VNI和三层VNI。

图 12-3　BGP EVPN 路由（Type2 路由）

主机 MAC 地址通告示意图如图 12-4 所示。

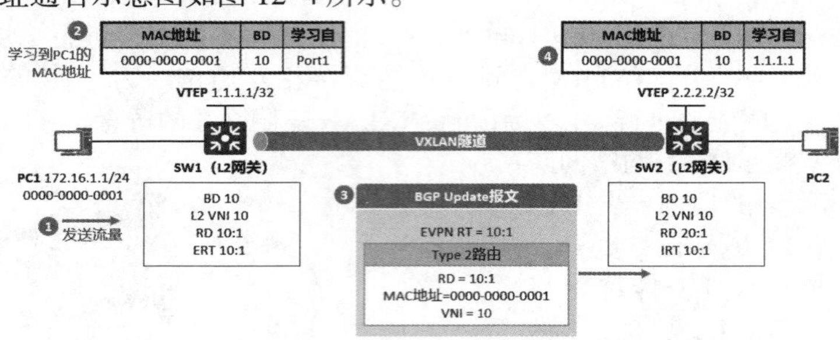

图 12-4　主机 MAC 地址通告示意图

同子网主机 MAC 地址通告过程如下：

PC1 产生数据流量并发往 SW1。

SW1 获知了 PC1 的 MAC 地址，它在 MAC 地址表中创建一个表项，记录该 MAC 地址、BD ID 及入接口。

SW1 根据该表项生成 BGP EVPN 路由并发送给 SW2。该路由携带本端 EVPN 实例的 RT 值（扩展团体属性）以及 Type2 路由（MAC 路由）。在 MAC 路由中，PC1 的 MAC 地址存放在 MAC Address 字段中，二层 VNI 存放在 MPLS Label1 字段中。

SW2 收到 SW1 发来的 BGP EVPN 路由后，首先检查该路由携带的 RT（类似于 MPLS VPN 中的 RT 的概念），如果与本端 EVPN 实例的入站 RT 相等，则接收该路由；否则丢弃。在接收该路由后，SW2 获得 PC1 的 MAC 地址、BD ID 和 SW1 上 VTEP IP 地址（MP_REACH_NLRI 中 Next hop network address 字段携带）的对应关系，并在本地的 MAC 表中生成 MAC 表项，其出接口需要根据下一跳进行迭代，最终迭代结果指向 SW1 的 VXLAN 隧道。

主机 ARP 通告示意图如图 12-5 所示。

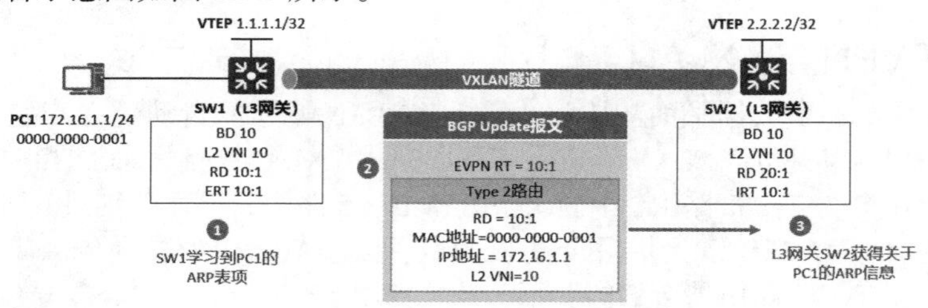

图 12-5　主机 ARP 通告示意图

MAC/IP 路由可以同时携带主机 MAC 地址 + 主机 IP 地址，因此该路由可以用于在 VTEP 之间传递主机 ARP 表项，实现主机 ARP 通告。其中，MAC Address 和 MAC Address Length 字段为主机 MAC 地址，IP Address 和 IP Address Length 字段为主机 IP 地址。此时的 MAC/IP 路由也称为 ARP 类型路由。主机 ARP 通告主要用于以下两种场景。

① ARP 广播抑制。当三层网关学习到其子网下的主机 ARP 时，生成主机信息（包含主机 IP 地址、主机 MAC 地址、二层 VNI、网关 VTEP IP 地址），然后通过传递 ARP 类型路由将主机信息同步到二层网关上。这样当二层网关再收到 ARP 请求时，先查找是否存在目的 IP 地址对应的主机信息，如果存在，则直接将 ARP 请求报文中的广播 MAC 地址替换为目的单播 MAC 地址，实现广播变单播，达到 ARP 广播抑制的目的。

②分布式网关场景下的虚拟机迁移。当一台虚拟机从当前网关迁移到另一个网关下之后，新网关学习到该虚拟机的 ARP（一般通过虚拟机发送免费 ARP 实现），并生成主机信息（包含主机 IP 地址、主机 MAC 地址、二层 VNI、网关 VTEP IP 地址），然后通过传递 ARP 类型路由将主机信息发送给虚拟机的原网关。原网关收到后，感知到虚拟机的位置发生变化，触发 ARP 探测，当探测不到原位置的虚拟机时，撤销原位置虚拟机的 ARP 和主机路由。

主机 ARP 通告主要在 VXLAN 集中式网关 +BGP EVPN 场景下使用，在 BGP EVPN 中，向对等体通告 ARP 路由或者 IRB 路由为互斥选项，只能配置其中一种路由对外发布，一般在 VXLAN 集中式网关 +BGP EVPN 场景下选择发布 ARP 路由，而且在 VXLAN 分布式网关 +BGP EVPN 场景下选择发布 IRB 路由。

主机 IP 路由通告示意图如图 12-6 所示。

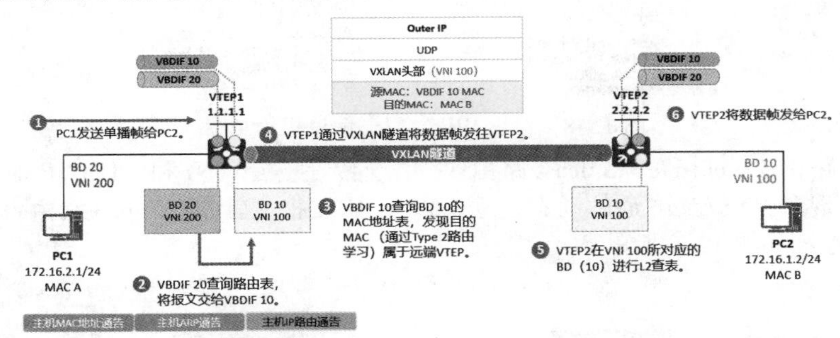

图 12-6　主机 IP 路由通告示意图

分布式网关组网中的 VTEP 设备既是 L2 网关，又是 L3 网关。在该组网下跨子网通信的实现方式并不唯一，根据接收报文的 VTEP（Ingress VTEP）处理方式不同，可以划分为非对称 IRB（Asymmetric Integrated Routing and Bridging）转发和对称 IRB（Symmetric Integrated Routing and Bridging）转发。

非对称 IRB 转发：Ingress VTEP 同时执行 L3、L2 查表转发，Egress VTEP 只需要进行 L2 查表、转发，因为 Ingress、Egress 所执行的操作不一致，被称为非对称转发。

对称 IRB 转发：Ingress VTEP、Egress VTEP 都执行 L3 查表转发。

相比较于非对称 IRB 转发，新增了一个 IP VPN 实例以及其所绑定的 L3 VNI 概念（非对称 IRB 转发时 VTEP 之间传输报文的 VXLAN 头部中的 VNI 值为 L2 VNI），VBDIF 接口需要绑定 IP VPN 实例，此时该 VBDIF 接口的路由学习、数据转发都被限制在该 IP VPN 实例中，与 MPLS VPN 类似。

IRB 路由通告过程示意图如图 12-7 所示。

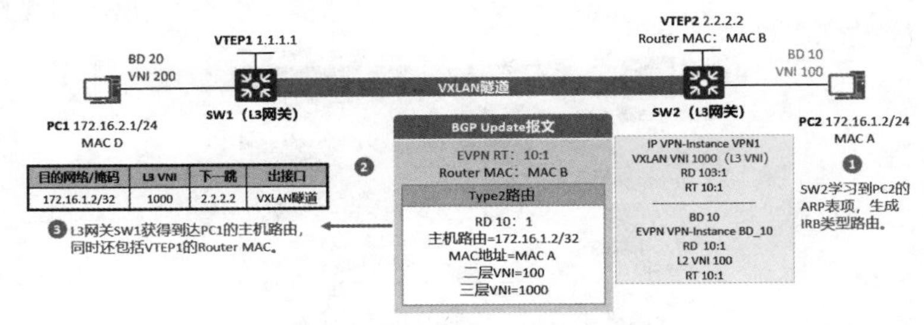

图 12-7　IRB 路由通告过程示意图

在配置对称 IRB 转发时，需要在 EDGE 设备上配置 IP VPN 实例和 EVPN 实例。对称 IRB 转发通信过程示意图如图 12-8 所示，本端设备的 BD 域中 EVPN 实例配置了一个出方向 RT 为 11：1，需要与对端设备的 IP VPN 实例的如方向 RT 匹配，用于生成主机路由。

对称 IRB 转发通信过程：

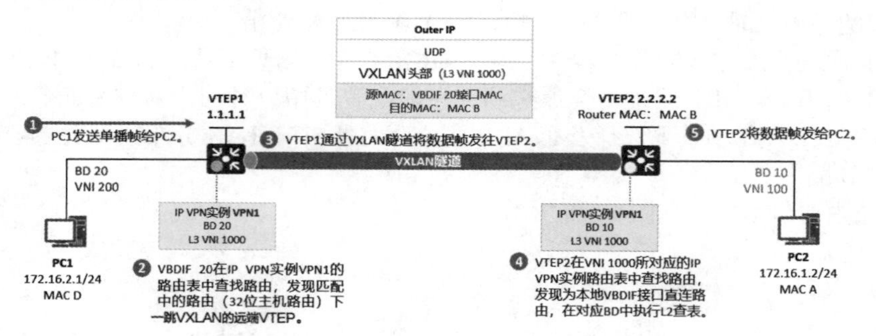

图 12-8　对称 IRB 转发通信过程示意图

（2）Type3 路由（Inclusive Multicast 路由）：用于传递二层 VNI 和 VTEP IP 地址信息，实现 VTEP 的自动发现和 VXLAN 隧道的动态建立，以及 BUM 报文转发。Type3 路由自动发现和建立 VXLAN 隧道示意图如图 12-9 所示。

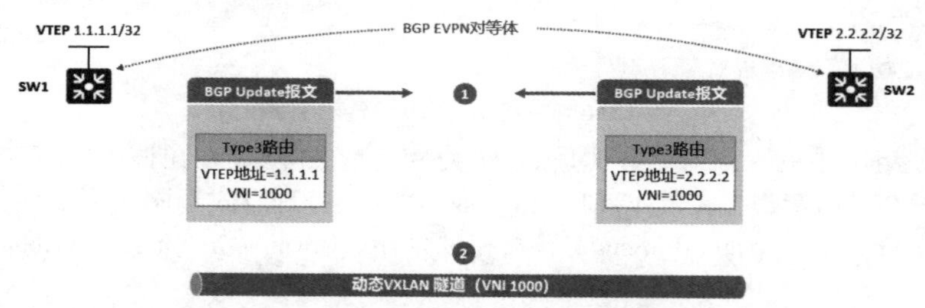

图 12-9　Type3 路由自动发现和建立 VXLAN 隧道示意图

VTEP 通过 Type3 路由互相传递二层 VNI 和 VTEP IP 地址信息。如果对端 VTEP IP 地址是三层路由可达的，则建立一条到对端的 VXLAN 隧道。同时，如果对端 VNI 与本端相同，则创建一个头端复制列表，用于后续 BUM 报文转发。

（3）Type5 路由（IP 前缀路由）：用于主机 MAC 地址 /ARP/IP 路由通告，外部网络路由通告。Type5 路由通告示意图如图 12-10 所示。

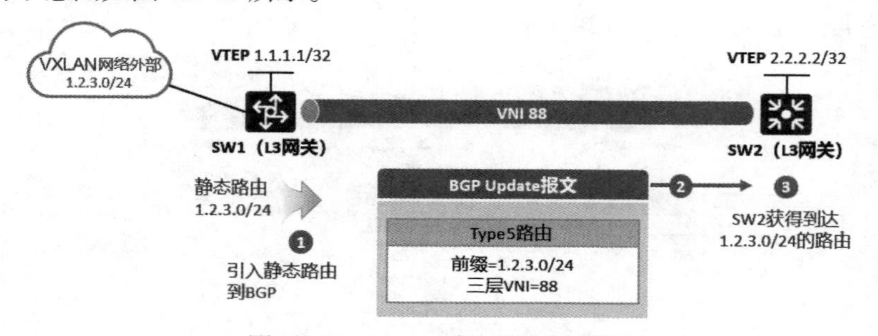

图 12-10　Type5 路由通告示意图

12.2 VXLAN 实验

12.2.1 配置 VXLAN 实现相同网段互访（静态方式）

1. 实验目的

在 edge1 和 edge2 配置静态 VXLAN，实现相同网段的 PC 能够互访。

2. 实验拓扑

配置 VXLAN 实现相同网段互访（静态方式）的实验拓扑如图 12-11 所示。

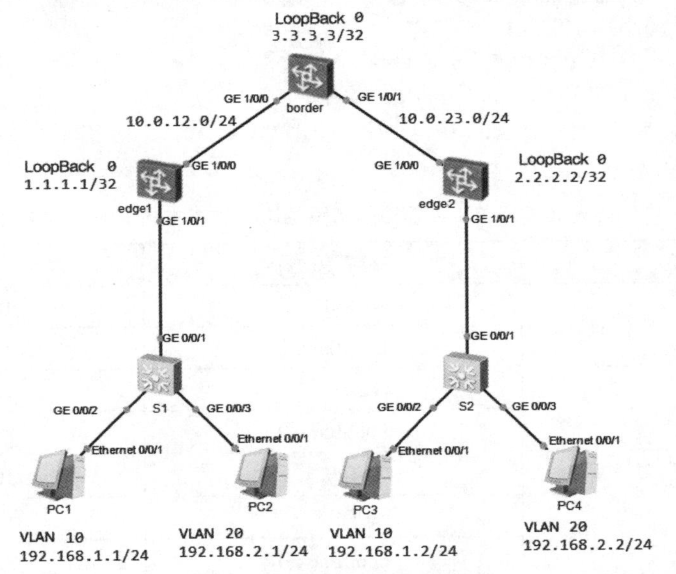

图 12-11　配置 VXLAN 实现相同网段互访（静态方式）实验拓扑

3. 实验步骤

步骤 1：S1 和 S2 的 VLAN 基本配置。

S1 的配置：

```
[S1]vlan batch 10 20
[S1]interface GigabitEthernet0/0/1
[S1-GigabitEthernet0/0/1]port link-type trunk
[S1-GigabitEthernet0/0/1]port trunk allow-pass vlan 10 20
[S1-GigabitEthernet0/0/1]q
[S1]interface GigabitEthernet0/0/2
[S1-GigabitEthernet0/0/2]port link-type access
[S1-GigabitEthernet0/0/2]port default vlan 10
[S1-GigabitEthernet0/0/2]q
[S1]interface GigabitEthernet0/0/3
[S1-GigabitEthernet0/0/3]port link-type access
[S1-GigabitEthernet0/0/3]port default vlan 20
```

S2 的配置：

```
[S2]vlan batch 10 20
```

```
[S2]interface GigabitEthernet0/0/1
[S2-GigabitEthernet0/0/1]port link-type trunk
[S2-GigabitEthernet0/0/1]port trunk allow-pass vlan 10 20
[S2-GigabitEthernet0/0/1]q
[S2]interface GigabitEthernet0/0/2
[S2-GigabitEthernet0/0/2]port link-type access
[S2-GigabitEthernet0/0/2]port default vlan 10
[S2-GigabitEthernet0/0/2]q
[S2]interface GigabitEthernet0/0/3
[S2-GigabitEthernet0/0/3]port link-type access
[S2-GigabitEthernet0/0/3]port default vlan 20
```

步骤 2：配置 border 和 EDGE 的 underlay 网络。

（1）配置 border 和 EDGE 节点的 IPv4 地址。配置 VXLAN 实现相同网段互访（静态方式）IP 地址规划表见表 12-1。

表 12-1　配置 VXLAN 实现相同网段互访（静态方式）IP 地址规划表

设备名称	接口编号	IP 地址
edge1	G1/0/0	10.0.12.2/24
	LoopBack 0	1.1.1.1/32
edge2	G1/0/0	10.0.23.2/24
	LoopBack 0	2.2.2.2/32
border	G1/0/0	10.0.12.1/24
	G1/0/1	10.0.23.1/24
	LoopBack 0	3.3.3.3/32

ENSP 的 CE 设备接口默认为二层口，并且物理状态为 down，需要配置 IP 执行以下操作（以 edge1 的 G1/0/0 接口为例）：

```
[edge1]interface G1/0/0
[edge1-GE1/0/0]undo shutdown
[edge1-GE1/0/0]undo portswitch
[edge1-GE1/0/0]ip address 10.0.12.2 24
```

（2）配置 underlay 网络的 OSPF 协议。

border 的配置：

```
[border]ospf 1
[border-ospf-1]area 0.0.0.0
[border-ospf-1-area-0.0.0.0]network 3.3.3.3 0.0.0.0
[border-ospf-1-area-0.0.0.0]network 10.0.12.0 0.0.0.255
[border-ospf-1-area-0.0.0.0]network 10.0.23.0 0.0.0.255
```

edge1 的配置：

```
[edge1]ospf 1
[edge1-ospf-1]area 0.0.0.0
[edge1-ospf-1-area-0.0.0.0]network 1.1.1.1 0.0.0.0
```

```
[edge1-ospf-1-area-0.0.0.0]network 10.0.12.0 0.0.0.255
```

edege2 的配置：

```
[edge2]ospf 1
[edge2-ospf-1]area 0.0.0.0
[edge2-ospf-1-area-0.0.0.0]network 2.2.2.2 0.0.0.0
[edge2-ospf-1-area-0.0.0.0]network 10.0.23.0 0.0.0.255
```

查看路由表：

```
[edge1]display ip routing-table
Proto: Protocol        Pre: Preference
Route Flags: R - relay, D - download to fib, T - to vpn-instance, B - black hole route
------------------------------------------------------------------------------
Routing Table : _public_
         Destinations : 11       Routes : 11
 Destination/Mask    Proto    Pre   Cost  Flags    NextHop      Interface
       1.1.1.1/32    Direct   0     0     D        127.0.0.1    LoopBack0
       2.2.2.2/32    OSPF     10    2     D        10.0.12.1    GE1/0/0
       3.3.3.3/32    OSPF     10    1     D        10.0.12.1    GE1/0/0
     10.0.12.0/24    Direct   0     0     D        10.0.12.2    GE1/0/0
     10.0.12.2/32    Direct   0     0     D        127.0.0.1    GE1/0/0
   10.0.12.255/32    Direct   0     0     D        127.0.0.1    GE1/0/0
     10.0.23.0/24    OSPF     10    2     D        10.0.12.1    GE1/0/0
      127.0.0.0/8    Direct   0     0     D        127.0.0.1    InLoopBack0
     127.0.0.1/32    Direct   0     0     D        127.0.0.1    InLoopBack0
127.255.255.255/32   Direct   0     0     D        127.0.0.1    InLoopBack0
255.255.255.255/32   Direct   0     0     D        127.0.0.1    InLoopBack0
```

通过以上输出可以看出，设备之间已经学习到了各自的环回口路由。

步骤 3：配置 EDGE 节点的 BD，将 VNI 绑定到 BD 中。BD 为本地的概念，在 VXLAN 中进行数据传递时需要通过 VNI 来体现。

edge1 的配置：

```
[edge1]bridge-domain 10
[edge1-bd10]vxlan vni 10
[edge1-bd10]q
[edge1]bridge-domain 20
[edge1-bd20]vxlan vni 20
```

edge2 的配置：

```
[edge2]bridge-domain 10
[edge2-bd10]vxlan vni 10
[edge2-bd10]q
[edge2]bridge-domain 20
[edge2-bd20]vxlan vni 20
```

步骤 4：配置 EDGE 节点的子接口，将不同的子接口绑定到不同的 BD。

edge1 的配置：

```
[edge1]interface GE1/0/1
```

```
[edge1-GE1/0/1]undo shutdown                    //默认情况下的物理接口为 down
[edge1-GE1/0/1]q
[edge1]interface GE1/0/1.10 mode l2             //配置子接口为二层子接口
[edge1-GE1/0/1.10]encapsulation dot1q vid 10    //配置子接口能够处理 VLAN 10 的数据
[edge1-GE1/0/1.10]bridge-domain 10              //将子接口绑定到 BD
[edge1-GE1/0/1.10]q
[edge1]interface GE1/0/1.20 mode l2
[edge1-GE1/0/1.20]encapsulation dot1q vid 20
[edge1-GE1/0/1.20]bridge-domain 20
```

edge2 的配置：

```
[edge2]interface GE1/0/1
[edge2-GE1/0/1]undo shutdown
[edge2-GE1/0/1]q
[edge2-GE1/0/1]interface GE1/0/1.10 mode l2
[edge2-GE1/0/1.10]encapsulation dot1q vid 10
[edge2-GE1/0/1.10]bridge-domain 10
[edge2-GE1/0/1.10]q
[edge2-]interface GE1/0/1.20 mode l2
[edge2-GE1/0/1.20]encapsulation dot1q vid 20
[edge2-GE1/0/1.20]bridge-domain 20
```

步骤 5：配置 NVE、指定 VXLAN 的头尾复制功能，以及静态的 VXLAN 隧道。

edge1 的配置：

```
[edge1]interface Nve1                           //创建 NVE 接口
[edge1-Nve1]source 1.1.1.1                      //指定 VTEP 地址为 1.1.1.1
[edge1-Nve1]vni 10 head-end peer-list 2.2.2.2  //配置头端复制列表
[edge1-Nve1]vni 20 head-end  peer-list  2.2.2.2
```

头端是指 VXLAN 隧道的入节点；复制是指当 VXLAN 隧道的入节点收到一份 BUM 报文后，需要将其复制多份并发送给列表中的所有 VTEP；头端复制列表是指用于指导 VXLAN 隧道的入节点进行 BUM 报文复制和发送的远端 VTEP 的 IP 地址列表。

edge2 的配置：

```
[edge2]interface Nve1
[edge2-Nve1]source 2.2.2.2
[edge2-Nve1]vni 10 head-end peer-list 1.1.1.1
[edge2-Nve1]vni 20 head-end peer-list 1.1.1.1
```

查看 VXLAN 隧道的建立情况：

```
[edge1]display  vxlan  tunnel
Number of vxlan tunnel : 1
Tunnel ID   Source                Destination           State  Type     Uptime
------------------------------------------------------------------------------
4026531841  1.1.1.1               2.2.2.2               up     static   00:01:41
```

步骤 6：测试。

PC1 访问 PC3 的测试图如图 12-12 所示。

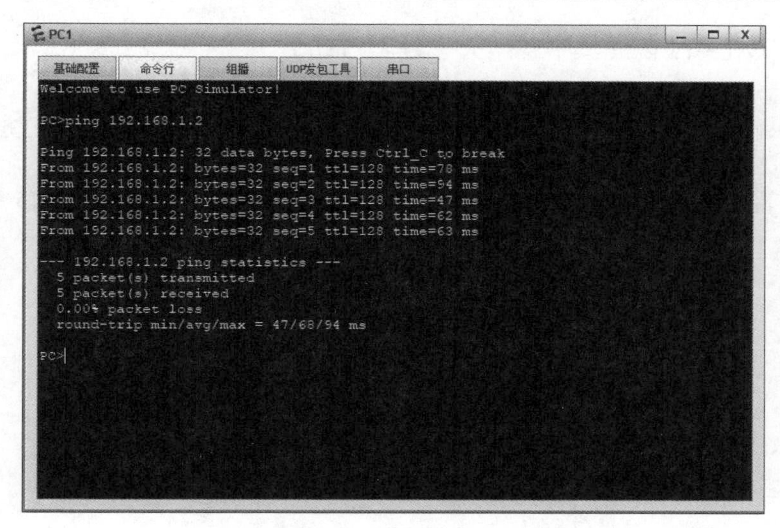

图 12-12 PC1 访问 PC3 的测试图

PC2 访问 PC4 的测试图如图 12-13 所示。

图 12-13 PC2 访问 PC4 的测试图

以 PC1 访问 PC3 的报文为例，在 edge1 的 G1/0/0 接口抓包，查看报文信息，结果如图 12-14 所示。

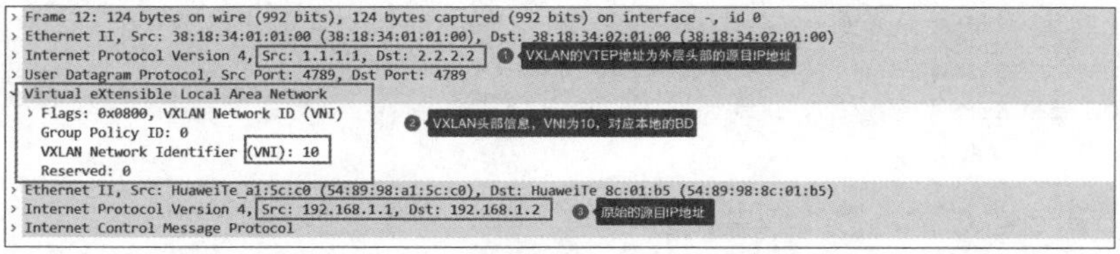

```
> Frame 12: 124 bytes on wire (992 bits), 124 bytes captured (992 bits) on interface -, id 0
> Ethernet II, Src: 38:18:34:01:01:00 (38:18:34:01:01:00), Dst: 38:18:34:02:01:00 (38:18:34:02:01:00)
> Internet Protocol Version 4, Src: 1.1.1.1, Dst: 2.2.2.2      ❶ VXLAN的VTEP地址为外层头部的源目IP地址
> User Datagram Protocol, Src Port: 4789, Dst Port: 4789
v Virtual eXtensible Local Area Network
  > Flags: 0x0800, VXLAN Network ID (VNI)                      ❷ VXLAN头部信息，VNI为10，对应本地的BD
    Group Policy ID: 0
    VXLAN Network Identifier (VNI): 10
    Reserved: 0
> Ethernet II, Src: HuaweiTe_a1:5c:c0 (54:89:98:a1:5c:c0), Dst: HuaweiTe_8c:01:b5 (54:89:98:8c:01:b5)
> Internet Protocol Version 4, Src: 192.168.1.1, Dst: 192.168.1.2    ❸ 原始的源目IP地址
> Internet Control Message Protocol
```

图 12-14 在 edge1 的 G1/0/0 接口的抓包结果

12.2.2 配置VXLAN创建集中式网关（静态方式）

扫一扫，看视频

1. 实验目的

配置border为集中式网关，不同网段通过border作为网关进行互访。

2. 实验拓扑

配置VXLAN创建集中式网关（静态方式）的实验拓扑如图12-15所示。

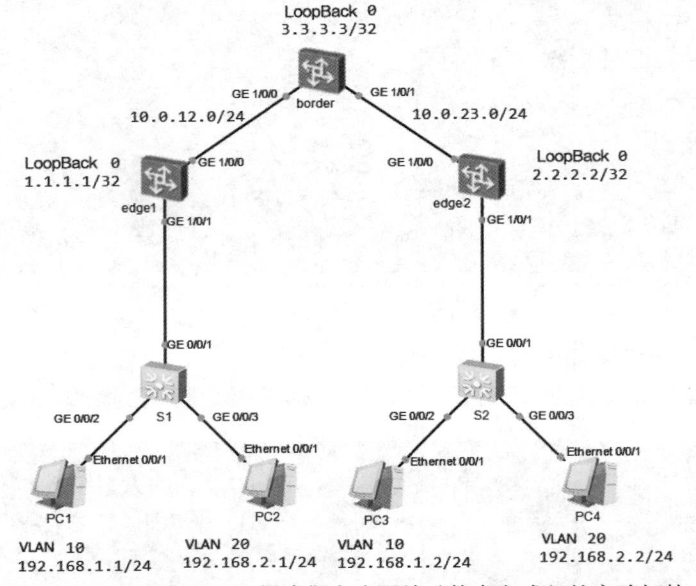

图12-15　配置VXLAN创建集中式网关（静态方式）的实验拓扑

3. 实验步骤

步骤1：S1和S2的VLAN基本配置。

S1的配置：

```
[S1]vlan batch 10 20
[S1]interface GigabitEthernet0/0/1
[S1-GigabitEthernet0/0/1]port link-type trunk
[S1-GigabitEthernet0/0/1]port trunk allow-pass vlan 10 20
[S1-GigabitEthernet0/0/1]q
[S1]interface GigabitEthernet0/0/2
[S1-GigabitEthernet0/0/2]port link-type access
[S1-GigabitEthernet0/0/2]port default vlan 10
[S1-GigabitEthernet0/0/2]q
[S1]interface GigabitEthernet0/0/3
[S1-GigabitEthernet0/0/3]port link-type access
[S1-GigabitEthernet0/0/3]port default vlan 20
```

S2的配置：

```
[S2]vlan batch 10 20
[S2]interface GigabitEthernet0/0/1
```

```
[S2-GigabitEthernet0/0/1]port link-type trunk
[S2-GigabitEthernet0/0/1]port trunk allow-pass vlan 10 20
[S2-GigabitEthernet0/0/1]q
[S2]interface GigabitEthernet0/0/2
[S2-GigabitEthernet0/0/2]port link-type access
[S2-GigabitEthernet0/0/2]port default vlan 10
[S2-GigabitEthernet0/0/2]q
[S2]interface GigabitEthernet0/0/3
[S2-GigabitEthernet0/0/3]port link-type access
[S2-GigabitEthernet0/0/3]port default vlan 20
```

步骤 2：配置 border 和 EDGE 的 underlay 网络。

（1）配置 border 和 EDGE 节点的 IPv4 地址。配置 VXLAN 创建集中式网关（静态方式）IP 地址规划表见表 12-2。

表 12-2　配置 VXLAN 创建集中式网关（静态方式）IP 地址规划表

设备名称	接口编号	IP 地址
edge1	G1/0/0	10.0.12.2/24
	LoopBack 0	1.1.1.1/32
edge2	G1/0/0	10.0.23.2/24
	LoopBack 0	2.2.2.2/32
border	G1/0/0	10.0.12.1/24
	G1/0/1	10.0.23.1/24
	LoopBack 0	3.3.3.3/32

ENSP 的 CE 设备接口默认为二层口，并且物理状态为 down，需要配置 IP 执行以下操作（下面以 edge1 的 G1/0/0 接口为例）：

```
[edge1]interface G1/0/0
[edge1-GE1/0/0]undo shutdown
[edge1-GE1/0/0]undo portswitch
[edge1-GE1/0/0]ip address  10.0.12.2 24
```

（2）配置 underlay 网络的 OSPF 协议。

border 的配置：

```
[border]ospf 1
[border-ospf-1]area 0.0.0.0
[border-ospf-1-area-0.0.0.0]network 3.3.3.3 0.0.0.0
[border-ospf-1-area-0.0.0.0]network 10.0.12.0 0.0.0.255
[border-ospf-1-area-0.0.0.0]network 10.0.23.0 0.0.0.255
```

edge1 的配置：

```
[edge1]ospf 1
[edge1-ospf-1]area 0.0.0.0
[edge1-ospf-1-area-0.0.0.0]network 1.1.1.1 0.0.0.0
[edge1-ospf-1-area-0.0.0.0]network 10.0.12.0 0.0.0.255
```

edge2 的配置：

```
[edge2]ospf 1
[edge2-ospf-1]area 0.0.0.0
[edge2-ospf-1-area-0.0.0.0]network 2.2.2.2 0.0.0.0
[edge2-ospf-1-area-0.0.0.0]network 10.0.23.0 0.0.0.255
```

查看路由表：

```
[edge1]display ip routing-table
Proto: Protocol        Pre: Preference
Route Flags: R - relay, D - download to fib, T - to vpn-instance, B - black hole route
------------------------------------------------------------------------------
Routing Table : _public_
         Destinations : 11        Routes : 11
  Destination/Mask      Proto     Pre      Cost     Flags   NextHop        Interface
        1.1.1.1/32      Direct    0        0        D       127.0.0.1      LoopBack0
        2.2.2.2/32      OSPF      10       2        D       10.0.12.1      GE1/0/0
        3.3.3.3/32      OSPF      10       1        D       10.0.12.1      GE1/0/0
      10.0.12.0/24      Direct    0        0        D       10.0.12.2      GE1/0/0
      10.0.12.2/32      Direct    0        0        D       127.0.0.1      GE1/0/0
    10.0.12.255/32      Direct    0        0        D       127.0.0.1      GE1/0/0
      10.0.23.0/24      OSPF      10       2        D       10.0.12.1      GE1/0/0
       127.0.0.0/8      Direct    0        0        D       127.0.0.1      InLoopBack0
      127.0.0.1/32      Direct    0        0        D       127.0.0.1      InLoopBack0
127.255.255.255/32      Direct    0        0        D       127.0.0.1      InLoopBack0
255.255.255.255/32      Direct    0        0        D       127.0.0.1      InLoopBack0
```

通过以上输出可以看出，设备之间已经学习到了各自的环回口路由。

步骤 3： 配置 EDGE 节点及 border 节点的 BD，将 VNI 绑定到 BD 中。

border 的配置：

```
[border]bridge-domain 10
[border-bd10]vxlan vni 10
[border-bd10]q
[border]bridge-domain 20
[border-bd20]vxlan vni 20
```

edge1 的配置：

```
[edge1]bridge-domain 10
[edge1-bd10]vxlan vni 10
[edge1-bd10]q
[edge1]bridge-domain 20
[edge1-bd20]vxlan vni 20
```

edge2 的配置：

```
[edge2]bridge-domain 10
[edge2-bd10]vxlan vni 10
[edge2-bd10]q
[edge2]bridge-domain 20
[edge2-bd20]vxlan vni 20
```

步骤 4: 配置 EDGE 节点的子接口，将不同的子接口绑定到不同的 BD 中。

edge1 的配置:

```
[edge1]interface GE1/0/1
[edge1-GE1/0/1]undo shutdown
[edge1-GE1/0/1]q
[edge1]interface GE1/0/1.10 mode l2
[edge1-GE1/0/1.10]encapsulation dot1q vid 10
[edge1-GE1/0/1.10]bridge-domain 10
[edge1-GE1/0/1.10]q
[edge1]interface GE1/0/1.20 mode l2
[edge1-GE1/0/1.20]encapsulation dot1q vid 20
[edge1-GE1/0/1.20]bridge-domain 20
```

edge2 的配置:

```
[edge2]interface GE1/0/1
[edge2-GE1/0/1]undo shutdown
[edge2-GE1/0/1]q
[edge2]interface GE1/0/1.10 mode l2
[edge2-GE1/0/1.10]encapsulation dot1q vid 10
[edge2-GE1/0/1.10]bridge-domain 10
[edge2-GE1/0/1.10]q
[edge2]interface GE1/0/1.20 mode l2
[edge2-GE1/0/1.20]encapsulation dot1q vid 20
[edge2-GE1/0/1.20]bridge-domain 20
[edge2-GE1/0/1.20]q
```

步骤 5: 配置 EDGE 节点和 border 节点的 NVE，指定 VXLAN 的头尾复制功能，EDGE 节点和 border 节点用于指定静态的 VXLAN 隧道。

border 的配置:

```
[border]interface Nve1
[border-Nve1]source 3.3.3.3
[border-Nve1]vni 10 head-end peer-list 1.1.1.1
[border-Nve1]vni 10 head-end peer-list 2.2.2.2
[border-Nve1]vni 20 head-end peer-list 1.1.1.1
[border-Nve1]vni 20 head-end peer-list 2.2.2.2
```

edge1 的配置:

```
[edge1]interface Nve1
[edge1-Nve1]source 1.1.1.1
[edge1-Nve1]vni 10 head-end peer-list 3.3.3.3
[edge1-Nve1]vni 20 head-end peer-list 3.3.3.3
```

edge2 的配置:

```
[edge2]interface Nve1
[edge2-Nve1]source 2.2.2.2
[edge2-Nve1]vni 10 head-end peer-list 3.3.3.3
[edge2-Nve1]vni 20 head-end peer-list 3.3.3.3
```

查看 border 的 VXLAN 隧道：

```
[border]display vxlan tunnel
Number of vxlan tunnel : 2
Tunnel ID   Source              Destination          State  Type     Uptime
---------------------------------------------------------------------------
4026531841  3.3.3.3             2.2.2.2              up     static   00:12:21
4026531842  3.3.3.3             1.1.1.1              up     static   00:12:19
```

步骤 6： 配置 INTERFACE VBDIF10 接口的网关 IP 地址。

```
[border]interface Vbdif10
[border-Vbdif10]ip address 192.168.1.254 255.255.255.0
[border-Vbdif10]q
[border]interface Vbdif20
[border-Vbdif20]ip address 192.168.2.254 255.255.255.0
```

查看接口状态，保证 VBDIF 接口为 up 状态：

```
[border]display ip int brief
Interface            IP Address/Mask       Physical    Protocol    VPN
GE1/0/0              10.0.12.1/24          up          up          --
GE1/0/1              10.0.23.1/24          up          up          --
LoopBack0            3.3.3.3/32            up          up(s)       --
MEth0/0/0            unassigned            up          down        --
NULL0                unassigned            up          up(s)       --
Vbdif10              192.168.1.254/24      up          up          --
Vbdif20              192.168.2.254/24      up          up          --
```

由于本实验中的 ENSP 模拟器出现了 bug，因此无法测试结果。

12.2.3 配置 VXLAN 实现相同网段互访（BGP EVPN 方式）

1. 实验目的

通过 BGP EVPN 方式实现 VXLAN 隧道的自动建立，并实现相同网段的互相访问。

2. 实验拓扑

配置 VXLAN 实现相同网段互访（BGP EVPN 方式）的实验拓扑如图 12-16 所示。

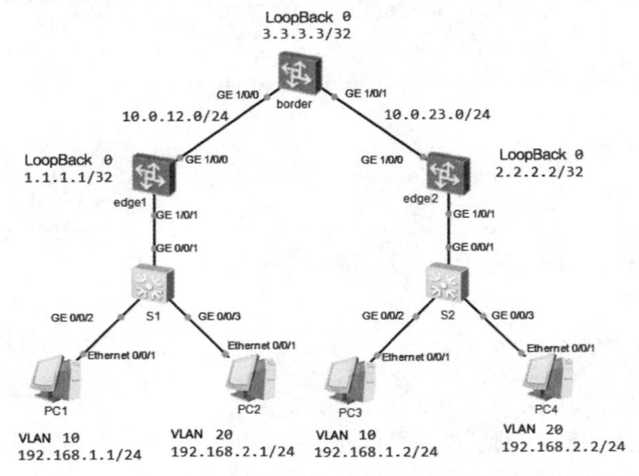

图 12-16 配置 VXLAN 实现相同网段互访（BGP EVPN 方式）的实验拓扑

3. 实验步骤

步骤 1： S1 和 S2 的 VLAN 基本配置。

S1 的配置：

```
[S1]vlan batch 10 20
[S1]interface GigabitEthernet0/0/1
[S1-GigabitEthernet0/0/1]port link-type trunk
[S1-GigabitEthernet0/0/1]port trunk allow-pass vlan 10 20
[S1-GigabitEthernet0/0/1]q
[S1]interface GigabitEthernet0/0/2
[S1-GigabitEthernet0/0/2]port link-type access
[S1-GigabitEthernet0/0/2]port default vlan 10
[S1-GigabitEthernet0/0/2]q
[S1]interface GigabitEthernet0/0/3
[S1-GigabitEthernet0/0/3]port link-type access
[S1-GigabitEthernet0/0/3]port default vlan 20
```

S2 的配置：

```
[S2]vlan batch 10 20
[S2]interface GigabitEthernet0/0/1
[S2-GigabitEthernet0/0/1]port link-type trunk
[S2-GigabitEthernet0/0/1]port trunk allow-pass vlan 10 20
[S2-GigabitEthernet0/0/1]q
[S2]interface GigabitEthernet0/0/2
[S2-GigabitEthernet0/0/2]port link-type access
[S2-GigabitEthernet0/0/2]port default vlan 10
[S2-GigabitEthernet0/0/2]q
[S2]interface GigabitEthernet0/0/3
[S2-GigabitEthernet0/0/3]port link-type access
[S2-GigabitEthernet0/0/3]port default vlan 20
```

步骤 2： 配置 border 和 EDGE 的 underlay 网络。

（1）配置 border 和 EDGE 节点的 IPv4 地址。配置 VXLAN 实现相同网段互访（BGP EVPN 方式）IP 地址规划表见表 12-3。

表 12-3　配置 VXLAN 实现相同网段互访（BGP EVPN 方式）IP 地址规划表

设备名称	接口编号	IP 地址
edge1	G1/0/0	10.0.12.2/24
	LoopBack 0	1.1.1.1/32
edge2	G1/0/0	10.0.23.2/24
	LoopBack 0	2.2.2.2/32
border	G1/0/0	10.0.12.1/24
	G1/0/1	10.0.23.1/24
	LoopBack 0	3.3.3.3/32

ENSP 的 CE 设备接口默认为二层口，并且物理状态为 down，需要配置 IP 执行以下操作（以 edge1 的 G1/0/0 接口为例）：

```
[edge1]interface G1/0/0
[edge1-GE1/0/0]undo shutdown
[edge1-GE1/0/0]undo portswitch
[edge1-GE1/0/0]ip address 10.0.12.2 24
```

（2）配置 underlay 网络的 OSPF 协议。

border 的配置：

```
[border]ospf 1
[border-ospf-1]area 0.0.0.0
[border-ospf-1-area-0.0.0.0]network 3.3.3.3 0.0.0.0
[border-ospf-1-area-0.0.0.0]network 10.0.12.0 0.0.0.255
[border-ospf-1-area-0.0.0.0]network 10.0.23.0 0.0.0.255
```

edge1 的配置：

```
[edge1]ospf 1
[edge1-ospf-1]area 0.0.0.0
[edge1-ospf-1-area-0.0.0.0]network 1.1.1.1 0.0.0.0
[edge1-ospf-1-area-0.0.0.0]network 10.0.12.0 0.0.0.255
```

edge2 的配置：

```
[edge2]ospf 1
[edge2-ospf-1]area 0.0.0.0
[edge2-ospf-1-area-0.0.0.0]network 2.2.2.2 0.0.0.0
[edge2-ospf-1-area-0.0.0.0]network 10.0.23.0 0.0.0.255
```

查看路由表：

```
[edge1]display ip routing-table
Proto: Protocol        Pre: Preference
Route Flags: R - relay, D - download to fib, T - to vpn-instance, B - black hole route
------------------------------------------------------------------------------
Routing Table : _public_
         Destinations : 11       Routes : 11
Destination/Mask    Proto   Pre  Cost  Flags    NextHop      Interface
      1.1.1.1/32    Direct  0    0     D        127.0.0.1    LoopBack0
      2.2.2.2/32    OSPF    10   2     D        10.0.12.1    GE1/0/0
      3.3.3.3/32    OSPF    10   1     D        10.0.12.1    GE1/0/0
    10.0.12.0/24    Direct  0    0     D        10.0.12.2    GE1/0/0
    10.0.12.2/32    Direct  0    0     D        127.0.0.1    GE1/0/0
  10.0.12.255/32    Direct  0    0     D        127.0.0.1    GE1/0/0
    10.0.23.0/24    OSPF    10   2     D        10.0.12.1    GE1/0/0
     127.0.0.0/8    Direct  0    0     D        127.0.0.1    InLoopBack0
    127.0.0.1/32    Direct  0    0     D        127.0.0.1    InLoopBack0
127.255.255.255/32  Direct  0    0     D        127.0.0.1    InLoopBack0
255.255.255.255/32  Direct  0    0     D        127.0.0.1    InLoopBack0
```

通过以上输出可以看出，设备之间已经学习到了各自的环回口路由。

步骤 3：配置 EDGE 节点的 BD，将 VNI 绑定到 BD 中。

edge1 的配置：

```
[edge1]bridge-domain 10
```

```
[edge1-bd10]vxlan vni 10
[edge1-bd10]q
[edge1]bridge-domain 20
[edge1-bd20]vxlan vni 20
```

edge2 的配置：

```
[edge2]bridge-domain 10
[edge2-bd10]vxlan vni 10
[edge2-bd10]q
[edge2]bridge-domain 20
[edge2-bd20]vxlan vni 20
```

步骤 4：配置 EDGE 节点的子接口，将不同的子接口绑定到不同的 BD 中。

edge1 的配置：

```
[edge1]interface GE1/0/1
[edge1-GE1/0/1]undo shutdown
[edge1-GE1/0/1]q
[edge1]interface GE1/0/1.10 mode l2
[edge1-GE1/0/1.10]encapsulation dot1q vid 10
[edge1-GE1/0/1.10]bridge-domain 10
[edge1-GE1/0/1.10]q
[edge1]interface GE1/0/1.20 mode l2
[edge1-GE1/0/1.20]encapsulation dot1q vid 20
[edge1-GE1/0/1.20]bridge-domain 20
```

edge2 的配置：

```
[edge2]interface GE1/0/1
[edge2-GE1/0/1]undo shutdown
[edge2-GE1/0/1]q
[edge2]interface GE1/0/1.10 mode l2
[edge2-GE1/0/1.10]encapsulation dot1q vid 10
[edge2-GE1/0/1.10]bridge-domain 10
[edge2-GE1/0/1.10]q
[edge2]interface GE1/0/1.20 mode l2
[edge2-GE1/0/1.20]encapsulation dot1q vid 20
[edge2-GE1/0/1.20]bridge-domain 20
[edge2-GE1/0/1.20]q
```

步骤 5：开启 EVPN 功能，并且配置 BGP 的 MP-BGP 邻居关系。

edge1 的配置：

```
[edge1]evpn-overlay enable                         //开启 EVPN 功能
[edge1]bgp 100
[edge1-bgp]router-id 1.1.1.1
[edge1-bgp]peer 2.2.2.2 as-number 100
[edge1-bgp]peer 2.2.2.2 connect-interface LoopBack0
[edge1-bgp]l2vpn-family evpn                       //进入 EVPN 地址族
[edge1-bgp-af-evpn]peer 2.2.2.2 enable             //使能 EVPN 邻居关系
Warning: This operation will reset the peer session. Continue? [Y/N]:y
```

edge2 的配置：

```
[edge2]evpn-overlay enable
[edge2]bgp 100
[edge2-bgp]router-id 2.2.2.2
[edge2-bgp]peer 1.1.1.1 as-number 100
[edge2-bgp]peer 1.1.1.1 connect-interface LoopBack0
[edge2-bgp]l2vpn-family evpn
[edge2-bgp-af-evpn]peer 1.1.1.1 enable
Warning: This operation will reset the peer session. Continue? [Y/N]:y
```

查看 BGP EVPN 的邻居关系：

```
[edge1]display  bgp evpn peer
 BGP local router ID          : 1.1.1.1
 Local AS number              : 100
 Total number of peers        : 1
 Peers in established state : 1
  Peer          V   AS    MsgRcvd   MsgSent   OutQ   Up/Down    State         PrefRcv
  2.2.2.2       4   100   5         5         0      00:01:17   Established   0
```

步骤 6：在 NVE 中指定使用 BGP 协议完成 VXLAN 的隧道建立。

edge1 的配置：

```
[edge1]interface Nve1
[edge1-Nve1]source 1.1.1.1
[edge1-Nve1]vni 10 head-end peer-list protocol bgp     //通过 BGP EVPN 自动建立 VXLAN
                                                       //隧道
[edge1-Nve1]vni 20 head-end peer-list protocol bgp
```

edge2 的配置：

```
[edge2]interface Nve1
[edge2-Nve1]source 2.2.2.2
[edge2-Nve1]vni 10 head-end peer-list protocol bgp
[edge2-Nve1]vni 20 head-end peer-list protocol bgp
```

步骤 7：配置 EVPN 实例并且绑定到 BD。

edge1 的配置：

```
[edge1]bridge-domain 10
[edge1-bd10]evpn
[edge1-bd10-evpn]route-distinguisher 10:10              //配置 RD 值
[edge1-bd10-evpn]vpn-target 10:10 export-extcommunity //配置 RT 值控制 EVPN 路由的
                                                       //收发
[edge1-bd10-evpn]vpn-target 10:10 import-extcommunity
[edge1-bd10-evpn]q
[edge1-bd10]q
[edge1]bridge-domain 20
[edge1-bd20]evpn
[edge1-bd20-evpn]route-distinguisher 20:20
[edge1-bd20-evpn]vpn-target 20:20 export-extcommunity
[edge1-bd20-evpn]vpn-target 20:20 import-extcommunity
```

edge2 的配置：

```
[edge2]bridge-domain 10
[edge2-bd10]evpn
[edge2-bd10-evpn]route-distinguisher 10:10
[edge2-bd10-evpn]vpn-target 10:10 export-extcommunity
[edge2-bd10-evpn]vpn-target 10:10 import-extcommunity
[edge2-bd10-evpn]q
[edge2-bd10]q
[edge2]bridge-domain 20
[edge2-bd20]evpn
[edge2-bd20-evpn]route-distinguisher 20:20
[edge2-bd20-evpn]vpn-target 20:20 export-extcommunity
[edge2-bd20-evpn]vpn-target 20:20 import-extcommunity
```

查看 VXLAN 隧道的建立情况：

```
[edge1]display vxlan tunnel
Number of vxlan tunnel : 1
Tunnel ID   Source          Destination      State    Type       Uptime
--------------------------------------------------------------------------------
4026531841  1.1.1.1         2.2.2.2          up       dynamic    00:00:58
```

通过以上输出可以看出，对于 edge1 和 edge2 建立的 VXLAN 隧道，Type 为 dynamic，表示此隧道为动态建立的隧道。

下面查看 BGP 的 EVPN 路由，如图 12-17 所示。其中，Inclusive Multicast Route 代表 Type3 的 EVPN 路由，该类型路由的主要作用是在 VXLAN 控制平面中用于 VTEP 的自动发现和 VXLAN 隧道的动态建立。

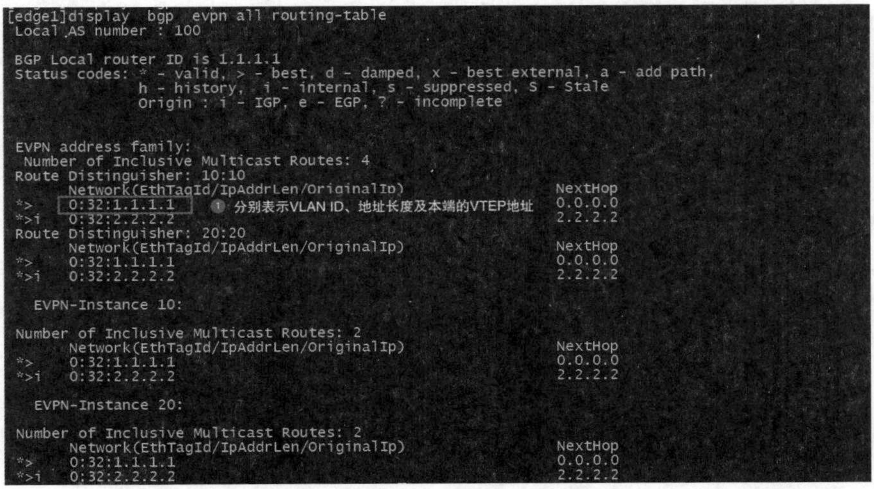

```
[edge1]display  bgp  evpn all routing-table
 Local AS number : 100

 BGP Local router ID is 1.1.1.1
 Status codes: * - valid, > - best, d - damped, x - best external, a - add path,
               h - history,  i - internal, s - suppressed, S - Stale
               Origin : i - IGP, e - EGP, ? - incomplete


 EVPN address family:
  Number of Inclusive Multicast Routes: 4
 Route Distinguisher: 10:10
       Network(EthTagId/IpAddrLen/OriginalIp)                NextHop
 *>    0:32:1.1.1.1     ① 分别表示VLAN ID、地址长度及本端的VTEP地址   0.0.0.0
 *>i   0:32:2.2.2.2                                          2.2.2.2
 Route Distinguisher: 20:20
       Network(EthTagId/IpAddrLen/OriginalIp)                NextHop
 *>    0:32:1.1.1.1                                          0.0.0.0
 *>i   0:32:2.2.2.2                                          2.2.2.2

   EVPN-Instance 10:

 Number of Inclusive Multicast Routes: 2
       Network(EthTagId/IpAddrLen/OriginalIp)                NextHop
 *>    0:32:1.1.1.1                                          0.0.0.0
 *>i   0:32:2.2.2.2                                          2.2.2.2

   EVPN-Instance 20:

 Number of Inclusive Multicast Routes: 2
       Network(EthTagId/IpAddrLen/OriginalIp)                NextHop
 *>    0:32:1.1.1.1                                          0.0.0.0
 *>i   0:32:2.2.2.2                                          2.2.2.2
```

图 12-17　edge1 的 EVPN 路由表

作为 BGP EVPN 对等体的 VTEP，通过 Inclusive Multicast 路由互相传递二层 VNI 和 VTEP IP 地址信息。

其中，Originating Router's IP Address 字段为本端 VTEP IP 地址，MPLS Label 字段为二层 VNI。

BGP 的 UPDATE 报文的抓包结果如图 12-18 所示。

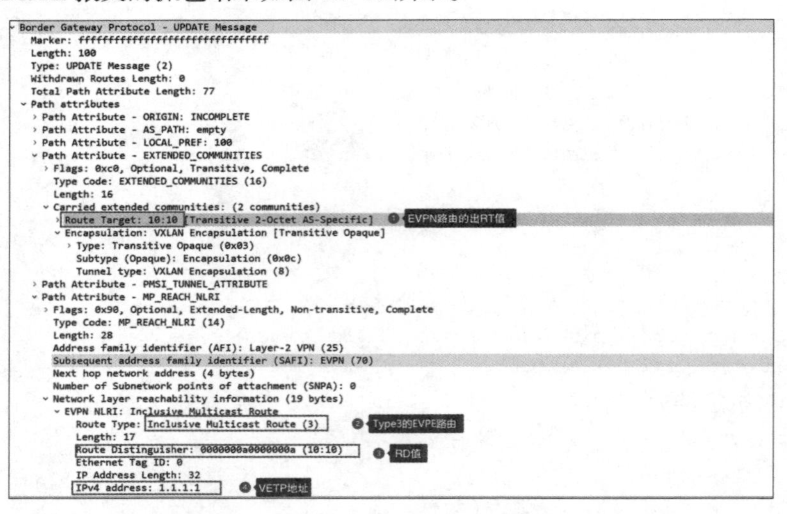

图 12-18　BGP 的 UPDATE 报文的抓包结果

由于本实验中的 ENSP 模拟器有 bug，因此无法测试结果。

12.2.4　配置 VXLAN 创建集中式网关（BGP EVPN 方式）

1. 实验目的

配置 border 为集中式网关，不同网段的互访通过 border 作为网关。

2. 实验拓扑

配置 VXLAN 创建集中式网关（BGP EVPN 方式）的实验拓扑如图 12-19 所示。

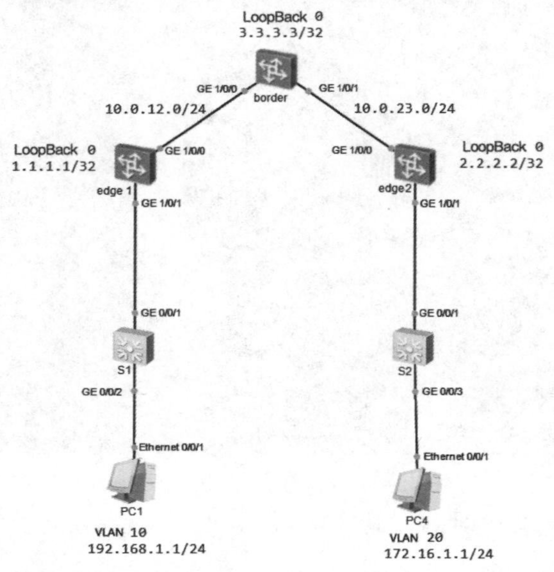

图 12-19　配置 VXLAN 创建集中式网关（BGP EVPN 方式）的实验拓扑

3. 实验步骤

步骤 1：S1 和 S2 的 VLAN 基本配置。

S1 的配置：

```
[S1]vlan batch 10 20
[S1]interface GigabitEthernet0/0/1
[S1-GigabitEthernet0/0/1]port link-type trunk
[S1-GigabitEthernet0/0/1]port trunk allow-pass vlan 10 20
[S1-GigabitEthernet0/0/1]q
[S1]interface GigabitEthernet0/0/2
[S1-GigabitEthernet0/0/2]port link-type access
[S1-GigabitEthernet0/0/2]port default vlan 10
[S1-GigabitEthernet0/0/2]q
```

S2 的配置：

```
[S2]vlan batch 10 20
[S2]interface GigabitEthernet0/0/1
[S2-GigabitEthernet0/0/1]port link-type trunk
[S2-GigabitEthernet0/0/1]port trunk allow-pass vlan 10 20
[S2-GigabitEthernet0/0/1]q
[S2]interface GigabitEthernet0/0/3
[S2-GigabitEthernet0/0/3]port link-type access
[S2-GigabitEthernet0/0/3]port default vlan 20
```

步骤 2：配置 border 和 EDGE 节点的 underlay 网络。

（1）配置 border 和 EDGE 节点的 IPv4 地址。配置 VXLAN 创建集中式网关（BGP EVPN 方式）IP 地址规划表见表 12-4。

表 12-4　配置 VXLAN 创建集中式网关（BGP EVPN 方式）IP 地址规划表

设备名称	接口编号	IP 地址
edge1	G1/0/0	10.0.12.2/24
	LoopBack 0	1.1.1.1/32
edge2	G1/0/0	10.0.23.2/24
	LoopBack 0	2.2.2.2/32
border	G1/0/0	10.0.12.1/24
	G1/0/1	10.0.23.1/24
	LoopBack 0	3.3.3.3/32

ENSP 的 CE 设备接口默认为二层口，并且物理状态为 down，配置 IP 需要执行以下操作（以 edge1 的 G1/0/0 接口为例）：

```
[edge1]interface G1/0/0
[edge1-GE1/0/0]undo shutdown
[edge1-GE1/0/0]undo portswitch
[edge1-GE1/0/0]ip address 10.0.12.2 24
```

（2）配置 underlay 网络的 OSPF 协议。

border 的配置：

```
[border]ospf 1
[border-ospf-1]area 0.0.0.0
[border-ospf-1-area-0.0.0.0]network 3.3.3.3 0.0.0.0
[border-ospf-1-area-0.0.0.0]network 10.0.12.0 0.0.0.255
[border-ospf-1-area-0.0.0.0]network 10.0.23.0 0.0.0.255
```

edge1 的配置：

```
[edge1]ospf 1
[edge1-ospf-1]area 0.0.0.0
[edge1-ospf-1-area-0.0.0.0]network 1.1.1.1 0.0.0.0
[edge1-ospf-1-area-0.0.0.0]network 10.0.12.0 0.0.0.255
```

edge2 的配置：

```
[edge2]ospf 1
[edge2-ospf-1]area 0.0.0.0
[edge2-ospf-1-area-0.0.0.0]network 2.2.2.2 0.0.0.0
[edge2-ospf-1-area-0.0.0.0]network 10.0.23.0 0.0.0.255
```

查看路由表：

```
[edge1]display ip routing-table
Proto: Protocol        Pre: Preference
Route Flags: R - relay, D - download to fib, T - to vpn-instance, B - black hole
route
------------------------------------------------------------------------------
Routing Table : _public_
         Destinations : 11        Routes : 11
Destination/Mask     Proto    Pre   Cost   Flags    NextHop        Interface
      1.1.1.1/32     Direct   0     0      D        127.0.0.1      LoopBack0
      2.2.2.2/32     OSPF     10    2      D        10.0.12.1      GE1/0/0
      3.3.3.3/32     OSPF     10    1      D        10.0.12.1      GE1/0/0
    10.0.12.0/24     Direct   0     0      D        10.0.12.2      GE1/0/0
    10.0.12.2/32     Direct   0     0      D        127.0.0.1      GE1/0/0
  10.0.12.255/32     Direct   0     0      D        127.0.0.1      GE1/0/0
    10.0.23.0/24     OSPF     10    2      D        10.0.12.1      GE1/0/0
     127.0.0.0/8     Direct   0     0      D        127.0.0.1      InLoopBack0
    127.0.0.1/32     Direct   0     0      D        127.0.0.1      InLoopBack0
127.255.255.255/32   Direct   0     0      D        127.0.0.1      InLoopBack0
255.255.255.255/32   Direct   0     0      D        127.0.0.1      InLoopBack0
```

通过以上输出可以看出，设备之间已经学习到了各自的环回口路由。

步骤 3: 配置 EDGE 节点及 border 节点的 BD，将 VNI 绑定到 BD 中。

border 的配置：

```
[border]bridge-domain 10
[border-bd10]vxlan vni 10
[border-bd10]q
```

```
[border]bridge-domain 20
[border-bd20]vxlan vni 20
```

edge1 的配置：

```
[edge1]bridge-domain 10
[edge1-bd10]vxlan vni 10
```

edge2 的配置：

```
[edge2]bridge-domain 20
[edge2-bd20]vxlan vni 20
```

步骤 4： 配置 EDGE 节点的子接口，将不同的子接口绑定到不同的 BD 中。

edge1 的配置：

```
[edge1]interface GE1/0/1
[edge1-GE1/0/1]undo shutdown
[edge1-GE1/0/1]q
[edge1]interface GE1/0/1.10 mode l2
[edge1-GE1/0/1.10]encapsulation dot1q vid 10
[edge1-GE1/0/1.10]bridge-domain 10
[edge1-GE1/0/1.10]q
```

edge2 的配置：

```
[edge2]interface GE1/0/1
[edge2-GE1/0/1]undo shutdown
[edge2-GE1/0/1]q
[edge2]interface GE1/0/1.20 mode l2
[edge2-GE1/0/1.20]encapsulation dot1q vid 20
[edge2-GE1/0/1.20]bridge-domain 20
[edge2-GE1/0/1.20]q
```

步骤 5： 配置 BD 的 VPN 实例，配置 RD 和 RT 值。

border 的配置：

```
[border]evpn-overlay enable
[border]bridge-domain 10
[border-bd10]evpn
[border-bd10-evpn]route-distinguisher 10:10
[border-bd10-evpn]vpn-target 10:10 export-extcommunity
[border-bd10-evpn]vpn-target 10:10 import-extcommunity
[border-bd10-evpn]q
[border-bd10]q
[border]bridge-domain 20
[border-bd20]evpn
[border-bd20-evpn]route-distinguisher 20:20
[border-bd20-evpn]vpn-target 20:20 export-extcommunity
[border-bd20-evpn]vpn-target 20:20 import-extcommunity
```

edge1 的配置：

```
[edge1]evpn-overlay enable
[edge1]bridge-domain 10
```

```
[edge1-bd10]evpn
[edge1-bd10-evpn]route-distinguisher 10:10
[edge1-bd10-evpn]vpn-target 10:10 export-extcommunity
[edge1-bd10-evpn]vpn-target 10:10 import-extcommunity
```

edge2 的配置：

```
[edge2]evpn-overlay enable
[edge2]bridge-domain 20
[edge2-bd20]evpn
[edge2-bd20-evpn]route-distinguisher 20:20
[edge2-bd20-evpn]vpn-target 20:20 export-extcommunity
[edge2-bd20-evpn]vpn-target 20:20 import-extcommunity
```

步骤 6：建立 BGP 的 EVPN 邻居关系。

border 的配置：

```
[border]bgp 100
[border-bgp]peer 1.1.1.1 as-number 100
[border-bgp]peer 1.1.1.1 connect-interface LoopBack0
[border-bgp]peer 2.2.2.2 as-number 100
[border-bgp]peer 2.2.2.2 connect-interface LoopBack0
[border-bgp]l2vpn-family evpn
[border-bgp-af-evpn]peer 1.1.1.1 enable
Warning: This operation will reset the peer session. Continue? [Y/N]:y
[border-bgp-af-evpn]peer 1.1.1.1 reflect-client
[border-bgp-af-evpn]peer 2.2.2.2 enable
Warning: This operation will reset the peer session. Continue? [Y/N]:y
[border-bgp-af-evpn]peer 2.2.2.2 reflect-client
```

edge1 的配置：

```
[edge1]bgp 100
[edge1-bgp]peer 3.3.3.3 as-number 100
[edge1-bgp]peer 3.3.3.3 connect-interface LoopBack0
[edge1-bgp]l2vpn-family evpn
[edge1-bgp-af-evpn]peer 3.3.3.3 enable
Warning: This operation will reset the peer session. Continue? [Y/N]:y
```

edge2 的配置：

```
[edge2]bgp 100
[edge2-bgp]peer 3.3.3.3 as-number 100
[edge2-bgp]peer 3.3.3.3 connect-interface LoopBack0
[edge2-bgp]l2vpn-family evpn
[edge2-bgp-af-evpn]peer 3.3.3.3 enable
Warning: This operation will reset the peer session. Continue? [Y/N]:y
```

步骤 7：配置 NVE 接口，自动建立 VXLAN 隧道。

border 的配置：

```
[border]interface Nve1
[border-Nve1]source 3.3.3.3
```

```
[border-Nve1]vni 10 head-end peer-list protocol bgp
[border-Nve1]vni 20 head-end peer-list protocol bgp
```

edge1 的配置：

```
[edge1]interface Nve1
[edge1-Nve1]source 1.1.1.1
[edge1-Nve1]vni 10 head-end peer-list protocol bgp
```

edge2 的配置：

```
[edge2]interface Nve1
[edge2-Nve1]source 2.2.2.2
[edge2-Nve1]vni 20 head-end peer-list protocol bgp
```

查看 VXLAN 隧道的建立情况：

```
[border]display vxlan tunnel
Number of vxlan tunnel : 2
Tunnel ID   Source         Destination      State     Type       Uptime
-----------------------------------------------------------------------------
4026531841  3.3.3.3        1.1.1.1          up        dynamic    00:00:33
4026531842  3.3.3.3        2.2.2.2          up        dynamic    00:00:18
```

步骤 8：配置 border 的网关 IP 地址。

```
[border]interface Vbdif10
[border-Vbdif10]ip address 192.168.1.254 255.255.255.0
[border-Vbdif10]q
[border]interface Vbdif20
[border-Vbdif20]ip address  172.16.1.254 255.255.255.0
```

测试连通性，PC1 ping PC4 的测试结果如图 12-20 所示。

图 12-20　PC1 ping PC4 的测试结果

12.2.5 配置 VXLAN 创建分布式网关（BGP EVPN 方式）

1. 实验目的

使用 EVPN 在 edge1 和 edge2 之间建立 VXLAN 隧道，EDGE 设备作为终端的网关。实现 PC1 和 PC2 之间能够互相通信。border 作为 edge1 的 EVPN 的反射器，负责 EVPN 路由的传递和数据的转发；AR1 作为去往外部网络的设备，配置 VXLAN 网络和外部网络的互通；edge1 作为去往外部网络的网关。

2. 实验拓扑

配置 VXLAN 创建分布式网关（BGP EVPN 方式）的实验拓扑如图 12-21 所示。

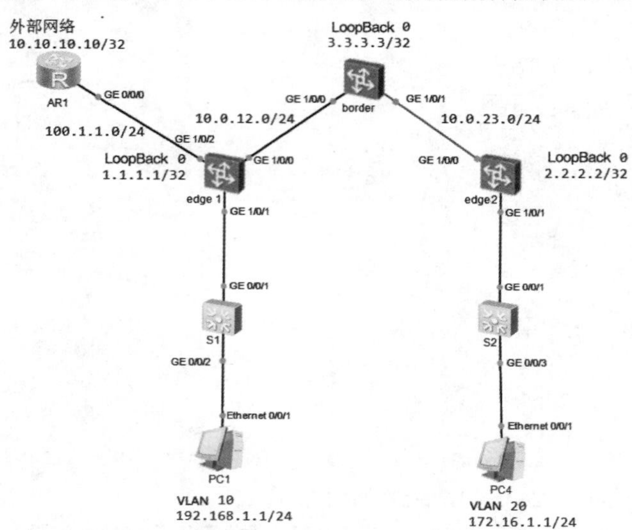

图 12-21 配置 VXLAN 创建分布式网关（BGP EVPN 方式）的实验拓扑

3. 实验步骤

步骤 1： S1 和 S2 的 VLAN 基本配置。

S1 的配置：

```
[S1]vlan batch 10 20
[S1]interface GigabitEthernet0/0/1
[S1-GigabitEthernet0/0/1]port link-type trunk
[S1-GigabitEthernet0/0/1]port trunk allow-pass vlan 10 20
[S1-GigabitEthernet0/0/1]q
[S1]interface GigabitEthernet0/0/2
[S1-GigabitEthernet0/0/2]port link-type access
[S1-GigabitEthernet0/0/2]port default vlan 10
[S1-GigabitEthernet0/0/2]q
```

S2 的配置：

```
[S2]vlan batch 10 20
[S2]interface GigabitEthernet0/0/1
[S2-GigabitEthernet0/0/1]port link-type trunk
```

```
[S2-GigabitEthernet0/0/1]port trunk allow-pass vlan 10 20
[S2-GigabitEthernet0/0/1]q
[S2]interface GigabitEthernet0/0/3
[S2-GigabitEthernet0/0/3]port link-type access
[S2-GigabitEthernet0/0/3]port default vlan 20
```

步骤 2： 配置 border 和 EDGE 节点的 underlay 网络。

（1）配置 border 和 EDGE 节点的 IPv4 地址。配置 VXLAN 创建分布式网关（BGP EVPN 方式）IP 地址规划表见表 12-5。

表 12-5　配置 VXLAN 创建分布式网关（BGP EVPN 方式）IP 地址规划表

设备名称	接口编号	IP 地址
edge1	G1/0/0	10.0.12.2/24
	LoopBack 0	1.1.1.1/32
edge2	G1/0/0	10.0.23.2/24
	LoopBack 0	2.2.2.2/32
border	G1/0/0	10.0.12.1/24
	G1/0/1	10.0.23.1/24
	LoopBack 0	3.3.3.3/32

ENSP 的 CE 设备接口默认为二层口，并且物理状态为 down，配置 IP 需要执行以下操作（以 edge1 的 G1/0/0 接口为例）：

```
[edge1]interface G1/0/0
[edge1-GE1/0/0]undo shutdown
[edge1-GE1/0/0]undo portswitch
[edge1-GE1/0/0]ip address 10.0.12.2 24
```

（2）配置 underlay 网络的 OSPF 协议。

border 的配置：

```
[border]ospf 1
[border-ospf-1]area 0.0.0.0
[border-ospf-1-area-0.0.0.0]network 3.3.3.3 0.0.0.0
[border-ospf-1-area-0.0.0.0]network 10.0.12.0 0.0.0.255
[border-ospf-1-area-0.0.0.0]network 10.0.23.0 0.0.0.255
```

edge1 的配置：

```
[edge1]ospf 1
[edge1-ospf-1]area 0.0.0.0
[edge1-ospf-1-area-0.0.0.0]network 1.1.1.1 0.0.0.0
[edge1-ospf-1-area-0.0.0.0]network 10.0.12.0 0.0.0.255
```

edge2 的配置：

```
[edge2]ospf 1
[edge2-ospf-1]area 0.0.0.0
[edge2-ospf-1-area-0.0.0.0]network 2.2.2.2 0.0.0.0
[edge2-ospf-1-area-0.0.0.0]network 10.0.23.0 0.0.0.255
```

查看路由表：

```
[edge1]display ip routing-table
Proto: Protocol        Pre: Preference
Route Flags: R - relay, D - download to fib, T - to vpn-instance, B - black hole route
------------------------------------------------------------------------------
Routing Table : _public_
         Destinations : 11       Routes : 11
  Destination/Mask     Proto   Pre  Cost    Flags   NextHop         Interface
        1.1.1.1/32     Direct  0    0       D       127.0.0.1       LoopBack0
        2.2.2.2/32     OSPF    10   2       D       10.0.12.1       GE1/0/0
        3.3.3.3/32     OSPF    10   1       D       10.0.12.1       GE1/0/0
      10.0.12.0/24     Direct  0    0       D       10.0.12.2       GE1/0/0
      10.0.12.2/32     Direct  0    0       D       127.0.0.1       GE1/0/0
    10.0.12.255/32     Direct  0    0       D       127.0.0.1       GE1/0/0
      10.0.23.0/24     OSPF    10   2       D       10.0.12.1       GE1/0/0
       127.0.0.0/8     Direct  0    0       D       127.0.0.1       InLoopBack0
      127.0.0.1/32     Direct  0    0       D       127.0.0.1       InLoopBack0
127.255.255.255/32     Direct  0    0       D       127.0.0.1       InLoopBack0
255.255.255.255/32     Direct  0    0       D       127.0.0.1       InLoopBack0
```

通过以上输出可以看出，设备之间已经学习到了各自的环回口路由。

步骤 3: 配置 EDGE 节点的 BD，将 VNI 绑定到 BD 中。

edge1 的配置：

```
[edge1]bridge-domain 10
[edge1-bd10]vxlan vni 10
```

edge2 的配置：

```
[edge2]bridge-domain 20
[edge2-bd20]vxlan vni 20
```

步骤 4: 配置 EDGE 节点的子接口，将不同的子接口绑定到不同的 BD 中。

edge1 的配置：

```
[edge1]interface GE1/0/1
[edge1-GE1/0/1]undo shutdown
[edge1-GE1/0/1]interface GE1/0/1.10 mode l2
[edge1-GE1/0/1.10]encapsulation dot1q vid 10
[edge1-GE1/0/1.10]bridge-domain 10
```

edge2 的配置：

```
[edge2]interface GE1/0/1
[edge2-GE1/0/1]undo shutdown
[edge2-GE1/0/1]interface GE1/0/1.20 mode l2
[edge2-GE1/0/1.20]encapsulation dot1q vid 20
[edge2-GE1/0/1.20]bridge-domain 20
```

步骤 5: 配置 EVPN 的邻居关系。

border 的配置：

```
[border]evpn-overlay enable
```

```
[border]bgp 100
[border-bgp]peer 1.1.1.1 as-number 100
[border-bgp]peer 1.1.1.1 connect-interface LoopBack0
[border-bgp]peer 2.2.2.2 as-number 100
[border-bgp]peer 2.2.2.2 connect-interface LoopBack0
[border-bgp]l2vpn-family evpn
[border-bgp-af-evpn]undo policy vpn-target
[border-bgp-af-evpn]peer 1.1.1.1 enable
Warning: This operation will reset the peer session. Continue? [Y/N]:y
[border-bgp-af-evpn]peer 1.1.1.1 advertise irb
[border-bgp-af-evpn]peer 1.1.1.1 reflect-client
[border-bgp-af-evpn]peer 2.2.2.2 enable
Warning: This operation will reset the peer session. Continue? [Y/N]:y
[border-bgp-af-evpn]peer 2.2.2.2 advertise irb
[border-bgp-af-evpn]peer 2.2.2.2 reflect-client
```

edge1 的配置：

```
[edge1]evpn-overlay enable
[edge1]bgp 100
[edge1-bgp]peer 3.3.3.3 as-number 100
[edge1-bgp]peer 3.3.3.3 connect-interface LoopBack0
[edge1-bgp]l2vpn-family evpn
[edge1-bgp-af-evpn]policy vpn-target
[edge1-bgp-af-evpn]peer 3.3.3.3 enable
Warning: This operation will reset the peer session. Continue? [Y/N]:y
[edge1-bgp-af-evpn]peer 3.3.3.3 advertise irb
```

edge2 的配置：

```
[edge2]evpn-overlay enable
[edge2]bgp 100
[edge2-bgp]peer 3.3.3.3 as-number 100
[edge2-bgp]peer 3.3.3.3 connect-interface LoopBack0
[edge2-bgp]l2vpn-family evpn
[edge2-bgp-af-evpn]policy vpn-target
[edge2-bgp-af-evpn]peer 3.3.3.3 enable
Warning: This operation will reset the peer session. Continue? [Y/N]:y
[edge2-bgp-af-evpn]peer 3.3.3.3 advertise irb
```

查看 BGP 的 EVPN 邻居关系：

```
[border]display bgp evpn peer
 BGP local router ID        : 10.0.12.1
 Local AS number            : 100
 Total number of peers      : 2
 Peers in established state : 2
  Peer         V    AS    MsgRcvd    MsgSent    OutQ    Up/Down     State        PrefRcv
  1.1.1.1      4    100   5          5          0       00:01:29    Established  0
  2.2.2.2      4    100   4          5          0       00:00:30    Established  0
```

步骤 6： 配置 BD 的 EVPN 实例，以及 RD 和 RT 值。

edge1 的配置：

```
[edge1]bridge-domain 10
[edge1-bd10]vxlan vni 10
[edge1-bd10]evpn
[edge1-bd10-evpn]route-distinguisher 10:10
[edge1-bd10-evpn]vpn-target 10:10 export-extcommunity //RT 10:10 用于两端 MAC
                                                     //Route(Type2 路由的发布和接收)
[edge1-bd10-evpn]vpn-target 11:1 export-extcommunity  //RT11: 1 与三层 VPN 实例的入
                                   // 方向 RT 对应，主要用于生成 VPN 实例的全局主机路由
[edge1-bd10-evpn]vpn-target 10:10 import-extcommunity
```

edge2 的配置：

```
[edge2-bd20]vxlan vni 20
[edge2-bd20]evpn
[edge2-bd20-evpn]route-distinguisher 10:11
[edge2-bd20-evpn]vpn-target 10:10 export-extcommunity
[edge2-bd20-evpn]vpn-target 11:1 export-extcommunity
[edge2-bd20-evpn]vpn-target 10:10 import-extcommunity
```

步骤 7： 配置 IP VPN 实例。

edge1 的配置：

```
[edge1]ip vpn-instance 1
[edge1-vpn-instance-1]ipv4-family
[edge1-vpn-instance-1-af-ipv4]route-distinguisher 10:10
[edge1-vpn-instance-1-af-ipv4]vpn-target 11:1 evpn /* 与 BD 的 EVPN 实例出方向 RT 一致，
当设备收到 Type2 的路由时，如果 RT 与 IP  VPN 实例的入方向 RT 一致，则收集其中的 ARP 信息，生成主
机路由，并且放在 VPN 实例路由表中 */
[edge1-vpn-instance-1-af-ipv4]q
[edge1-vpn-instance-1]vxlan vni 1000  // 配置三层 VNI，用于指定设备收到流量后将数据发
                                      // 往哪一个 VPN 实例，然后查表转发，主要用于业务隔离
```

edge2 的配置：

```
[edge2]ip vpn-instance 1
[edge2-vpn-instance-1]ipv4-family
[edge2-vpn-instance-1-af-ipv4]route-distinguisher 10:10
[edge2-vpn-instance-1-af-ipv4]vpn-target 11:1 evpn
[edge2-vpn-instance-1-af-ipv4]q
[edge2-vpn-instance-1]vxlan vni 1000  // 配置三层 VNI，用于指定设备收到流量后将数据发往
                                      // 哪一个 VPN 实例，然后查表转发
```

步骤 8： 创建 VBDIF 接口，作为终端设备的网关。

edge1 的配置：

```
[edge1]interface Vbdif10
[edge1-Vbdif10]ip binding vpn-instance 1
[edge1-Vbdif10]ip address 192.168.1.254 255.255.255.0
[edge1-Vbdif10]vxlan anycast-gateway enable
[edge1-Vbdif10]arp collect host enable
```

edge2 的配置：

```
[edge2]interface Vbdif20
[edge2-Vbdif20]ip binding vpn-instance 1
[edge2-Vbdif20]ip address 172.16.1.254 255.255.255.0
[edge2-Vbdif20]vxlan anycast-gateway enable
[edge2-Vbdif20]arp collect host enable
```

步骤 9： 配置 NVE 接口，使用 EVPN 自动创建 VXLAN 隧道。

edge1 的配置：

```
[edge1]interface Nve1
[edge1-Nve1]source 1.1.1.1
[edge1-Nve1]vni 10 head-end peer-list protocol bgp
```

edge2 的配置：

```
[edge2]interface Nve1
[edge2-Nve1]source 2.2.2.2
[edge2-Nve1]vni 20 head-end peer-list protocol bgp
```

查看 VXLAN 隧道的建立情况：

```
[edge1]display vxlan tunnel
Number of vxlan tunnel : 1
Tunnel ID      Source       Destination        State       Type          Uptime
---------------------------------------------------------------------------------------
4026531841     1.1.1.1      2.2.2.2            up          dynamic       00:00:36
```

步骤 10： 使用 PC1 访问网关，产生 Type2 的主机路由信息，如图 12-22 所示。

图 12-22　使用 PC1 访问网关

查看 edge1 的 BGP EVPN 路由表，如图 12-23 所示。

```
[edge1]display  bgp  evpn route-distinguisher 10:10  routing-table m
[edge1]display  bgp  evpn route-distinguisher 10:10  routing-table mac-route
 Route Distinguisher: 10:10

 Number of Mac Routes: 2
 BGP Local router ID is 10.0.12.2
 Status codes: * - valid, > - best, d - damped, x - best external, a - add path,
               h - history,  i - internal, s - suppressed, S - Stale
               Origin : i - IGP, e - EGP, ? - incomplete

       Network(EthTagId/MacAddrLen/MacAddr/IpAddrLen/IpAddr)  NextHop
 *>    0:48:5489-98a1-5cc0:32:192.168.1.1   ① PC1的ARP信息     0.0.0.0
 *>    0:48:707b-e837-654f:0:0.0.0.0                           0.0.0.0
```

图 12-23　edge1 的 BGP EVPN 路由表

此时 PC2 也访问网关，再次查看 BGP EVPN 路由表，如图 12-24 所示。

```
[edge1]display bgp  evpn  all  r
[edge1]display bgp  evpn  all  routing-table
 Local AS number : 100

 BGP Local router ID is 10.0.12.2
 Status codes: * - valid, > - best, d - damped, x - best external, a - add path,
               h - history,  i - internal, s - suppressed, S - Stale
               Origin : i - IGP, e - EGP, ? - incomplete


 EVPN address family:
  Number of Mac Routes: 4
 Route Distinguisher: 10:10
       Network(EthTagId/MacAddrLen/MacAddr/IpAddrLen/IpAddr)  NextHop
 *>    0:48:5489-98a1-5cc0:32:192.168.1.1                     0.0.0.0
 *>    0:48:707b-e837-654f:0:0.0.0.0                          0.0.0.0
 Route Distinguisher: 10:11
       Network(EthTagId/MacAddrLen/MacAddr/IpAddrLen/IpAddr)  NextHop
 *>i   0:48:5489-9846-4f51:32:172.16.1.1                      2.2.2.2
 *>i   0:48:707b-e863-5183:0:0.0.0.0                          2.2.2.2

   EVPN-Instance 10:              ① EDGE设备可以学习到两端的ARP路由信息

 Number of Mac Routes: 4
       Network(EthTagId/MacAddrLen/MacAddr/IpAddrLen/IpAddr)  NextHop
 *>i   0:48:5489-9846-4f51:32:172.16.1.1                      2.2.2.2
 *>    0:48:5489-98a1-5cc0:32:192.168.1.1                     0.0.0.0
 *>    0:48:707b-e837-654f:0:0.0.0.0                          0.0.0.0
 *>i   0:48:707b-e863-5183:0:0.0.0.0                          2.2.2.2

   EVPN-Instance __RD_1_10_10__:

 Number of Mac Routes: 1
       Network(EthTagId/MacAddrLen/MacAddr/IpAddrLen/IpAddr)  NextHop
 *>i   0:48:5489-9846-4f51:32:172.16.1.1                      2.2.2.2
```

② PC1的ARP信息，会携带RT为11:1，与IP VPN实例中的入方向RT一致，则会提取其中的主机IP地址，生成主机路由信息，放入全局的VPN实例路由表中

图 12-24　PC2 访问网关后 edge1 的 BGP EVPN 路由表

查看 PC2 的 ARP 路由详细信息，如图 12-25 所示。

```
[edge1]display  bgp  evpn  all  routing-table  mac-route 0:48:5489-9846-4f51:32:172.16.1.1


 BGP local router ID : 10.0.12.2
 Local AS number : 100
 Total routes of Route Distinguisher(10:11): 1
 BGP routing table entry information of 0:48:5489-9846-4f51:32:172.16.1.1:
 Label information (Received/Applied): 20 1000/NULL    ① 二层VNI为20，三层VNI为1000
 From: 3.3.3.3 (10.0.12.1)
 Route Duration: 0d00h28m12s
 Relay IP Nexthop: 10.0.12.1
 Relay Tunnel Out-Interface: VXLAN    ② 迭代进入VXLAN隧道
 Original nexthop: 2.2.2.2
 Qos information : 0x0
 Ext-Community: RT <10 : 10>, RT <11 : 1>, Tunnel Type <VxLan>, Router's MAC <707b-e863-5183>
 AS-path Nil, origin incomplete, localpref 100, pref-val 0, valid, internal, best, select, pre 25
 Originator: 10.0.23.2
 Cluster list: 10.0.12.1
 Route Type: 2 (MAC Advertisement Route)
 Ethernet Tag ID: 0, MAC Address/Len: 5489-9846-4f51/48, IP Address/Len: 172.16.1.1/32, ESI:0000.0000.0000.0000.0000
 Not advertised to any peer yet
```

③ RT为11:1，与IP VPN实例的一致，将生成主机路由信息并且加入全局的VPN实例路由表中

图 12-25　PC2 的 ARP 路由详细信息

查看 edge1 的 BGP VPN 实例 1 的路由表，可以发现生成了 172.16.1.1/32 的路由信息，如图 12-26 所示。

```
[edge1]display  bgp  vpnv4 vpn-instance 1 routing-table

 BGP Local router ID is 10.0.12.2
 Status codes: * - valid, > - best, d - damped, x - best external, a - add path,
               h - history,  i - internal, s - suppressed, S - Stale
               Origin : i - IGP, e - EGP, ? - incomplete
 RPKI validation codes: V - valid, I - invalid, N - not-found


 VPN-Instance 1, Router ID 10.0.12.2:
 Total Number of Routes: 1
       Network            NextHop                        MED        LocPrf    PrefVal Path/Ogn
 *>i   172.16.1.1/32      2.2.2.2                                    100        0       ?
```

图 12-26　edge1 的 BGP EVPN 实例 1 路由表

查看全局的 VPN 实例路由表，当网关收到去往 172.16.1.1/32 的目的 IP 地址时，将迭代到 VXLAN 隧道，如图 12-27 所示。

```
[edge1]display  ip routing-table vpn-instance 1
Proto: Protocol        Pre: Preference
Route Flags: R - relay, D - download to fib, T - to vpn-instance, B - black hole route
------------------------------------------------------------------------------
Routing Table : 1
         Destinations : 5        Routes : 5

Destination/Mask    Proto   Pre  Cost        Flags NextHop         Interface

     172.16.1.1/32  IBGP    255  0             RD  2.2.2.2         VXLAN    PC2的主机路由
    192.168.1.0/24  Direct  0    0             D   192.168.1.254   Vbdif10
  192.168.1.254/32  Direct  0    0             D   127.0.0.1       Vbdif10
  192.168.1.255/32  Direct  0    0             D   127.0.0.1       Vbdif10
255.255.255.255/32  Direct  0    0             D   127.0.0.1       InLoopBack0
```

图 12-27　edge1 的 VPN 实例全局路由表

PC1 访问 PC2 的测试结果如图 12-28 所示。测试之前先使用 PC 访问网关，确保网关上有两个 PC 的 Type2 ARP 信息。

图 12-28　PC1 访问 PC2 的测试结果

查看抓包结果，如图 12-29 所示。

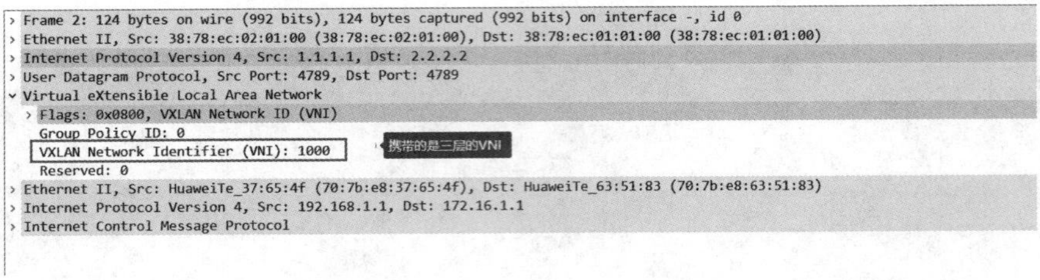

```
> Frame 2: 124 bytes on wire (992 bits), 124 bytes captured (992 bits) on interface -, id 0
> Ethernet II, Src: 38:78:ec:02:01:00 (38:78:ec:02:01:00), Dst: 38:78:ec:01:01:00 (38:78:ec:01:01:00)
> Internet Protocol Version 4, Src: 1.1.1.1, Dst: 2.2.2.2
> User Datagram Protocol, Src Port: 4789, Dst Port: 4789
v Virtual eXtensible Local Area Network
  > Flags: 0x0800, VXLAN Network ID (VNI)
    Group Policy ID: 0
    VXLAN Network Identifier (VNI): 1000        携带的是三层的VNI
    Reserved: 0
> Ethernet II, Src: HuaweiTe_37:65:4f (70:7b:e8:37:65:4f), Dst: HuaweiTe_63:51:83 (70:7b:e8:63:51:83)
> Internet Protocol Version 4, Src: 192.168.1.1, Dst: 172.16.1.1
> Internet Control Message Protocol
```

图 12-29　edge1 的 G1/0/0 接口的抓包结果

步骤 11： 配置 edge1 和 AR1 的互联接口及 IP 地址。

```
[edge1]interface GE1/0/2
[edge1-GE1/0/2]undo portswitch
[edge1-GE1/0/2]undo shutdown
[edge1-GE1/0/2]ip binding vpn-instance 1
Info: All IPv4 and IPv6 related configurations on this interface are removed.
[edge1-GE1/0/2]ip address 100.1.1.1 255.255.255.0
```

步骤 12： 配置去往外部网络的路由及回包路由。

edge1 的配置：

```
[edge1]ip route-static vpn-instance 1 10.10.10.10 32 100.1.1.2
```

AR1 的配置：

```
[AR1]ip route-static 172.16.1.0 24 100.1.1.1
[AR1]ip route-static 192.168.1.0 24 100.1.1.1 //配置回包路由
```

步骤 13： 将静态路由转换为 EVPN 的 Type5 路由并传递给 edge2 设备。

```
[edge1]bgp 100
[edge1-bgp]ipv4-family vpn-instance 1
[edge1-bgp-1]import-route static  //引入 VPN 实例 1 中的静态路由
[edge1-bgp-1]advertise l2vpn evpn //转换为 EVPN 的 Type5 路由
```

查看 BGP EVPN 的 Type5 的路由表，可以看到生成了 10.10.10.10/32 的路由信息。

```
[edge1]display bgp evpn all routing-table prefix-route
 Local AS number : 100
 BGP Local router ID is 10.0.12.2
 Status codes: * - valid, > - best, d - damped, x - best external, a - add path,
               h - history,  i - internal, s - suppressed, S - Stale
               Origin : i - IGP, e - EGP, ? - incomplete
 EVPN address family:
  Number of Ip Prefix Routes: 1
 Route Distinguisher: 10:10
       Network(EthTagId/IpPrefix/IpPrefixLen)                          NextHop
 *>    0:10.10.10.10:32                                                0.0.0.0
   EVPN-Instance __RD_1_10_10__:
 Number of Ip Prefix Routes: 1
       Network(EthTagId/IpPrefix/IpPrefixLen)                          NextHop
```

```
   *>    0:10.10.10.10:32                                       0.0.0.0
```

查看 edge2 的 VPN 实例 1 的路由表（可以看到通过 IBGP 学习到了 10.10.10.10/32 的路由信息，下一跳接口为 VXLAN）：

```
[edge2]display ip routing-table vpn-instance 1
Proto: Protocol        Pre: Preference
Route Flags: R - relay, D - download to fib, T - to vpn-instance, B - black hole
route
------------------------------------------------------------------------------
Routing Table : 1
         Destinations : 6        Routes : 6
Destination/Mask      Proto    Pre    Cost   Flags    NextHop         Interface
  10.10.10.10/32      IBGP     255    0      RD       1.1.1.1         VXLAN
   172.16.1.0/24      Direct   0      0      D        172.16.1.254    Vbdif20
 172.16.1.254/32      Direct   0      0      D        127.0.0.1       Vbdif20
 172.16.1.255/32      Direct   0      0      D        127.0.0.1       Vbdif20
  192.168.1.1/32      IBGP     255    0      RD       1.1.1.1         VXLAN
255.255.255.255/32    Direct   0      0      D        127.0.0.1       InLoopBack0
```

步骤 14：测试。

PC4 访问 10.10.10.10 的测试结果如图 12-30 所示。在 edge2 的 G1/0/0 接口的抓包结果如图 12-31 所示。

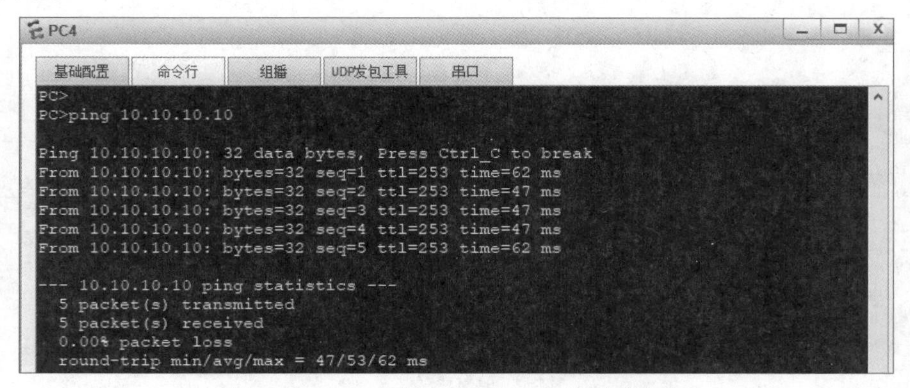

图 12-30　PC4 访问 10.10.10.10 的测试结果

```
> Frame 8: 124 bytes on wire (992 bits), 124 bytes captured (992 bits) on interface -, id 0
> Ethernet II, Src: 38:0f:86:03:01:00 (38:0f:86:03:01:00), Dst: 38:0f:86:02:01:01 (38:0f:86:02:01:01)
> Internet Protocol Version 4, Src: 2.2.2.2, Dst: 1.1.1.1    ③ 源目地址为本端和对端的VTEP地址
> User Datagram Protocol, Src Port: 4789, Dst Port: 4789
v Virtual eXtensible Local Area Network
  > Flags: 0x0800, VXLAN Network ID (VNI)
    Group Policy ID: 0
    VXLAN Network Identifier (VNI): 1000    ② 携带三层VNI，迭代到了VXLAN隧道
    Reserved: 0
> Ethernet II, Src: HuaweiTe_63:51:83 (70:7b:e8:63:51:83), Dst: HuaweiTe_37:65:4f (70:7b:e8:37:65:4f)
> Internet Protocol Version 4, Src: 172.16.1.2, Dst: 10.10.10.10    ① 原始数据包，说明流量已经发出去了
> Internet Control Message Protocol
```

图 12-31　在 edge2 的 G1/0/0 接口的抓包结果

PC1 无法 ping 通目标设备，因为模拟器存在问题。

第 13 章

准入控制技术

13.1　准入控制技术概述

NAC（Network Access Control，网络接入控制）通过对接入网络的客户端和用户的认证保证网络的安全，是一种“端到端”的安全技术。

NAC 包括三种认证方式：802.1x 认证、MAC 认证和 Portal 认证。由于这三种认证方式的认证原理不同，因此其各自适合的场景也有所差异。在实际应用中，用户可以根据场景部署某一种合适的认证方式，也可以部署几种认证方式组成的混合认证，混合认证的组合方式以设备实际支持为准。三种认证方式的对比表见表 13-1。

表 13-1　三种认证方式的对比表

对比项	802.1x 认证	MAC 认证	Portal 认证
适合场景	新建网络、用户集中、信息安全要求严格的场景	打印机、传真机等哑终端接入认证的场景	用户分散、用户流动性大的场景
客户端需求	需要	不需要	不需要
优点	安全性高	无须安装客户端	部署灵活
缺点	部署不灵活	须登记 MAC 地址，管理复杂	安全性不高

13.1.1　802.1x 认证

1. 802.1x 认证简介

802.1x 认证是一种基于端口的网络接入控制协议，即在接入设备的端口这一级验证用户身份并控制其访问权限。

扫一扫，看视频

802.1x 认证使用 EAPoL（Extensible Authentication Protocol over LAN，局域网可扩展认证协议）实现客户端、设备端和认证服务器之间认证信息的交换。

2. 802.1x 认证系统

如图 13-1 所示，802.1x 认证系统是典型的客户端 / 服务器结构，包括三个实体：客户端、接入设备和认证服务器。

图 13-1　802.1x 认证系统

客户端一般为一个用户终端设备，用户可以通过启动客户端软件发起 802.1x 认证。客户端必须支持局域网上的可扩展认证协议 EAPoL。

接入设备通常为支持 802.1x 协议的网络设备，它为客户端提供接入局域网的端口，该端口可以是物理端口，也可以是逻辑端口。

认证服务器用于实现对用户进行认证、授权和计费，通常为 RADIUS 服务器。

3. 802.1x 认证流程

根据接入设备对 802.1x 客户端发送的 EAPoL 报文处理机制的不同，可将认证方式分为 EAP 中继方式和 EAP 终结方式。

802.1x EAP 中继方式的认证流程如图 13-2 所示。

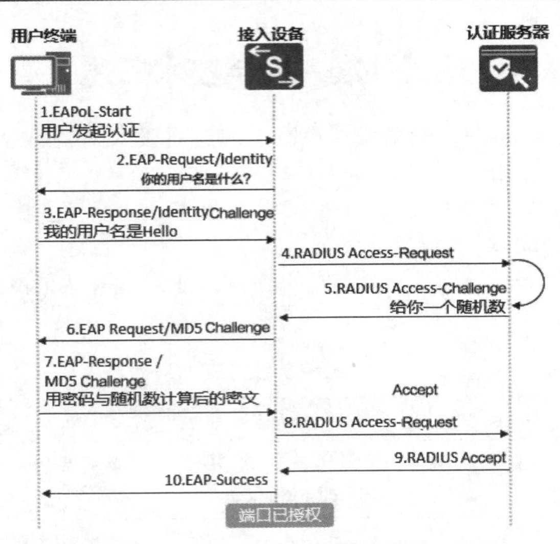

图 13-2　802.1x EAP 中继方式的认证流程

802.1x EAP 中继方式的认证流程如下：

（1）用户在访问外部网络时需要打开 802.1x 客户端程序，输入已经申请、登记过的用户名和密码，发起连接请求。此时，客户端程序将向设备端发出认证请求报文（EAPoL-Start），开始启动一次认证过程。

（2）设备端收到认证请求报文后，将发出一个 Identity 类型的请求报文（EAP-Request/Identity），要求用户的客户端程序发送输入的用户名。

（3）客户端程序响应设备端发出的请求，将用户名信息通过 Identity 类型的响应报文（EAP-Response/Identity）发送给设备端。

（4）设备端将客户端发送的响应报文中的 EAP 报文封装在 RADIUS 报文（RADIUS Access-Request）中发送给认证服务器进行处理。

（5）RADIUS 服务器收到设备端转发的用户名信息后，将该信息与数据库中的用户名列表对比，找到该用户名对应的密码信息，用随机生成的一个 MD5 Challenge 对密码进行加密处理，同时将此 MD5 Challenge 通过 RADIUS 报文发送给设备端。

（6）设备端将 RADIUS 服务器发送的 MD5 Challenge 转发给客户端。

（7）客户端收到由设备端传来的 MD5 Challenge 后，用该 Challenge 对密码部分进行加密处理，生成 EAP-Response/MD5 Challenge 报文，并发送给设备端。

（8）设备端将此 EAP-Response/MD5 Challenge 报文封装在 RADIUS 报文中发送给 RADIUS 服务器。

13.1.2　MAC 认证

扫一扫，看视频

1. MAC 认证简介

MAC 认证是一种基于端口和 MAC 地址对用户的网络访问权限进行控制的认证方法。以用户的 MAC 地址作为身份凭据到认证服务器进行认证。

默认时，交换机收到 DHCP/ARP/DHCPv6/ND 报文后均能触发对用户进行 MAC 认证的操作。支持通过配置使交换机收到任意的数据帧后触发 MAC 认证。

2. MAC 认证流程

MAC 认证流程如图 13-3 所示［以 PAP（Password Authentication Protocol，密码认证协议）方式为例］。

（1）接入设备首次检测到终端的 MAC 地址，进行 MAC 地址学习并触发 MAC 认证。

（2）接入设备生成一个随机值（MD5 挑战字），并对 MAC 认证用户的 MAC 地址、密码、随机值依次排列后经过 MD5 算法进行加密，然后将用户名、加密后的密码以及随机值封装在 RADIUS 认证请求报文中发送给 RADIUS 服务器，请求 RADIUS 服务器对该终端进行 MAC 认证。

（3）RADIUS 服务器使用收到的随机值对本地数据库中对应 MAC 认证用户的 MAC 地址、密码、随机值依次排列后进行加密（MD5 算法）处理，如果与设备发来的密码相同，则向设备发送认证接收报文，表示终端 MAC 认证成功，允许该终端访问网络。

（4）CHAP（Challenge Handshake Authentication Protocol，挑战握手认证协议）方式的 MAC 认证与 PAP 方式的 MAC 认证相比，不同之处在于是对 MAC 认证用户的 CHAP ID、MAC 地址、随机值依次排列后进行 MD5 算法加密。

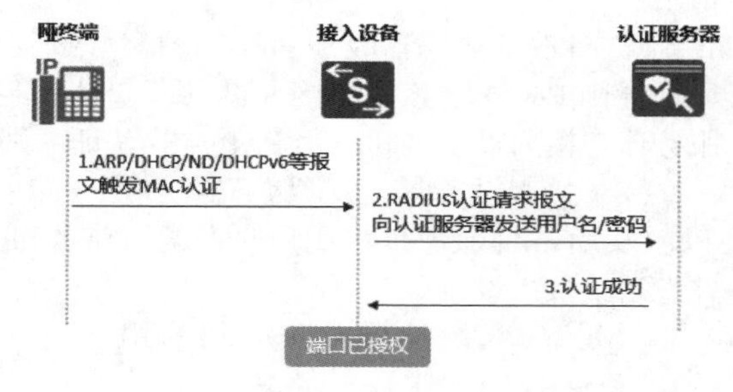

图 13-3　MAC 认证流程

13.1.3　Portal 认证

1. Portal 认证简介

Portal 认证也称为 Web 认证。用户可以通过在 Portal 认证页面中输入用户账号和密码信息，实现对终端用户身份的认证。

用户可以通过以下两种方式实现 Portal 认证页面访问。

（1）主动认证：用户通过浏览器主动访问 Portal 认证网站。

（2）重定向认证：用户输入的访问地址不是 Portal 认证网站地址，被接入设备强制访问 Portal 认证网站（通常称为重定向）。

2. Portal 认证流程

基于 Portal 协议的 Portal 认证流程如图 13-4 所示。

（1）在认证之前，客户端与接入设备之间会建立预连接，即客户端用户在认证成功之前已在接入设备上建立用户在线表项，并且只有部分网络访问权限。

（2）客户端发起 HTTP 连接请求。

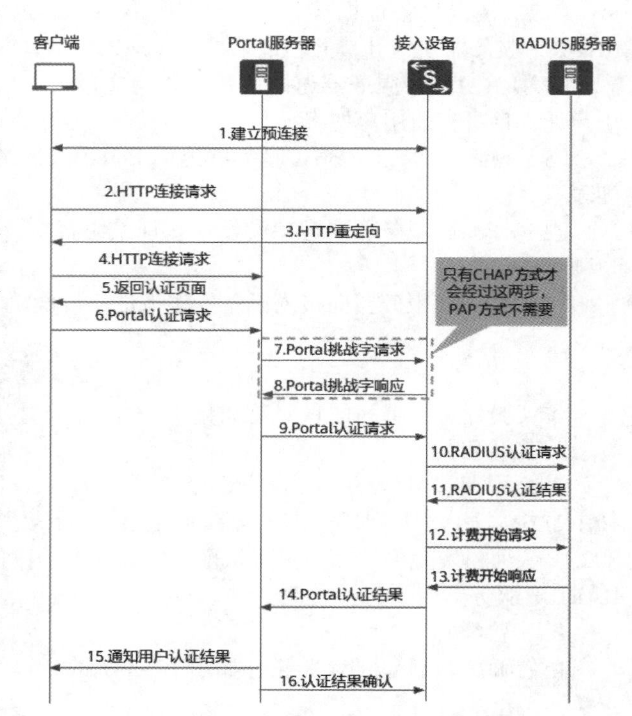

图 13-4　基于 Portal 协议的 Portal 认证流程

（3）接入设备收到 HTTP 连接请求报文时，如果是访问 Portal 服务器或免认证网络资源的 HTTP 报文，则接入设备允许其通过；如果是访问其他地址的 HTTP 报文，则接入设备将其 URL 地址重定向到 Portal 认证页面。

（4）客户端根据获得的 URL 地址向 Portal 服务器发起 HTTP 连接请求。

（5）Portal 服务器向客户端返回 Portal 认证页面。

（6）用户在 Portal 认证页面中输入用户名和密码后，客户端向 Portal 服务器发起 Portal 认证请求。

（7）Portal 服务器收到 Portal 认证请求后，如果 Portal 服务器与接入设备之间采用 CHAP 方式认证，则 Portal 服务器向接入设备发起 Portal 挑战字请求报文（REQ_CHALLENGE）；如果 Portal 服务器与接入设备之间采用 PAP 方式认证，则接入设备直接进行第（9）步。（可选。）

（8）接入设备向 Portal 服务器响应 Portal 挑战字应答报文（ACK_CHALLENGE）。（可选。）

（9）Portal 服务器将用户输入的用户名和密码封装在 Portal 认证请求报文（REQ_AUTH）中，并发送给接入设备。

（10）接入设备根据获取到的用户名和密码，向 RADIUS 服务器发送 RADIUS 认证请求（ACCESS-REQUEST）。

（11）RADIUS 服务器对用户名和密码进行认证。如果认证成功，则 RADIUS 服务器向接入设备发送认证接收报文（ACCESS-ACCEPT）；如果认证失败，则 RADIUS 服务器返回认证拒绝报文（ACCESS-REJECT）。由于 RADIUS 协议合并了认证和授权的过程，因此认证接收报文中也包含了用户的授权信息。

（12）接入设备根据接收到的认证结果接入 / 拒绝用户。如果允许用户接入，则接入设备向 RADIUS 服务器发送计费开始请求报文（ACCOUNTING-REQUEST）。

（13）RADIUS 服务器返回计费开始响应报文（ACCOUNTING-RESPONSE），并开始计费，将用户加入自身在线用户列表。

（14）接入设备向 Portal 服务器返回 Portal 认证结果（ACK_AUTH），并将用户加入在线用户列表。

（15）Portal 服务器向客户端发送认证结果报文，通知客户端认证成功，并将用户加入在线用户列表。

（16）Portal 服务器向接入设备发送认证应答确认（AFF_ACK_AUTH）。

13.2 准入控制实验

本实验通过配置 802.1x 认证讲解准入控制实验操作过程。

1. 实验目的

某公司拥有两个部门：市场部和人事部。其中，市场部和人事部的 IP 地址分别为 10.1.11.0/24、10.1.21.0/24 两个 IP 网段。市场部属于 VLAN 11，人事部属于 VLAN 21。现在需要在 SW2 上配置 802.1x 认证，实现终端用户只有认证成功后才能访问网络中的资源。（本实验使用真实设备，其中 RADIUS 服务器需要自行搭建。）

2. 实验拓扑

配置 802.1x 认证的实验拓扑如图 13-5 所示。

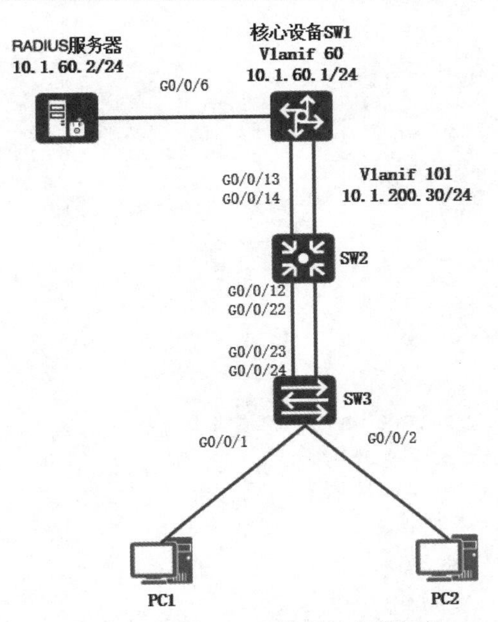

图 13-5　配置 802.1x 认证的实验拓扑

3. 实验步骤

步骤 1：配置交换机的链路聚合。

SW1 的配置：

```
[SW1]interface Eth-Trunk 1
[SW1-Eth-Trunk1]mode lacp
[SW1-Eth-Trunk1]trunkport g0/0/13
[SW1-Eth-Trunk1]trunkport g0/0/14
```

SW2 的配置：

```
[SW2]interface Eth-Trunk 1
[SW2-Eth-Trunk1]mode lacp
[SW2-Eth-Trunk1]trunkport g0/0/13
[SW2-Eth-Trunk1]trunkport g0/0/14
[SW2]interface Eth-Trunk 2
[SW2-Eth-Trunk2]mode lacp
[SW2-Eth-Trunk2]trunkport g0/0/12
[SW2-Eth-Trunk2]trunkport g0/0/22
```

SW3 的配置：

```
[SW3]interface Eth-Trunk 1
[SW3-Eth-Trunk1]mode lacp
[SW3-Eth-Trunk1]trunkport g0/0/23
[SW3-Eth-Trunk1]trunkport g0/0/24
```

步骤 2：创建 VLAN，配置 SW1 和 SW2 的互联 IP 地址以及终端的网关 IP 地址。

SW1 的配置：

```
[SW1]vlan batch 60 101
[SW1]interface Vlanif 60
[SW1-Vlanif60]ip address 10.1.60.254 24
[SW1]interface Vlanif 101
[SW1-Vlanif101]ip address 10.1.200.29 30
```

SW2 的配置：

```
[SW2]vlan batch 11 21 101
[SW2]interface Vlanif 11
[SW2-Vlanif10]ip address 10.1.11.254 24
[SW2-Vlanif10]q
[SW2]interface Vlanif 21
[SW2-Vlanif20]ip address 10.2.21.254 24
[SW2]interface Vlanif 101
[SW2-Vlanif101]ip address 10.1.200.30 30
```

步骤 3：配置交换机的链路类型。

SW1 的配置：

```
[SW1]interface g0/0/6
[SW1-GigabitEthernet0/0/6]port link-type access
[SW1-GigabitEthernet0/0/6]port default vlan 60
[SW1-GigabitEthernet0/0/6]q
[SW1]interface Eth-Trunk 1
[SW1-Eth-Trunk1]port link-type trunk
[SW1-Eth-Trunk1]port trunk allow-pass vlan 101
```

SW2 的配置：

```
[SW2]interface Eth-Trunk 1
[SW2-Eth-Trunk1]port link-type trunk
[SW2-Eth-Trunk1]port trunk allow-pass vlan 101
[SW2-Eth-Trunk1]q
[SW2]interface Eth-Trunk 2
[SW2-Eth-Trunk2]port link-type hybrid
[SW2-Eth-Trunk2]port hybrid tagged vlan 11 21
```

步骤 4：配置路由并开启 SW2 的 DHCP 服务。

SW1 的配置：

```
[SW1]ospf
[SW1-ospf-1]area 0
[SW1-ospf-1-area-0.0.0.0]network 10.1.60.254 0.0.0.0
[SW1-ospf-1-area-0.0.0.0]network 10.1.200.29 0.0.0.0
```

SW2 的配置：

```
[SW2]ospf
[SW2-ospf-1]area 0
[SW2-ospf-1-area-0.0.0.0]network 10.1.11.254 0.0.0.0
[SW2-ospf-1-area-0.0.0.0]network 10.1.21.254 0.0.0.0
```

```
[SW2-ospf-1-area-0.0.0.0]network 10.1.200.30 0.0.0.0
[SW2]dhcp enable
Info: The operation may take a few seconds. Please wait for a moment. done.
[SW2]interface Vlanif 11
[SW2-Vlanif10]dhcp select interface
[SW2-Vlanif10]q
[SW2]interface Vlanif 21
[SW2-Vlanif20]dhcp select interface
[SW2-Vlanif20]q
```

步骤 5: 配置 RADIUS 模板。

```
[SW2]radius-server template radius
//指定 RADIUS 认证的 IP 地址和端口号
[SW2-radius-radius]radius-server authentication 10.1.60.2 1812
//指定 RADIUS 计费的 IP 地址和端口号
[SW2-radius-radius]radius-server accounting 10.1.60.2 1813
//配置与 RADIUS 之间的密码
[SW2-radius-radius]radius-server shared-key cipher Huawei@123
[SW2-radius-radius]q
//指定 RADIUS 授权服务器的 IP 地址
[SW2]radius-server authorization 10.1.60.2 shared-key cipher Huawei@123
```

步骤 6: 创建认证、计费方法为 RADIUS 认证。

```
[SW2]aaa
[SW2-aaa]authentication-scheme radius                //创建认证模板
[SW2-aaa-authen-radius]authentication-mode radius    //认证模式为 RADIUS
[SW2-aaa-authen-radius]q
[SW2-aaa]accounting-scheme radius                    //创建计费模板
[SW2-aaa-accounting-radius]accounting-mode radius    //计费模式为 RADIUS
[SW2-aaa-accounting-radius]q
[SW2-aaa]domain huawei                               //创建认证域
[SW2-aaa-domain-huawei]authentication-scheme radius  //调用认证模板
[SW2-aaa-domain-huawei]accounting-scheme radius      //调用计费模板
[SW2-aaa-domain-huawei]radius-server radius
[SW2-aaa-domain-huawei]q
```

步骤 7: 配置认证模板。

```
[SW2]dot1x-access-profile name dot1x                       //创建 802.1x 认证模板
[SW2-dot1x-access-profile-dot1x]q
[SW2]mac-access-profile name mac                           //创建 MAC 认证模板
[SW2-mac-access-profile-mac]q
[SW2]authentication-profile name huawei                    //创建认证模板
[SW2-authen-profile-huawei]dot1x-access-profile dot1x      //调用 802.1x 认证模板
[SW2-authen-profile-huawei]mac-access-profile mac          //调用 MAC 认证模板
[SW2-authen-profile-huawei]access-domain huawei force      //配置强制使用认证域 huawei
[SW2-authen-profile-huawei]authentication dot1x-mac-bypass//配置旁路认证
[SW2-authen-profile-huawei]q
```

步骤 8： 配置地址池。

```
[SW2]vlan pool market                                   //创建市场部 VLAN 池
[SW2-vlan-pool-market]vlan 11
[SW2-vlan-pool-market]q
[SW2]vlan pool hr                                       //创建人事部 VLAN 池
[SW2-vlan-pool-hr]vlan 21
[SW2-vlan-pool-hr]q

[SW2]interface Eth-Trunk 2
[SW2-Eth-Trunk2]authentication-profile huawei    //接口调用认证模板
```

步骤 9： 配置 EAP 报文透传功能。

```
[SW3]l2protocol-tunnel user-defined-protocol dot1x protocol-mac 0180-c200-0003
group-mac 0100-0000-0002
[SW3]interface Eth-Trunk 1
[SW3-Eth-Trunk1]l2protocol-tunnel user-defined-protocol dot1x enable
[SW3]interface g0/0/1
[SW3-GigabitEthernet0/0/1]l2protocol-tunnel user-defined-protocol dot1x enable
```

4. 认证

（1）在 PC1 上进行 802.1x 认证，如图 13-6 所示。

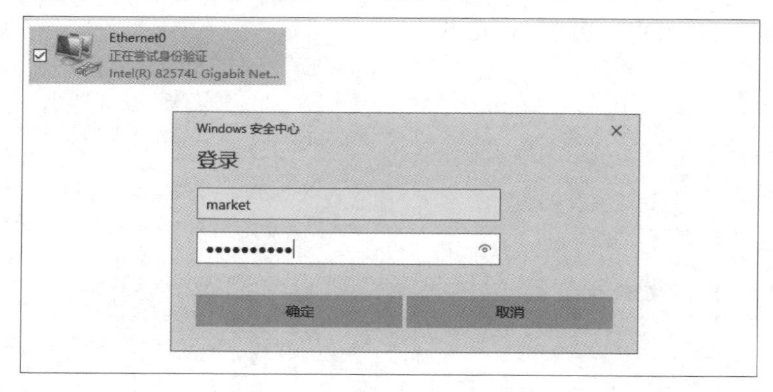

图 13-6　通过 PC1 进行认证登录

认证成功后，PC1 可以获取 VLAN 11 的 IP 地址，如图 13-7 所示。

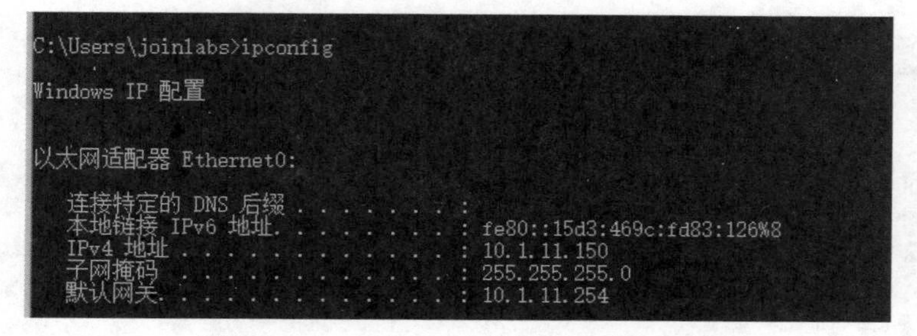

图 13-7　PC1 认证成功后获取的地址结果

（2）在 PC2 上进行 802.1x 认证，如图 13-8 所示。

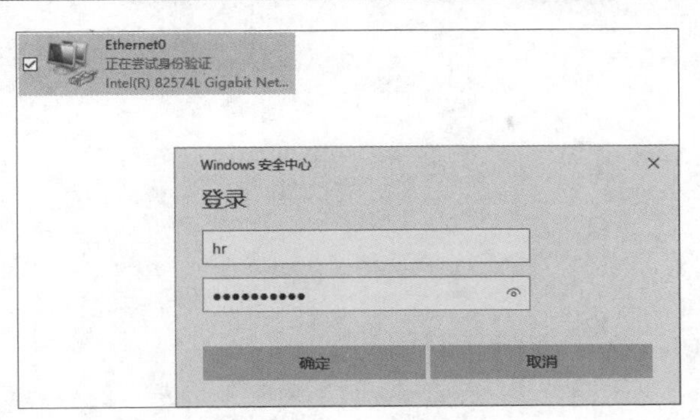

图 13-8　通过 PC2 进行认证登录

认证成功后，PC2 可以获取 VLAN 21 的 IP 地址，如图 13-9 所示。

图 13-9　PC2 认证成功后获取的地址结果

第 14 章

广域网 VPN 技术

14.1 VPN 技术概述

越来越多的企业在发展中需要在分支与总部之间进行内网通信。传统的分支与总部之间的内网通信需要租用专线（如 MPLS、传输专线等）。但是专线价格昂贵，对于中小型企业或跨国公司来说，成本较高。

由于 Internet 的发展，Internet 网络有了足够的带宽和覆盖范围，通过 Internet 建立分支与总部之间的内网通信的可行性越来越高，隧道就是在这种背景下被提出的。

通过各种隧道，分支与总部之间可以基于 Internet 建立企业网络。

14.1.1 GRE VPN

GRE（Generic Routing Encapsulation，通用路由封装）协议可以对某些网络层协议（如 IPX、ATM、IPv6、AppleTalk 等）的数据报文进行封装，使这些被封装的数据报文能够在另一个网络层协议（如 IPv4）中传输。

GRE 提供了将一种协议的报文封装在另一种协议报文中的机制，是一种三层隧道封装技术，使报文可以通过 GRE 隧道透明地传输，以解决异种网络的传输问题。

报文在 GRE 隧道中传输包括封装和解封装两个过程。如图 14-1 所示，如果 X 协议报文从 Ingress PE 向 Egress PE 传输，则封装在 Ingress PE 上完成，而解封装在 Egress PE 上进行。封装后的数据报文在网络中传输的路径称为 GRE 隧道。

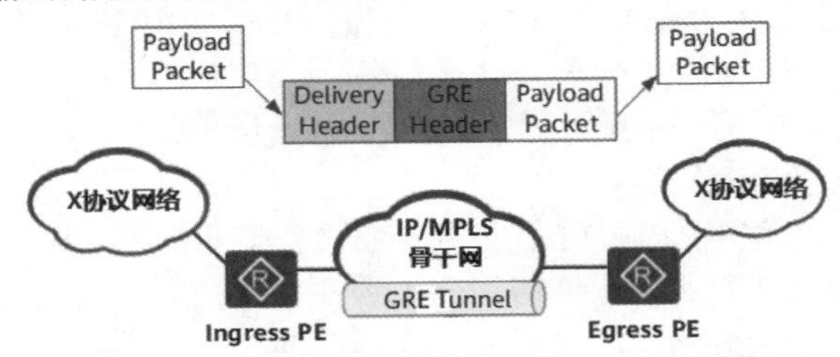

图 14-1　通过 GRE 隧道实现 X 协议报文互通组网图

（1）封装。

① Ingress PE 从连接 X 协议网络的接口接收到 X 协议报文后，首先将报文交由 X 协议处理。

② X 协议根据报文头中的目的地址在路由表或转发表中查找出接口，确定如何转发此报文。如果发现出接口是 GRE Tunnel 接口，则对报文进行 GRE 封装，即添加 GRE 头。

③根据骨干网传输协议为 IP，给报文加上 IP 头。IP 头的源地址就是隧道源地址，目的地址就是隧道目的地址。

④根据该 IP 头的目的地址（隧道目的地址），在骨干网路由表中查找相应的出接口并发送报文，然后封装后的报文将在该骨干网中传输。

（2）解封装，解封装过程和封装过程相反。

① Egress PE 从 GRE Tunnel 接口收到该报文，如果分析 IP 头发现报文的目的地址为本设备，则 Egress PE 去掉 IP 头后交给 GRE 协议处理。

② GRE 协议剥掉 GRE 报头，获取 X 协议报文，再交由 X 协议对此数据报文进行后续的转发处理。

14.1.2 IPSec VPN

IPSec（IP Security）协议族是IETF制定的一系列安全协议，它为端到端IP报文交互提供了基于密码学、可互操作、高质量的安全保护机制。

通过对数据加密、认证，IPSec使数据能够在Internet网络上安全传输。

IPSec VPN技术可以和多种VPN技术结合使用，使企业互联更加灵活安全。

1. IPSec协议体系

IPSec不是一个单独的协议，如图14-2所示。它给出了IP网络上数据安全的一整套体系结构，包括ESP（Encapsulating Security Payload，封装安全载荷）、AH（Authentication Header，认证头）、IKE（Internet Key Exchange，Internet密钥交换）等协议。

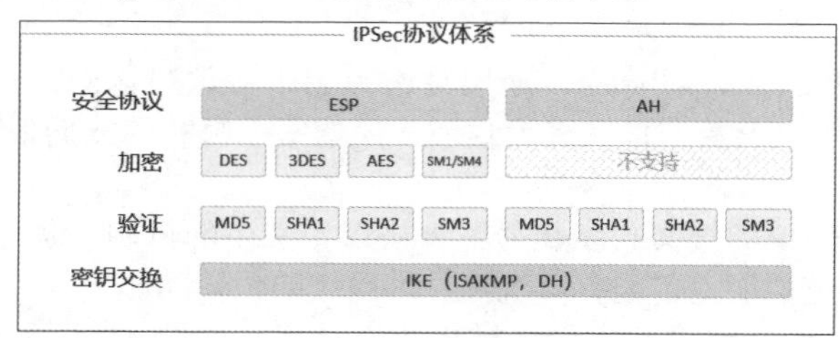

图14-2　IPSec协议体系

IPSec使用AH和ESP两种安全协议来传输和封装数据，提供认证或加密等安全服务。

AH和ESP协议提供的安全功能依赖于协议采用的验证、加密算法。AH仅支持认证功能，不支持加密功能。ESP支持认证和加密功能。

安全协议提供认证或加密等安全服务时需要存在密钥。

密钥交换的方式有以下两种。

（1）带外共享密钥：在发送、接收设备上手工配置静态的加密、验证密钥。双方通过带外共享的方式（如通过电话或邮件方式）保证密钥的一致性。这种方式的缺点是可扩展性差，在点到多点组网中配置密钥的工作量成倍增加。另外，为提升网络安全性，需要周期性修改密钥，这种方式下也很难实施。

（2）通过IKE协议自动协商密钥：IKE建立在Internet安全联盟和密钥管理协议ISAKMP定义的框架上，采用DH（Diffie-Hellman）算法在不安全的网络上安全地分发密钥。这种方式配置简单，可扩展性好，特别是在大型动态的网络环境下，此优点更加突出。同时，通信双方通过交换密钥材料来计算共享的密钥，即使第三方截获了双方用于计算密钥的所有交换数据，也无法计算出真正的密钥。

2. IPSec基本原理

如图14-3所示，IPSec隧道建立过程中需要协商IPSec SA（Security Association，安全联盟），IPSec SA一般通过IKE协商生成。

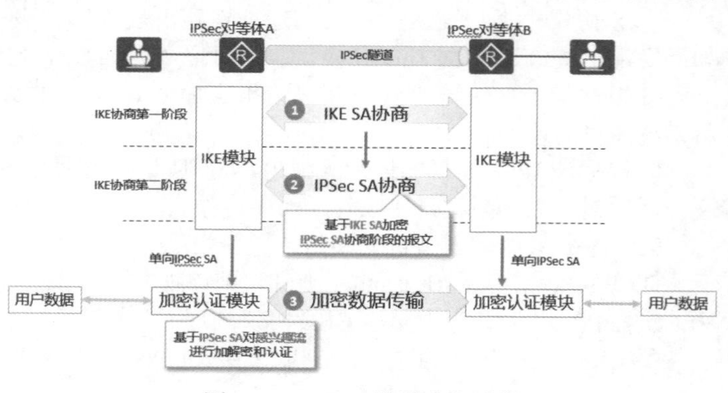

图14-3　IPSec隧道建立过程

SA 由一个三元组来唯一标识，这个三元组包括 SPI（Security Parameter Index，安全参数索引）、目的 IP 地址和使用的安全协议号（AH 或 ESP）。其中，SPI 是为唯一标识 SA 而生成的一个 32 比特的数值，它在 AH 和 ESP 头中传输。在手工配置 SA 时，需要手工指定 SPI 的取值。使用 IKE 协商产生 SA 时，SPI 将随机生成。

SA 是单向的逻辑连接，因此两个 IPSec 对等体之间的双向通信最少需要建立两个 SA 来分别对两个方向的数据流进行安全保护。

IKE 作为密钥协商协议存在两个版本：IKEv1 和 IKEv2。下面以 IKEv1 为例进行介绍。

IKEv1 协商阶段 1 的目的是建立 IKE SA。IKE SA 建立后，对等体间的所有 ISAKMP 消息都将通过加密和验证，这条安全通道可以保证 IKEv1 第二阶段的协商能够安全进行。IKE SA 是一个双向的逻辑连接，两个 IPSec 对等体间只建立一个 IKE SA。

IKEv1 协商阶段 2 的目的就是建立用于安全传输数据的 IPSec SA，并为数据传输衍生出密钥。该阶段使用 IKEv1 协商阶段 1 中生成的密钥对 ISAKMP 消息的完整性和身份进行验证，并对 ISAKMP 消息进行加密，故保证了交换的安全性。

IKE 协商成功意味着双向的 IPSec 隧道已经建立，可以通过 ACL 方式或者安全框架方式定义 IPSec 感兴趣流（需要被 IPSec 保护的数据流），符合感兴趣流流量特征的数据都将被送入 IPSec 隧道进行处理。

14.1.3　L2TP VPN

L2TP（Layer 2 Tunneling Protocol，第二层隧道协议）协议结合了 L2F 协议和 PPTP 协议的优点，是 IETF 有关第二层隧道协议的工业标准。L2TP 是 VPDN（Virtual Private Dial-up Network，虚拟私有拨号网）隧道协议的一种，它扩展了 PPP（Point-to-Point Protocol，点到点协议）的应用，是一种应用于远程办公场景中为出差员工远程访问企业内网资源提供接入服务的重要 VPN 技术。

L2TP 的基本架构如图 14-4 所示。

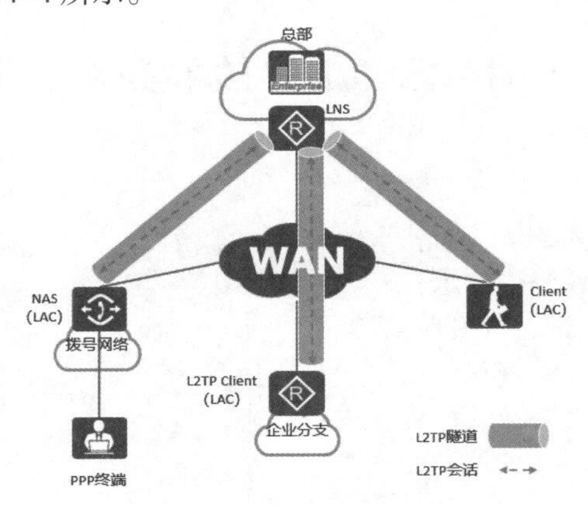

14-4　L2TP 的基本架构

（1）NAS：NAS（Network Access Server，网络接入服务器）主要由 ISP 维护，用于连接拨号网络。

（2）LAC：LAC，即 L2TP 访问集中器，是交换网络上具有 PPP 和 L2TP 协议处理能力的设备。

（3）LNS：LNS 是 LAC 的对端设备，即 LAC 和 LNS 之间建立了 L2TP 隧道。

（4）隧道和会话：L2TP 隧道在 LAC 和 LNS 之间建立，一对 LAC 和 LNS 可以建立多个 L2TP 隧道，一个 L2TP 隧道可以包含多个 L2TP 会话。

L2TP 的工作过程如图 14-5 所示。

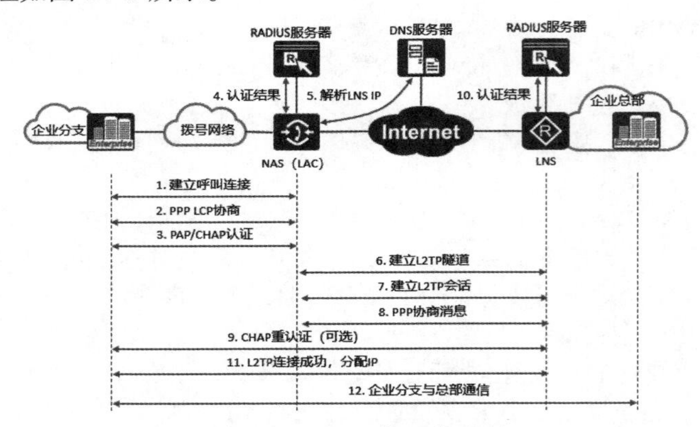

图 14-5　L2TP 的工作过程

（1）建立 L2TP 隧道连接。当 LAC 收到远程用户的 PPP 协商请求时，LAC 向 LNS 发起 L2TP 隧道请求。LAC 和 LNS 之间通过 L2TP 的控制消息协商隧道 ID、隧道认证等内容，协商成功后则建立起一条 L2TP 隧道，由隧道 ID 进行标识。

（2）建立 L2TP 会话连接。如果 L2TP 隧道已存在，则在 LAC 和 LNS 之间通过 L2TP 的控制消息来协商会话 ID 等内容；否则先建立 L2TP 隧道连接。会话中携带了 LAC 的 LCP 协商信息和用户认证信息，LNS 对收到的信息认证通过后，则通知 LAC 会话建立成功。L2TP 会话连接由会话 ID 进行标识。

（3）传输 PPP 报文。L2TP 会话建立成功后，PPP 终端将数据报文发送至 LAC，LAC 根据 L2TP 隧道和会话 ID 等信息进行 L2TP 报文封装，并发送到 LNS，LNS 进行 L2TP 解封装处理，根据路由转发表发送至目的主机，完成报文的传输。

14.2　VPN 技术实验

14.2.1　配置 IPSec VPN

1. 实验目的

AR1 和 AR3 作为公司的出口设备，AR2 作为运营商设备，要求在 AR1 和 AR3 之间部署 IPSec VPN 实现总部 PC1 和分支 PC2 的私有网段之间跨公网进行互相访问。

扫一扫，看视频

2. 实验拓扑

配置 IPSec VPN 的实验拓扑如图 14-6 所示。

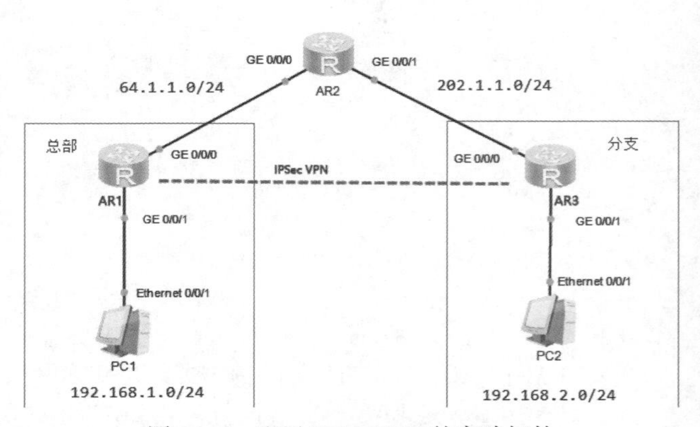

图 14-6　配置 IPSec VPN 的实验拓扑

3. 实验步骤

步骤 1：配置 IP 地址。配置 IPSec VPN IP 地址规划表见表 14-1。

表 14-1 配置 IPSec VPN IP 地址规划表

设备名称	接口编号	IP 地址
AR1	G0/0/0	64.1.1.1/24
	G0/0/1	192.168.1.254/24
AR2	G0/0/0	64.1.1.2/24
	G0/0/1	202.1.1.2/24
AR3	G0/0/0	202.1.1.1/24
	G0/0/1	192.168.2.254/24

步骤 2：配置出口设备去往公网以及对端私网的路由条目，此处可直接使用一条默认路由代替。

AR1 的配置：

```
[AR1]ip route-static 0.0.0.0 0 64.1.1.2
```

AR3 的配置：

```
[AR3]ip route-static 0.0.0.0 0 202.1.1.2
```

步骤 3：使用 ACL 匹配私有网络互访的流量（需要被 IPSec VPN 加密的流量）。

AR1 的配置：

```
[AR1]acl 3000
[AR1-acl-adv-3000]rule permit ip source 192.168.1.0 0.0.0.255 destination
192.168.2.0 0.0.0.255                //匹配 PC1 访问 PC2 的流量
```

AR3 的配置：

```
[AR3]acl 3000
[AR3-acl-adv-3000]rule permit ip source 192.168.2.0 0.0.0.255 destination
192.168.1.0 0.0.0.255                //匹配 PC2 访问 PC1 的流量
```

步骤 4：配置 IKE 的安全提议。

AR1 的配置：

```
[AR1]ike proposal 1
[AR1-ike-proposal-1]q
```

AR3 的配置：

```
[AR3]ike proposal 1
[AR3-ike-proposal-1]q
```

查看创建好的安全提议内容：

```
[AR1]display ike proposal number 1
-------------------------------------------
 IKE Proposal: 1
   Authentication method      : pre-shared    //认证方法为预共享密钥
   Authentication algorithm   : SHA1          //认证算法为 SHA1
   Encryption algorithm       : DES-CBC       //加密算法为 DES-CBC
   DH group                   : MODP-768      //DH 密钥交换参数为 MODP-768
   SA duration                : 86400         //IKE SA 的生存周期
   PRF                        : PRF-HMAC-SHA
```

步骤 5：配置 IKE 对等体。

AR1 的配置：

```
[AR1]ike peer hcie v2
[AR1-ike-peer-hcie]remote-address 202.1.1.1 //配置为分支出口设备的公网 IP 地址
[AR1-ike-peer-hcie]ike-proposal 1
[AR1-ike-peer-hcie]pre-shared-key cipher huawei@123
```

AR3 的配置：

```
[AR3]ike peer hcie v2
[AR3-ike-peer-hcie]pre-shared-key cipher huawei@123
[AR3-ike-peer-hcie]ike-proposal 1
[AR3-ike-peer-hcie]remote-address 64.1.1.1
```

步骤 6：配置 IPSec 安全提议。

AR1 的配置：

```
[AR1]ipsec proposal hcie
[AR1-ipsec-proposal-hice]q
```

AR3 的配置：

```
[AR3]ipsec proposal hcie
[AR3-ipsec-proposal-hcie]q
```

查看 IPSec 安全提议：

```
[AR1]display ipsec  proposal
Number of proposals: 1
IPSec proposal name: hcie
 Encapsulation mode: Tunnel                              //封装模式为隧道模式
 Transform         : esp-new                             //封装协议为 ESP
 ESP protocol      : Authentication MD5-HMAC-96          //ESP 采用的认证算法
                     Encryption     DES                  //ESP 采用的加密算法
```

步骤 7：配置 IPSec 安全策略。

AR1 的配置：

```
[AR1]ipsec policy huawei 10 isakmp
[AR1-ipsec-policy-isakmp-huawei-10]proposal hcie       //关联 IPSec 安全提议 HCIE
[AR1-ipsec-policy-isakmp-huawei-10]security acl 3000  //关联需要加密的流量
[AR1-ipsec-policy-isakmp-huawei-10]ike-peer hcie       //关联 IKE 的对等体
```

AR2 的配置：

```
[AR3]ipsec policy huawei 10 isakmp
[AR3-ipsec-policy-isakmp-huawei-10]security acl 3000
[AR3-ipsec-policy-isakmp-huawei-10]ike-peer hcie
[AR3-ipsec-policy-isakmp-huawei-10]proposal hcie
```

步骤 8：在接口应用 IPSec 安全策略。

AR1 的配置：

```
[AR1]interface g0/0/0
[AR1-GigabitEthernet0/0/0]ipsec policy huawei
```

AR3 的配置：

```
[AR3]interface g0/0/0
[AR3-GigabitEthernet0/0/0]ipsec policy huawei
```

步骤 9： 测试。

PC2 访问 PC1 的测试结果如图 14-7 所示。

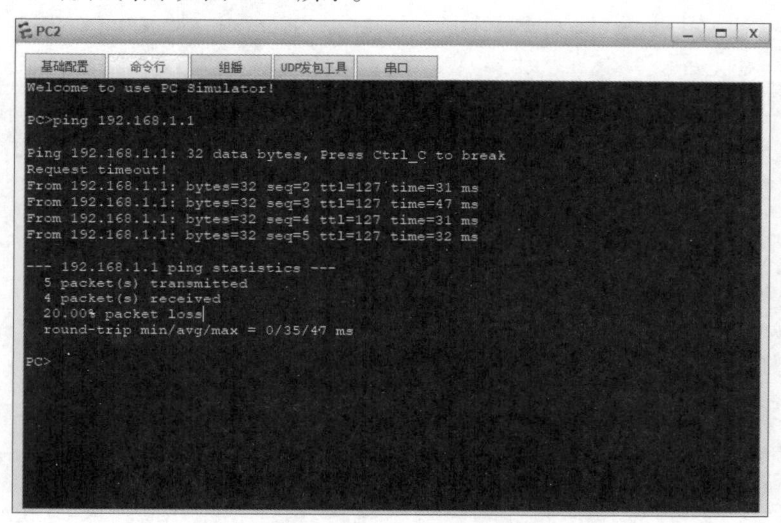

图 14-7　PC2 访问 PC1 的测试结果

查看设备安全联盟的建立情况：

```
[AR1]display ike sa v2
   Conn-ID        Peer              VPN       Flag(s)            Phase
  ---------------------------------------------------------------------
   4              202.1.1.1         0         RD                 2
   3              202.1.1.1         0         RD                 1

  Flag Description:
  RD--READY   ST--STAYALIVE   RL--REPLACED   FD--FADING   TO--TIMEOUT
  HRT--HEARTBEAT   LKG--LAST KNOWN GOOD SEQ NO.   BCK--BACKED UP
```

以上输出为 IKE 安全联盟的状态，RD 表示建立成功。

查看 IPSec 安全联盟的建立情况：

```
[AR1]display ipsec sa

===============================
Interface: GigabitEthernet0/0/0
 Path MTU: 1500
===============================

  -----------------------------
  IPSec policy name: "huawei"
```

```
Sequence number  : 10
Acl Group        : 3000
Acl rule         : 5
Mode             : ISAKMP
-----------------------------
  Connection ID      : 4
  Encapsulation mode: Tunnel                            //封装模式为隧道模式
  Tunnel local       : 64.1.1.1                         //本端的 IP 地址
  Tunnel remote      : 202.1.1.1                        //对端的 IP 地址
  Flow source        : 192.168.1.0/255.255.255.0 0/0 //加密流量的源 IP 地址
  Flow destination   : 192.168.2.0/255.255.255.0 0/0 //加密流量的目的 IP 地址
  Qos pre-classify   : Disable

  [Outbound ESP SAs] 出方向采用 ESP 协议的 IPSec SA 信息
    SPI: 1628093829 (0x610abd85)
    Proposal: ESP-ENCRYPT-DES-64 ESP-AUTH-MD5
    SA remaining key duration (bytes/sec): 1887298560/3536
    Max sent sequence-number: 9                          //接收到加密报文的数量
    UDP encapsulation used for NAT traversal: N

  [Inbound ESP SAs] 入方向采用 ESP 协议的 IPSec SA 信息
    SPI: 3380830098 (0xc9835f92)
    Proposal: ESP-ENCRYPT-DES-64 ESP-AUTH-MD5
    SA remaining key duration (bytes/sec): 1887436260/3536
    Max received sequence-number: 9                      //发送报文的数量
    Anti-replay window size: 32
    UDP encapsulation used for NAT traversal: N
```

14.2.2 配置 L2TP VPN

扫一扫，看视频

1. 实验目的

LNS 作为公司的出口设备，要求在 LNS 设备上配置 L2TP VPN 实现外部出差人员能够通过认证接入公司内部网络。Cloud1 为模拟出差人员，需要在本地使用计算机进行桥接。

2. 实验拓扑

配置 L2TP VPN 的实验拓扑如图 14-8 所示。

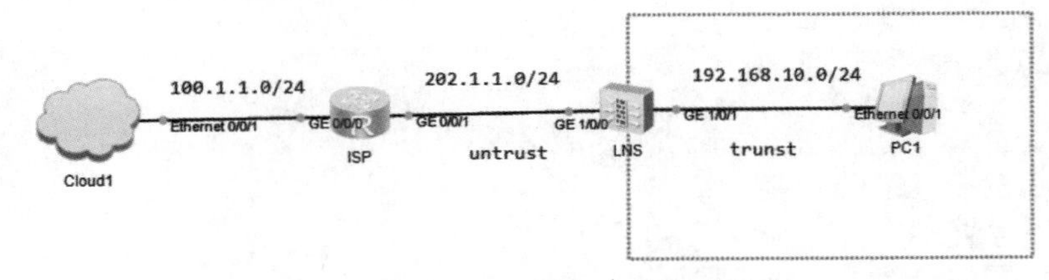

图 14-8　配置 L2TP VPN 的实验拓扑

3. 实验步骤

步骤 1：配置设备的接口 IP 地址。配置 L2TP VPN IP 地址规划表见表 14-2。

表 14-2　配置 L2TP VPN IP 地址规划表

设备名称	接口编号	IP 地址
ISP	G0/0/0	100.1.1.1/24
	G0/0/1	202.1.1.2/24
LNS	G1/0/0	202.1.1.1/24
	G1/0/1	192.168.10.254/24
Cloud1	E0/0/1	100.1.1.2/24

步骤 2: 将防火墙的接口加入到对应的安全区域。

LNS 的配置:

```
[LNS]firewall zone  untrust
[LNS-zone-untrust]add interface GigabitEthernet1/0/0
[LNS]firewall zone  trust
[LNS-zone-trust]add  interface g1/0/1
```

步骤 3: 配置 LNS 的默认路由。

```
[LNS]ip route-static 0.0.0.0 0.0.0.0 202.1.1.2
```

步骤 4: 配置本地计算机的路由，目的网段指向 202.1.1.0/24，下一跳为 100.1.1.1。目的是让本地计算机能够访问到防火墙的出口。

打开本地计算机中的 PowerShell，如图 14-9 所示。在本地计算机中添加临时路由，如图 14-10 所示。

图 14-9　打开本地计算机中的 PowerShell

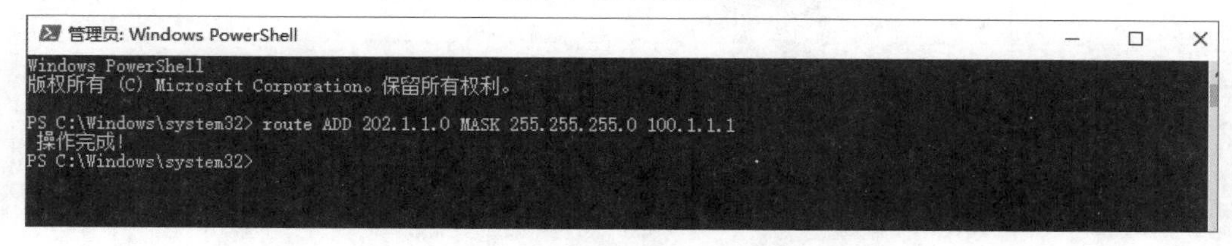

图 14-10　在本地计算机中添加临时路由

步骤 5: 在防火墙的出口开启 HTTPS 的功能。

```
[LNS]interface g1/0/0
[LNS-GigabitEthernet1/0/0]service-manage https permit
```

在浏览器中输入 https://202.1.1.1:8443_进入防火墙的 Web 界面。通过本地计算机登录 ESNP 防火墙，如图 14–11 所示。

图 14–11　通过本地计算机登录 ESNP 防火墙

步骤 6： 配置 L2TP VPN。

（1）创建 L2TP 组，如图 14–12 所示。

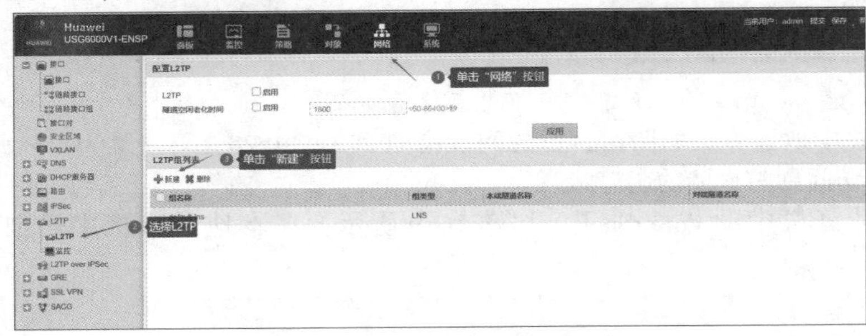

图 14–12　创建 L2TP 组

（2）配置 L2TP 组参数，如图 14–13~ 图 14–18 所示。

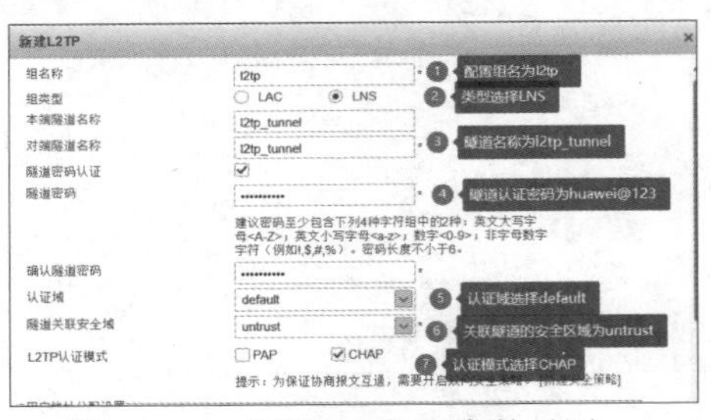

图 14–13　配置 L2TP 组参数（1）

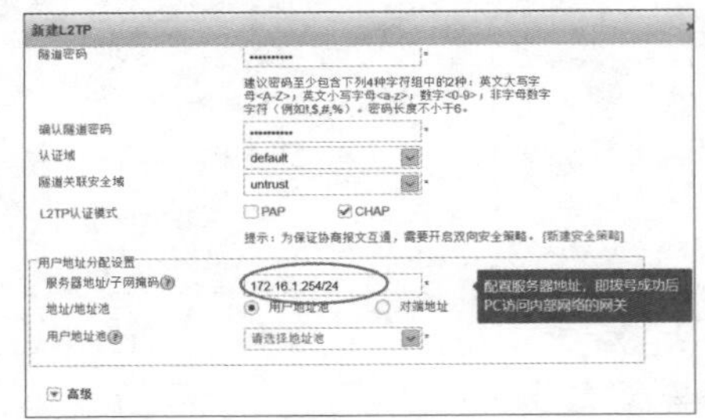

图 14–14　配置 L2TP 组参数（2）

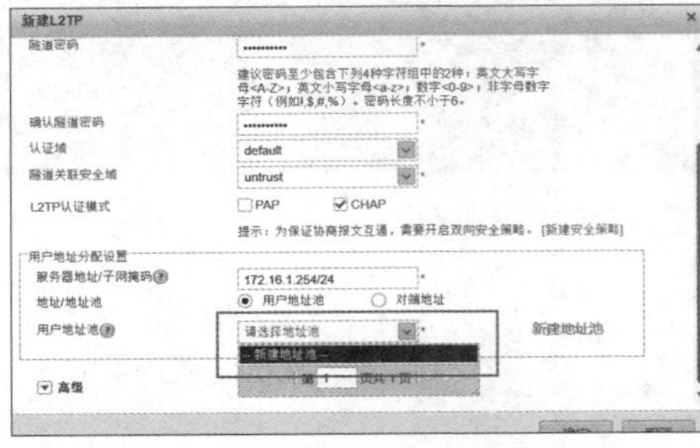

图 14–15　配置 L2TP 组参数（3）

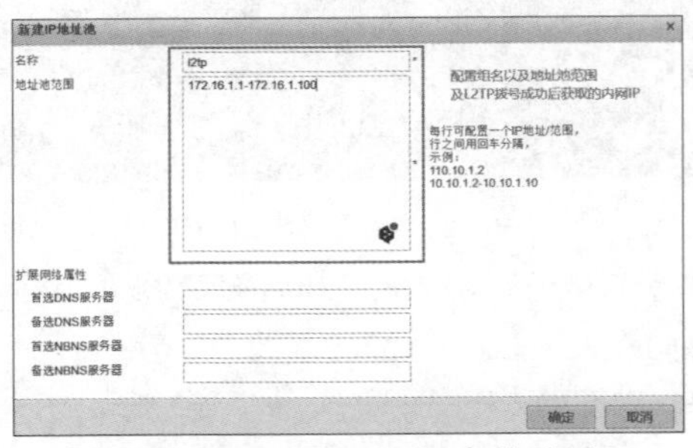

图 14–16　配置 L2TP 组参数（4）

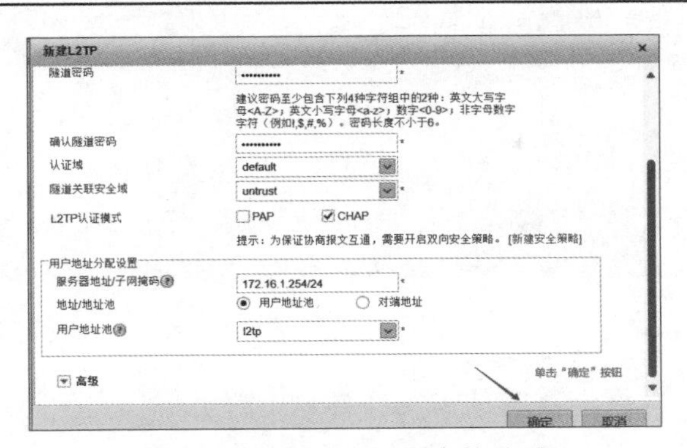

图 14–17　配置 L2TP 组参数（5）

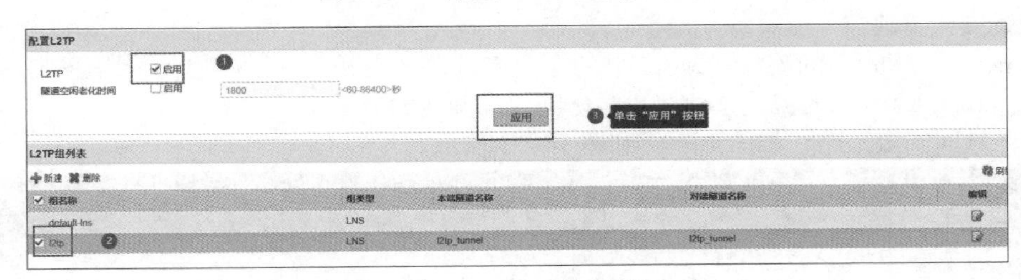

图 14–18　配置 L2TP 组参数（6）

4. 创建用户名及密码

下面创建 L2TP 的用户名及密码，如图 14–19~ 图 14–21 所示。

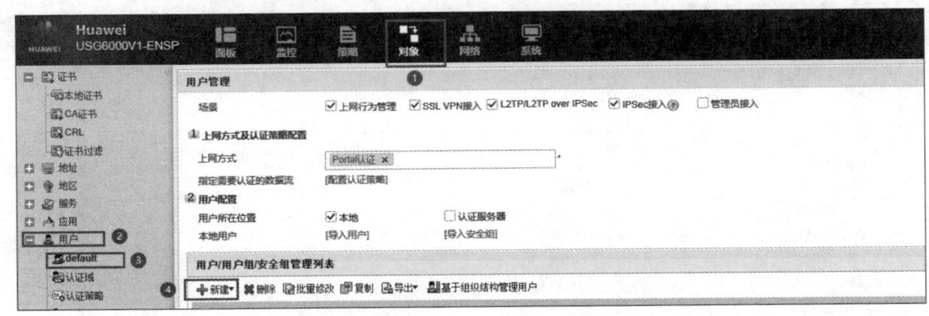

图 14–19　创建 L2TP 的用户名及密码（1）

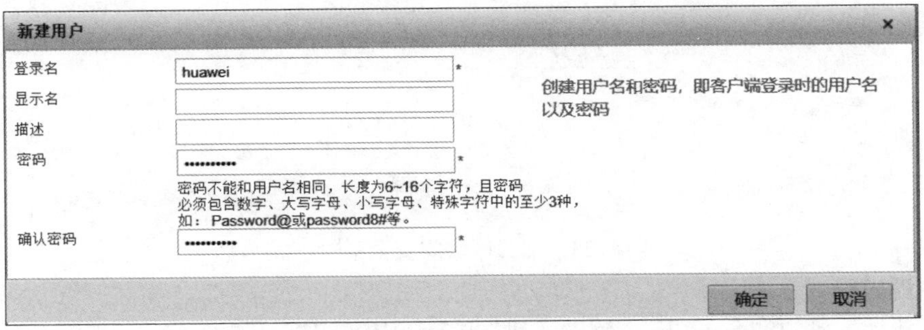

新建用户

登录名　huawei

显示名

描述

密码

密码不能和用户名相同，长度为6~16个字符，且密码必须包含数字、大写字母、小写字母、特殊字符中的至少3种，如：Password@或password8#等。

确认密码

创建用户名和密码，即客户端登录时的用户名以及密码

确定　取消

图 14–20　创建 L2TP 的用户名及密码（2）

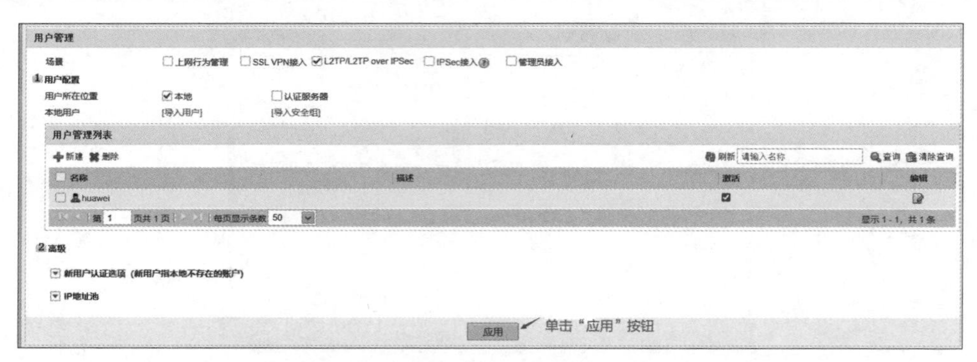

图 14-21　创建 L2TP 的用户名及密码（3）

5. 配置安全策略

放行 untrust 到 local 的流量，即 L2TP 的协商流量。

放行 untrust 到 trust 的流量，即客户端访问内部网络的流量。

配置区域间安全策略，如图 14-22 和图 14-23 所示。

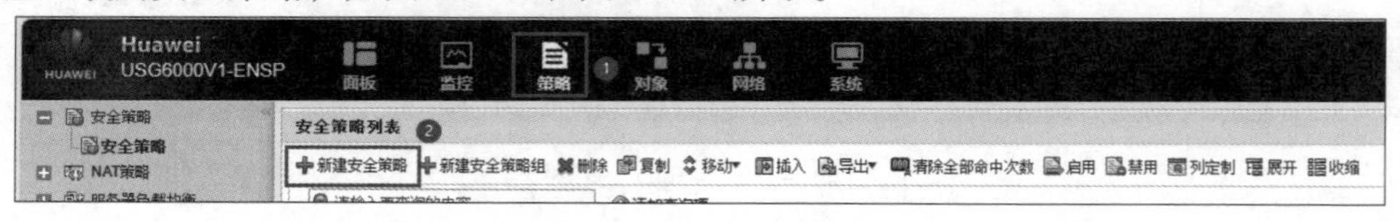

图 14-22　配置区域间安全策略（1）

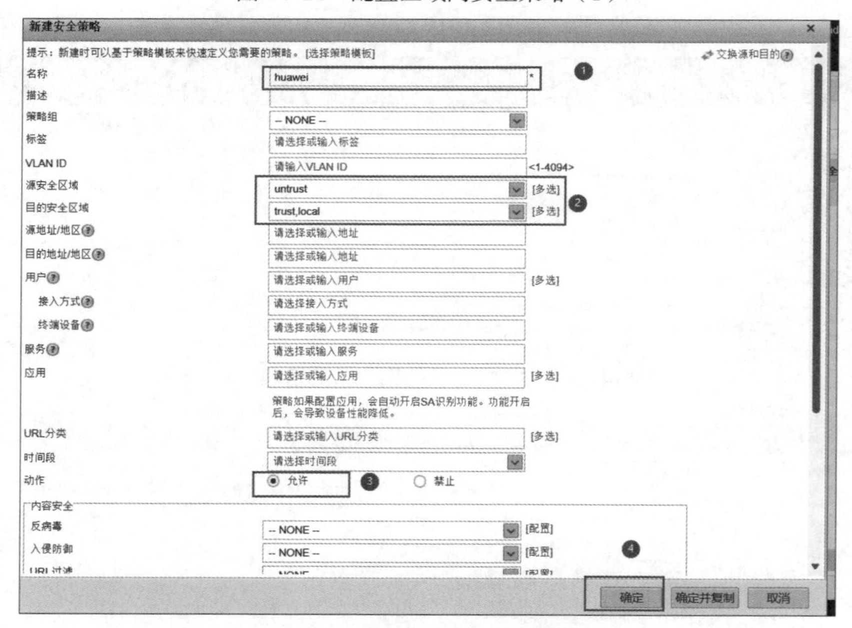

图 14-23　配置区域间安全策略（2）

6. 客户端设置

（1）打开计算机中的 SecoClient 软件，建立 L2TP VPN 连接，如图 14-24 所示。

图 14-24　使用 SecoClient 软件建立 L2TP VPN 连接（1）

（2）设置 L2TP VPN 连接参数，如图 14-25 和图 14-26 所示。

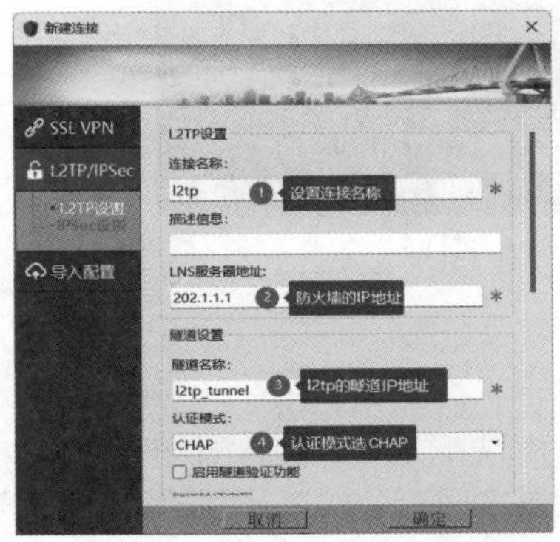

图 14-25　设置 L2TP VPN 连接参数（1）

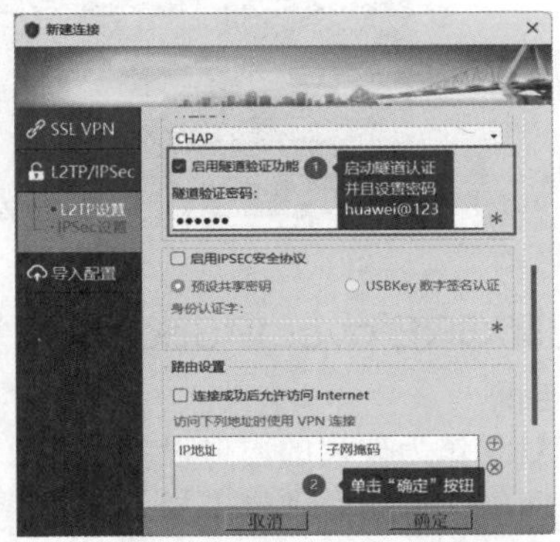

图 14-26　设置 L2TP VPN 连接参数（2）

（3）客户端登录，使用 SecoClient 软件建立 L2TP VPN 连接，如图 14-27 所示。

（4）输入登录的用户名和密码，如图 14-28 所示。

图 14-27　使用 SecoClient 软件建立 L2TP VPN 连接（2）

图 14-28　输入登录的用户名和密码

按 Win+R 组合键，输入 cmd 命令，按 Enter 键，打开命令行窗口，接着输入 ipconfig 命令后按 Enter 键，Windows 系统会获取 172.16.1.0/24 网段的 IP 地址，如图 14-29 所示。

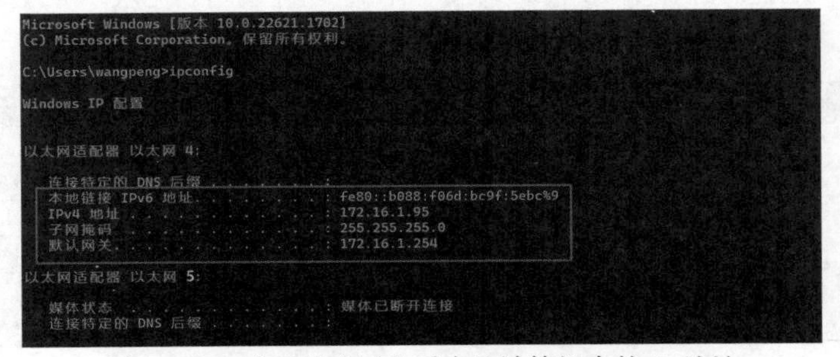

图 14-29　登录成功后查看本地计算机中的 IP 地址

测试网络的连通性，发现其可以访问内部网络，如图 14-30 所示。

```
C:\Users\wangpeng>ping 192.168.10.254

正在 Ping 192.168.10.254 具有 32 字节的数据:
来自 192.168.10.254 的回复: 字节=32 时间=8ms TTL=255
来自 192.168.10.254 的回复: 字节=32 时间=6ms TTL=255
来自 192.168.10.254 的回复: 字节=32 时间=10ms TTL=255

192.168.10.254 的 Ping 统计信息:
    数据包: 已发送 = 3，已接收 = 3，丢失 = 0 (0% 丢失)，
往返行程的估计时间(以毫秒为单位):
    最短 = 6ms，最长 = 10ms，平均 = 8ms
Control-C
^C
C:\Users\wangpeng>ping 192.168.10.1

正在 Ping 192.168.10.1 具有 32 字节的数据:
请求超时。
来自 192.168.10.1 的回复: 字节=32 时间=9ms TTL=127
来自 192.168.10.1 的回复: 字节=32 时间=7ms TTL=127
来自 192.168.10.1 的回复: 字节=32 时间=10ms TTL=127

192.168.10.1 的 Ping 统计信息:
    数据包: 已发送 = 4，已接收 = 3，丢失 = 1 (25% 丢失)，
往返行程的估计时间(以毫秒为单位):
    最短 = 7ms，最长 = 10ms，平均 = 8ms
```

图 14-30　测试网络的连通性

14.2.3　配置 GRE over IPSec VPN

1. 实验目的

AR1 和 AR3 作为公司的出口设备，AR2 作为运营商设备，要求在 AR1 和 AR3 之间部署 GRE over IPSec VPN，以实现总部 PC1 和分支 PC2 的私有网段之间跨公网进行互相访问。

2. 实验拓扑

配置 GRE over IPSec VPN 的实验拓扑如图 14-31 所示。

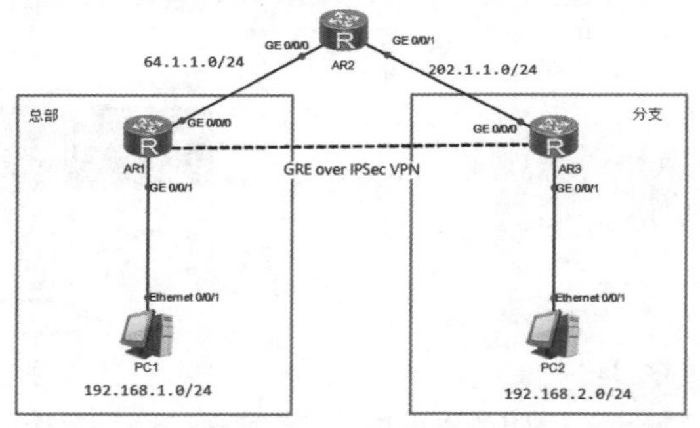

图 14-31　配置 GRE over IPSec VPN 的实验拓扑

3. 实验步骤

步骤 1：配置 IP 地址。配置 GRE over IPSec VPN IP 地址规划表见表 14-3。

表 14-3　配置 GRE over IPSec VPN IP 地址规划表

设备名称	接口编号	IP 地址
AR1	G0/0/0	64.1.1.1/24
	G0/0/1	192.168.1.254/24
AR2	G0/0/0	64.1.1.2/24
	G0/0/1	202.1.1.2/24
AR3	G0/0/0	202.1.1.1/24
	G0/0/1	192.168.2.254/24

步骤 2：配置出口设备去往公网以及对端私网的路由条目，此处可以直接使用一条默认路由代替。

AR1 的配置：

```
[AR1]ip route-static 0.0.0.0 0 64.1.1.2
```

AR3 的配置：

```
[AR3]ip route-static 0.0.0.0 0 202.1.1.2
```

步骤 3：配置 GRE 隧道。

AR1 的配置：

```
[AR1]interface Tunnel0/0/0
[AR1-Tunnel0/0/0]ip address 13.1.1.1 255.255.255.0
[AR1-Tunnel0/0/0]tunnel-protocol gre
[AR1-Tunnel0/0/0]source 64.1.1.1
[AR1-Tunnel0/0/0]destination 202.1.1.1
```

AR3 的配置：

```
[AR3]interface Tunnel0/0/0
[AR3-Tunnel0/0/0]ip address 13.1.1.3 255.255.255.0
[AR3-Tunnel0/0/0]tunnel-protocol gre
[AR3-Tunnel0/0/0]source 202.1.1.1
[AR3-Tunnel0/0/0]destination 64.1.1.1
```

查看隧道的建立情况（隧道状态为 UP 即可）：

```
[AR1]display interface Tunnel 0/0/0
Tunnel0/0/0 current state : UP
Line protocol current state : UP
Last line protocol up time : 2023-11-15 14:21:48 UTC-08:00
Description:HUAWEI, AR Series, Tunnel0/0/0 Interface
Route Port,The Maximum Transmit Unit is 1500
Internet Address is 13.1.1.1/24
Encapsulation is TUNNEL, loopback not set
Tunnel source 64.1.1.1 (GigabitEthernet0/0/0), destination 202.1.1.1
Tunnel protocol/transport GRE/IP, key disabled
keepalive disabled
Checksumming of packets disabled
Current system time: 2023-11-15 14:23:20-08:00
    300 seconds input rate 0 bits/sec, 0 packets/sec
    300 seconds output rate 0 bits/sec, 0 packets/sec
    0 seconds input rate 0 bits/sec, 0 packets/sec
    0 seconds output rate 0 bits/sec, 0 packets/sec
    0 packets input,  0 bytes
    0 input error
    0 packets output,  0 bytes
    0 output error
    Input bandwidth utilization  : --
Output bandwidth utilization : --
```

步骤 4： 使用 GRE 隧道建立 OSPF 邻居，让总部和分支能学习到各自的私网路由。

AR1 的配置：

```
[AR1]ospf 1
[AR1-ospf-1]area 0.0.0.0
[AR1-ospf-1-area-0.0.0.0]network 13.1.1.0 0.0.0.255
[AR1-ospf-1-area-0.0.0.0]network 192.168.1.0 0.0.0.255
```

AR3 的配置：

```
[AR3]ospf 1
[AR3-ospf-1]area 0.0.0.0
[AR3-ospf-1-area-0.0.0.0]network 13.1.1.0 0.0.0.255
[AR3-ospf-1-area-0.0.0.0]network 192.168.2.0 0.0.0.255
```

查看邻居的建立情况：

```
[AR1]display ospf peer brief

         OSPF Process 1 with Router ID 64.1.1.1
                  Peer Statistic Information
 ----------------------------------------------------------------------------
 Area Id          Interface                      Neighbor id          State
 0.0.0.0          Tunnel0/0/0                    202.1.1.1            Full
 ----------------------------------------------------------------------------
```

通过以上输出可以看出，已经通过 GRE 隧道建立了 OSPF 的邻居关系。

查看路由的学习情况：

```
[AR1]display ip routing-table protocol ospf
Route Flags: R - relay, D - download to fib
------------------------------------------------------------------------------
Public routing table : OSPF
         Destinations : 1        Routes : 1

OSPF routing table status : <Active>
         Destinations : 1        Routes : 1

Destination/Mask    Proto   Pre    Cost   Flags     NextHop       Interface

  192.168.2.0/24    OSPF    10     1563   D         13.1.1.3      Tunnel0/0/0
```

通过以上输出可以看出，AR1 学习到了分支的 192.168.2.0/24 的路由信息，下一跳为 Tunnel 0/0/0，此时分支也可以学习到总部的路由，查看方式同理。

步骤 5： 配置 ACL，匹配需要加密的流量，由于总部和分支之间的互访流量都需要通过 GRE 隧道实现互访，因此 ACL 仅需匹配隧道的源目 IP 地址即可。

AR1 的配置：

```
[AR1]acl number 3000
[AR1-acl-adv-3000]rule 5 permit ip source 64.1.1.1 0 destination 202.1.1.1 0
                                                          // 匹配通过 GRE 隧道的流量
```

AR2 的配置：

```
[AR2]acl number 3000
[AR2-acl-adv-3000]rule 5 permit ip source 202.1.1.1 0 destination 64.1.1.1 0
```

步骤 6：配置 IKE 安全提议。

AR1 的配置：

```
[AR1]ike proposal 1
[AR1-ike-proposal-1]q
```

AR3 的配置：

```
[AR3]ike proposal 1
[AR3-ike-proposal-1]q
```

步骤 7：配置 IKE 对等体。

AR1 的配置：

```
[AR1]ike peer huawei v2
[AR1-ike-peer-huawei]pre-shared-key simple huawei
[AR1-ike-peer-huawei]ike-proposal 1
[AR1-ike-peer-huawei]remote-address 202.1.1.1
```

AR3 的配置：

```
[AR3]ike peer huawei v2
[AR3-ike-peer-huawei]pre-shared-key simple huawei
[AR3-ike-peer-huawei]ike-proposal 1
[AR3-ike-peer-huawei]remote-address 64.1.1.1
```

步骤 8：配置 IPSec 安全提议。

AR1 的配置：

```
[AR1]ipsec proposal huawei
[AR1-ipsec-proposal-huawei]q
```

AR3 的配置：

```
[AR3]ipsec proposal huawei
[AR3-ipsec-proposal-huawei]q
```

步骤 9：配置 IPSec 安全策略。

AR1 的配置：

```
[AR1]ipsec policy huawei 10 isakmp
[AR1-ipsec-policy-isakmp-huawei-10]security acl 3000
[AR1-ipsec-policy-isakmp-huawei-10]ike-peer huawei
[AR1-ipsec-policy-isakmp-huawei-10]proposal huawei
```

AR3 的配置：

```
[AR3]ipsec policy huawei 10 isakmp
[AR3-ipsec-policy-isakmp-huawei-10]security acl 3000
[AR3-ipsec-policy-isakmp-huawei-10]ike-peer huawei
[AR3-ipsec-policy-isakmp-huawei-10]proposal huawei
```

步骤 10: 在接口调用 IPSec 安全策略。

AR1 的配置：

```
[AR1]interface GigabitEthernet0/0/0
[AR1-GigabitEthernet0/0/0]ipsec policy huawei
```

AR3 的配置：

```
[AR3]interface GigabitEthernet0/0/0
[AR3-GigabitEthernet0/0/0]ipsec policy huawei
```

PC1 访问 PC2 的测试图如图 14-32 所示。

图 14-32　PC1 访问 PC2 的测试图

第 15 章

Segment Routing

15.1 Segment Routing 概述

Segment Routing（段路由）是一种基于源路由理念设计的在网络上转发数据包的协议。Segment Routing MPLS 是指基于 MPLS 转发平面的 Segment Routing，下文简称 Segment Routing。Segment Routing 将网络路径分成一个一个的段，并且为这些段和网络中的转发节点分配段标识 ID。通过对段和网络中的转发节点进行有序排列，就可以得到一条转发路径。

Segment Routing 将代表转发路径的段序列编码在数据包头部，随数据包传输。接收端收到数据包后，对段序列进行解析，如果段序列的顶部段标识是本节点，则弹出该标识，然后进行下一步处理；如果不是本节点，则使用 ECMP（Equal Cost Multiple Path，等价多路径）方式将数据包转发到下一个节点。

Segment Routing 技术将带来以下好处。

（1）简化 MPLS 网络的控制平面。Segment Routing 使用控制器或 IGP 集中算路和分发标签，不再需要 RSVP-TE、LDP 等隧道协议。Segment Routing 可以直接应用于 MPLS 架构，转发平面没有变化。

（2）提供高效 TI-LFA（Topology-Independent Loop-Free Alternate，拓扑无关的无环路备份）FRR 算法保护，实现路径故障的快速恢复。在 Segment Routing 技术的基础上结合 RLFA（Remote Loop-Free Alternate，远端无环路备份）FRR 算法，形成高效的 TI-LFA FRR 算法。TI-LFA FRR 算法支持任意拓扑的节点和链路保护，能够弥补传统隧道保护技术的不足。

（3）Segment Routing 技术更具有网络容量扩展能力。传统 MPLS TE 是一种面向连接的技术，为了维护连接状态，节点间需要发送和处理大量报文，设备控制层面压力大。Segment Routing 仅在头节点对报文进行标签操作，即可任意控制业务路径，中间节点不需要维护路径信息，设备控制层面压力小。

此外，Segment Routing 技术的标签数量为全网节点数 + 本地邻接数，只与网络规模相关，与隧道数量和业务规模无关。

（4）更好地向 SDN 网络平滑演进。Segment Routing 技术基于源路由理念而设计，通过源节点即可控制数据包在网络中的转发路径。配合集中算路模块，即可灵活简便地实现路径控制与调整。

Segment Routing 同时支持传统网络和 SDN 网络，兼容现有设备，保障现有网络平滑演进到 SDN 网络，而不是颠覆现有网络。

15.1.1 Segment Routing 基本概念

1. Segment ID

Segment ID（SID）用于标识 Segment，其格式取决于具体的技术实现。例如，Segment ID 可以使用 MPLS 标签、MPLS 标签空间中的索引、IPv6 地址。

R1 访问 R8 的流量（Segment 示意图）如图 15-1 所示，其中包含以下三个指令。

（1）指令 1（400）：沿着支持 ECMP 的最短路径到达 R4。

（2）指令 2（1046）：沿着 R4 的 GE0/0/2 接口转发数据包。

（3）指令 3（800）：沿着支持 ECMP 的最短路径到达 R8。

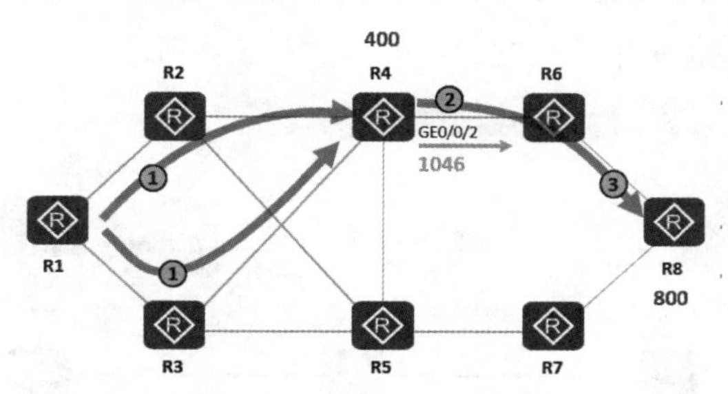

图 15-1　Segment ID 示意图

2. 源路由

源节点选择一条路径并在报文中压入一个有序的 Segment List，网络中的其他节点按照报文封装的 Segment List 进行转发。Segment List 是由一个或多个 SID 构成的有序列表，源路由示意图如图 15-2 所示。

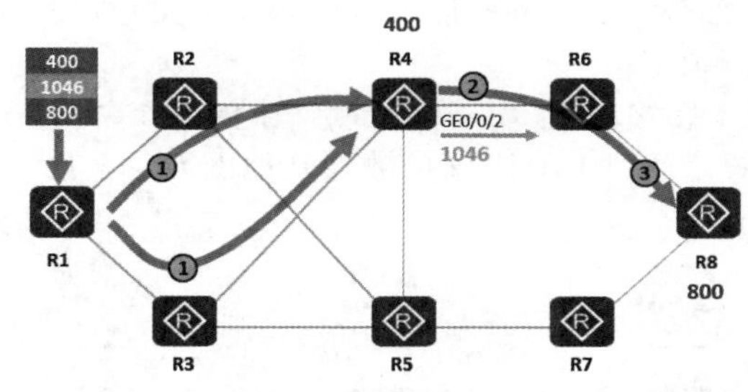

图 15-2　源路由示意图

3. Segment 分类

Segment 可以分为以下三类。

（1）Prefix Segment（前缀段）：用于标识网络中的某个目的地址前缀，必须手动配置，可以通过 IGP 协议扩散到其他网元，全局可见，全局有效。其中，Node Segment 是特殊的 Prefix Segment。前缀段 SID 示意图如图 15-3 所示。

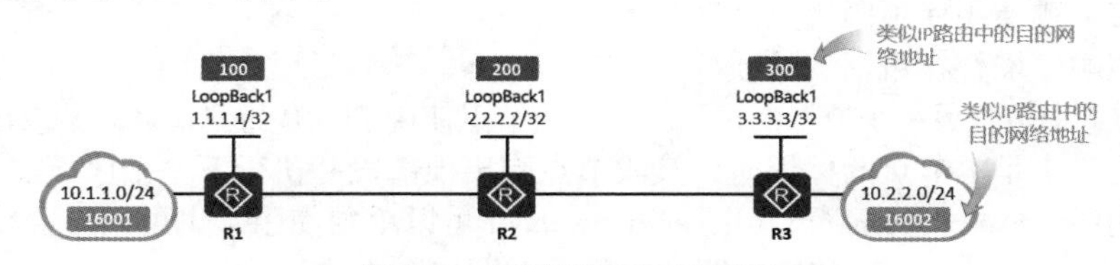

图 15-3　前缀段 SID 示意图

Prefix Segment 通过 Prefix SID（Segment ID）标识。Prefix SID 是发布端通告的 SRGB（Segment Routing Global Block，段路由全局块）范围内的偏移值，接收端会根据自己的 SRGB 计算实际标签值，用于生成 MPLS 转发表项。Node Segment 是特殊的 Prefix Segment，用于标识特定的节点（Node）。

在节点的 LoopBack 接口下配置 IP 地址作为前缀，这个节点的 Prefix SID 实际就是 Node SID。

（2）SRGB：用户指定的为 Segment Routing MPLS 预留的全局标签集合。

每台设备通过扩展的路由协议通告自己的 SRGB。

节点通过扩展的路由协议通告前缀 SID 索引（Index）后，各台设备分别根据 SRGB 计算入站及出站 SID。

在实际部署中，建议设备采用统一的 SRGB，如图 15-4 所示。

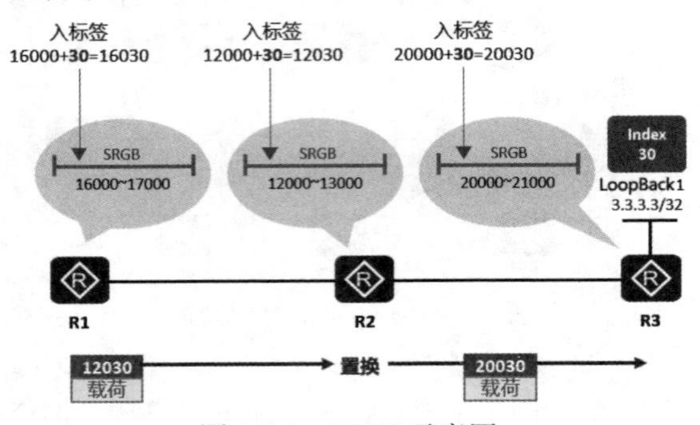

图 15-4　SRGB 示意图

Segment Routing 要求前缀 SID 全局有效。

在 MPLS 中，设备的一部分标签空间可能被其他协议（如 LDP）占用，因此需要指定明确的空间用于 Segment Routing 全局标签。

（3）Adjacency Segment（邻接段）：用于标识网络中的某个邻接，可以通过源节点使用协议动态分配，也可以手动配置。通过 IGP 协议扩散到其他网元，全局可见，本地有效。Adjacency Segment 示意图如图 15-5 所示。

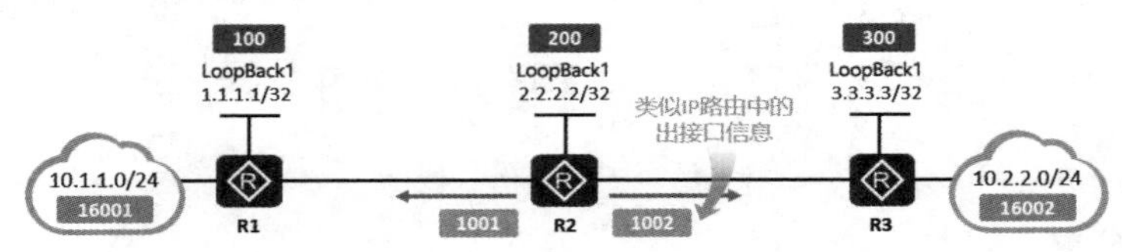

图 15-5　Adjacency Segment 示意图

15.1.2　SID 的使用

按序组合前缀 SID 和邻接 SID 可以构建出网络内的任何路径。

路径中的每一跳使用栈顶段信息来区分下一跳。

段信息按照顺序堆叠在数据头的顶部。

当栈顶段信息中包含另一个节点的标识时，接收节点使用 ECMP 的方式将数据包转发到下一跳。

当栈顶段信息中是本节点的标识时，接收节点弹出顶部段并执行下一个段所需的任务。

在实际应用中，Prefix Segment、Adjacency Segment 可以单独使用，也可以结合使用。

场景 1：基于 Prefix Segment 的转发路径，其示意图如图 15-6 所示。

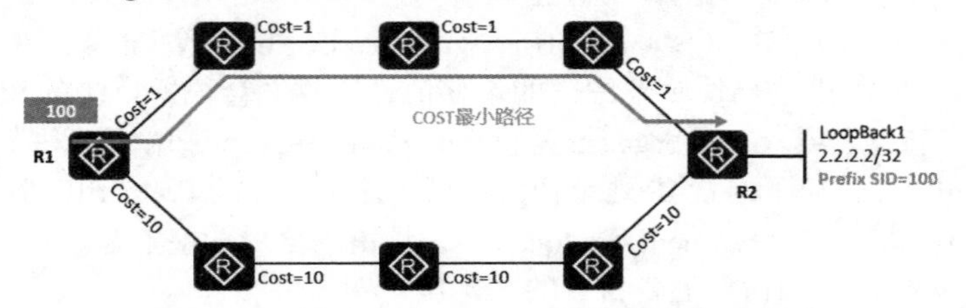

图 15-6　基于 Prefix Segment 的转发路径示意图

基于 Prefix Segment 的转发路径是由 IGP 通过 SPF（Shortest Path First，最短路径优先）算法计

算得出的。

（1）通过 IGP 扩散之后，整个 IGP 域的所有设备学习到 R2 的 Prefix SID（100）。

（2）以 R1 为例（其他设备类似），它通过 SPF 算法计算出一条到达 R2 的最短路径。

场景 2：基于 Adjacency Segment 的转发路径，其示意图如图 15-7 所示。

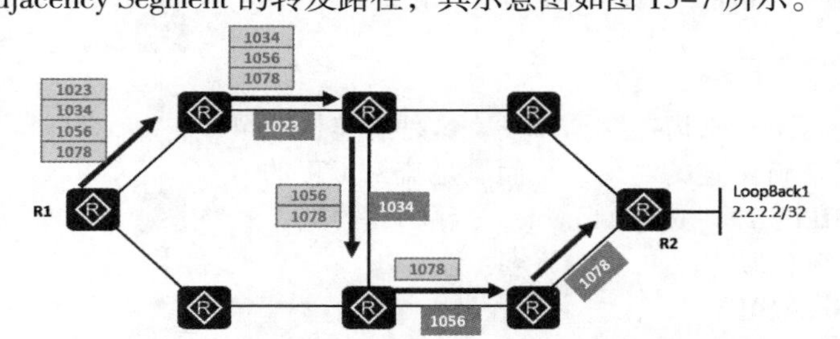

图 15-7　基于 Adjacency Segment 的转发路径示意图

首先给网络中的每个邻接 SID 分配一个 Adjacency Segment，然后在头节点定义一个包含多个 Adjacency Segment 的 Segment List。

场景 3：基于 Adjacency Segment + Node Segment 的转发路径，其示意图如图 15-8 所示。

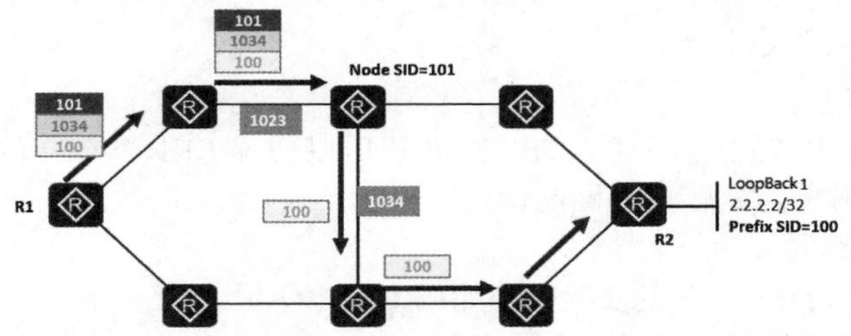

图 15-8　基于 Adjacency Segment + Node Segment 的转发路径示意图

将 Adjacency Segment 和 Node Segment 结合，Adjacency Segment 可以强制整条路径包含某一个邻接。而对于 Node Segment，节点可以使用 SPF 算法计算最短路径，也可以负载分担。

15.1.3　SR MPLS BE

使用 SID 来指导设备基于最短路径进行数据转发，这种工作机制称为 SR MPLS BE（Best Effort，尽力服务），其示意图如图 15-9 所示。

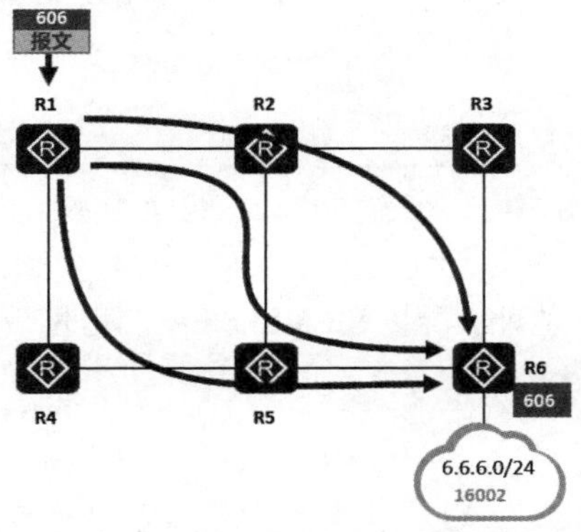

图 15-9　SR MPLS BE 示意图

例如，在本例中，使用R6的Node SID 606可以指导数据沿着去往R6的最短路径来转发数据，该最短路径是基于路由协议计算得出的，并且支持等价路径。

SR MPLS BE是一种替代“LDP+IGP方案”的新方案。

15.1.4 SR MPLS TE

使用多个SID进行组合来指导数据转发，这种工作机制可以对数据的转发路径进行一定的约束，从而满足流量工程的需求，因此被称为SR MPLS TE（Traffic Engine，流量工程）。

SID有以下三种组合形式。

（1）使用多个Node SID。

（2）使用多个Adjacency SID。

（3）使用Node SID与Adjacency SID组合，如图15-10所示。

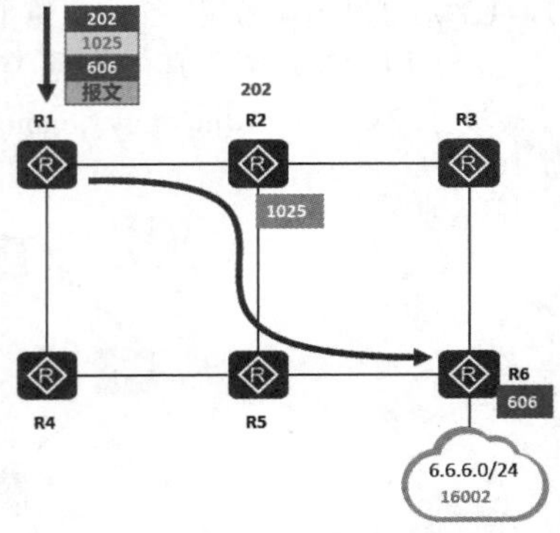

图15-10　SR MPLS TE示意图

15.2 Segment Routing实验

15.2.1 配置SR MPLS BE

扫一扫，看视频

1. 实验目的

在PE1和PE3之间建立MP-BGP邻居以传递CE1和CE2的私网路由，并且使用SR MPLS BE的方式传递私网流量。

2. 实验拓扑

配置SR MPLS BE的实验拓扑如图15-11所示。

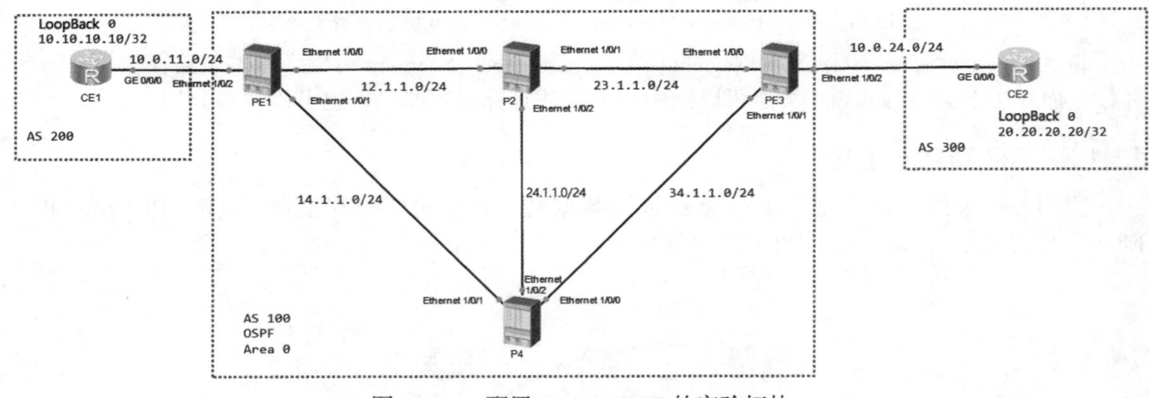

图15-11　配置SR MPLS BE的实验拓扑

3. 实验步骤

步骤1： 配置设备接口的IP地址。配置SR MPLS BE IP地址规划表见表15-1。

表 15-1 配置 SR MPLS BE IP 地址规划表

设备名称	接口编号	IP 地址
PE1	Ethernet 1/0/0	12.1.1.1/24
	Ethernet 1/0/1	14.1.1.1/24
	LoopBack 0	1.1.1.1/32
	Ethernet 1/0/2	10.0.11.1/24
P2	Ethernet 1/0/0	12.1.1.2/24
	Ethernet 1/0/1	23.1.1.2/24
	LoopBack 0	2.2.2.2/32
PE3	Ethernet 1/0/0	23.1.1.3/24
	Ethernet 1/0/1	34.1.1.3/24
	LoopBack 0	3.3.3.3/24
	Ethernet 1/0/2	10.0.24.1/24
P4	Ethernet 1/0/0	34.1.1.4/24
	Ethernet 1/0/1	14.1.1.4/24
	LoopBack 0	4.4.4.4/32
CE1	G0/0/0	10.0.11.2/24
	LoopBack 0	10.10.10.10/32
CE2	G0/0/0	10.0.24.2/24
	LoopBack 0	20.20.20.20/32

步骤 2：配置 AS 100 内的 IGP 协议。

PE1 的配置：

```
[PE1]ospf
[PE1-ospf-1]area 0
[PE1-ospf-1-area-0.0.0.0]network 12.1.1.0 0.0.0.255
[PE1-ospf-1-area-0.0.0.0]network 14.1.1.0 0.0.0.255
[PE1-ospf-1-area-0.0.0.0]network 1.1.1.1 0.0.0.0
```

P2 的配置：

```
[P2]ospf 1
[P2-ospf-1]area 0
[P2-ospf-1-area-0.0.0.0]network 2.2.2.2 0.0.0.0
[P2-ospf-1-area-0.0.0.0]network 12.1.1.0 0.0.0.255
[P2-ospf-1-area-0.0.0.0]network 23.1.1.0 0.0.0.255
```

PE3 的配置：

```
[PE3]ospf 1
[PE3-ospf-1]area 0
[PE3-ospf-1-area-0.0.0.0]network 3.3.3.3 0.0.0.0
[PE3-ospf-1-area-0.0.0.0]network 23.1.1.0 0.0.0.255
[PE3-ospf-1-area-0.0.0.0]network 34.1.1.0 0.0.0.255
```

P4 的配置：

```
[P4]ospf 1
[P4-ospf-1]area 0
[P4-ospf-1-area-0.0.0.0]network 4.4.4.4 0.0.0.0
[P4-ospf-1-area-0.0.0.0]network 34.1.1.0 0.0.0.255
[P4-ospf-1-area-0.0.0.0]network 14.1.1.0 0.0.0.255
```

查看设备的路由学习情况，保证 AS 100 内的设备能学习到每台设备的环回口路由，此处以 PE1 为例：

```
[PE1]display ip routing-table protocol ospf
Route Flags: R - relay, D - download to fib, T - to vpn-instance, B - black hole
route
------------------------------------------------------------------------------
_public_ Routing Table : OSPF
         Destinations : 8        Routes : 9
OSPF routing table status : <Active>
         Destinations : 5        Routes : 6
Destination/Mask    Proto    Pre   Cost  Flags      NextHop         Interface
      2.2.2.2/32    OSPF     10    1     D          12.1.1.2        Ethernet1/0/0
      3.3.3.3/32    OSPF     10    2     D          14.1.1.4        Ethernet1/0/1
                    OSPF     10    2     D          12.1.1.2        Ethernet1/0/0
      4.4.4.4/32    OSPF     10    1     D          14.1.1.4        Ethernet1/0/1
     23.1.1.0/24    OSPF     10    2     D          12.1.1.2        Ethernet1/0/0
     34.1.1.0/24    OSPF     10    2     D          14.1.1.4        Ethernet1/0/1

OSPF routing table status : <Inactive>
         Destinations : 3        Routes : 3
Destination/Mask    Proto    Pre   Cost  Flags      NextHop         Interface
      1.1.1.1/32    OSPF     10    0                1.1.1.1         LoopBack0
     12.1.1.0/24    OSPF     10    1                12.1.1.1        Ethernet1/0/0
     14.1.1.0/24    OSPF     10    1                14.1.1.1        Ethernet1/0/1
```

步骤 3： 配置 AS 100 内的设备，开启 MPLS，并且配置 MPLS LSR-ID。

PE1 的配置：

```
[PE1]mpls lsr-id 1.1.1.1
[PE1]mpls
Info: Mpls starting, please wait... OK!
[PE1-mpls]
```

P2 的配置：

```
[P2]mpls lsr-id 2.2.2.2
[P2]mpls
Info: Mpls starting, please wait... OK!
```

PE3 的配置：

```
[PE3]mpls lsr-id 3.3.3.3
[PE3]mpls
Info: Mpls starting, please wait... OK!
```

P4 的配置：

```
[P4]mpls lsr-id 4.4.4.4
[P4]mpls
Info: Mpls starting, please wait... OK!
```

步骤 4：开启 SR 功能，建立 SR LSP。

PE1 的配置：

```
[PE1]segment-routing                         //全局开启 SR 功能
[PE1-segment-routing]q
[PE1-segment-routing]ospf 1
[PE1-ospf-1]opaque-capability enable        //使能 OSPF 的 OPAQUE-LSA 能力
[PE1-ospf-1]segment-routing mpls            //使能 OSPF 对应拓扑的 Segment Routing 功能
[PE1-ospf-1]segment-routing global-block 16000 16999//配置 SRGB 的范围
                                                     //（每台设备的范围空间需要一致）

[PE1-ospf-1]q
[PE1]interface LoopBack0
[PE1-LoopBack0]ospf prefix-sid index 1     //配置 Loopback 接口 IP 地址的 Prefix SID
```

在 Segment Routing 技术中，使用 IGP 协议来收集整个网络的拓扑信息，并为每台路由器分发标签。因此，在 IGP 协议下使能 Segment Routing 功能是必要的配置。如果 IGP 协议采用 OSPF，则需要在 OSPF 视图下执行 segment-routing mpls 命令。

创建 LoopBack 接口后可以为该接口配置 IP 地址，这样 LoopBack 接口的 IP 地址才可以对外发布。但是只有在 LoopBack 接口上配置 32 位掩码的 IP 地址时，Prefix SID 才能生效，并且只对 LoopBack 接口的主 IP 地址生效。

如果 Prefix SID 值超过 SRGB 的范围，则 Prefix SID 不会被发布。

P2 的配置：

```
[P2]segment-routing
[P2-segment-routing]q
[P2]ospf 1
[P2-ospf-1]opaque-capability enable
[P2-ospf-1]segment-routing mpls
[P2-ospf-1]segment-routing global-block 16000 16999
[P2-ospf-1]q
[P2]interface LoopBack0
[P2-LoopBack0]ospf prefix-sid index 2
```

PE3 的配置：

```
[PE3]segment-routing
[PE3-segment-routing]q
[PE3]ospf 1
[PE3-ospf-1]opaque-capability enable
[PE3-ospf-1]segment-routing mpls
[PE3-ospf-1]segment-routing global-block 16000 16999
[PE3-ospf-1]q
[PE3]interface LoopBack0
[PE3-LoopBack0]ospf prefix-sid index 3
```

P4 的配置：

```
[P4]segment-routing
[P4-segment-routing]q
[P4]ospf 1
[P4-ospf-1]opaque-capability enable
[P4-ospf-1]segment-routing mpls
[P4-ospf-1]segment-routing global-block 16000 16999
[P4-ospf-1]q
[P4]interface LoopBack0
[P4-LoopBack0]ospf prefix-sid index 4
```

通过查看隧道建立情况可以看到，PE1 已经和其他几台设备建立了 SRBE 的 LSP。

```
[PE1]display tunnel-info all
Tunnel ID                    Type                  Destination                Status
-------------------------------------------------------------------------------------
0x000000002900000042         srbe-lsp              2.2.2.2                    UP
0x000000002900000043         srbe-lsp              3.3.3.3                    UP
0x000000002900000045         srbe-lsp              4.4.4.4                    UP
```

查看 OSPF 的 10 类 LSA：

```
[PE1]display ospf lsdb opaque-area self-originate
          OSPF Process 1 with Router ID 12.1.1.1
                          Area: 0.0.0.0
                  Link State Database
  Type       : Opq-Area
  Ls id      : 4.0.0.0
  Adv rtr    : 12.1.1.1
  Ls age     : 215
  Len        : 44
  Options    :  E
  seq#       : 80000002
  chksum     : 0xd58
  Opaque Type: 4
  Opaque Id: 0
  Router-Information LSA TLV information:
    SR-Algorithm TLV:
      Algorithm: SPF
    SID/Label Range TLV:
      Range Size: 1000
      SID/Label Sub-TLV:
        Label: 16000  //本设备的 SRGB。其中，16000 代表起始的标签；Range Size:1000 代表
                      //SRGB 的标签空间大小为 1000，标签取值应为 16000~16999

  Type       : Opq-Area
  Ls id      : 7.0.0.0
  Adv rtr    : 12.1.1.1
  Ls age     : 220
  Len        : 44
```

```
Options    :  E
seq#       : 80000001
chksum     : 0x505a
Opaque Type: 7
Opaque Id: 0
OSPFv2 Extended Prefix Opaque LSA TLV information:
  OSPFv2 Extended Prefix TLV:
    Route Type: Intra-Area
    AF: IPv4-Unicast
    Flags: 0x40 (-|N|-|-|-|-|-|-)
    Prefix: 1.1.1.1/32 // 环回口路由
    Prefix SID Sub-TLV:
      Flags: 0x00 (-|-|-|-|-|-|-|-)
      MT ID: 0
      Algorithm: SPF
      Index: 1 // 设备在环回口设置的偏移值为 1，结合 SRGB 的标签范围，那么此设备的 Node
             //ID 为 16001

Type       : Opq-Area
Ls id      : 8.0.0.2
Adv rtr    : 12.1.1.1
Ls age     : 105
Len        : 52
Options    :  E
seq#       : 80000001
chksum     : 0x7e5e
Opaque Type: 8
Opaque Id: 2
OSPFv2 Extended Link Opaque LSA TLV information:
  OSPFv2 Extended Link TLV:
    Link Type: TransNet
    Link ID: 12.1.1.1
    Link Data: 12.1.1.1
    LAN Adj-SID Sub-TLV:
      Flags: 0x60 (-|V|L|-|-|-|-|-)
      MT ID: 0
      Weight: 0
      Neighbor ID: 12.1.1.2
      Label: 48080 //ospf 为 12.1.1.1 分配的 Adj-SID 为 48020
Type       : Opq-Area
Ls id      : 8.0.0.3
Adv rtr    : 12.1.1.1
Ls age     : 49
Len        : 52
Options    :  E
seq#       : 80000001
chksum     : 0x3c96
Opaque Type: 8
```

```
  Opaque Id: 3
  OSPFv2 Extended Link Opaque LSA TLV information:
OSPFv2 Extended Link TLV:
      Link Type: TransNet
      Link ID: 14.1.1.1
      Link Data: 14.1.1.1
      LAN Adj-SID Sub-TLV:
        Flags: 0x60 (-|V|L|-|-|-|-|-)
        MT ID: 0
        Weight: 0
        Neighbor ID: 14.1.1.4
        Label: 48081 //OSPF 为 14.1.1.1 分配的 ADJ-SID 为 48021
```

步骤 5：配置 VPN 实例。

PE1 的配置：

```
[PE1]ip vpn-instance 1
[PE1-vpn-instance-1]ipv4-family
[PE1-vpn-instance-1-af-ipv4]route-distinguisher 100:1
[PE1-vpn-instance-1-af-ipv4]vpn-target 100:1 export-extcommunity
[PE1-vpn-instance-1-af-ipv4]vpn-target 100:1 import-extcommunity
```

PE3 的配置：

```
[PE3]ip vpn-instance 1
[PE3-vpn-instance-1]ipv4-family
[PE3-vpn-instance-1-af-ipv4]route-distinguisher 100:2
[PE3-vpn-instance-1-af-ipv4]vpn-target 100:1 export-extcommunity
[PE3-vpn-instance-1-af-ipv4]vpn-target 100:1 import-extcommunity
```

步骤 6：建立 MP-BGP 邻居关系。

PE1 的配置：

```
[PE1]bgp 100
[PE1-bgp]peer 3.3.3.3 as-number 100
[PE1-bgp]peer 3.3.3.3 connect-interface LoopBack0
[PE1-bgp]ipv4-family vpnv4
[PE1-bgp-af-vpnv4]peer 3.3.3.3 enable
Warning: This operation will reset the peer session. Continue? [Y/N]:y
```

PE3 的配置：

```
[PE3]bgp 100
[PE3-bgp]peer 1.1.1.1 as-number 100
[PE3-bgp]peer 1.1.1.1 connect-interface LoopBack0
[PE3-bgp]ipv4-family vpnv4
[PE3-bgp-af-vpnv4]peer 1.1.1.1 enable
Warning: This operation will reset the peer session. Continue? [Y/N]:y
```

步骤 7：配置 PE 和 CE 的 BGP 邻居关系。

PE1 的配置：

```
[PE1]interface Ethernet1/0/2
[PE1-Ethernet1/0/2]ip binding vpn-instance 1
```

```
[PE1-Ethernet1/0/2]ip address 10.0.11.1 255.255.255.0
[PE1-Ethernet1/0/2]q
[PE1]bgp 100
[PE1-bgp]ipv4-family vpn-instance 1
[PE1-bgp-1]peer 10.0.11.2 as-number 200
```

PE3 的配置：

```
[PE3]interface Ethernet1/0/2
[PE3-Ethernet1/0/2]ip binding vpn-instance 1
[PE3-Ethernet1/0/2]ip address 10.0.24.1 255.255.255.0
[PE3-Ethernet1/0/2]q
[PE3]bgp 100
[PE3-bgp]ipv4-family vpn-instance 1
[PE3-bgp-1]peer 10.0.24.2 as-number 300
```

CE1 的配置：

```
[CE1]bgp 200
[CE1-bgp]peer 10.0.11.1 as-number 100
[CE1-bgp]network 10.10.10.10 255.255.255.255
```

CE2 的配置：

```
[CE2]bgp 300
[CE2-bgp]peer 10.0.24.1 as-number 100
[CE2-bgp]network 20.20.20.20 255.255.255.255
```

查看 VPNv4 的路由信息：

```
[PE1]display bgp vpnv4 all routing-table
 BGP Local router ID is 12.1.1.1
 Status codes: * - valid, > - best, d - damped, x - best external, a - add path,
               h - history,  i - internal, s - suppressed, S - Stale
               Origin : i - IGP, e - EGP, ? - incomplete
 RPKI validation codes: V - valid, I - invalid, N - not-found
 Total number of routes from all PE: 2
 Route Distinguisher: 100:1
        Network            NextHop        MED      LocPrf      PrefVal      Path/Ogn
 *>     10.10.10.10/32     10.0.11.2      0                    0            200i
 Route Distinguisher: 100:2
        Network            NextHop        MED      LocPrf      PrefVal      Path/Ogn
 *>i    20.20.20.20/32     3.3.3.3        0        100         0            300i

 VPN-Instance 1, Router ID 12.1.1.1:
 Total Number of Routes: 2
        Network            NextHop        MED      LocPrf      PrefVal      Path/Ogn
 *>     10.10.10.10/32     10.0.11.2      0                    0            200i
 *>i    20.20.20.20/32     3.3.3.3        0        100         0            300i
```

通过以上输出可以看出，能够正常学习到对端的路由信息。

查看 MPLS 隧道，以下输出表示，PE3 设备为本端 CE2 设备的 20.20.20.20/32 路由分配了私网标签 48120。

```
[PE3]display  mpls lsp
-------------------------------------------------------------------------------
                 LSP Information: BGP LSP
-------------------------------------------------------------------------------
FEC                      In/Out Label          In/Out IF              Vrf Name
20.20.20.20/32           48120/NULL            -/-                    1
```

步骤 8：配置隧道选择 SR LSP。

PE1 的配置：

```
[PE1]tunnel-policy 1
Info: New tunnel-policy is configured.
[PE1-tunnel-policy-1]tunnel select-seq sr-lsp load-balance-number 2
[PE1-tunnel-policy-1]q
[PE1]ip vpn-instance 1
[PE1-vpn-instance-1]ipv4-family
[PE1-vpn-instance-1-af-ipv4]tnl-policy 1
```

PE3 的配置：

```
[PE3]tunnel-policy 1
Info: New tunnel-policy is configured.
[PE3-tunnel-policy-1]tunnel select-seq sr-lsp load-balance-number 2
[PE3-tunnel-policy-1]q
[PE3]ip vpn-instance 1
[PE3-vpn-instance-1]ipv4-family
[PE3-vpn-instance-1-af-ipv4]tnl-policy 1
```

步骤 9：测试。

```
<CE1>ping -a 10.10.10.10 20.20.20.20
  PING 20.20.20.20: 56  data bytes, press CTRL_C to break
    Reply from 20.20.20.20: bytes=56 Sequence=1 ttl=252 time=30 ms
    Reply from 20.20.20.20: bytes=56 Sequence=2 ttl=252 time=30 ms
    Reply from 20.20.20.20: bytes=56 Sequence=3 ttl=252 time=30 ms
    Reply from 20.20.20.20: bytes=56 Sequence=4 ttl=252 time=30 ms
    Reply from 20.20.20.20: bytes=56 Sequence=5 ttl=252 time=20 ms
  --- 20.20.20.20 ping statistics ---
    5 packet(s) transmitted
    5 packet(s) received
    0.00% packet loss
round-trip min/avg/max = 20/28/30 ms
```

在 PE1 的 E0/0/0 接口或 E0/0/1 接口抓包查看结果，如图 15-12 所示。

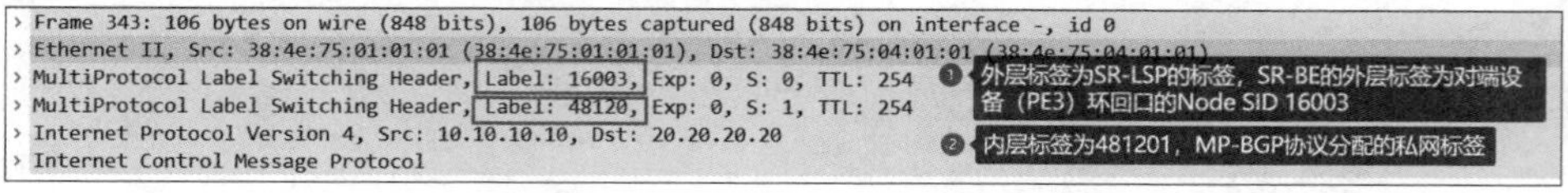

```
> Frame 343: 106 bytes on wire (848 bits), 106 bytes captured (848 bits) on interface -, id 0
> Ethernet II, Src: 38:4e:75:01:01:01 (38:4e:75:01:01:01), Dst: 38:4e:75:04:01:01 (38:4e:75:04:01:01)
> MultiProtocol Label Switching Header, Label: 16003, Exp: 0, S: 0, TTL: 254
> MultiProtocol Label Switching Header, Label: 48120, Exp: 0, S: 1, TTL: 254
> Internet Protocol Version 4, Src: 10.10.10.10, Dst: 20.20.20.20
> Internet Control Message Protocol
```

❶ 外层标签为SR-LSP的标签，SR-BE的外层标签为对端设备（PE3）环回口的Node SID 16003

❷ 内层标签为481201，MP-BGP协议分配的私网标签

图 15-12　PE1 的抓包结果

通过以上输出可以看出，内层标签为 48120，外层标签为 16003。其中，内层标签由 BGP 分配；外层标签由 OSPF 分配，其具体信息可以通过查看以下路由表得出。

```
[PE1]display ip routing-table vpn-instance 1 20.20.20.20 verbose
Route Flags: R - relay, D - download to fib, T - to vpn-instance, B - black hole
route
------------------------------------------------------------------------------
Routing Table : 1
Summary Count : 1
Destination: 20.20.20.20/32
     Protocol: IBGP                  Process ID: 0
   Preference: 255                         Cost: 0
      NextHop: 3.3.3.3                Neighbour: 3.3.3.3
        State: Active Adv Relied            Age: 00h06m27s
          Tag: 0                       Priority: low
        Label: 48120                    QoSInfo: 0x0
   IndirectID: 0x10000A4               Instance:
 RelayNextHop: 12.1.1.2               Interface: Ethernet1/0/0
     TunnelID: 0x000000002900000043       Flags: RD
 RelayNextHop: 14.1.1.4               Interface: Ethernet1/0/1
     TunnelID: 0x000000002900000043       Flags: RD
```

通过查看VPN实例中20.20.20.20/32的路由表可知，流量将迭代到隧道中，隧道ID为0x000000002900000043。

查看隧道信息：

```
[PE1]display tunnel-info all
Tunnel ID                  Type                 Destination             Status
----------------------------------------------------------------------------------
0x000000002900000042       srbe-lsp             2.2.2.2                 UP
0x000000002900000043       srbe-lsp             3.3.3.3                 UP
0x000000002900000045       srbe-lsp             4.4.4.4                 UP
```

通过以上输出可以看出，隧道ID为0x000000002900000043的目的端是IP地址为3.3.3.3的设备，即PE3。

```
[PE3]display ospf lsdb opaque-area self-originate
          OSPF Process 1 with Router ID 23.1.1.3
                          Area: 0.0.0.0
                    Link State Database
  Type       : Opq-Area
  Ls id      : 4.0.0.0
  Adv rtr    : 23.1.1.3
  Ls age     : 962
  Len        : 44
  Options    :  E
  seq#       : 80000003
  chksum     : 0x9bbb
  Opaque Type: 4
  Opaque Id: 0
  Router-Information LSA TLV information:
    SR-Algorithm TLV:
      Algorithm: SPF
```

```
   SID/Label Range TLV:
     Range Size: 1000
     SID/Label Sub-TLV:
       Label: 16000 //此处表示PE3的SRGB的范围为16000~16999
 Type         : Opq-Area
 Ls id        : 7.0.0.0
 Adv rtr      : 23.1.1.3
 Ls age       : 967
 Len          : 44
 Options      : E
 seq#         : 80000002
 chksum       : 0x771b
 Opaque Type: 7
 Opaque Id: 0
 OSPFv2 Extended Prefix Opaque LSA TLV information:
   OSPFv2 Extended Prefix TLV:
     Route Type: Intra-Area
     AF: IPv4-Unicast
     Flags: 0x40 (-|N|-|-|-|-|-|-)
     Prefix: 3.3.3.3/32
     Prefix SID Sub-TLV:
       Flags: 0x00 (-|-|-|-|-|-|-|-)
       MT ID: 0
       Algorithm: SPF
       Index: 3 //PE3的环回口偏移值为3，那么Node SID为16003
```

15.2.2 配置 SR MPLS TE

扫一扫，看视频

1. 实验目的

通过运营商网络配置SR MPLS TE，使CE1和CE2之间的互访流量通过PE1→P2→P4→PE3。

2. 实验拓扑

配置 SR MPLS TE 的实验拓扑如图 15-13 所示。

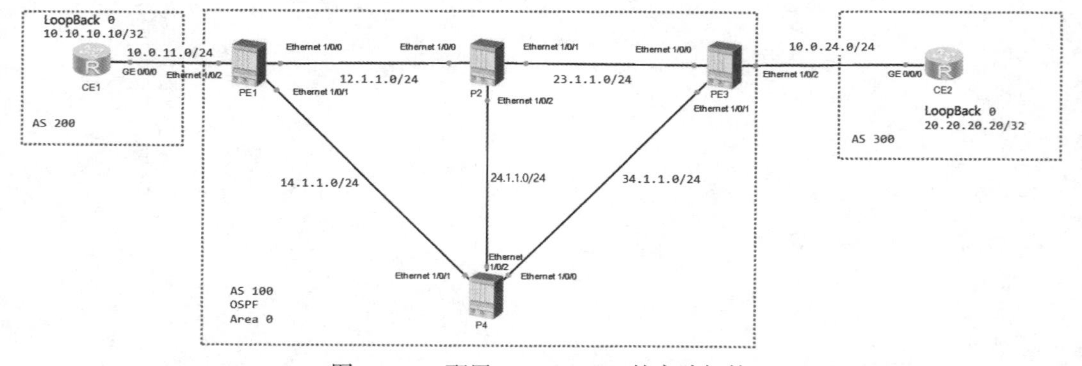

图 15-13 配置 SR MPLS TE 的实验拓扑

3. 实验步骤

步骤 1：配置设备接口的 IP 地址。配置 SR MPLS TE IP 地址规划表见表 15-2。

表 15-2　配置 SR MPLS TE IP 地址规划表

设备名称	接口编号	IP 地址
PE1	Ethernet 1/0/0	12.1.1.1/24
	Ethernet 1/0/1	14.1.1.1/24
	LoopBack 0	1.1.1.1/32
	Ethernet 1/0/2	10.0.11.1/24
P2	Ethernet 1/0/0	12.1.1.2/24
	Ethernet 1/0/1	23.1.1.2/24
	LoopBack 0	2.2.2.2/32
	Ethernet 1/0/2	24.1.1.2/24
PE3	Ethernet 1/0/0	23.1.1.3/24
	Ethernet 1/0/1	34.1.1.3/24
	LoopBack 0	3.3.3.3/24
	Ethernet 1/0/2	10.0.24.1/24
P4	Ethernet 1/0/0	34.1.1.4/24
	Ethernet 1/0/1	14.1.1.4/24
	LoopBack 0	4.4.4.4/32
	Ethernet 1/0/2	24.1.1.4/24
CE1	G0/0/0	10.0.11.2/24
	LoopBack 0	10.10.10.10/32
CE2	G0/0/0	10.0.24.2/24
	LoopBack 0	20.20.20.20/32

步骤 2：配置 AS 100 内的 IGP 协议。

PE1 的配置：

```
[PE1]isis 1
[PE1-isis-1]is-level level-2
[PE1-isis-1]cost-style wide
[PE1-isis-1]network-entity 49.0001.0000.0000.0001.00
[PE1-isis-1]q
[PE1]interface Ethernet1/0/0
```

```
[PE1-Ethernet1/0/0]isis enable 1
[PE1-Ethernet1/0/0]q
[PE1]interface Ethernet1/0/1
[PE1-Ethernet1/0/1]isis enable 1
[PE1-Ethernet1/0/1]q
[PE1]interface LoopBack0
[PE1-LoopBack0]isis enable 1
```

P2 的配置：

```
[P2]isis 1
[P2-isis-1]is-level level-2
[P2-isis-1]cost-style wide
[P2-isis-1]network-entity 49.0001.0000.0000.0002.00
[P2-isis-1]q
[P2]interface Ethernet1/0/0
[P2-Ethernet1/0/0]isis enable 1
[P2-Ethernet1/0/0]q
[P2]interface Ethernet1/0/1
[P2-Ethernet1/0/1]isis enable 1
[P2-Ethernet1/0/1]interface Ethernet1/0/2
[P2-Ethernet1/0/2]isis enable 1
[P2-Ethernet1/0/2]q
[P2]interface LoopBack0
[P2-LoopBack0]isis enable 1
```

PE3 的配置：

```
[PE3]isis 1
[PE3-isis-1]is-level level-2
[PE3-isis-1]cost-style wide
[PE3-isis-1]network-entity 49.0001.0000.0000.0003.00
[PE3-isis-1]q
[PE3]interface Ethernet1/0/0
[PE3-Ethernet1/0/0]isis enable 1
[PE3-Ethernet1/0/0]q
[PE3]interface Ethernet1/0/1
[PE3-Ethernet1/0/1]isis enable 1
[PE3-Ethernet1/0/1]q
[PE3]interface LoopBack0
[PE3-LoopBack0]isis enable 1
```

P4 的配置：

```
[P4]isis 1
[P4-isis-1]is-level level-2
[P4-isis-1]cost-style wide
[P4-isis-1]network-entity 49.0001.0000.0000.0004.00
[P4-isis-1]q
[P4]interface Ethernet1/0/0
[P4-Ethernet1/0/0]isis enable 1
```

```
[P4-Ethernet1/0/0]q
[P4]interface Ethernet1/0/1
[P4-Ethernet1/0/1]isis enable 1
[P4-Ethernet1/0/1]q
[P4]interface Ethernet1/0/2
[P4-Ethernet1/0/2]isis enable 1
[P4-Ethernet1/0/2]q
[P4]interface LoopBack0
[P4-LoopBack0]isis enable 1
```

步骤 3: 开启 AS 100 内的 MPLS 功能。

PE1 的配置:

```
[PE1]mpls lsr-id 1.1.1.1
[PE1]mpls
[PE1-mpls]mpls te
```

P2 的配置:

```
[P2]mpls lsr-id 2.2.2.2
[P2]mpls
[P2-mpls]mpls te
```

PE3 的配置:

```
[PE3]mpls lsr-id 3.3.3.3
[PE3]mpls
[PE3-mpls]mpls te
```

P4 的配置

```
[P4]mpls lsr-id 4.4.4.4
[P4]mpls
[P4-mpls]mpls te
```

步骤 4: 开启 SR 功能，配置 IS-IS 的 SRGB，并且为设备的环回口路由配置 Node SID。

PE1 的配置:

```
[PE1]segment-routing
[PE1-segment-routing]q
[PE1]Isis
[PE1-isis-1]segment-routing mpls
[PE1-isis-1]segment-routing global-block 16000 16999
[PE1-isis-1]q
[PE1]interface LoopBack0
[PE1-LoopBack0]isis prefix-sid index 1
```

P2 的配置:

```
[P2]segment-routing
[P2-segment-routing]q
[P2]Isis
[P2-isis-1]segment-routing mpls
[P2-isis-1]segment-routing global-block 16000 16999
[P2-isis-1]q
```

```
[P2]interface LoopBack0
[P2-LoopBack0]isis enable 1
[P2-LoopBack0]isis prefix-sid index 2
```

PE3 的配置：

```
[PE3]segment-routing
[PE3-segment-routing]q
[PE3]Isis
[PE3-isis-1]segment-routing mpls
[PE3-isis-1]segment-routing global-block 16000 16999
[PE3-isis-1]q
[PE3]interface LoopBack0
[PE3-LoopBack0]isis enable 1
[PE3-LoopBack0]isis prefix-sid index 3
```

P4 的配置：

```
[P4]segment-routing
[P4-segment-routing]q
[P4]Isis
[P4-isis-1]segment-routing mpls
[P4-isis-1]segment-routing global-block 16000 16999
[P4-isis-1]q
[P4]interface LoopBack0
[P4-LoopBack0]isis enable 1
[P4-LoopBack0]isis prefix-sid index 4
```

步骤 5：配置 SR 的显示路径。

（1）配置 P2 和 P4 之间互联链路的邻接端 SID。

P2 的配置：

```
[P2]segment-routing
[P2-segment-routing]ipv4 adjacency local-ip-addr 24.1.1.2 remote-ip-addr 24.1.1.4
sid 330000
```

查看设备的邻接端 SID，其中 330000 是通过上述操作静态配置的，其他的三个是通过 IS-IS 自动分配的：

```
[P2]display segment-routing adjacency mpls forwarding
            Segment Routing Adjacency MPLS Forwarding Information
Label     Interface          NextHop          Type         MPLSMtu           Mtu
--------------------------------------------------------------------------------
48020     Eth1/0/0           12.1.1.1         ISIS-V4      ---              1500
48021     Eth1/0/1           23.1.1.3         ISIS-V4      ---              1500
48022     Eth1/0/2           24.1.1.4         ISIS-V4      ---              1500
330000    Eth1/0/2           24.1.1.4         STATIC-V4    ---              1500
```

P4 的配置：

```
[P4]segment-routing
[P4-segment-routing]ipv4 adjacency local-ip-addr 24.1.1.4 remote-ip-addr 24.1.1.2
sid 330001
```

查看设备的邻接端 SID，其中 330001 是通过上述操作静态配置的，其他的三个是通过 IS-IS 自动分配的。

```
[P4]display segment-routing adjacency mpls forwarding
           Segment Routing Adjacency MPLS Forwarding Information
Label      Interface           NextHop           Type         MPLSMtu            Mtu
-------------------------------------------------------------------------------------
48020      Eth1/0/1            14.1.1.1          ISIS-V4        ---              1500
48021      Eth1/0/2            24.1.1.2          ISIS-V4        ---              1500
48022      Eth1/0/0            34.1.1.3          ISIS-V4        ---              1500
330001     Eth1/0/2            24.1.1.2          STATIC-V4      ---              1500
```

（2）在 IS-IS 中开启 MPLS TE 功能。

PE1 的配置：

```
[PE1]Isis
[PE1-isis-1]traffic-eng level-2
```

P2 的配置：

```
[P2]Isis
[P2-isis-1]traffic-eng level-2
```

PE3 的配置：

```
[PE3]Isis
[PE3-isis-1]traffic-eng level-2
```

P4 的配置：

```
[P4]Isis
[P4-isis-1]traffic-eng level-2
```

（3）配置显示路径。

PE1 的配置：

```
[PE1]explicit-path pe1-pe3
[PE1-explicit-path-pe1-pe3]next sid label 16002 type prefix
[PE1-explicit-path-pe1-pe3]next sid label 330000 type adjacency
[PE1-explicit-path-pe1-pe3]next sid label 16003 type prefix
[PE1-explicit-path-pe1-pe3]q
[PE1]interface Tunnel1
[PE1-Tunnel1]ip address unnumbered interface LoopBack0
[PE1-Tunnel1]tunnel-protocol mpls te
[PE1-Tunnel1]destination 3.3.3.3
[PE1-Tunnel1]mpls te signal-protocol segment-routing
[PE1-Tunnel1]mpls te tunnel-id 1
[PE1-Tunnel1]mpls te path explicit-path pe1-pe3
```

PE3 的配置：

```
[PE3]explicit-path pe3-pe1
[PE3-explicit-path-pe3-pe1]next sid label 16004 type prefix
[PE3-explicit-path-pe3-pe1]next sid label 330001 type adjacency
[PE3-explicit-path-pe3-pe1]next sid label 16001 type prefix
```

```
[PE3-explicit-path-pe3-pe1]
[PE3]interface Tunnel1
[PE3-Tunnel1]ip address unnumbered interface LoopBack0
[PE3-Tunnel1]tunnel-protocol mpls te
[PE3-Tunnel1]destination 1.1.1.1
[PE3-Tunnel1]mpls te signal-protocol segment-routing
[PE3-Tunnel1]mpls te tunnel-id 1
[PE3-Tunnel1]mpls te path explicit-path pe3-pe1
[PE1]display  tunnel-info all
Tunnel ID                    Type                Destination              Status
----------------------------------------------------------------------------------
0x000000000300000001         sr-te               3.3.3.3                  UP
0x000000002900000004         srbe-lsp            4.4.4.4                  UP
0x000000002900000005         srbe-lsp            2.2.2.2                  UP
0x000000002900000006         srbe-lsp            3.3.3.3                  UP
```

步骤6: 配置PE之间的MP-BGP邻居。

PE1的配置:

```
[PE1]bgp 100
[PE1-bgp]peer 3.3.3.3 as-number 100
[PE1-bgp]peer 3.3.3.3 connect-interface LoopBack0
[PE1-bgp]ipv4-family vpnv4
[PE1-bgp-af-vpnv4]policy vpn-target
[PE1-bgp-af-vpnv4]peer 3.3.3.3 enable
```

PE3的配置:

```
[PE3]bgp 100
[PE3-bgp]peer 1.1.1.1 as-number 100
[PE3-bgp]peer 1.1.1.1 connect-interface LoopBack0
[PE3-bgp]ipv4-family vpnv4
[PE3-bgp-af-vpnv4]policy vpn-target
[PE3-bgp-af-vpnv4]peer 1.1.1.1 enable
```

步骤7: 配置PE和CE之间的BGP邻居。

PE1的配置:

```
[PE1]ip vpn-instance 1
[PE1-vpn-instance-1]ipv4-family
[PE1-vpn-instance-1-af-ipv4]route-distinguisher 100:1
[PE1-vpn-instance-1-af-ipv4]vpn-target 100:1 export-extcommunity
[PE1-vpn-instance-1-af-ipv4]vpn-target 100:1 import-extcommunity
[PE1-vpn-instance-1-af-ipv4]q
[PE1-vpn-instance-1]interface Ethernet1/0/2
[PE1-Ethernet1/0/2]ip binding vpn-instance 1
[PE1-Ethernet1/0/2]ip address 10.0.11.1 255.255.255.0
[PE1-Ethernet1/0/2]q
[PE1]bgp 100
[PE1-bgp]ipv4-family vpn-instance 1
[PE1-bgp-1]peer 10.0.11.2 as-number 200
```

PE3 的配置：

```
[PE3]ip vpn-instance 1
[PE3-vpn-instance-1]ipv4-family
[PE3-vpn-instance-1-af-ipv4]route-distinguisher 100:2
[PE3-vpn-instance-1-af-ipv4]vpn-target 100:1 export-extcommunity
[PE3-vpn-instance-1-af-ipv4]vpn-target 100:1 import-extcommunity
[PE3-vpn-instance-1-af-ipv4]q
[PE3-vpn-instance-1]
[PE3-vpn-instance-1]interface Ethernet1/0/2
[PE3-Ethernet1/0/2]ip binding vpn-instance 1
[PE3-Ethernet1/0/2]ip address 10.0.24.1 255.255.255.0
[PE3-Ethernet1/0/2]q
[PE3]bgp 100
[PE3-bgp]ipv4-family vpn-instance 1
[PE3-bgp-1]peer 10.0.24.2 as-number 300
```

CE1 的配置：

```
[CE1]bgp 200
[CE1-bgp]peer 10.0.11.1 as-number 100
[CE1-bgp]network 10.10.10.10 255.255.255.255
```

CE2 的配置：

```
[CE2-bgp]peer 10.0.24.1 as-number 100
[CE2-bgp]network 20.20.20.20 255.255.255.255
```

步骤 8: 配置隧道选择 SR-TE。

PE1 的配置：

```
[PE1]tunnel-policy 1
[PE1-tunnel-policy-1]tunnel select-seq sr-te load-balance-number 1
[PE1-tunnel-policy-1]q
[PE1]ip vpn-instance 1
[PE1-vpn-instance-1]ipv4-family
[PE1-vpn-instance-1-af-ipv4]tnl-policy 1
```

PE3 的配置：

```
[PE3]tunnel-policy 1
[PE3-tunnel-policy-1]tunnel select-seq sr-te load-balance-number 1
[PE3-tunnel-policy-1]q
[PE3]ip vpn-instance 1
[PE3-vpn-instance-1]ipv4-family
[PE3-vpn-instance-1-af-ipv4]tnl-policy 1
```

步骤 9: 测试。

```
[CE1]ping -a 10.10.10.10 20.20.20.20
  PING 20.20.20.20: 56  data bytes, press CTRL_C to break
    Reply from 20.20.20.20: bytes=56 Sequence=1 ttl=250 time=40 ms
    Reply from 20.20.20.20: bytes=56 Sequence=2 ttl=250 time=30 ms
    Reply from 20.20.20.20: bytes=56 Sequence=3 ttl=250 time=30 ms
```

```
    Reply from 20.20.20.20: bytes=56 Sequence=4 ttl=250 time=30 ms
    Reply from 20.20.20.20: bytes=56 Sequence=5 ttl=250 time=40 ms
  --- 20.20.20.20 ping statistics ---
    5 packet(s) transmitted
    5 packet(s) received
    0.00% packet loss
round-trip min/avg/max = 30/34/40 ms
```

在 PE1 的 E1/0/0 接口的抓包结果如图 15-14 所示。

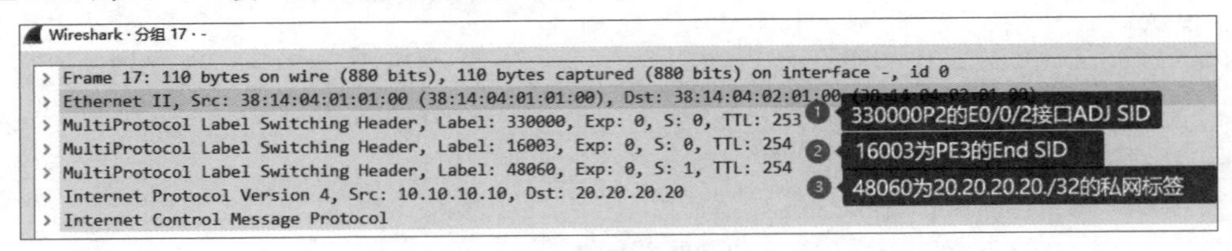

图 15-14　在 PE1 的 E1/0/0 接口的抓包结果

第16章

SRv6

16.1 SRv6 概述

SRv6（Segment Routing IPv6，基于 IPv6 转发平面的段路由）是一种基于源路由理念设计的在网络上转发 IPv6 数据包的协议。SRv6 通过在 IPv6 报文中插入一个 SRH（Segment Routing Header，路由扩展头），然后在 SRH 中压入一个显式的 IPv6 地址栈，并由中间节点不断更新目的地址和偏移地址栈的操作来完成逐跳转发。

SRv6 技术具有以下优点。

（1）简化网络配置，更简易地实现 VPN。SRv6 基于 IPv6 转发，不使用 MPLS 技术，完全兼容现有 IPv6 网络。中间 Transit 节点可以不支持 SRv6，按照正常路由转发含有 SRH 的 IPv6 报文。

（2）提供高保护率的 FRR 保护能力。在 SRv6 技术的基础上结合 RLFA FRR 算法，形成高效的 TI-LFA FRR 算法，原理上支持任意拓扑保护，能够弥补传统隧道保护技术的不足。

（3）便于 IPv6 转发路径的流量调优。各种服务类型的 SID 搭配使用，头节点可以灵活规划显式路径，调整对应的业务流量。

16.1.1 SRv6 原理简介

SRv6 在头节点上通过将数据压入 SRH 来指导数据转发，其原理示意图如图 16-1 所示。

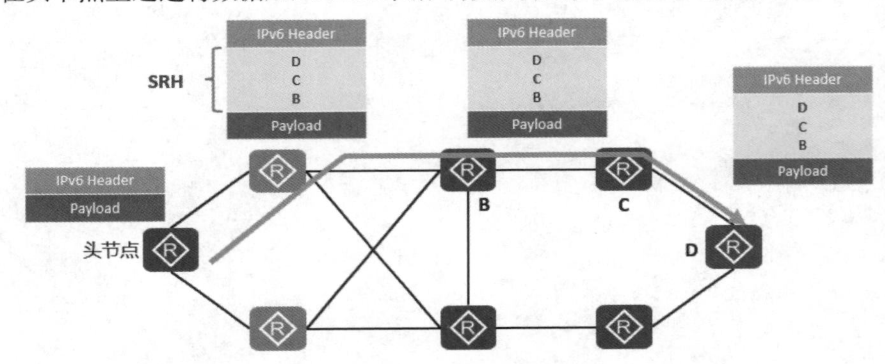

图 16-1　SRv6 原理示意图

SRv6 报文没有改变原有 IPv6 报文的封装结构，SRv6 报文仍旧是 IPv6 报文，普通的 IPv6 设备也可以识别，所以 SRv6 是 Native IPv6 技术。

SRv6 的 Native IPv6 特质使 SRv6 设备能够和普通 IPv6 设备共同组网，对现有网络具有更好的兼容性。

16.1.2 SRv6 基本概念

（1）IPv6 SRH。RFC 8754 中定义了 IPv6 SRH 标准，在 IPv6 报文增加一个 SRH。IPv6 SRH 格式如图 16-2 所示。

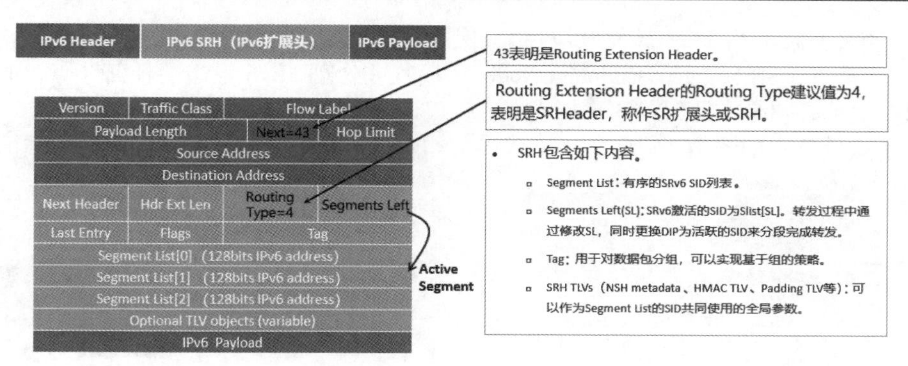

图 16–2　IPv6 SRH 格式

（2）SRv6 Segment。SRv6 Segment 是 IPv6 地址形式，通常也可以称为 SRv6 SID（Segment Identifier）。

如图 16–3 所示，SRv6 SID 由 Locator、Function 和 Arguments 三部分组成，格式为 Locator:Function:Arguments。Length(L+F+A) ≤ 128，当长度和小于 128 时，保留位用 0 补齐。

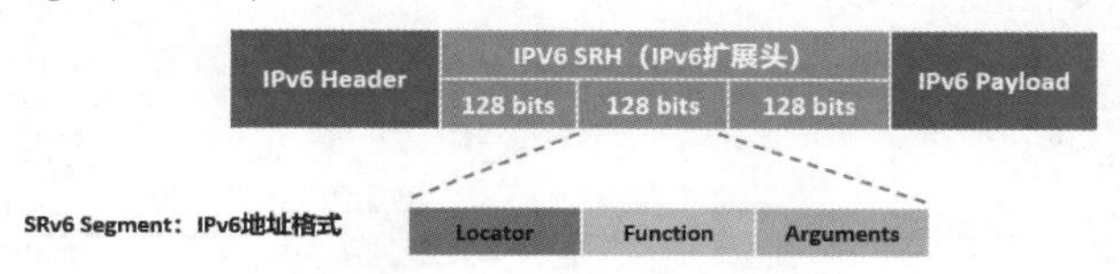

图 16–3　SRv6 Segment 示意图

如果没有 Arguments 字段，则格式是 Locator:Function。Locator 占据 IPv6 地址的高比特位，Function 占据 IPv6 地址的剩余部分。

① SRv6 Segment: Locator。如图 16–4 所示，Locator 是网络拓扑中的一个网络节点的标识，用于路由和转发报文到该节点，实现网络指令的可寻址。

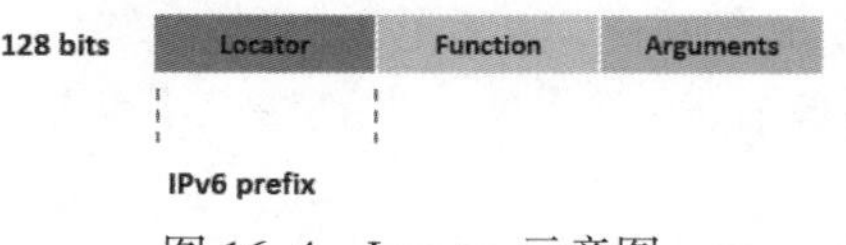

图 16–4　Locator 示意图

Locator 标识的位置信息有两个重要的属性：可路由和可聚合。节点配置 Locator 之后，系统会生成一条 Locator 网段路由，并且通过 IGP 在 SR 域内扩散。网络内其他节点通过 Locator 网段路由就可以定位到本节点，同时本节点发布的所有 SRv6 SID 也都可以通过该条 Locator 网段路由到达。

② SRv6 Segment: Function & Arguments。如图 16–5 所示，Function 用于标识该指令要执行的转发动作。在 SRv6 网络编程中，不同的转发行为由不同的 Function 来标识。例如，在 RFC 中定义了公认的 End、End.X、End.DX4、End.DX6 等。

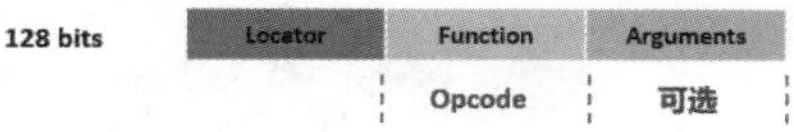

128 bits　Locator　Function　Arguments
Opcode　可选

图 16–5　Function & Arguments 示意图

（3）SRv6 节点。RFC8754 中定义了 SRv6 有以下三种类型的节点。

源节点（SR Source Node）：生成 SRv6 报文的源节点。

中转节点（Transit Node）：转发 SRv6 报文但不进行 SRv6 处理的 IPv6 节点。

Endpoint 节点（SR Segment Endpoint Node）：接收并处理 SRv6 报文的任何节点，其中该报文的 IPv6 目标地址必须是本地配置的 SID 或本地接口地址。

SRv6 节点示意图如图 16-6 所示。

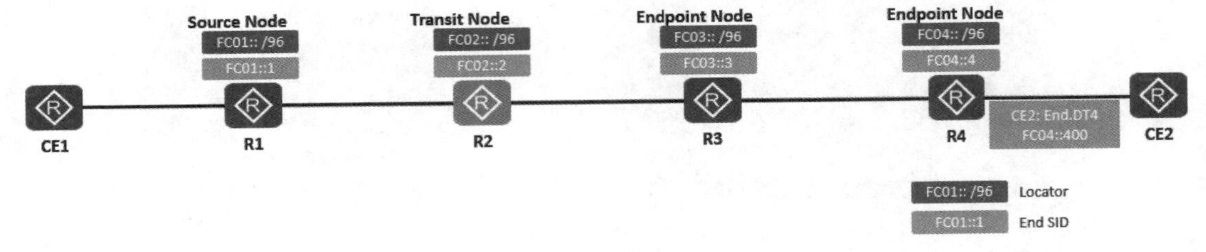

图 16-6　SRv6 节点示意图

①源节点。源节点将数据包引导到 SRv6 Segment List 中，如果 SRv6 Segment List 只包含单个 SID，并且无须在 SRv6 报文中添加信息或 TLV 字段，则 SRv6 报文的目的地址字段设置为该 SID。

源节点可以是生成 IPv6 报文且支持 SRv6 的主机，也可以是 SRv6 域的边缘设备。SRv6 源节点工作示意图如图 16-7 所示。

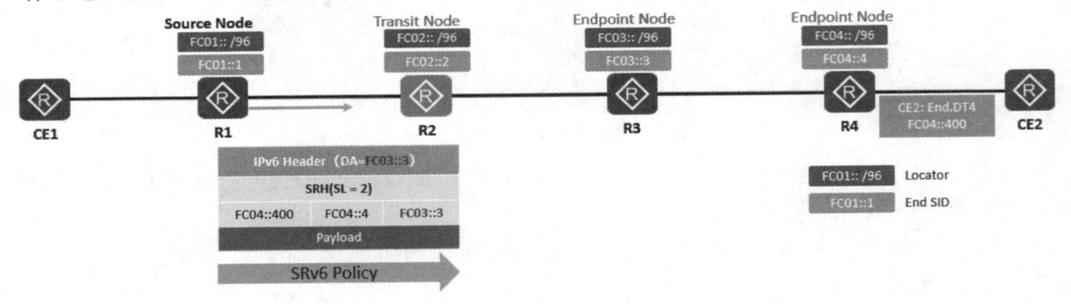

图 16-7　SRv6 源节点工作示意图

②中转节点。中转节点是指在 SRv6 报文转发路径上不参与 SRv6 处理的 IPv6 节点，中转节点只执行普通的 IPv6 报文转发。

节点在收到 SRv6 报文后解析报文的 IPv6 目的地址字段。如果 IPv6 目的地址既不是本地配置的 SRv6 SID，也不是本地接口地址，则节点将 SRv6 报文当作普通 IPv6 报文查询路由表执行转发，不处理 SRH。

中转节点可以是普通的 IPv6 节点，也可以是支持 SRv6 的节点，如图 16-8 所示。

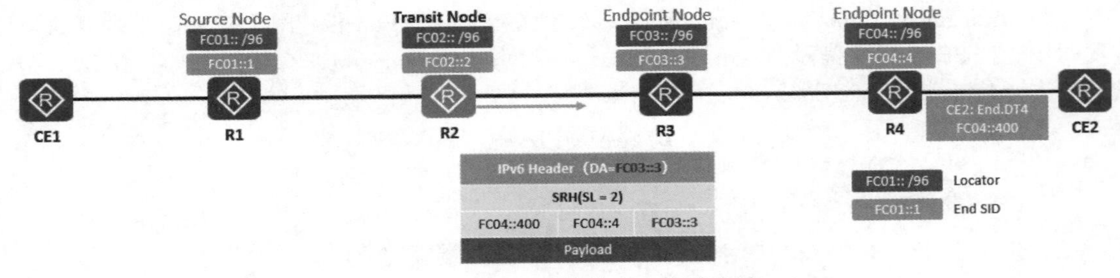

图 16-8　SRv6 中转节点工作示意图

③ Endpoint 节点。在 SRv6 报文转发过程中，节点接收报文的 IPv6 目的地址是本地配置的 SID，则节点被称为 Endpoint 节点。

例如，R3 使用报文的 IPv6 目的地址 FC03::3 查找 Local SID 表，命中到 End SID，然后 R3 将报文 SL 减 1，同时将 SL=1 的 SID 作为报文目标 IPv6 地址，并查路由表转发出去。

Endpoint 节点在数据的转发路径上可以有多个，每个 Endpoint 节点都会为报文提供转发、封装和解封装等服务。

SRv6 Endpoint 节点工作示意图如图 16-9 所示。

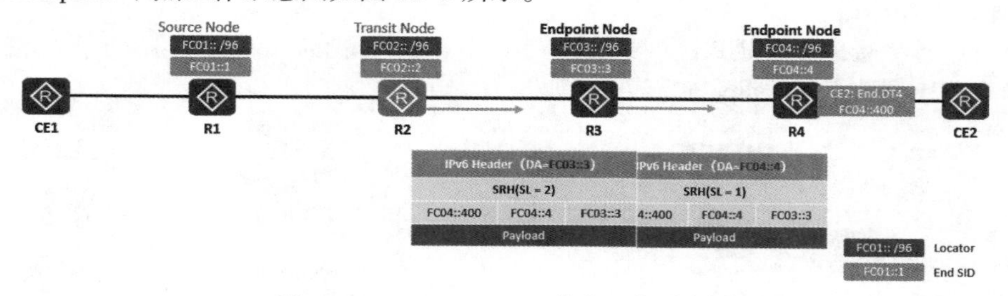

图 16-9　SRv6 Endpoint 节点工作示意图

Endpoint 节点行为见表 16-1。

表 16-1　Endpoint 节点行为

类　型	功能描述	发布协议	类　型
End	Endpoint SID，用于标识网络中的某个目的节点。对应的转发动作（Function）是：更新 IPv6 DA，查找 IPv6 FIB 进行报文转发	IGP	路径 SID
End.X	三层交叉连接的 Endpoint SID，用于标识网络中的某条链路。对应的转发动作是：更新 IPv6 DA，从 End.X SID 绑定的出接口转发报文	IGP	路径 SID
End.DT4	PE 类型的 Endpoint SID，用于标识网络中的某个 IPv4 VPN 实例。对应的转发动作是：解封装报文，并且查找 IPv4 VPN 实例路由表转发。End.DT4 SID 在 L3VPNv4 场景中使用，等价于 IPv4 VPN 的标签	BGP	业务 SID
End.DT6	PE 类型的 Endpoint SID，用于标识网络中的某个 IPv6 VPN 实例。对应的转发动作是：解封装报文，并且查找 IPv6 VPN 实例路由表转发。End.DT6 SID 在 L3VPNv6 场景中使用，等价于 IPv6 VPN 的标签	BGP	业务 SID
End.DX4	PE 类型的三层交叉连接的 Endpoint SID，用于标识网络中的某个 IPv4 CE。对应的转发动作是：解封装报文，并且将解封后的 IPv4 报文在该 SID 绑定的三层接口上转发。End.DX4 SID 在 L3VPNv4 场景中使用，等价于连接到 CE 的邻接标签	BGP	业务 SID
End.DX6	PE 类型的三层交叉连接的 Endpoint SID，用于标识网络中的某个 IPv6 CE。对应的转发动作是：解封装报文，并且将解封后的 IPv6 报文在该 SID 绑定的三层接口上转发。End.DX6 SID 在 L3VPNv6 场景中使用，等价于连接到 CE 的邻接标签	BGP	业务 SID

16.1.3 SRv6 转发模式

SRv6 有以下两种转发模式：SRv6 Policy 和 SRv6 BE。其中，SRv6 Policy 可以实现流量工程，配合控制器可以更好地响应业务的差异化需求，做到业务驱动网络；SRv6 BE 是一种简化的 SRv6 实现，正常情况下不含有 SRH，只能提供尽力而为的转发。

1. SRv6 Locator 信息扩散

无论是基于 SRv6 BE 还是基于 SRv6 Policy 转发流量，路由器都需要有 SRv6 Locator 的相关路由信息才能转发 SRv6 报文。

SRv6 节点通常通过扩展 IGP（扩展 OSPFv3 或 IS-IS），将 Locator 相关路由扩散到网络节点上（包括源节点、中转节点和 Endpoint 节点），如图 16-10 所示。

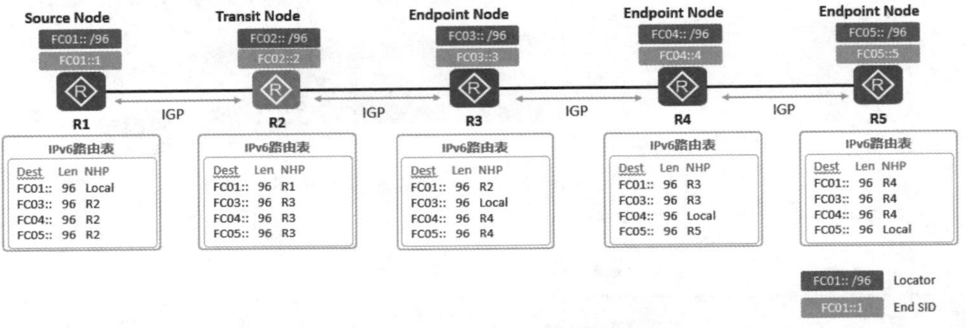

图 16-10　SRv6 Locator 信息扩散示意图

2. SRv6 BE 转发过程

传统 MPLS 有 LDP 和 RSVP-TE 两种控制协议。其中，LDP 方式不支持流量工程能力，而是利用 IGP 算路结果建立 LDP LSP 指导转发。在 SRv6 中，也有类似的方式，只不过 SRv6 仅使用一个业务 SID 来指引报文在 IP 网络中进行尽力而为的转发，这种方式就是 SRv6 BE。

SRv6 BE 依据 IGP 开销计算转发路径，其转发过程示意图如图 16-11 所示。

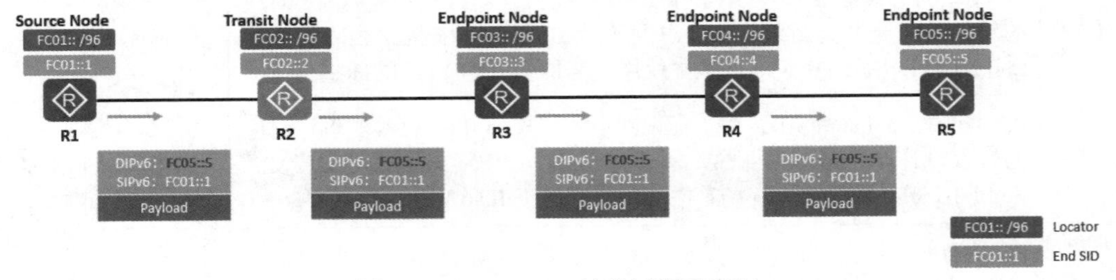

图 16-11　SRv6 BE 转发过程示意图

3. SRv6 Policy 转发过程

在 SRv6 转发过程中每经过一个 SRv6 Endpoint 节点，Segments Left（SL）字段减 1，IPv6 报文头中的目的 IPv6 地址变换一次。Segments Left 和 Segment List 字段共同决定 IPv6 DA 信息。

与 SR-MPLS 不同，SRv6 SRH 是从下到上逆序操作的，SRH 中的 Segment 在经过节点后也不会被弹出。因此 SRv6 报头可以进行路径回溯。

SRv6 Policy 是一种在 SRv6 技术基础上发展的新的引流技术，SRv6 Policy 转发过程示意图如图 16-12 所示。

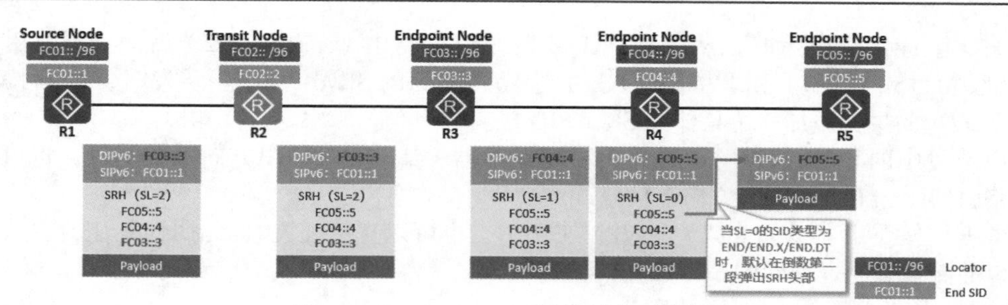

图 16-12 SRv6 Policy 转发过程示意图

SRv6 Policy 路径表示为指定路径的段列表（Segment List），称为 SID 列表（Segment ID List）。每个 SID 列表是从源到目的地的端到端路径，并指示网络中的设备遵循指定的路径，而不是遵循 IGP 计算的最短路径。

如果数据包被导入 SRv6 Policy，SID 列表由头端添加到数据包上，网络的其余设备执行 SID 列表中嵌入的指令。

4. 标识 SRv6 Policy

一个 SRv6 Policy 由一个元组标识，即 <Headend, Color, Endpoint>。

对于一个指定的源节点，SRv6 Policy 由 <Color, Endpoint> 标识。

- 头端（Headend）：SRv6 Policy 生成的节点，一般是全局唯一的 IP 地址。
- 颜色（Color）：32 比特扩展团体属性，用于标识某一种业务意图（如低延时）。
- 尾端（Endpoint）：SRv6 Policy 的目的地址，一般是全局唯一的 IPv6 地址。

在特定头端，Color 和 Endpoint 用于标识 SRv6 Policy 转发路径，如图 16-13 所示。

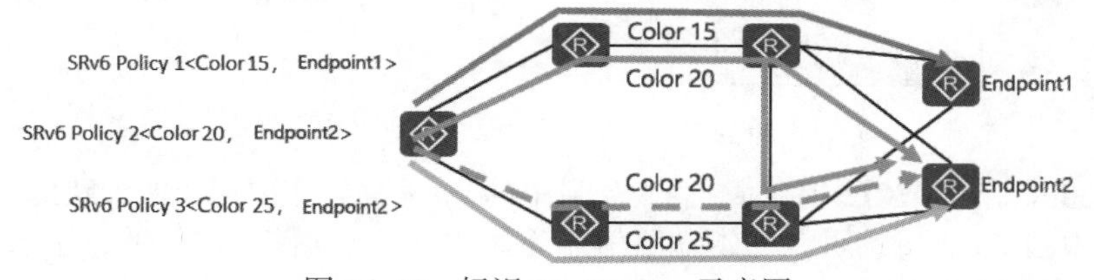

图 16-13 标识 SRv6 Policy 示意图

5. SRv6 Policy 基于 Color 引流

SRv6 Policy 可以基于路由 Color 引流。基于 Color 引流是直接基于路由的扩展团体属性 Color 和目的地址迭代到 SRv6 Policy，如图 16-14 所示。

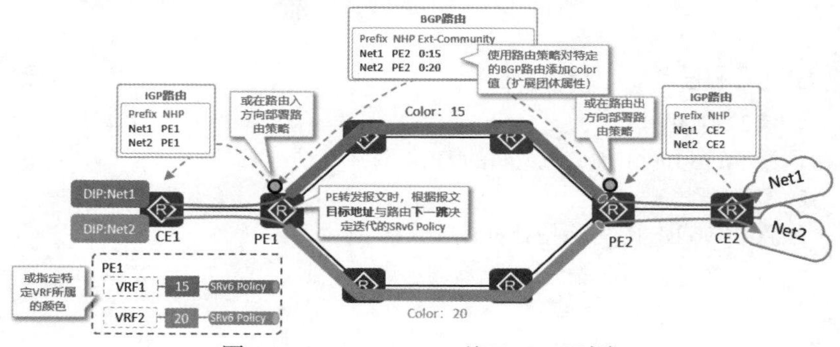

图 16-14 SRv6 Policy 基于 Color 引流

在 PE1 上配置隧道策略，然后当 PE1 接收到 BGP 路由（Net1、Net2）后，根据路由的扩展团体属性（0:15、0:20）和下一跳（PE2）迭代到不同的 SRv6 Policy。转发时，为到特定网段（Net1、Net2）的报文添加一个具体的 SRv6 SID 栈。

路由条目中的 Color 属性可以在路由始发路由器（如 PE2）的出方向上被修改，也可以在路由接收路由器（如 PE1）的入方向上被修改。

在数据始发路由器（如 PE1）的 VPN 实例中，也可以直接配置 Color 属性，让所有从该 VPN 实例发出的流量都通过相同的 SRv6 Policy 进行转发。

6. SRv6 Policy 基于 DSCP 引流

对于去往同一个目的地址的不同业务流量（如 HTTP、FTP 等），基于 Color 引流会导致所有流量都迭代到同一条转发路径，无法精细化控制流量。因此，需要使用基于 DSCP 引流的方式将流量迭代到不同的转发路径，如图 16–15 所示。

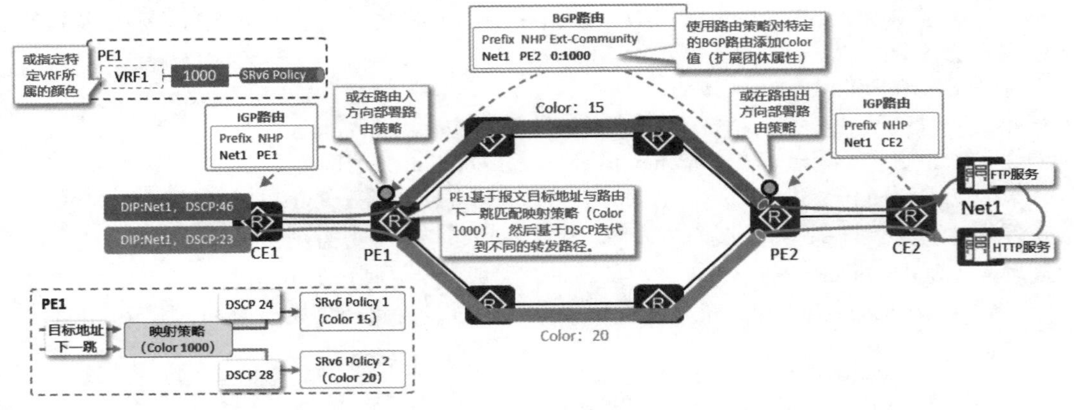

图 16–15　SRv6 Policy 基于 DSCP 引流

当 SRv6 Policy 基于 DSCP 引流时，路由条目中的 Color 属性主要用于匹配映射策略。

路由条目中的 Color 属性可以在本端路由器（如 PE2）发送路由前修改，也可以在对端路由器（如 PE1）接收路由后修改。

在数据始发路由器（如 PE1）的 VPN 实例中，也可以直接配置 Color 属性，让所有从该 VPN 实例发出的流量都通过相同的 SRv6 Policy 进行转发。

16.2　SRv6 实验

16.2.1　配置 SRv6 BE

扫一扫，看视频

1. 实验目的

通过 SRv6 BE 实现 CE 之间的 IPv4 网络互访。

2. 实验拓扑

配置 SRv6 BE 的实验拓扑如图 16–16 所示。

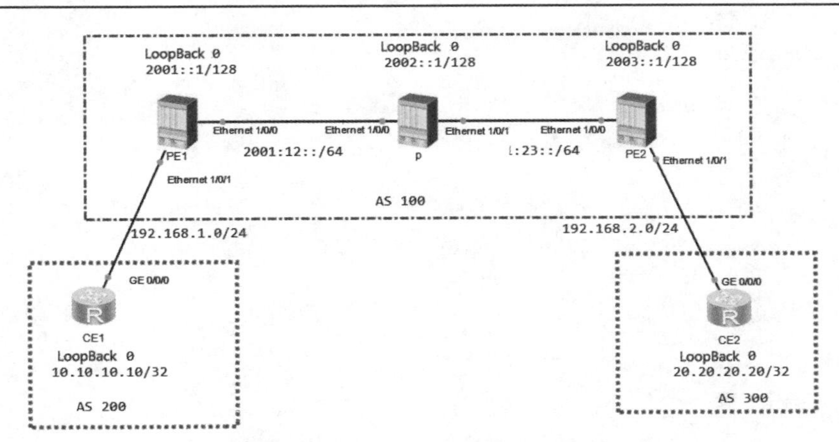

图 16-16 配置 SRv6 BE 的实验拓扑

3. 实验步骤

步骤 1： 配置设备的接口 IP 地址。配置 SRv6 BE IP 地址规划表见表 16-2。

表 16-2 配置 SRv6 BE IP 地址规划表

设备名称	接口编号	IPv4/IPv6 地址
PE1	Ethernet 1/0/0	2001:12::1/64
	LoopBack 0	2001::1/128
	Ethernet 1/0/0	192.168.1.1/24
P	Ethernet 1/0/1	2001:12::2/64
	LoopBack 0	2002::1/128
	Ethernet 1/0/1	2001:23::2/64
PE2	Ethernet 1/0/0	2001:23::3/64
	LoopBack 0	2003::1/128
	Ethernet 1/0/1	192.168.2.1/24
CE1	G0/0/0	192.168.1.2/24
	LoopBack 0	10.10.10.10/32
CE2	G0/0/0	192.168.2.2/24
	LoopBack 0	20.20.20.20/32

步骤 2： 配置 ISP 网络的 IGP 协议（IS-IS IPv6）。

PE1 的配置：

```
[PE1]isis 1
[PE1-isis-1]cost-style wide
[PE1-isis-1]network-entity 49.0001.0000.0000.0001.00
[PE1-isis-1]ipv6 enable topology ipv6
[PE1-isis-1]
[PE1-isis-1]interface Ethernet1/0/0
[PE1-Ethernet1/0/0]isis ipv6 enable 1
```

```
[PE1-Ethernet1/0/0]q
[PE1]interface LoopBack0
[PE1-LoopBack0]isis ipv6 enable 1
```

P 的配置：

```
[P]isis 1
[P-isis-1]cost-style wide
[P-isis-1]network-entity 49.0001.0000.0000.0002.00
[P-isis-1]ipv6 enable topology ipv6
[P-isis-1]q
[P]interface LoopBack0
[P-LoopBack0]isis ipv6 enable 1
[P-LoopBack0]q
[P]interface Ethernet1/0/0
[P-Ethernet1/0/0]isis ipv6 enable 1
[P-Ethernet1/0/0]q
[P]interface Ethernet1/0/1
[P-Ethernet1/0/1]isis ipv6 enable 1
[P-Ethernet1/0/1]q
```

PE2 的配置：

```
[PE2]isis 1
[PE2-isis-1]cost-style wide
[PE2-isis-1]network-entity 49.0001.0000.0000.0003.00
[PE2-isis-1]ipv6 enable topology ipv6
[PE2-isis-1]q
[PE2]interface LoopBack0
[PE2-LoopBack0]isis ipv6 enable 1
[PE2-LoopBack0]q
[PE2]interface Ethernet1/0/0
[PE2-Ethernet1/0/0]isis ipv6 enable 1
```

查看设备的环回口路由 IS-IS 是否能相互学习到：

```
[PE1]display isis route
                         Route information for ISIS(1)
                         -----------------------------
                        ISIS(1) Level-1 Forwarding Table
                        --------------------------------
 IPv6 Dest.      ExitInterface       NextHop                Cost     Flags
--------------------------------------------------------------------------------
2001::1/128      Loop0               Direct                 0        D/-/L/-
2001:12::/64     Eth1/0/0            Direct                 10       D/-/L/-
2001:23::/64     Eth1/0/0            FE80::3A14:4FF:FE02:100 20      A/-/L/-
2002::1/128      Eth1/0/0            FE80::3A14:4FF:FE02:100 10      A/-/L/-
2003::1/128      Eth1/0/0            FE80::3A14:4FF:FE02:100 20      A/-/L/-
     Flags: D-Direct, A-Added to URT, L-Advertised in LSPs, S-IGP Shortcut,
            U-Up/Down Bit Set, LP-Local Prefix-Sid
```

```
                      ISIS(1) Level-2 Forwarding Table
                      --------------------------------
 IPv6 Dest.       ExitInterface       NextHop                    Cost      Flags
--------------------------------------------------------------------------------
2001::1/128       Loop0               Direct                     0         D/-/L/-
2001:12::/64      Eth1/0/0            Direct                     10        D/-/L/-
2001:23::/64      -                   -                          20        -/-/-/-
2002::1/128       -                   -                          10        -/-/-/-
2003::1/128       -                   -                          20        -/-/-/-
     Flags: D-Direct, A-Added to URT, L-Advertised in LSPs, S-IGP Shortcut,
            U-Up/Down Bit Set, LP-Local Prefix-Sid
```

步骤 3: 在 PE 设备上配置 VPN 实例，在 PE 设备与 CE 设备之间建立 BGP 邻居关系。

PE1 的配置：

```
[PE1]ip vpn-instance vpn1
[PE1-vpn-instance-vpn1]ipv4-family
[PE1-vpn-instance-vpn1-af-ipv4]route-distinguisher 1:1
[PE1-vpn-instance-vpn1-af-ipv4]vpn-target 1:1 export-extcommunity
[PE1-vpn-instance-vpn1-af-ipv4]vpn-target 1:1 import-extcommunity
[PE1-vpn-instance-vpn1-af-ipv4]q
[PE1-vpn-instance-vpn1]interface Ethernet1/0/1
[PE1-Ethernet1/0/1]ip binding vpn-instance vpn1
Info: All IPv4 and IPv6 related configurations on this interface are removed.
[PE1-Ethernet1/0/1]ip address 192.168.1.1 255.255.255.0
[PE1-Ethernet1/0/1]q
[PE1]bgp 100
[PE1-bgp]ipv4-family vpn-instance vpn1
[PE1-bgp-vpn1]peer 192.168.1.2 as-number 200 //配置与 CE 建立 BGP 邻居
```

PE2 的配置：

```
[PE2]ip vpn-instance vpn1
[PE2-vpn-instance-vpn1]ipv4-family
[PE2-vpn-instance-vpn1-af-ipv4]route-distinguisher 1:1
[PE2-vpn-instance-vpn1-af-ipv4]vpn-target 1:1 export-extcommunity
[PE2-vpn-instance-vpn1-af-ipv4]vpn-target 1:1 import-extcommunity
[PE2-vpn-instance-vpn1-af-ipv4]q
[PE2-vpn-instance-vpn1]interface Ethernet1/0/1
[PE2-Ethernet1/0/1]ip binding vpn-instance vpn1
[PE2-Ethernet1/0/1]ip address 192.168.2.1 255.255.255.0
[PE2-Ethernet1/0/1]q
[PE2]bgp 100
[PE2-bgp]ipv4-family vpn-instance vpn1
[PE2-bgp-vpn1]peer 192.168.2.2 as-number 300 //配置与 CE 建立 BGP 邻居
```

CE1 的配置：

```
[CE1]bgp 200
[CE1-bgp]peer 192.168.1.1 as-number 100
[CE1-bgp]network 10.10.10.10 255.255.255.255
```

CE2 的配置：

```
[CE2]bgp 300
[CE2-bgp]peer 192.168.2.1 as-number 100
[CE2-bgp]network 20.20.20.20 255.255.255.255
```

配置完成后，在 PE 上可以看到 CE 设备 VPNv4 的路由：

```
[PE1-bgp]display bgp vpnv4 all routing-table
 BGP Local router ID is 0.0.0.0
 Status codes: * - valid, > - best, d - damped, x - best external, a - add path,
               h - history,  i - internal, s - suppressed, S - Stale
               Origin : i - IGP, e - EGP, ? - incomplete
 RPKI validation codes: V - valid, I - invalid, N - not-found
 Total number of routes from all PE: 1
 Route Distinguisher: 1:1
        Network            NextHop        MED      LocPrf      PrefVal        Path/Ogn
 *>     10.10.10.10/32     192.168.1.2    0                    0              200i

 VPN-Instance vpn1, Router ID 192.168.1.1:
 Total Number of Routes: 1
        Network            NextHop        MED      LocPrf      PrefVal        Path/Ogn
 *>     10.10.10.10/32     192.168.1.2    0                    0              200i

[PE2]display bgp vpnv4 all routing-table
 BGP Local router ID is 0.0.0.0
 Status codes: * - valid, > - best, d - damped, x - best external, a - add path,
               h - history,  i - internal, s - suppressed, S - Stale
               Origin : i - IGP, e - EGP, ? - incomplete
 RPKI validation codes: V - valid, I - invalid, N - not-found
 Total number of routes from all PE: 1
 Route Distinguisher: 1:1
        Network            NextHop        MED      LocPrf      PrefVal        Path/Ogn
 *>     20.20.20.20/32     192.168.2.2    0                    0              300i

 VPN-Instance vpn1, Router ID 192.168.2.1:
 Total Number of Routes: 1
        Network            NextHop        MED      LocPrf      PrefVal        Path/Ogn
 *>     20.20.20.20/32     192.168.2.2    0                    0              300i
```

通过以上输出可以看出，PE1 可以学习到 CE1 的路由，PE2 可以学习到 CE2 的路由。

步骤 4： 配置 PE1 和 PE2 的 VPNv4 邻居关系。

PE1 的配置：

```
[PE1]bgp 100
[PE1-bgp]router-id 1.1.1.1//一定要配置 Router ID，BGP4+ 无法自动生成 Router ID
[PE1-bgp]peer 2003::1 as-number 100
[PE1-bgp]peer 2003::1 connect-interface LoopBack0
[PE1-bgp]ipv4-family vpnv4
[PE1-bgp-af-vpnv4]peer 2003::1 enable
```

```
Warning: This operation will reset the peer session. Continue? [Y/N]:y
```

PE2 的配置：

```
[PE2]bgp 100
[PE2-bgp]router-id 2.2.2.2
[PE2-bgp]peer 2001::1 as-number 100
[PE2-bgp]peer 2001::1 connect-interface LoopBack0
[PE2-bgp]ipv4-family vpnv4
[PE2-bgp-af-vpnv4]peer 2001::1 enable
Warning: This operation will reset the peer session. Continue? [Y/N]:y
```

查看 PE1 的 VPNv4 邻居及路由表：

```
[PE1]display bgp vpnv4 all peer
 BGP local router ID : 1.1.1.1
 Local AS number : 100
 Total number of peers : 2                 Peers in established state : 2
  Peer   V      AS  MsgRcvd          MsgSent    OutQ    Up/Down      State         PrefRcv
  2003::1       4   100   4          5          0       00:00:25     Established   1

  Peer of IPv4-family for vpn instance :
  VPN-Instance vpn1, Router ID 1.1.1.1:
  Peer   V      AS  MsgRcvd          MsgSent    OutQ    Up/Down      State         PrefRcv
  192.168.1.2   4   200   3          5          0       00:00:59     Established   1
```

通过以上输出可以看出，PE1 和 PE2 已经建立好 MP-BGP 的邻居关系。

```
[PE1]dis bgp vpnv4 all routing-table
 BGP Local router ID is 1.1.1.1
 Status codes: * - valid, > - best, d - damped, x - best external, a - add path,
               h - history,  i - internal, s - suppressed, S - Stale
               Origin : i - IGP, e - EGP, ? - incomplete
 RPKI validation codes: V - valid, I - invalid, N - not-found
 Total number of routes from all PE: 2
 Route Distinguisher: 1:1
        Network           NextHop        MED      LocPrf     PrefVal       Path/Ogn
 *>     10.10.10.10/32    192.168.1.2    0                   0             200i
 *>i    20.20.20.20/32    2003::1        0        100        0             300i

 VPN-Instance vpn1, Router ID 1.1.1.1:
 Total Number of Routes: 2
        Network           NextHop        MED      LocPrf     PrefVal       Path/Ogn
 *>     10.10.10.10/32    192.168.1.2    0                   0             200i
i       20.20.20.20/32    2003::1        0        100        0             300i
```

通过以上输出可以看出，PE1 已经学习到对端的 VPNv4 路由 20.20.20.20/32，但是此路由在 VPN 实例 VPN1 中为无效路由。

```
[PE1]display bgp vpnv4 vpn-instance vpn1 routing-table 20.20.20.20
 BGP local router ID : 1.1.1.1
 Local AS number : 100
 VPN-Instance vpn1, Router ID 1.1.1.1:
```

```
 Paths:   1 available, 0 best, 0 select, 0 best-external, 0 add-path
 BGP routing table entry information of 20.20.20.20/32:
 Route Distinguisher: 1:1
 Remote-Cross route
 Label information (Received/Applied): 48120/NULL
 From: 2003::1 (2.2.2.2)
 Route Duration: 0d00h03m24s
 Relay Tunnel Out-Interface:
 Original nexthop: 2003::1
 Qos information : 0x0
 Ext-Community: RT <1 : 1>
  AS-path 300, origin igp, MED 0, localpref 100, pref-val 0, internal, pre 255,
invalid for nexthop tunnel unreachable
 Not advertised to any peer yet
```

无效路由的原因是隧道不可达，接下来配置 SRv6 的隧道，用于传递 CE 之间的数据。

步骤 5：配置 SRv6 BE。

（1）开启 SRv6 功能，并且配置 SRv6 的 Locator。

PE1 的配置：

```
[PE1]segment-routing ipv6 //使能 SRv6 功能
//配置 SRv6 VPN 封装的源地址，一般使用 BGP 更新源地址
[PE1-segment-routing-ipv6]encapsulation source-address 2001::1
//配置静态 Locator 的前缀为 2001:ABCD::/64，取值范围为后 32 位
[PE1-segment-routing-ipv6]locator pe1 ipv6-prefix 2001:ABCD:: 64 static 32
```

PE2 的配置：

```
[PE2]segment-routing ipv6
[PE2-segment-routing-ipv6]encapsulation source-address 2003::1
[PE2-segment-routing-ipv6]locator PE2 ipv6-prefix 2003:ABCD:: 64 static 32
```

查看配置结果：

```
[PE1]display segment-routing ipv6 locator pe1 verbose

                        Locator Configuration Table
                        ---------------------------

LocatorName  : pe1                                  LocatorID    : 1
IPv6Prefix   : 2001:ABCD::                          PrefixLength : 64
StaticLength : 32                                   Reference    : 0
Default      : N                                    ArgsLength   : 0
AutoSIDBegin : 2001:ABCD::1:0:0
AutoSIDEnd   : 2001:ABCD::FFFF:FFFF:FFFF:FFFF //此段表示 Locator 的范围

Total Locator(s): 1
```

（2）配置 IS-IS 的 SRv6 功能。

PE1 的配置：

```
[PE1]isis 1
```

```
[PE1-isis-1]segment-routing ipv6 locator pe1 //使能IS-IS的SRv6功能和IS-IS通告Locator
                                             //的前缀路由
```

PE2 的配置：

```
[PE2]isis 1
[PE2-isis-1]segment-routing ipv6 locator PE2
```

查看 IPv6 路由表：

```
[PE1]display ipv6 routing-table protocol isis
_public_ Routing Table : IS-IS
Summary Count : 7
IS-IS routing table status : <Active>
Summary Count : 5
Destination   : 2001:23::                          PrefixLength : 64
NextHop       : FE80::3A14:4FF:FE02:100            Preference   : 15
Cost          : 20                                 Protocol     : ISIS-L1
RelayNextHop  : ::                                 TunnelID     : 0x0
Interface     : Ethernet1/0/0                      Flags        : D
Destination   : 2001:ABCD::                        PrefixLength : 64
NextHop       : ::                                 Preference   : 15
Cost          : 0                                  Protocol     : ISIS-L1
RelayNextHop  : ::                                 TunnelID     : 0x0
Interface     : NULL0                              Flags        : DB
Destination   : 2002::1                            PrefixLength : 128
NextHop       : FE80::3A14:4FF:FE02:100            Preference   : 15
Cost          : 10                                 Protocol     : ISIS-L1
RelayNextHop  : ::                                 TunnelID     : 0x0
Interface     : Ethernet1/0/0                      Flags        : D
Destination   : 2003::1                            PrefixLength : 128
NextHop       : FE80::3A14:4FF:FE02:100            Preference   : 15
Cost          : 20                                 Protocol     : ISIS-L1
RelayNextHop  : ::                                 TunnelID     : 0x0
Interface     : Ethernet1/0/0                      Flags        : D
Destination   : 2003:ABCD::                        PrefixLength : 64
NextHop       : FE80::3A14:4FF:FE02:100            Preference   : 15
Cost          : 20                                 Protocol     : ISIS-L1
RelayNextHop  : ::                                 TunnelID     : 0x0
Interface     : Ethernet1/0/0                      Flags        : D
```

通过以上输出可以看出，PE1 已经通过 IS-IS 学习到 PE2 的 Locator 的路由，即 2003:ABCD::/64。

（3）配置 BGP 的 SRv6 功能。

PE1 的配置：

```
[PE1]bgp 100
[PE1-bgp]ipv4-family vpnv4
[PE1-bgp-af-vpnv4]peer 2003::1 prefix-sid //向对端PE通告路由时携带SID属性
[PE1-bgp-af-vpnv4]ipv4-family vpn-instance vpn1
[PE1-bgp-vpn1]segment-routing ipv6 best-effort//使能根据路由携带的SID属性进行私网路由迭代
[PE1-bgp-vpn1]segment-routing ipv6 locator pe1//使能VPN私网路由携带SID属性
```

PE2 的配置：

```
[PE2]bgp 100
[PE2-bgp]ipv4-family vpnv4
[PE2-bgp-af-vpnv4]peer 2001::1 prefix-sid
[PE2-bgp-af-vpnv4]ipv4-family vpn-instance vpn1
[PE2-bgp-vpn1]segment-routing ipv6 locator PE2
[PE2-bgp-vpn1]segment-routing ipv6 best-effort
```

查看由 BGP 生成的 End-DT4 SID：

```
[PE2]display segment-routing ipv6 local-sid end-dt4 forwarding
                    My Local-SID End.DT4 Forwarding Table
                    -------------------------------------
SID         : 2003:ABCD::1:0:3C/128                    FuncType : End.DT4
VPN Name    : vpn1                                     VPN ID   : 2
LocatorName: PE2                                       LocatorID: 1
Total SID(s): 1
```

本地生成的 SID 类似于 MPLS 中的私网标签，用于决定收到此 DIP 时，发往哪一个 VPN 实例。

在 PE1 上查看 20.20.20.20/32 的 VPNv4 路由详细信息：

```
[PE1]display bgp vpnv4 vpn-instance vpn1 routing-table 20.20.20.20
 BGP local router ID : 1.1.1.1
 Local AS number : 100
 VPN-Instance vpn1, Router ID 1.1.1.1:
 Paths:   1 available, 1 best, 1 select, 0 best-external, 0 add-path
 BGP routing table entry information of 20.20.20.20/32:
 Route Distinguisher: 1:1
 Remote-Cross route
 Label information (Received/Applied): 3/NULL
 From: 2003::1 (2.2.2.2)
 Route Duration: 0d00h03m53s
 Relay IP Nexthop: FE80::3A14:4FF:FE02:100
 Relay IP Out-Interface: Ethernet1/0/0
 Relay Tunnel Out-Interface:
 Original nexthop: 2003::1
 Qos information : 0x0
 Ext-Community: RT <1 : 1>
 Prefix-sid: 2003:ABCD::1:0:3C //PE2 生成的 end-dt4 SID
 AS-path 300, origin igp, MED 0, localpref 100, pref-val 0, valid, internal,
best, select, pre 255, IGP cost 20
 Advertised to such 1 peers:
192.168.1.2
```

在 CE1 上 ping CE2：

```
[CE1]ping -a 10.10.10.10 20.20.20.20
  PING 20.20.20.20: 56  data bytes, press CTRL_C to break
    Reply from 20.20.20.20: bytes=56 Sequence=1 ttl=253 time=40 ms
    Reply from 20.20.20.20: bytes=56 Sequence=2 ttl=253 time=30 ms
    Reply from 20.20.20.20: bytes=56 Sequence=3 ttl=253 time=40 ms
```

```
    Reply from 20.20.20.20: bytes=56 Sequence=4 ttl=253 time=30 ms
Reply from 20.20.20.20: bytes=56 Sequence=5 ttl=253 time=30 ms
```

在 PE1 的 Ethernet 1/0/0 接口抓包查看结果，如图 16-17 所示。

```
> Frame 20: 166 bytes on wire (1328 bits), 166 bytes captured (1328 bits) on interface -, id 0
> Ethernet II, Src: 38:4e:75:02:01:00 (38:4e:75:02:01:00), Dst: 38:4e:75:03:01:00 (38:4e:75:03:01:00)
v Internet Protocol Version 6, Src: 2001::1, Dst: 2003:abcd::1:0:3c    ❶ 使用IPv6封装私网的IPv6数据
    0110 .... = Version: 6
  > .... 1111 1111 .... .... .... .... .... = Traffic Class: 0xff (DSCP: Unknown, ECN: CE)
    .... .... .... 0000 0000 0000 0000 0001 = Flow Label: 0x00001
    Payload Length: 92
    Next Header: Destination Options for IPv6 (60)
    Hop Limit: 254
    Source: 2001::1
    Destination: 2003:abcd::1:0:3c    ❷ 目的IP为对端VPNv4邻居发布过来的Prefix-SID
    [Source Teredo Server IPv4: 0.0.0.0]
    [Source Teredo Port: 65535]
    [Source Teredo Client IPv4: 255.255.255.254]
  > Destination Options for IPv6
> Internet Protocol Version 4, Src: 10.10.10.10, Dst: 20.20.20.20
> Internet Control Message Protocol
> Exablaze trailer, Device: 8, Port: 255, Timestamp: 08:00:0.000000000000
```

图 16-17　PE1 的 Ethernet 1/0/0 接口的抓包结果

16.2.2　配置 SRv6 Policy

1. 实验目的

通过在 PE 之间建立 SRv6 Policy 来实现两端的 CE 设备访问对端的 IP 网段 10.10.10.10/32（使用显示路径 PE1→P1→PE2 进行转发），访问对端的 IP 网段 20.20.20.10/32（使用显示路径 PE2→P1→PE1 进行转发）。访问对端的 IP 网段 10.10.10.20/32（使用显示路径 PE1→P2→PE2 进行转发），访问对端的 IP 网段 20.20.20.20/32（使用显示路径 PE2→P1→PE1 进行转发）。

扫一扫，看视频

需要注意的是，本实验需要使用华为 ENSP Pro 完成。

2. 实验拓扑

配置 SRv6 Policy 的实验拓扑如图 16-18 所示。

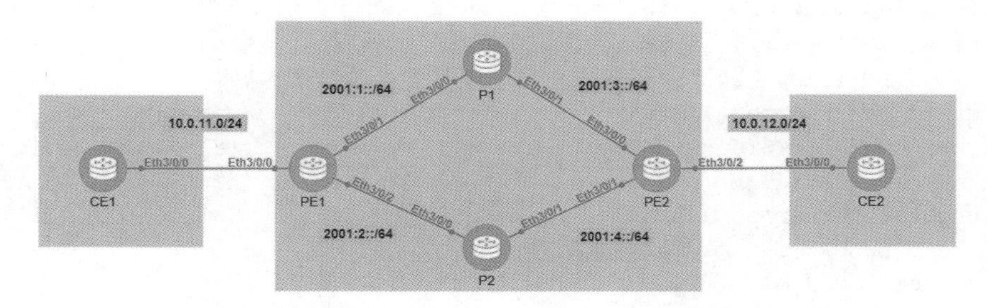

图 16-18　配置 SRv6 Policy 的实验拓扑

3. 实验步骤

步骤 1： 配置设备的 IP 地址。配置 SRv6 Policy IP 地址规划表见表 16-3。

表 16-3　配置 SRv6 Policy IP 地址规划表

设备名称	接口编号	IPv4/IPv6 地址
PE1	Ethernet 3/0/0	10.0.11.1/24
	Ethernet 3/0/1	2001:1::1/64

续表

设备名称	接口编号	IPv4/IPv6 地址
PE1	Ethernet 3/0/2	2001:2::1/64
	LoopBack 0	1::1/128
PE2	Ethernet 3/0/0	2001:3::2/64
	Ethernet 3/0/1	2001:4::2/64
	Ethernet 3/0/2	10.0.12.1/24
	LoopBack 0	2::2/128
P1	Ethernet 3/0/0	2001:1::2/64
	Ethernet 3/0/1	2001:3::1/64
	LoopBack 0	3::3/128
P2	Ethernet 3/0/0	2001:2::2/64
	Ethernet 3/0/1	2001:4::1/64
	LoopBack 0	4::4/128
CE1	Ethernet 3/0/0	10.0.11.2/24
	LoopBack 0	10.10.10.10/32
	LoopBack 1	10.10.10.20/32
CE2	Ethernet 3/0/0	10.0.12.2/24
	LoopBack 0	20.20.20.10/32
	LoopBack 1	20.20.20.20/32

步骤 2：配置骨干网络的 IS-IS-IPv6。

PE1 的配置：

```
[PE1]isis
[PE1-isis-1]cost-style wide
[PE1-isis-1]network-entity 49.0001.0000.0000.0001.00
[PE1-isis-1]ipv6 enable topology ipv6
[PE1-isis-1]is-level level-2
[PE1]interface Ethernet 3/0/1
[PE1-Ethernet3/0/1]isis ipv6 enable
[PE1-Ethernet3/0/1]q
[PE1]interface Ethernet 3/0/2
[PE1-Ethernet3/0/2]isis ipv6 enable
[PE1-Ethernet3/0/2]q
[PE1]interface lo0
[PE1-LoopBack0]isis ipv6 enable
```

PE2 的配置：

```
[PE2]isis 1
[PE2-isis-1]is-level level-2
[PE2-isis-1]cost-style wide
[PE2-isis-1]network-entity 49.0001.0000.0000.0002.00
[PE2-isis-1]ipv6 enable topology ipv6
[PE2-isis-1]q
[PE2]interface Ethernet3/0/0
```

```
[PE2-Ethernet3/0/0]isis ipv6 enable 1
[PE2-Ethernet3/0/0]q
[PE2]interface Ethernet3/0/1
[PE2-Ethernet3/0/1]isis ipv6 enable 1
[PE2-Ethernet3/0/1]q
[PE2]interface loopback 0
[PE2-LoopBack0]isis ipv6 enable 1
```

P1 的配置:

```
[P1]isis 1
[P1-isis-1]is-level level-2
[P1-isis-1]cost-style wide
[P1-isis-1]network-entity 49.0001.0000.0000.0003.00
[P1-isis-1]ipv6 enable topology ipv6
[P1-isis-1]q
[P1]interface Ethernet3/0/0
[P1-Ethernet3/0/0]isis ipv6 enable 1
[P1-Ethernet3/0/0]q
[P1]interface Ethernet3/0/1
[P1-Ethernet3/0/1]isis ipv6 enable 1
[P1-Ethernet3/0/1]q
[P1]interface loopback 0
[P1-LoopBack0]isis ipv6 enable 1
```

P2 的配置:

```
[P2]isis 1
[P2-isis-1]is-level level-2
Info: IS level Changed. IS-IS process 1 will be reset.
[P2-isis-1]cost-style wide
Info: Cost style Changed. IS-IS process 1 will be reset.
[P2-isis-1]network-entity 49.0001.0000.0000.0004.00
[P2-isis-1]ipv6 enable topology ipv6
[P2-isis-1]q
[P2]interface Ethernet3/0/0
[P2-Ethernet3/0/0]isis ipv6 enable 1
[P2-Ethernet3/0/0]q
[P2]interface Ethernet3/0/1
[P2-Ethernet3/0/1]isis ipv6 enable 1
[P2-Ethernet3/0/1]q
[P2]interface loopback 0
[P2-LoopBack0]isis ipv6 enable 1
[P2-LoopBack0]q
```

步骤 3: 配置 PE1 和 PE2 之间的 MP-BGP 邻居关系。

PE1 的配置:

```
[PE1]bgp 100
[PE1-bgp]router-id 1.1.1.1
[PE1-bgp]peer 2::2 as-number 100
[PE1-bgp]peer 2::2 connect-interface LoopBack 0
```

```
[PE1-bgp-af-vpnv4]peer 2::2 enable
Warning: This operation will reset the peer session. Continue? [Y/N]:y
```

PE2 的配置：

```
[PE2]bgp 100
[PE2-bgp]router-id 2.2.2.2
[PE2-bgp]peer 1::1 as-number  100
[PE2-bgp]peer 1::1 connect-interface LoopBack 0
[PE2-bgp]ipv4-family vpnv4
[PE2-bgp-af-vpnv4]peer 1::1 enable
Warning: This operation will reset the peer session. Continue? [Y/N]:y
```

查看 VPNv4 邻居关系：

```
[PE1]display  bgp vpnv4 all  peer
 BGP local router ID : 1.1.1.1
 Local AS number : 100
 Total number of peers : 1                Peers in established state : 1
  Peer          V   AS    MsgRcvd     MsgSent    OutQ    Up/Down     State          PrefRcv
  2::2          4   100   3           4          0       00:00:41    Established    0
```

通过以上输出可以看出，PE1 和 PE2 已经建立了 MP-BGP 的邻居关系。

步骤 4：在骨干网络使能 SR 功能，并且建立 SRv6 Policy 路径。

（1）配置设备的 SRv6 功能及接口的 End-x SID，用于后续 SRv6 Policy 的显示路径。

PE1 的配置：

```
[PE1]segment-routing ipv6
[PE1-segment-routing-ipv6]encapsulation source-address 1::1
[PE1-segment-routing-ipv6]locator PE1 ipv6-prefix 2001:100:: 64 static 32
                                                  //配置设备的 Locator 并且命名为 PE1
[PE1-segment-routing-ipv6-locator]opcode ::1 end-x interface  Ethernet 3/0/1
nexthop 2001:1::2 psp //配置 PE1 的 Ethernet 3/0/1 接口的 End-x SID 为 2001:100::1
[PE1-segment-routing-ipv6-locator]opcode ::2 end-x interface  Ethernet 3/0/2
nexthop 2001:2::2 psp //配置 PE1 的 Ethernet 3/0/2 接口的 End-x SID 为 2001:100::2
[PE1-segment-routing-ipv6-locator]opcode ::100 end psp //配置设备的 End-x SID 为
                                                          //2001:100::100
```

PE2 的配置：

```
[PE2]segment-routing ipv6
[PE2-segment-routing-ipv6]encapsulation source-address  2::2
[PE2-segment-routing-ipv6]locator PE2 ipv6-prefix 2001:200:: 64 static  32
[PE2-segment-routing-ipv6-locator]opcode ::1 end-x interface Ethernet 3/0/0
nexthop 2001:3::1 psp
[PE2-segment-routing-ipv6-locator]opcode ::1 end-x interface Ethernet 3/0/1
nexthop 2001:4::1 psp
[PE2-segment-routing-ipv6-locator]opcode ::100 end psp
```

P1 的配置：

```
[P1]segment-routing ipv6
[P1-segment-routing-ipv6]encapsulation source-address 3::3
[P1-segment-routing-ipv6]locator P1ipv6-prefix 2001:300:: 64 static 32
```

```
[P1-segment-routing-ipv6-locator]opcode ::1 end-x interface Ethernet3/0/0 nexthop
2001:1::1 psp
[P1-segment-routing-ipv6-locator]opcode ::2 end-x interface Ethernet3/0/1 nexthop
2001:3::2 psp
```

P2 的配置：

```
[P2]segment-routing ipv6
[P2-segment-routing-ipv6]encapsulation source-address 4::4
[P2-segment-routing-ipv6]locator P2 ipv6-prefix 2001:400:: 64 static 32
[P2-segment-routing-ipv6-locator]opcode ::1 end-x interface Ethernet3/0/0 nexthop
2001:2::1 psp
[P2-segment-routing-ipv6-locator]opcode ::2 end-x interface Ethernet3/0/1 nexthop
2001:4::2 psp
```

查看设备的 End-x SID，以 PE1 为例：

```
[PE1]display segment-routing ipv6 local-sid end-x forwarding
                        My Local-SID End.X Forwarding Table
                        ------------------------------------
SID         : 2001:100::1/128                          FuncType    : End.X
Flavor      : PSP                                      SidCompress : NO
LocatorName : PE1                                      LocatorID   : 1
ProtocolType: STATIC                                    ProcessID  : --
UpdateTime  : 2023-11-23 15:10:23.074
TeFrrFlags  : --                                     DelayTimerRemain: -
NextHop     :                 Interface :              ExitIndex:
2001:1::2                     Eth3/0/1                 0x00000006
SID         : 2001:100::2/128                         FuncType     : End.X
Flavor      : PSP                                     SidCompress  : NO
LocatorName : PE1                                     LocatorID    : 1
ProtocolType: STATIC                                  ProcessID    : --
UpdateTime  : 2023-11-23 15:10:46.934
TeFrrFlags  : --                                     DelayTimerRemain: -
NextHop     :                 Interface :               ExitIndex:
2001:2::2                     Eth3/0/2                  0x00000007
```

通过以上输出可以看出，PE1 的 Ethernet 3/0/1 接口的 End-x SID 为 2001:100::1，PE1 的 Ethernet 3/0/2 接口的 End-x SID 为 2001:100::2。

（2）使能 IS-IS 的 SRv6 功能。

PE1 的配置：

```
[PE1]isis
[PE1-isis-1]segment-routing ipv6 locator PE1 auto-sid-disable //使能SRv6功能，并
//且关闭协议自动为接口分配End-x SID，因为本实验中的每条链路的End-x SID都是手动静态配置的
```

PE2 的配置：

```
[PE2]isis
[PE2-isis-1]segment-routing ipv6 locator PE2 auto-sid-disable
```

P1 的配置：

```
[P1]isis
```

```
[P1-isis-1]segment-routing ipv6 locator P1 auto-sid-disable
```

P2 的配置：

```
[P2]isis
[P2-isis-1]segment-routing ipv6 locator PE1 auto-sid-disable
```

（3）使能与指定 IPv6 对等体之间交换 IPv4 Prefix SID 信息。

PE1 的配置：

```
[PE1]bgp 100
[PE1-bgp]ipv4-family vpnv4
[PE1-bgp-af-vpnv4]peer 2::2 prefix-sid
```

PE2 的配置：

```
[PE2]bgp 100
[PE2-bgp]ipv4-family vpnv4
[PE2-bgp-af-vpnv4]peer 1::1 prefix-sid
```

步骤 5： 配置设备的 SRv6 Policy，在 PE1 上配置显示路径 PE1 → P1 → PE2，Color 为 101；配置显示路径 PE1 → P2 → PE2，Color 为 102。在 PE2 上配置显示路径 PE2 → P1 → PE1，Color 为 101；配置显示路径 PE2 → P2 → PE1，Color 为 102。

PE1 的配置：

```
[PE1]segment-routing ipv6
[PE1-segment-routing-ipv6]segment-list list1
[PE1-segment-routing-ipv6-segment-list-list1]index 5 sid ipv6 2001:100::1
[PE1-segment-routing-ipv6-segment-list-list1]index 10 sid ipv6 2001:300::2
    //配置显示路径，第一跳为 PE1 的 Ethernet 3/0/1 接口，第二跳为 P1 的 Ethernet 3/0/1 接口
[PE1-segment-routing-ipv6-segment-list-list1]q
[PE1-segment-routing-ipv6]segment-list list2
[PE1-segment-routing-ipv6-segment-list-list2]index 5 sid ipv6 2001:100::2
[PE1-segment-routing-ipv6-segment-list-list2]index 10 sid ipv6 2001:400::2
[PE1-segment-routing-ipv6-segment-list-list2]q
    //配置显示路径，第一跳为 PE1 的 Ethernet 3/0/2 接口，第二跳为 P2 的 Ethernet 3/0/1 接口
[PE1-segment-routing-ipv6]srv6-te policy PE1-P1-PE2 endpoint 2::2 color 101
[PE1-segment-routing-ipv6-policy-PE1-P1-PE2]candidate-path preference 100
[PE1-segment-routing-ipv6-policy-PE1-P1-PE2-path]segment-list list1
[PE1-segment-routing-ipv6-policy-PE1-P1-PE2-path]q
    //配置 SRv6 隧道策略，命名为 PE1-P1-PE2，配置策略 Color 值为 101。调用显示路径 list1
[PE1-segment-routing-ipv6]srv6-te policy PE1-P2-PE2 endpoint 2::2 color 102
[PE1-segment-routing-ipv6-policy-PE1-P2-PE2]candidate-path preference 100
[PE1-segment-routing-ipv6-policy-PE1-P2-PE2-path]segment-list list2
    //配置 SRv6 隧道策略，命名为 PE1-P2-PE2，配置策略 Color 值为 102。调用显示路径 list1
```

PE2 的配置：

```
[PE2]segment-routing ipv6
[PE2-segment-routing-ipv6]segment-list list1
[PE2-segment-routing-ipv6-segment-list-list1]index 5 sid ipv6 2001:200::1
[PE2-segment-routing-ipv6-segment-list-list1]index 10 sid ipv6 2001:300::1
[PE2-segment-routing-ipv6-segment-list-list1]q
```

```
[PE2-segment-routing-ipv6]segment-list list2
[PE2-segment-routing-ipv6-segment-list-list2]index 5 sid ipv6 2001:200::2
[PE2-segment-routing-ipv6-segment-list-list2]index 10 sid ipv6 2001:400::1
[PE2-segment-routing-ipv6-segment-list-list2]q
[PE2-segment-routing-ipv6]srv6-te policy PE2-P1-PE1 endpoint 1::1 color 101
[PE2-segment-routing-ipv6-policy-PE2-P1-PE1]candidate-path preference 100
[PE2-segment-routing-ipv6-policy-PE2-P1-PE1-path]segment-list list1
[PE2-segment-routing-ipv6-policy-PE2-P1-PE1-path]q
[PE2-segment-routing-ipv6-policy-PE2-P1-PE1]srv6-te policy PE2-P2-PE1 endpoint
1::1 color 102
[PE2-segment-routing-ipv6-policy-PE2-P2-PE1]candidate-path preference 100
[PE2-segment-routing-ipv6-policy-PE2-P2-PE1-path]segment-list list2
```

查看 SRv6 隧道状态，以 PE1 为例：

```
[PE1]display tunnel-info all
Tunnel ID                   Type                Destination          Status
-------------------------------------------------------------------------------
0x000000003400000001        srv6tepolicy        2001:200::100        UP
0x000000003400000002        srv6tepolicy        2001:200::100        UP
```

通过以上输出可以看出，PE1 生成了两条到 PE2 的 SRv6 隧道，并且状态都为 UP。

步骤 6: 配置 PE 和 CE 之间的 BGP 邻居关系。

PE1 的配置：

```
[PE1]ip vpn-instance vpn1
[PE1-vpn-instance-vpn1]route-distinguisher 100:1        //创建 VPN 实例 vpn1
[PE1-vpn-instance-vpn1-af-ipv4]vpn-target 1:1 both
[PE1]interface Ethernet 3/0/0
[PE1-Ethernet3/0/0]ip binding vpn-instance  vpn1        //接口加入 VPN 实例 vpn1
[PE1-Ethernet3/0/0]ip address 10.0.11.1 24
[PE1]bgp 100
[PE1-bgp]ipv4-family vpn-instance vpn1
[PE1-bgp-vpn1]peer 10.0.11.2 as-number 200              //与 CE1 建立 BGP 邻居关系
[PE1-bgp-vpn1]segment-routing ipv6 traffic-engineer best-effort  //设置 SRv6 的逃生邻居
                                                                 //为 SRv6 BE
[PE1-bgp-vpn1]segment-routing ipv6 locator PE1
```

CE1 的配置：

```
[CE1]bgp 200
[CE1-bgp]peer 10.0.11.1 as-number 100
[CE1-bgp]network 10.10.10.10 32
[CE1-bgp]network 10.10.10.20 32
```

PE2 的配置：

```
[PE2]ip vpn-instance vpn1
[PE2-vpn-instance-vpn1]route-distinguisher 100:2
[PE2-vpn-instance-vpn1-af-ipv4]vpn-target 1:1 both
[PE2]interface  Ethernet 3/0/2
[PE2-Ethernet3/0/2]ip binding vpn-instance vpn1
[PE2-Ethernet3/0/2]ip address 10.0.12.1 24
```

```
[PE2]bgp 100
[PE2-bgp]ipv4-family vpn-instance vpn1
[PE2-bgp-vpn1]peer 10.0.12.2 as-number  300
[PE2-bgp-vpn1]segment-routing ipv6 traffic-engineer best-effort
[PE2-bgp-vpn1]segment-routing ipv6 locator PE2
```

CE2 的配置：

```
[CE2]bgp 300
[CE2-bgp]peer 10.0.12.1 as-number 100
[CE2-bgp]network 20.20.20.10 32
[CE2-bgp]network 20.20.20.20 32
```

步骤 7： 配置隧道策略，引入私网流量，将 10.10.10.10/32 以及 20.20.20.10/32 的路径染色为 101，将 10.10.10.20/32 以及 20.20.20.20/32 的路径染色为 102。

（1）创建策略，为后续的路由染色。

PE1 的配置：

```
[PE1]ip ip-prefix 1 permit 20.20.20.10 32
[PE1]ip ip-prefix 2 permit 20.20.20.20 32
[PE1]route-policy color permit node 10
[PE1-route-policy]if-match ip-prefix 1
[PE1-route-policy]apply extcommunity color 0:101
[PE1-route-policy]q
[PE1]route-policy color permit node 20
[PE1-route-policy]if-match ip-prefix 2
[PE1-route-policy]apply extcommunity color 0:102
```

PE2 的配置：

```
[PE2]ip ip-prefix 1 permit 10.10.10.10 32
[PE2]ip ip-prefix 2 permit 10.10.10.20 32
[PE2]route-policy color permit node 10
[PE2-route-policy]if-match ip-prefix 1
[PE2-route-policy]apply extcommunity color 0:101
[PE2-route-policy]q
[PE2]route-policy color permit node 20
[PE2-route-policy]if-match ip-prefix 2
[PE2-route-policy]apply extcommunity color 0:102
```

（2）应用路由策略，将从对端 VPNv4 邻居接收的路由按照策略进行染色。

PE1 的配置：

```
[PE1]bgp 100
[PE1-bgp]ipv4-family vpnv4
[PE1-bgp-af-vpnv4]peer 2::2 route-policy color import
[PE1]tunnel-policy p1
[PE1-tunnel-policy-p1]tunnel select-seq ipv6 srv6-te-policy load-balance-number 1
[PE1-tunnel-policy-p1]q
[PE1]ip vpn-instance vpn1
[PE1-vpn-instance-vpn1]ipv4-family
[PE1-vpn-instance-vpn1-af-ipv4]tnl-policy p1
```

PE2 的配置：

```
[PE2]bgp 100
[PE2-bgp]ipv4-family vpnv4
[PE2-bgp-af-vpnv4]peer 1::1 route-policy color import
[PE2]tunnel-policy p1
[PE2-tunnel-policy-p1]tunnel select-seq ipv6 srv6-te-policy load-balance-number 1
[PE2]ip vpn-instance vpn1
[PE2-vpn-instance-vpn1-af-ipv4]tnl-policy  p1
```

查看 CE 设备的路由学习情况：

```
[CE1]display bgp routing-table
 BGP Local router ID is 10.0.11.2
 Status codes: * - valid, > - best, d - damped, x - best external, a - add path,
               h - history,  i - internal, s - suppressed, S - Stale
               Origin : i - IGP, e - EGP, ? - incomplete
 RPKI validation codes: V - valid, I - invalid, N - not-found
 Total Number of Routes: 4
        Network           NextHop        MED      LocPrf        PrefVal       Path/Ogn
 *>     10.10.10.10/32    0.0.0.0        0                      0             i
 *>     10.10.10.20/32    0.0.0.0        0                      0             i
 *>     20.20.20.10/32    10.0.11.1                             0             100 300i
 *>     20.20.20.20/32    10.0.11.1                             0             100 300i
```

查看 PE 设备接收对端路由的情况，以 PE1 为例：

```
[PE1]display bgp vpnv4 all routing-table 20.20.20.10
 BGP local router ID : 1.1.1.1
 Local AS number : 100
 Total routes of Route Distinguisher(100:2): 1
 BGP routing table entry information of 20.20.20.10/32:
 Label information (Received/Applied): 3/NULL
 From: 2::2 (2.2.2.2)
 Route Duration: 0d00h00m56s
 Relay IP Nexthop: FE80::3A02:FF:FE11:300
 Relay IP Out-Interface: Ethernet3/0/1
 Relay Tunnel Out-Interface:
 Original nexthop: 2::2
 Qos information : 0x0
 Ext-Community: RT <1 : 1>, Color <0 : 101> //路由的 Color 值为 101
 Prefix-sid: 2001:200::1:0:3
 AS-path 300, origin igp, MED 0, localpref 100, pref-val 0, valid, internal,
best, select, pre 255, IGP cost 20
 Not advertised to any peer yet
 VPN-Instance vpn1, Router ID 1.1.1.1:
 Total Number of Routes: 1
 BGP routing table entry information of 20.20.20.10/32:
 Route Distinguisher: 100:2
 Remote-Cross route
 Label information (Received/Applied): 3/NULL
```

```
 From: 2::2 (2.2.2.2)
 Route Duration: 0d00h00m56s
 Relay IP Nexthop: FE80::3A02:FF:FE11:300
 Relay IP Out-Interface: Ethernet3/0/1
 Relay Tunnel Out-Interface:
 Original nexthop: 2::2
 Qos information : 0x0
 Ext-Community: RT <1 : 1>, Color <0 : 101>
 Prefix-sid: 2001:200::1:0:3
 AS-path 300, origin igp, MED 0, localpref 100, pref-val 0, valid, internal, best,
select, pre 255, IGP cost 20
 Advertised to such 1 peers:
10.0.11.2
[PE1]display bgp vpnv4 all routing-table 20.20.20.20
 BGP local router ID : 1.1.1.1
 Local AS number : 100
 Total routes of Route Distinguisher(100:2): 1
 BGP routing table entry information of 20.20.20.20/32:
 Label information (Received/Applied): 3/NULL
 From: 2::2 (2.2.2.2)
 Route Duration: 0d00h01m01s
 Relay IP Nexthop: FE80::3A02:FF:FE11:300
 Relay IP Out-Interface: Ethernet3/0/1
 Relay Tunnel Out-Interface:
 Original nexthop: 2::2
 Qos information : 0x0
 Ext-Community: RT <1 : 1>, Color <0 : 102> //路由的Color值为102
 Prefix-sid: 2001:200::1:0:3
 AS-path 300, origin igp, MED 0, localpref 100, pref-val 0, valid, internal,
best, select, pre 255, IGP cost 20
 Not advertised to any peer yet
 VPN-Instance vpn1, Router ID 1.1.1.1:
 Total Number of Routes: 1
 BGP routing table entry information of 20.20.20.20/32:
 Route Distinguisher: 100:2
 Remote-Cross route
 Label information (Received/Applied): 3/NULL
 From: 2::2 (2.2.2.2)
 Route Duration: 0d00h01m01s
 Relay IP Nexthop: FE80::3A02:FF:FE11:300
 Relay IP Out-Interface: Ethernet3/0/1
 Relay Tunnel Out-Interface:
 Original nexthop: 2::2
 Qos information : 0x0
 Ext-Community: RT <1 : 1>, Color <0 : 102>
 Prefix-sid: 2001:200::1:0:3
 AS-path 300, origin igp, MED 0, localpref 100, pref-val 0, valid, internal, best,
select, pre 255, IGP cost 20
```

```
 Advertised to such 1 peers:
   10.0.11.2
```

查看 SRv6 隧道的情况，以 PE1 为例：

```
[PE1]display srv6-te policy
PolicyName : PE1-P1-PE2   //隧道名称
Color                 : 101 //颜色为101     Endpoint             : 2001:200::100
TunnelId              : 1                   Binding SID          : -
TunnelType            : SRv6-TE Policy      DelayTimerRemain     : -
Policy State          : Up                  State Change Time    : 2023-11-23 16:19:21
Admin State           : Up                  Traffic Statistics   : Disable
Backup Hot-Standby    : Disable             BFD                  : Disable
Interface Index       : -                   Interface Name       : -
Interface State       : -                   Encapsulation Mode : Insert
Candidate-path Count: 1

 Candidate-path Preference : 100
 Path State           : Active              Path Type            : Primary
 Protocol-Origin      : Configuration(30)   Originator           : 0, 0.0.0.0
 Discriminator        : 100                 Binding SID          : -
 GroupId              : 1                   Policy Name          : PE1-P1-PE2
 Template ID          : 0                   Path Verification    : Disable
 DelayTimerRemain     : -                   Network Slice ID     : -
 Segment-List Count : 1
 Segment-List         : list1
 Segment-List ID      : 1                   XcIndex              : 1
 List State           : Up                  DelayTimerRemain     : -
 Verification State   : -                   SuppressTimeRemain : -
 PMTU                 : 9600                Active PMTU          : 9600
 Weight               : 1                   BFD State            : -
 Network Slice ID     : -
 Binding SID          : -
 Reverse Binding SID: -
 SID :
        2001:100::1
        2001:300::2
//显示路径的SID
PolicyName : PE1-P2-PE2   //隧道名称
Color                  : 102//颜色为102          Endpoint             : 2001:200::100
TunnelId               : 2                       Binding SID          : -
TunnelType             : SRv6-TE Policy          DelayTimerRemain     : -
Policy State           : Up                      State Change Time    : 2023-11-23 16:20:01
Admin State            : Up                      Traffic Statistics   : Disable
Backup Hot-Standby     : Disable                 BFD                  : Disable
Interface Index        : -                       Interface Name       : -
Interface State        : -                       Encapsulation Mode: Insert
Candidate-path Count : 1
```

```
Candidate-path Preference : 100
Path State             : Active              Path Type            : Primary
Protocol-Origin        : Configuration(30)   Originator           : 0, 0.0.0.0
Discriminator          : 100                 Binding SID          : -
GroupId                : 2                   Policy Name          : PE1-P2-PE2
Template ID            : 0                   Path Verification    : Disable
DelayTimerRemain       : -                   Network Slice ID     : -
Segment-List Count     : 1
Segment-List           : list2
Segment-List ID        : 2                   XcIndex              : 2
List State             : Up                  DelayTimerRemain     : -
Verification State     : -                   SuppressTimeRemain: -
PMTU                   : 9600                Active PMTU          : 9600
Weight                 : 1                   BFD State            : -
Network Slice ID       : -
Binding SID            : -
Reverse Binding SID : -
SID :
        2001:100::2
        2001:400::2
// 显示路径的 SID
```

通过以上输出可以看出，PE1 接收路由为 20.20.20.10/32 路由将染色为 101，而显示路径 PE1 → P1 → PE2 的 Color 值也为 101，PE1 在发送目的 IP 地址为 20.20.20.10/32 的流量时，会将流量迭代进入颜色为 101 的 SRv6 Policy 隧道。PE1 接收路由为 20.20.20.20/32 的路由将染色为 102，而显示路径 PE1 → P2 → PE2 的 Color 值也为 102，PE1 在发送目的 IP 地址为 20.20.20.20/32 的流量时，会将流量迭代进入颜色为 102 的 SRv6 Policy 隧道。

步骤 8：测试。CE1 测试结果如图 16-19 所示。

```
[CE1]ping -a 10.10.10.10 20.20.20.10
  PING 20.20.20.10: 56  data bytes, press CTRL_C to break
    Reply from 20.20.20.10: bytes=56 Sequence=1 ttl=252 time=36 ms
    Reply from 20.20.20.10: bytes=56 Sequence=2 ttl=252 time=34 ms
    Reply from 20.20.20.10: bytes=56 Sequence=3 ttl=252 time=34 ms
    Reply from 20.20.20.10: bytes=56 Sequence=4 ttl=252 time=34 ms
    Reply from 20.20.20.10: bytes=56 Sequence=5 ttl=252 time=40 ms

  --- 20.20.20.10 ping statistics ---
    5 packet(s) transmitted
    5 packet(s) received
    0.00% packet loss
    round-trip min/avg/max = 34/35/40 ms

[CE1]ping -a 10.10.10.20 20.20.20.20
  PING 20.20.20.20: 56  data bytes, press CTRL_C to break
    Reply from 20.20.20.20: bytes=56 Sequence=1 ttl=252 time=32 ms
    Reply from 20.20.20.20: bytes=56 Sequence=2 ttl=252 time=39 ms
    Reply from 20.20.20.20: bytes=56 Sequence=3 ttl=252 time=31 ms
    Reply from 20.20.20.20: bytes=56 Sequence=4 ttl=252 time=42 ms
    Reply from 20.20.20.20: bytes=56 Sequence=5 ttl=252 time=35 ms

  --- 20.20.20.20 ping statistics ---
    5 packet(s) transmitted
    5 packet(s) received
    0.00% packet loss
    round-trip min/avg/max = 31/35/42 ms
```

图 16-19　CE1 测试结果

第 17 章

网络自动化

17.1 网络自动化概述

网络自动化是指通过工具实现网络自动化部署和运维，逐步降低对人的依赖程度。

业界有很多实现网络自动化的开源工具，如Ansible、SaltStack、Puppet、Chef和Python等。这些工具通过SSH连接到设备实现批量化的操作和管理，实现了初级网络自动化。

初级网络自动化基于CLI方式管理网络，其痛点在于网络设备返回的是非结构化数据（文本回显），不利于计算机处理。网络自动化发展的基础需求是设备提供结构化的数据，这可以极大地推进网络自动化的进程。设备开放NETCONF/RESTCONF接口，提供XML或者JSON格式的数据类型。

17.1.1 Paramiko简介

Paramiko是Python实现SSHv2协议的模块，它支持口令认证和公钥认证两种方式，可以实现安全的远程命令执行、文件传输等功能。

工程师可以基于Paramiko模块编写Python代码，实现SSH相关功能。Paramiko使用场景示意图如图17-1所示。

图17-1　Paramiko使用场景示意图

Paramiko组件架构如图17-2所示，最常用的两个类为SSHClient类和SFTPClient类，分别提供了SSH和SFTP功能。

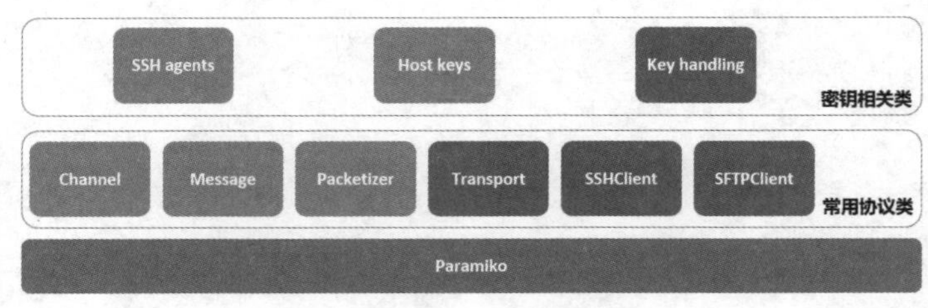

图17-2　Paramiko组件架构

下面简单介绍Paramiko的常用类。

（1）Channel类：该类用于创建在SSH Transport上的安全通道。

（2）Message类：SSH Message是字节流。该类对字符串、整数、布尔值和无限精度整数（Python中称为long）的某些组合进行编码。

（3）Packetizer类：数据包处理类。

（4）Transport类：该类用于在现有套接字或类套接字对象上创建一个Transport会话对象。

（5）SSHClient类：SSHClient类是与SSH服务器会话的高级表示。该类集成了Transport、Channel和SFTPClient类。

（6）SFTPClient类：该类通过一个打开的SSH Transport会话创建SFTP会话通道并执行远程文件操作。

Paramiko 使用流程如图 17-3 所示。

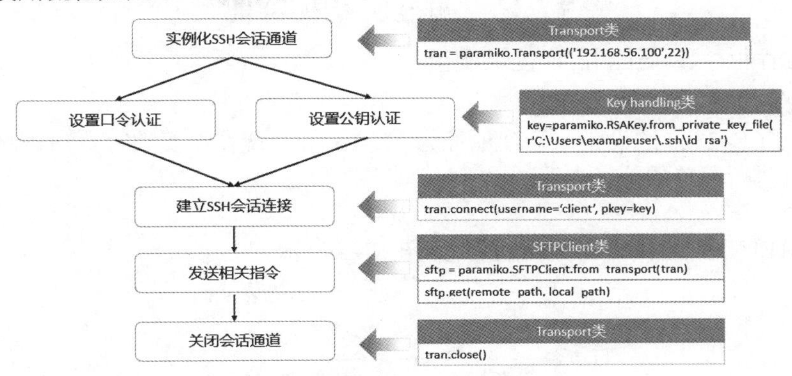

图 17-3　Paramiko 使用流程

17.1.2　NETCONF 简介

NETCONF（Network Configuration Protocol，网络配置协议）提供了一套管理网络设备的机制。用户可以使用这套机制增加、修改、删除网络设备的配置，获取网络设备的配置和状态信息。

NETCONF 协议框架示意图如图 17-4 所示。

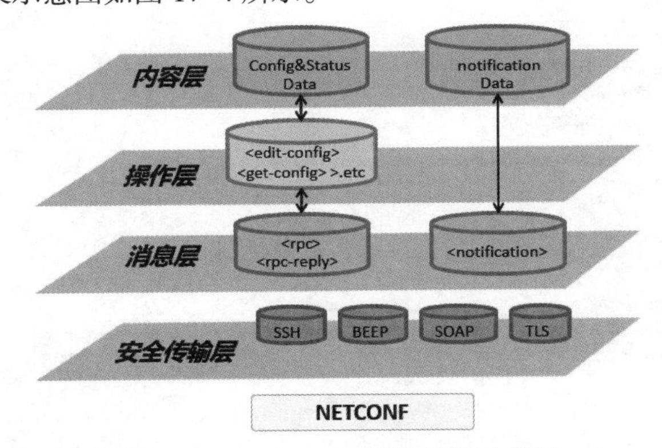

图 17-4　NETCONF 协议框架示意图

NETCONF 协议在概念上可以划分为以下 4 层。

（1）安全传输（Secure Transport）层：为客户端和服务器之间交互提供通信路径。目前华为使用 SSH 协议作为 NETCONF 协议的承载协议。

（2）消息（Messages）层：提供一种简单的不依赖安全传输层的 RPC 请求和回应机制。客户端首先把 RPC 请求内容封装在一个 <rpc> 元素内，然后发送给服务器；服务器首先把请求处理结果封装在一个 <rpc-reply> 元素内，然后回应给客户端。

（3）操作（Operations）层：定义一组基本的操作，作为 RPC 的调用方法，这些操作组成了 NETCONF 基本能力。

（4）内容（Content）层：描述了网络管理所涉及的配置数据，而这些数据依赖于各制造商设备。目前主流的数据模型有 Schema 模型和 YANG 模型等。

17.2 编程自动化实验

17.2.1 Paramiko STELNET 登录设备

1. 实验目的

公司有一台 CE12800 的设备，管理地址为 172.16.1.2，现在需要编写自动化脚本，通过 SSH 登录到设备上并进行简单的信息查看。

2. 实验拓扑

Paramiko STELNET 登录设备的实验拓扑如图 17-5 所示。

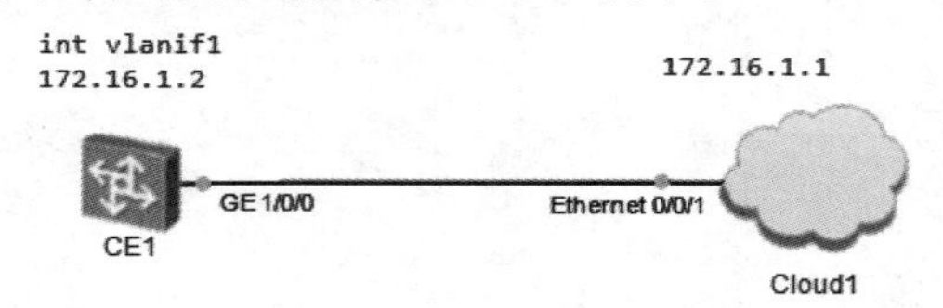

图 17-5 Paramiko STELNET 登录设备的实验拓扑

3. 实验步骤

步骤 1：将本地计算机和 ENSP 的设备进行桥接，桥接配置如图 17-6 所示。

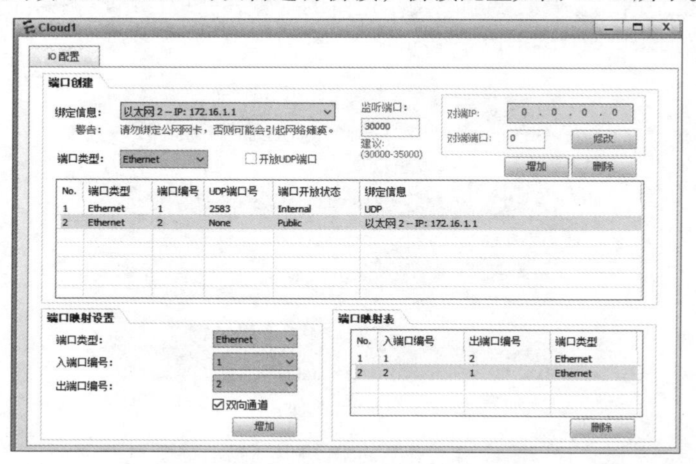

图 17-6 桥接配置

步骤 2：配置交换机的 IP 地址。

```
[HUAWEI]system-view immediately
[HUAWEI]sysname CE1
[CE1]interface  Vlanif 1
[CE1-Vlanif1]ip address 172.16.1.2 24
[CE1-Vlanif1]quit
[CE1]interface  g1/0/0
[CE1-GE1/0/0]undo  shutdown
```

测试本地的 Cmd 窗口与 CE1 设备的连通性：

```
C:\Users\xxx>ping 172.16.1.2
```

```
正在 Ping 172.16.1.2 具有 32 字节的数据：
来自 172.16.1.2 的回复：字节=32 时间=19 ms TTL=255
来自 172.16.1.2 的回复：字节=32 时间=7 ms TTL=255
来自 172.16.1.2 的回复：字节=32 时间=5 ms TTL=255
来自 172.16.1.2 的回复：字节=32 时间=7 ms TTL=255
172.16.1.2 的 Ping 统计信息：
    数据包：已发送 = 4，已接收 = 4，丢失 = 0 (0% 丢失)，
往返行程的估计时间(以毫秒为单位)：
最短 = 5 ms，最长 = 19 ms，平均 = 9 ms
```

步骤 3： 配置 CE1 的 SSH 登录。

（1）创建 SSH 登录的账号：

```
[CE1]aaa
[CE1-aaa]local-user python password cipher Huawei@123
[CE1-aaa]local-user python user-group manage-ug
[CE1-aaa]local-user python service-type ssh
[CE1-aaa]local-user python level 3
```

（2）在 CE1 设备配置 SSH 用户的认证方式和服务类型：

```
[CE1]ssh user python
[CE1]ssh user python authentication-type password
[CE1]ssh user python service-type stelnet
```

（3）配置 VTY 的登录方式以及开启 STELNET 服务：

```
[CE1]stelnet server enable
Info: Succeeded in starting the STelnet server.
[CE1]user-interface vty 0 4
[CE1-ui-vty0-4]authentication-mode aaa
[CE1-ui-vty0-4]protocol inbound ssh
[CE1-ui-vty0-4]user privilege level 3
[CE1-ui-vty0-4]q
```

使用 Shell 工具查看是否能够登录到 CE1 设备，如图 17-7 所示。

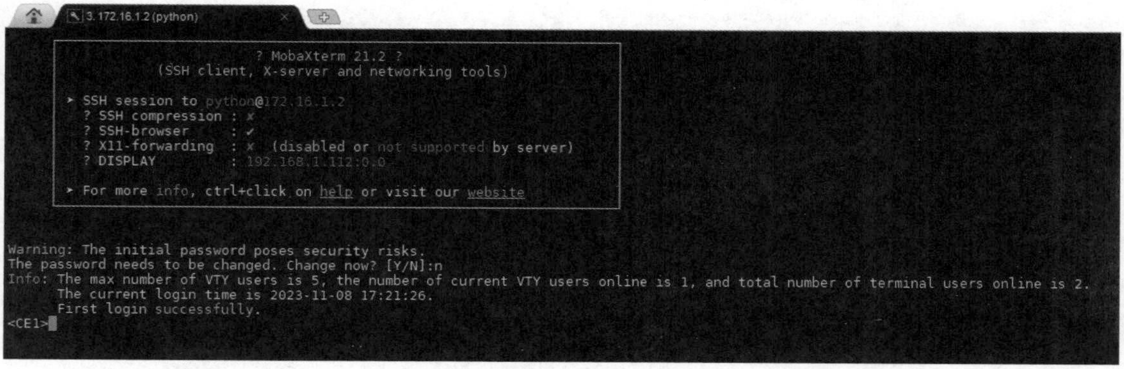

```
3. 172.16.1.2 (python)

                    ? MobaXterm 21.2 ?
          (SSH client, X-server and networking tools)

  ➤ SSH session to python@172.16.1.2
    ? SSH compression : ✘
    ? SSH-browser     : ✔
    ? X11-forwarding  : ✘  (disabled or not supported by server)
    ? DISPLAY         : 192.168.1.112:0.0

  ➤ For more info, ctrl+click on help or visit our website

Warning: The initial password poses security risks.
The password needs to be changed. Change now? [Y/N]:n
Info: The max number of VTY users is 5, the number of current VTY users online is 1, and total number of terminal users online is 2.
      The current login time is 2023-11-08 17:21:26.
      First login successfully.
<CE1>
```

图 17-7 SSH 登录 CE1

步骤 4： 编写 Python 代码。

完整代码如下：

```
import paramiko
import time
ssh_user = 'python'
ssh_pass = 'Huawei@123'
service_IP = '172.16.1.2'
client = paramiko.SSHClient()
client.set_missing_host_key_policy(paramiko.AutoAddPolicy)
client.connect(hostname=service_IP,username=ssh_user,password=ssh_pass)
shell = client.invoke_shell()
shell.send('n\n')
shell.send('screen-length 0 temporary\n')
shell.send('display  current-configuration\n')
time.sleep(3)
dis_cu = shell.recv(999999).decode()
shell.send('display  version\n')
time.sleep(3)
dis_ve = shell.recv(999999).decode()
print(dis_cu)
print('______________________________')
print(dis_ve)
print('______________________________')
client.close()
```

步骤 5： 编译器执行，如图 17-8 所示。

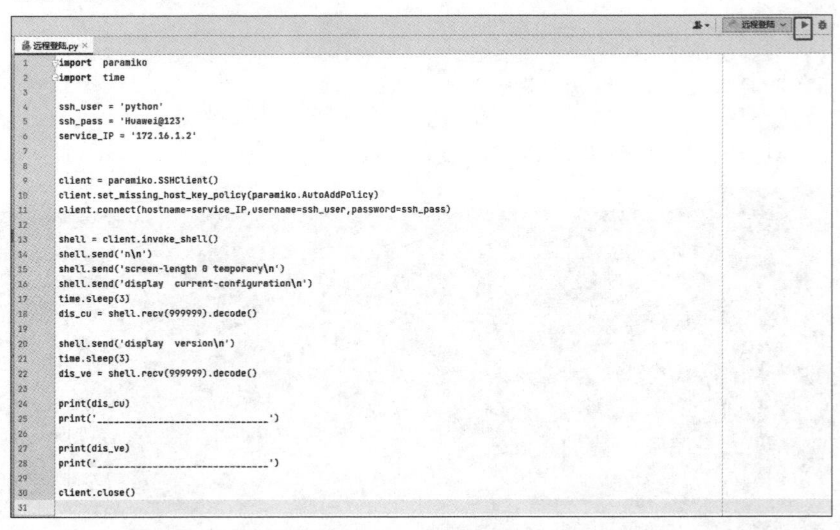

图 17-8　编译器执行

步骤 6： 查看输出结果。

```
Warning: The initial password poses security risks.
```

```
The password needs to be changed. Change now? [Y/N]:n
Info: The max number of VTY users is 5, the number of current VTY users online is 1,
and total number of terminal users online is 2.
      The current login time is 2023-11-08 17:44:26.
      The last login time is 2023-11-08 17:34:00 from 172.16.1.1 through SSH.
<CE1>screen-length 0 temporary
Info: The configuration takes effect on the current user terminal interface only.
<CE1>display current-configuration
!Software Version V800R011C00SPC607B607
!Last configuration was updated at 2023-11-08 17:33:12+00:00 by SYSTEM
automatically
#
sysname CE1
#
device board 17 board-type CE-MPUB
device board 1 board-type CE-LPUE
#
aaa
 local-user python password irreversible-cipher $1c$1}y.#(<UOQ$zxCQM:&Z(CZ$k"$,0p
K,V(2oKQ9T;726;t3}zG8U$
 local-user python service-type ssh
 local-user python level 3
 local-user python user-group manage-ug
 #
 authentication-scheme default
 #
 authorization-scheme default
 #
 accounting-scheme default
 #
 domain default
 #
 domain default_admin
#
interface Vlanif1
 ip address 172.16.1.2 255.255.255.0
#
interface MEth0/0/0
 undo shutdown
#
interface GE1/0/0
 undo shutdown
#
interface GE1/0/1
 shutdown
#
interface GE1/0/2
```

```
 shutdown
#
interface GE1/0/3
 shutdown
#
interface GE1/0/4
 shutdown
#
interface GE1/0/5
 shutdown
#
interface GE1/0/6
 shutdown
#
interface GE1/0/7
 shutdown
#
interface GE1/0/8
 shutdown
#
interface GE1/0/9
 shutdown
#
interface NULL0
#
stelnet server enable
ssh user python
ssh user python authentication-type password
ssh user python service-type stelnet
ssh authorization-type default aaa
#
ssh server cipher aes256_gcm aes128_gcm aes256_ctr aes192_ctr aes128_ctr aes256_
cbc aes128_cbc 3des_cbc
#
ssh server dh-exchange min-len 1024
#
ssh client cipher aes256_gcm aes128_gcm aes256_ctr aes192_ctr aes128_ctr aes256_
cbc aes128_cbc 3des_cbc
#
user-interface con 0
#
user-interface vty 0 4
 authentication-mode aaa
 user privilege level 3
 protocol inbound ssh
#
vm-manager
#
```

```
return
<CE1>
____________________________________
display  version
Huawei Versatile Routing Platform Software
VRP (R) software, Version 8.180 (CE12800 V800R011C00SPC607B607)
Copyright (C) 2012-2018 Huawei Technologies Co., Ltd.
HUAWEI CE12800 uptime is 0 day, 0 hour, 38 minutes
SVRP Platform Version 1.0
<CE1>
____________________________________

进程已结束，退出代码 0
```

下面对以上 Python 代码进行解析。

（1）导入库：

```
import paramiko
import time
```

导入 paramiko 和 time 两个库，如果没有安装 paramiko 库，可以使用编译工具进行安装，以 PyCharm 为例。

单击 PyCharm 的文件中的“设置”按钮，选择“Python 解释器”，单击“+”按钮，添加 Python 库，如图 17-9 所示。

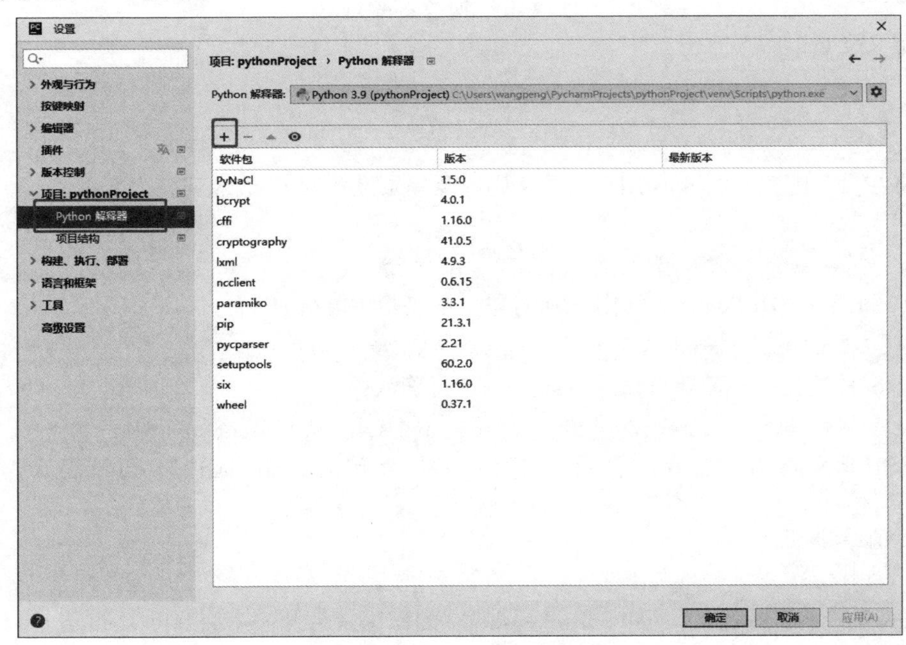

图 17-9　添加 Python 库

搜索 paramiko，单击安装软件包，如图 17-10 所示。

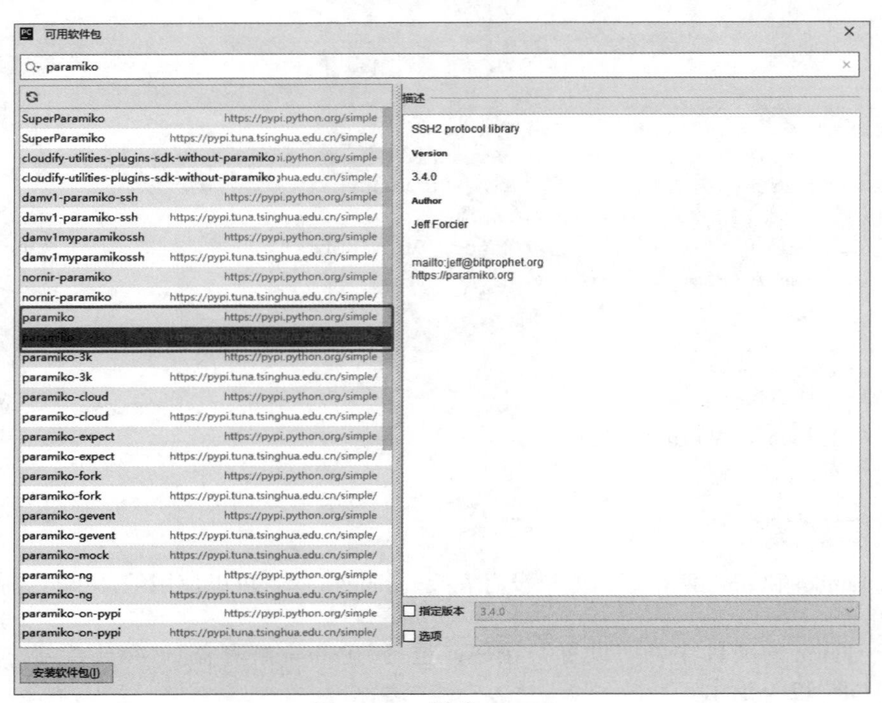

图 17-10　搜索 Paramiko

（2）定义变量：

```
ssh_user = 'python'
ssh_pass = 'Huawei@123'
service_IP = '172.16.1.2'
```

分别将在设备配置好的 SSH 用户名、密码以及登录设备的 IP 地址定义为变量。

（3）建立 SSH 会话连接：

```
client = paramiko.SSHClient()
```

使用 Paramiko SSHClient() 实例化 SSH 对象。本实验赋值给 client：

```
client.set_missing_host_key_policy(paramiko.AutoAddPolicy)
```

新建立 SSH 连接时不需要再输入 yes 或 no 进行确认：

```
client.connect(hostname=service_IP,username=ssh_user,password=ssh_pass)
```

目的 SSH 服务器为 172.16.1.2，用户名为 python，密码为 Huawei@123，以密码认证方式进行用户认证。

（4）赋值给 shell：

```
shell = client.invoke_shell()
```

调用 invoke_shell() 赋值给 shell。invoke_shell() 的作用是打开一个交互的 shell 会话。该会话为一个逻辑通道 channel，建立在 SSH 会话连接上。

```
shell.send('n\n')
shell.send('screen-length 0 temporary\n')
```

```
shell.send('display current-configuration\n')
time.sleep(3)
```

以上代码的具体说明如下：

shell.send 表示在命令行输入相应的指令。

shell.send('n\n') 表示登录到设备时无须再次修改密码。

shell.send('screen-length 0 temporary\n') 表示取消分屏。

shell.send('display current-configuration\n') 表示查看设备的当前配置。

time.sleep(3) 表示等待 3 秒。

```
dis_cu = shell.recv(999999).decode()
```

invoke_shell() 已经创建了一个 channel 逻辑通道。此前所有的输入 / 输出的过程信息都在此 channel 中。用户可以获取这个 channel 中的所有信息，并显示到 Python 编译器。

```
shell.send('display version\n')
time.sleep(3)
dis_ve = shell.recv(999999).decode()
```

查看设备的版本信息，并且获取回显信息：

```
print(dis_cu)
print('______________________________')
print(dis_ve)
print('______________________________')
```

输出回显信息：

```
client.close()
```

关闭会话连接。

17.2.2 Paramiko SFTP 文件传输

1. 实验目的

公司有一台 CE12800 的设备，管理地址为 172.16.1.2，现在需要编写自动化脚本，通过 SFTP 实现简单的上传下载操作。

2. 实验拓扑

Paramiko SFTP 文件传输的实验拓扑如图 17-11 所示。

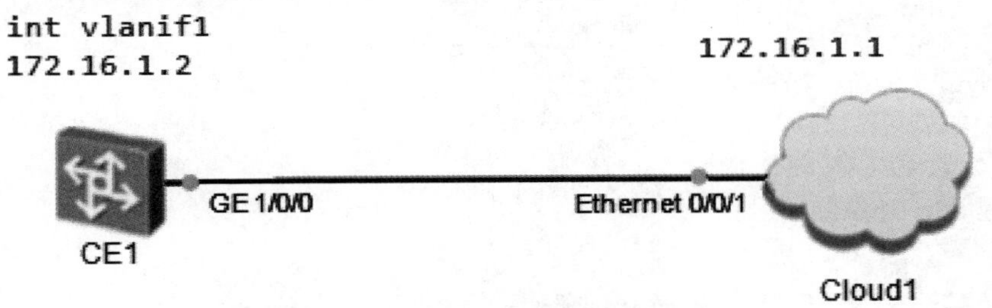

图 17-11 Paramiko SFTP 文件传输的实验拓扑

3. 实验步骤

步骤 1： 将本地计算机和 ENSP 的设备进行桥接，桥接配置图如图 17-12 所示。

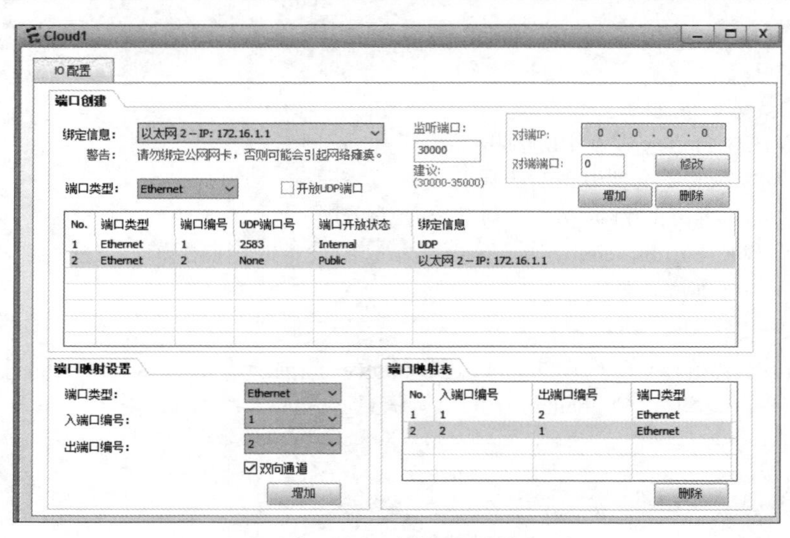

图 17-12　桥接配置图

步骤 2：配置交换机的 IP 地址。

```
[HUAWEI]system-view immediately
[HUAWEI]sysname CE1
[CE1]interface Vlanif 1
[CE1-Vlanif1]ip address 172.16.1.2 24
[CE1-Vlanif1]quit
[CE1]interface g1/0/0
[CE1-GE1/0/0]undo shutdown
```

测试本地的 Cmd 窗口与 CE1 设备的连通性：

```
C:\Users\xxx>ping 172.16.1.2
正在 Ping 172.16.1.2 具有 32 字节的数据：
来自 172.16.1.2 的回复：字节 =32 时间 =19 ms TTL=255
来自 172.16.1.2 的回复：字节 =32 时间 =7 ms TTL=255
来自 172.16.1.2 的回复：字节 =32 时间 =5 ms TTL=255
来自 172.16.1.2 的回复：字节 =32 时间 =7 ms TTL=255
172.16.1.2 的 Ping 统计信息：
    数据包：已发送 = 4，已接收 = 4，丢失 = 0 (0% 丢失)，
往返行程的估计时间(以毫秒为单位)：
最短 = 5 ms，最长 = 19 ms，平均 = 9 ms
```

步骤 3：配置 CE1 的 SSH 登录。

（1）创建 SSH 登录的账号：

```
[CE1]aaa
[CE1-aaa]local-user python password cipher Huawei@123
[CE1-aaa]local-user python user-group manage-ug
[CE1-aaa]local-user python service-type ssh
[CE1-aaa]local-user python level 3
```

（2）在 CE1 设备配置 SSH 用户的认证方式和服务类型：

```
[CE1]ssh user python
[CE1]ssh user python authentication-type password
[CE1]ssh user python service-type sftp
[CE1]ssh user python sftp-directory cfcard:
[CE1]ssh authentication-type default password
[CE1]sftp server enable
[CE1]user-interface vty 0 4
[CE1-ui-vty0-4]authentication-mode aaa
[CE1-ui-vty0-4]protocol inbound ssh
[CE1-ui-vty0-4]user privilege level 3
[CE1-ui-vty0-4]q
```

步骤 4：编写 Python 代码。

完整代码如下：

```
import paramiko
tran = paramiko.Transport(('172.16.1.2',22))
tran.connect(username='python',password='Huawei@123')
sftp = paramiko.SFTPClient.from_transport(tran)
local_path = r'F:\test\test.cfg'
remote_path = '/vrpcfg.cfg'
sftp.get(remote_path,local_path)
print('get file succeed')
sftp.put(local_path,'/test.cfg')
print('put file succeed')
tran.close()
```

步骤 5：编译器执行，如图 17–13 所示。

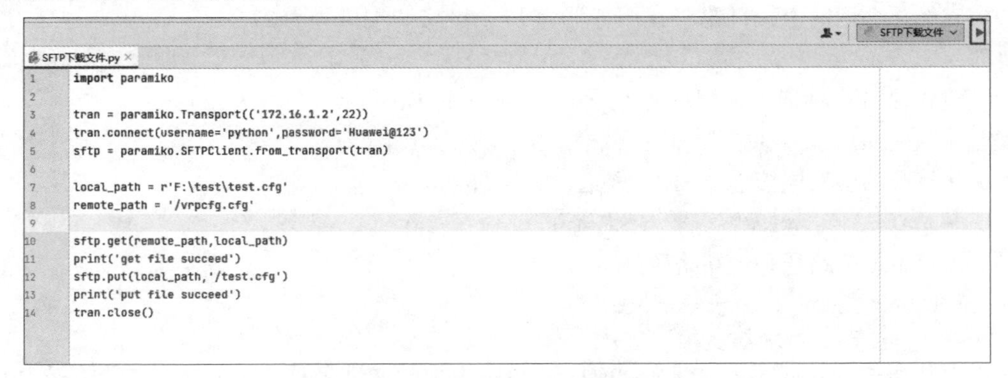

图 17–13　编译器执行

步骤 6：查看执行结果。

```
get file succeed
put file succeed
```

在本地计算机的 F:\test 目录下查看是否有备份的 test.cfg 文件，如图 17–14 所示。

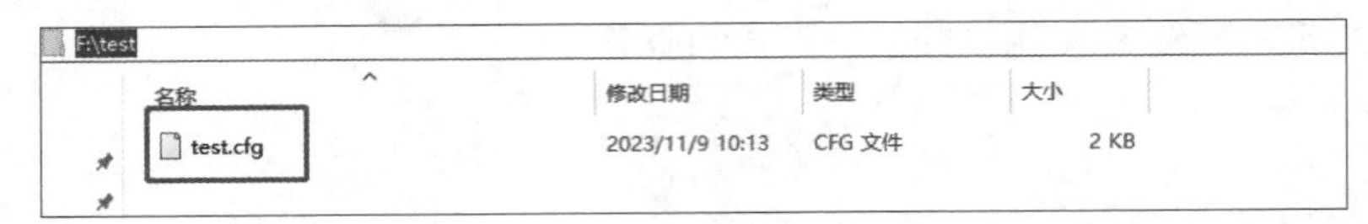

图 17-14　查看本地文件

在 CE1 设备上查看是否上传了 test.cfg 文件：

```
<CE1>dir
Directory of cfcard:/
  Idx   Attr    Size(Byte)  Date           Time        FileName
  0     dr-x    -           Nov 09 2023    10:03:28    $_checkpoint
  1     dr-x    -           Nov 09 2023    09:36:26    $_install_mod
  2     dr-x    -           Nov 09 2023    09:37:01    $_license
  3     dr-x    -           Nov 09 2023    09:37:06    $_security_info
  4     dr-x    -           Nov 09 2023    09:37:03    $_system
  5     -rw-    0           Nov 09 2023    09:36:26    CE12800
  6     -rw-    104         Nov 09 2023    09:36:26    VRPV800R011C00SPC607B607D0203_
s12800.cc
  7     -rw-    251         Nov 09 2023    09:36:26    device.sys
  8     -rw-    1,718       Nov 09 2023    10:13:31    test.cfg
  9     -rw-    1,718       Nov 09 2023    09:36:26    vrpcfg.cfg
```

下面对以上 Python 代码进行解析。

（1）导入库：

```
import paramiko
```

导入 paramiko 库，本章会使用 Transport 类和 SFTPClient 类。Transport 类用于建立 SFTP 的会话通道，SFTPClient 类用于上传和下载文件。

（2）实例化会话通道，目的 SSH 服务器为 172.16.1.2，端口号为 22：

```
tran = paramiko.Transport(('172.16.1.2',22))
```

（3）建立 SSH 会话通道，用户名为 python，密码为 Huawei@123：

```
tran.connect(username='python',password='Huawei@123')
```

（4）从打开的会话连接创建 SFTP 通道，赋值给 sftp：

```
sftp = paramiko.SFTPClient.from_transport(tran)
```

（5）指定本地的路径和远端路径：

```
local_path = r'F:\test\test.cfg'
remote_path = '/vrpcfg.cfg'
```

在以上代码中，local_path = r'F:\test\test.cfg' 表示本地的文件名称为 test.cfg，文件路径为 F:\test；remote_path = '/vrpcfg.cfg' 表示 SFTP 服务器上的文件路径和名称为 /vrpcfg.cfg。

（6）执行下载文件操作：

```
sftp.get(remote_path,local_path)
print('get file succeed')
```

将 SFTP 服务器上的 /vrpcfg.cfg 下载到本地路径 F:\test 上，并命名为 test.cfg，下载完成后，输出 get file succeed 提示语句。

（7）执行下载文件操作：

```
sftp.put(local_path,'/test.cfg')
print('put file succeed')
```

将本地路径上的 F:\test\test.cfg 上传到 SFTP 服务器的根路径下。

（8）关闭会话连接：

```
tran.close()
```

17.2.3 Paramiko 综合实验

1. 实验目的

公司有一台 CE12800 的设备，管理地址为 172.16.1.2，现在需要编写自动化脚本，使用 STELNET 实现设备的自动保存配置文件，使用 SFTP 实现设备的文件下载。

2. 实验拓扑

Paramiko 综合实验的实验拓扑如图 17-15 所示。

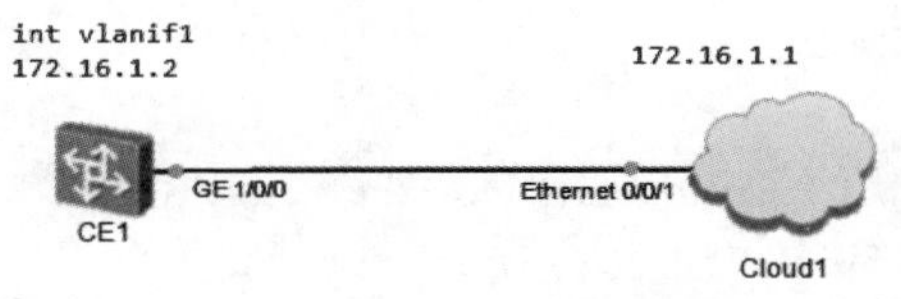

图 17-15 Paramiko 综合实验的实验拓扑

3. 实验步骤

步骤 1： 将本地计算机和 ENSP 的设备进行桥接，桥接示意图如图 17-16 所示。

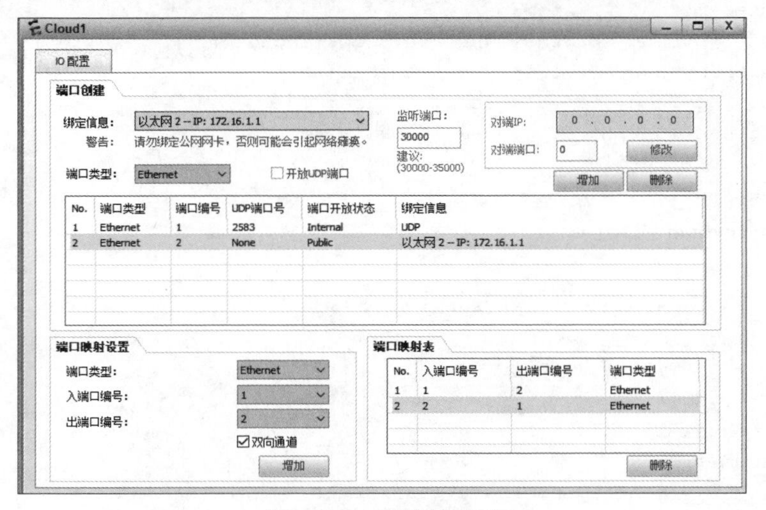

图 17-16 桥接示意图

步骤 2： 配置交换机的 IP 地址。

```
[HUAWEI]system-view immediately
[HUAWEI]sysname CE1
[CE1]interface Vlanif 1
[CE1-Vlanif1]ip address 172.16.1.2 24
```

```
[CE1-Vlanif1]quit
[CE1]interface g1/0/0
[CE1-GE1/0/0]undo shutdown
```

测试本地的Cmd窗口与CE1设备的连通性：

```
C:\Users\xxx>ping 172.16.1.2
正在 Ping 172.16.1.2 具有 32 字节的数据：
来自 172.16.1.2 的回复：字节=32 时间=19 ms TTL=255
来自 172.16.1.2 的回复：字节=32 时间=7 ms TTL=255
来自 172.16.1.2 的回复：字节=32 时间=5 ms TTL=255
来自 172.16.1.2 的回复：字节=32 时间=7 ms TTL=255
172.16.1.2 的 Ping 统计信息：
    数据包：已发送 = 4，已接收 = 4，丢失 = 0 (0% 丢失)，
往返行程的估计时间（以毫秒为单位）：
最短 = 5 ms，最长 = 19 ms，平均 = 9 ms
```

步骤3： 配置CE1的SSH登录。

（1）创建SSH登录的账号：

```
[CE1]aaa
[CE1-aaa]local-user python password cipher Huawei@123
[CE1-aaa]local-user python user-group manage-ug
[CE1-aaa]local-user python service-type ssh
[CE1-aaa]local-user python level 3
```

（2）在CE1设备配置SSH用户的认证方式和服务类型：

```
[CE1]ssh user python
[CE1]ssh user python authentication-type password
[CE1]ssh user python service-type sftp stelnet
[CE1]ssh user python sftp-directory cfcard:
[CE1]ssh authentication-type default password
[CE1]sftp server enable
[CE1]stelnet server enable
```

步骤4： 编写Python代码：

完整代码如下：

```
from paramiko import SSHClient,AutoAddPolicy
from time import  sleep
service_ip = '172.16.1.2'
ssh_user = 'python'
ssh_pass = 'Huawei@123'
class datacom():
    def __init__(self,ip,username,password):
        self.ip = ip
        self.username = username
        self.password = password
        self.ssh = self.ssh_connect()
    def ssh_connect(self):
        ssh = SSHClient()
        ssh.set_missing_host_key_policy(AutoAddPolicy)
```

```
        ssh.connect(self.ip,username=self.username,password=self.password)
        return ssh
    def save_config(self):
        shell = self.ssh.invoke_shell()
        shell.send('n\n')
        sleep(2)
        shell.send('save  CE1_save.zip\n')
        sleep(2)
        shell.send('y\n')
        sleep(2)
        shell.send('dir\n')
        sleep(2)
        dis_file = shell.recv(999999).decode()
        print(dis_file)
        self.ssh.close()
    def down_file(self):
        remotename = 'CE1_save.zip'
        localname = r'F:\test\CE1_save.zip'
        self.ssh.open_sftp().get(remotename,localname)
        print('get file succeed')
        self.ssh.close()
if __name__ =='__main__':
    Joinlabs = datacom(service_ip,ssh_user,ssh_pass)
    Joinlabs.save_config()
    Joinlabs = datacom(service_ip,ssh_user,ssh_pass)
    Joinlabs.down_file()
```

步骤 5: 编译器执行，如图 17–17 所示。

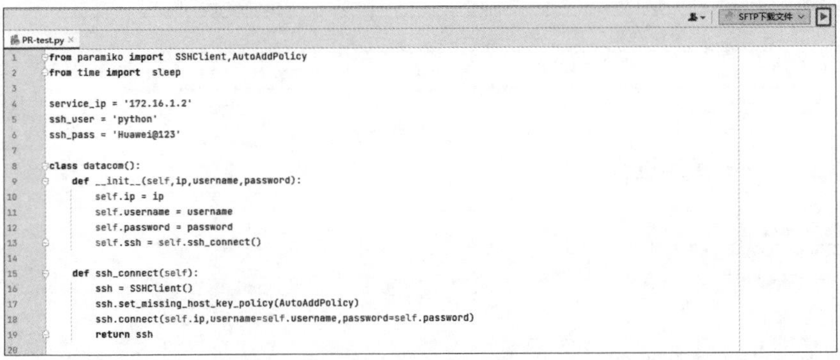

```
PR-test.py
1   from paramiko import  SSHClient,AutoAddPolicy
2   from time import  sleep
3
4   service_ip = '172.16.1.2'
5   ssh_user = 'python'
6   ssh_pass = 'Huawei@123'
7
8   class datacom():
9       def __init__(self,ip,username,password):
10          self.ip = ip
11          self.username = username
12          self.password = password
13          self.ssh = self.ssh_connect()
14
15      def ssh_connect(self):
16          ssh = SSHClient()
17          ssh.set_missing_host_key_policy(AutoAddPolicy)
18          ssh.connect(self.ip,username=self.username,password=self.password)
19          return ssh
20
```

图 17–17 编译器执行

步骤 6: 查看输出结果。

```
Warning: The initial password poses security risks.
The password needs to be changed. Change now? [Y/N]:n
Info: The max number of VTY users is 5, the number of current VTY users online is 1,
and total number of terminal users online is 2.
    The current login time is 2023-11-09 14:48:19.
```

```
        The last login time is 2023-11-09 11:52:29 from 172.16.1.1 through SSH.
<CE1>save  CE1_save.zip
Warning: Are you sure to save the configuration to cfcard:/CE1_save.zip? [Y/N]:y
Now saving the current configuration to the slot 17
Info: Save the configuration successfully.
<CE1>dir
Directory of cfcard:/
  Idx   Attr    Size(Byte)   Date          Time        FileName
  0     dr-x    -            Nov 09 2023   10:03:28    $_checkpoint
  1     dr-x    -            Nov 09 2023   09:36:26    $_install_mod
  2     dr-x    -            Nov 09 2023   09:37:01    $_license
  3     dr-x    -            Nov 09 2023   09:37:06    $_security_info
  4     dr-x    -            Nov 09 2023   09:37:03    $_system
  5     -rw-    0            Nov 09 2023   09:36:26    CE12800
  6     -rw-    866          Nov 09 2023   14:48:23    CE1_save.zip
  7     -rw-    104          Nov 09 2023   09:36:26    VRPV800R011C00SPC607B607D0203_
s12800.cc
  8     -rw-    2,893        Nov 09 2023   14:48:23    device.sys
  9     -rw-    1,718        Nov 09 2023   10:13:31    test.cfg
  10    -rw-    1,718        Nov 09 2023   09:36:26    vrpcfg.cfg
8,388,608 KB total (6,224,796 KB free)
<CE1>
get file succeed
```

通过以上输出可以看出，设备的文件系统多了 CE1_save.zip 的保存配置文件，并且执行结果输出了 get file succeed，表示文件下载成功，在本地计算机上查看 F:\test 路径中是否存在 CE1_save.zip，如图 17-18 所示。

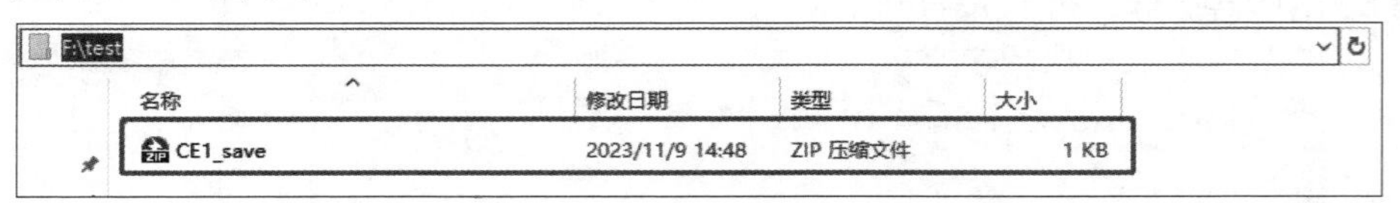

图 17-18　查看本地文件

下面对以上 Python 代码进行解析。

（1）导入库：

```
from paramiko import SSHClient,AutoAddPolicy
from time import sleep
```

（2）定义变量：

```
service_ip = '172.16.1.2'
ssh_user = 'python'
ssh_pass = 'Huawei@123'
```

将 SSH 登录需要用到的 IP 地址、用户名和密码定义为变量。

（3）定义类：

```
class datacom():
```

定义类 datacom，类名为 datacom。在此类中定义 ssh_connect()、save_config() 和 down_file() 三个方法。其中，ssh_connect() 用于建立 SSH 连接；save_config() 用于保存设备配置；down_file() 用于

下载设备的文件。

（4）定义构造函数：

```
def __init__(self,ip,username,password):
    self.ip = ip
    self.username = username
    self.password = password
    self.ssh = self.ssh_connect()
```

构造方法 __init__（__ 在实际输入时是两个下画线）用于创建实例对象，每当创建一个类的实例对象时，Python 解释器都会自动调用该构造方法来初始化对象的某些属性。

def __init__(self,ip,username,password) 表示在调用类 datacom() 时所需填入的 IP 地址、用户名和密码。其中，self.ip、self.username、self.password 可以在同一个类的函数下进行传参的动作。self.ssh = self.ssh_connect() 表示调用接下来要定义的 ssh_connect() 方法。

（5）定义 ssh_connect() 方法，用于建立 SSH 连接：

```
def ssh_connect(self):
    ssh = SSHClient()
    ssh.set_missing_host_key_policy(AutoAddPolicy)
    ssh.connect(self.ip,username=self.username,password=self.password)
    return ssh
```

在以上代码中，ssh.connect(self.ip,username=self.username,password=self.password) 表示当 SSH 登录网络设备时，需要输入 IP 地址、用户名和密码三个参数。

（6）定义 save_config() 方法，用于保存设备配置：

```
def save_config(self):
    shell = self.ssh.invoke_shell() # 调用构造函数中的 self.ssh，而 self.ssh 就是用
                                    # 于建立 SSH 连接的，并且开启交互式会话
    shell.send('n\n') # 输入命令 no ，登录时设备提示修改密码，此时输入 n 代表不修改密码
    sleep(2)
    shell.send('save  CE1_save.zip\n') # 输入命令 save CE1_save.zip，将保存的文件命名
                                       # 为 CE1_save.zip
    sleep(2)
    shell.send('y\n')
    sleep(2)
    shell.send('dir\n')
    sleep(2)
    dis_file = shell.recv(999999).decode()
    print(dis_file)
    self.ssh.close()
```

（7）定义 down_file() 方法，用于下载配置文件：

```
def down_file(self):
remotename = 'CE1_save.zip'
localname = r'F:\test\CE1_save.zip'
self.ssh.open_sftp().get(remotename,localname)
print('get file succeed')
self.ssh.close()
```

（8）定义主函数，顺序执行：

```
if __name__ =='__main__':
    Joinlabs = datacom(service_ip,ssh_user,ssh_pass)
    Joinlabs.save_config()
    Joinlabs = datacom(service_ip,ssh_user,ssh_pass)
    Joinlabs.down_file()
```

17.2.4 NETCONF 配置实验

1. 实验目的

公司有一台 CE12800 的设备，管理地址为 172.16.1.2，现在需要编写自动化脚本，通过 SSH 登录到设备上配置 NETCONF 协议的用户名、密码以及 NETCONF 服务，并且通过 NETCONF 协议将设备的 LoopBack 0 接口的 IP 地址配置为 1.1.1.1/32。

2. 实验拓扑

NETCONF 配置实验的实验拓扑如图 17-19 所示。

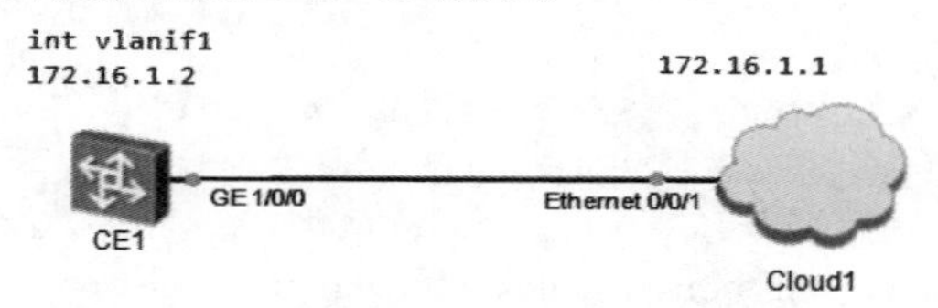

图 17-19　NETCONF 配置实验的实验拓扑

3. 实验步骤

步骤 1： 将本地计算机和 ENSP 的设备进行桥接，桥接示意图如图 17-20 所示。

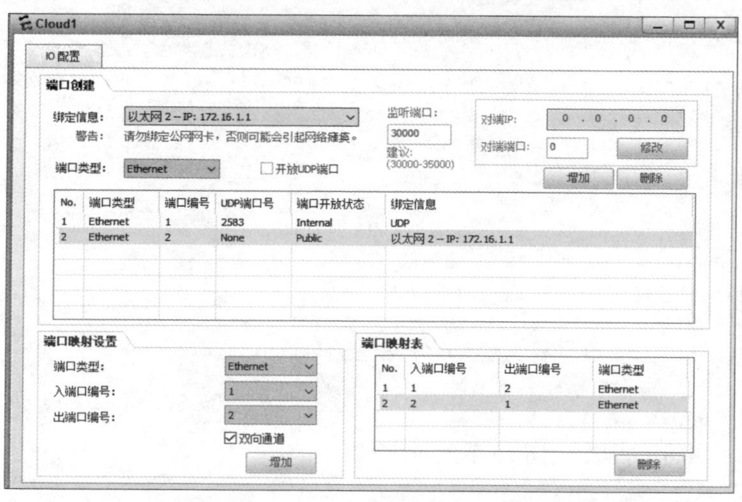

图 17-20　桥接示意图

步骤 2： 配置交换机的 IP 地址。

```
[HUAWEI]system-view immediately
[HUAWEI]sysname CE1
[CE1]interface Vlanif 1
```

```
[CE1-Vlanif1]ip address 172.16.1.2 24
[CE1-Vlanif1]quit
[CE1]interface g1/0/0
[CE1-GE1/0/0]undo  shutdown
```

测试本地的 Cmd 窗口与 CE1 设备的连通性：

```
C:\Users\xxx>ping 172.16.1.2
正在 Ping 172.16.1.2 具有 32 字节的数据：
来自 172.16.1.2 的回复：字节 =32 时间 =19 ms TTL=255
来自 172.16.1.2 的回复：字节 =32 时间 =7 ms TTL=255
来自 172.16.1.2 的回复：字节 =32 时间 =5 ms TTL=255
来自 172.16.1.2 的回复：字节 =32 时间 =7 ms TTL=255
172.16.1.2 的 Ping 统计信息：
    数据包：已发送 = 4，已接收 = 4，丢失 = 0 (0% 丢失)，
往返行程的估计时间(以毫秒为单位)：
最短 = 5 ms，最长 = 19 ms，平均 = 9 ms
```

步骤 3： 配置 CE1 的 SSH 登录。

（1）创建 SSH 登录的账号：

```
[CE1]aaa
[CE1-aaa]local-user python password cipher Huawei@123
[CE1-aaa]local-user python user-group manage-ug
[CE1-aaa]local-user python service-type ssh
[CE1-aaa]local-user python level 3
```

（2）在 CE1 设备配置 SSH 用户的认证方式和服务类型：

```
[CE1]ssh user python
[CE1]ssh user python authentication-type password
[CE1]ssh user python service-type stelnet
```

（3）配置 VTY 的登录方式并开启 STELNET 服务：

```
[CE1]stelnet server enable
Info: Succeeded in starting the STelnet server.
[CE1]user-interface vty 0 4
[CE1-ui-vty0-4]authentication-mode aaa
[CE1-ui-vty0-4]protocol inbound ssh
[CE1-ui-vty0-4]user privilege level 3
[CE1-ui-vty0-4]q
```

步骤 4： 编写 Python 代码。

完整代码如下：

```
from paramiko import SSHClient,AutoAddPolicy
from ncclient import manager
from ncclient import operations
from time import  sleep
service_ip = '172.16.1.2'
ssh_user = 'python'
ssh_pass = 'Huawei@123'
```

```
netconf_user = 'netconf'
netconf_pass = 'Huawei@123'
port = '830'
XMLS = '''<config>
        <ethernet  xmlns="http://www.huawei.com/netconf/vrp"  content-version="1.0"
format-version="1.0">
        <ethernetIfs>
          <ethernetIf operation="merge">
            <ifName>GE1/0/2</ifName>
            <l2Enable>disable</l2Enable>
          </ethernetIf>
        </ethernetIfs>
      </ethernet>
      <ifm xmlns="http://www.huawei.com/netconf/vrp" content-version="1.0" format-
version="1.0">
        <interfaces>
          <interface operation="merge">
            <ifName>Loopback0</ifName>
            <ifDescr>Config by NETCONF</ifDescr>
            <ifmAm4>
              <am4CfgAddrs>
                <am4CfgAddr operation="create">
                  <subnetMask>255.255.255.255</subnetMask>
                  <addrType>main</addrType>
                  <ifIpAddr>1.1.1.1</ifIpAddr>
                </am4CfgAddr>
              </am4CfgAddrs>
            </ifmAm4>
          </interface>
        </interfaces>
      </ifm>
    </config>'''
class datacom():
    def __init__(self,ip,username,password):
        self.ip = ip
        self.username = username
        self.password = password
        self.ssh = self.ssh_connect()
    def ssh_connect(self):
        ssh = SSHClient()
        ssh.set_missing_host_key_policy(AutoAddPolicy)
        ssh.connect(self.ip,username=self.username,password=self.password)
        return ssh
    def ssh_config(self):
        shell = self.ssh.invoke_shell()
        shell.send('n\n')
        f = open("config_netconf.txt")
        cmd = f.readlines()
```

```
        for i in cmd:
            shell.send(i)
            sleep(2)
        dis_this = shell.recv(999999).decode()
        print(dis_this)
        self.ssh.close()
def netconf_connect(host, port, user, password):
    return manager.connect(host=host,
                          port=port,
                          username=user,
                          password=password,
                          hostkey_verify = False,
                          device_params={'name': "huawei"},
                          allow_agent = False,
                          look_for_keys = False)
if __name__ == '__main__':
    joinlabs = datacom(service_ip,ssh_user,ssh_pass)
    joinlabs.ssh_config()
    netconf = netconf_connect(service_ip,port,netconf_user,netconf_pass)
    netconf.edit_config(target='running',config=XMLS)
```

步骤5: 编译器执行，如图 17-21 所示。

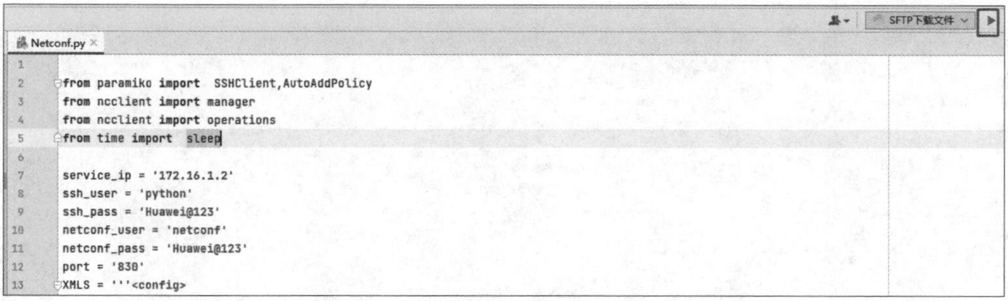

图 17-21 编译器执行

步骤6: 查看输出结果。

```
Warning: The initial password poses security risks.
The password needs to be changed. Change now? [Y/N]:n
Info: The max number of VTY users is 5, the number of current VTY users online is 1,
and total number of terminal users online is 1.
      The current login time is 2023-11-09 20:09:21.
      The last login time is 2023-11-09 19:40:14 from 172.16.1.1 through SSH.
<CE1>system-view immediately
Enter system view, return user view with return command.
[CE1]aaa
[CE1-aaa]local-user netconf password irreversible-cipher Huawei@123
Info: A new user is added.
[CE1-aaa]local-user netconf service-type ssh
[CE1-aaa]local-user netconf level 3
```

```
[CE1-aaa]quit
[CE1]ssh user netconf authentication-type password
Info: Succeeded in adding a new SSH user.
[CE1]ssh user netconf service-type snetconf
[CE1]snetconf server enable
Info: Succeeded in starting the SNETCONF server on SSH port 22.
[CE1]netconf
[CE1-netconf]protocol inbound ssh port 830
Info: Succeeded in starting the ssh port 830 service.
[CE1-netconf]quit
[CE1]
进程已结束，退出代码 0
```

登录 CE1 查看 LoopBack 0 接口配置：

```
[CE1]interface  LoopBack 0
[CE1-LoopBack0]dis this
interface LoopBack0
 description Config by NETCONF
 ip address 1.1.1.1 255.255.255.255
```

下面对以上 Python 代码进行解析。

（1）在 PyCharm 的本项目的 Python 文件的同一目录下创建 txt 文档，用于配置 NETCONF 服务。创建 txt 文档示意图如图 17-22 所示。

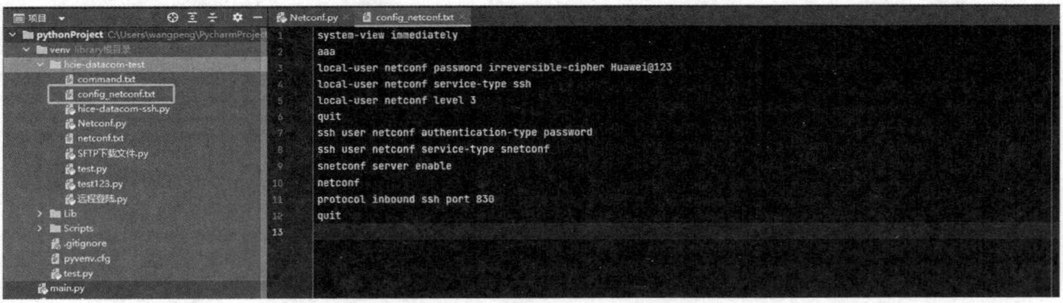

图 17-22　创建 txt 文档示意图

```
system-view immediately
aaa
local-user netconf password irreversible-cipher Huawei@123
local-user netconf service-type ssh
local-user netconf level 3
quit
ssh user netconf authentication-type password
ssh user netconf service-type snetconf
snetconf server enable
netconf
protocol inbound ssh port 830
quit
```

（2）导入库：

```
from paramiko import SSHClient,AutoAddPolicy
from ncclient import manager
from ncclient import operations
from time import sleep
```

导入 paramiko 库用于 SSH 远程登录设备进行配置，导入 ncclinet 库用于 NETCONF 的连接和配置。

（3）定义变量：

```
service_ip = '172.16.1.2'
ssh_user = 'python'
ssh_pass = 'Huawei@123'
netconf_user = 'netconf'
netconf_pass = 'Huawei@123'
port = '830'
```

将需要登录设备的 IP 地址、SSH 登录用户名、密码以及 NETCONF 的用户名和密码定义为变量。

（4）构建 XML 配置文件：

```
XMLS = '''<config>
        <ethernet xmlns="http://www.huawei.com/netconf/vrp" content-version="1.0"
format-version="1.0">
        <ethernetIfs>
          <ethernetIf operation="merge">
            <ifName>GE1/0/2</ifName>
            <l2Enable>disable</l2Enable>
          </ethernetIf>
        </ethernetIfs>
      </ethernet>
      <ifm xmlns="http://www.huawei.com/netconf/vrp" content-version="1.0" format-
version="1.0">
        <interfaces>
          <interface operation="merge">
            <ifName>Loopback0</ifName>
            <ifDescr>Config by NETCONF</ifDescr>
            <ifmAm4>
              <am4CfgAddrs>
                <am4CfgAddr operation="create">
                  <subnetMask>255.255.255.255</subnetMask>
                  <addrType>main</addrType>
                  <ifIpAddr>1.1.1.1</ifIpAddr>
                </am4CfgAddr>
              </am4CfgAddrs>
            </ifmAm4>
          </interface>
        </interfaces>
      </ifm>
    </config>'''
```

NETCONF 通过 XML 文件传递配置信息。XML 是一种非常常用的文本格式，可以使用“<>”

不断嵌套展开数据。完整的 NETCONF 会话有安全传输层、消息层、操作层和内容层。在当前 XML 配置文件中传递的信息仅包含操作层和内容层。

本实验中的 XML 文件用于将 LoopBack 0 接口的 IP 地址改为 1.1.1.1/32。

（5）定义构造函数：

```
class datacom():
    def __init__(self,ip,username,password):
        self.ip = ip
        self.username = username
        self.password = password
        self.ssh = self.ssh_connect()
```

（6）定义 def ssh_connect(): 方法，用于建立 SSH 连接，登录网络设备：

```
    def ssh_connect(self):
        ssh = SSHClient()
        ssh.set_missing_host_key_policy(AutoAddPolicy)
        ssh.connect(self.ip,username=self.username,password=self.password)
        return ssh
```

（7）定义 def ssh_config(): 方法，用于登录设备，进行配置：

```
    def ssh_config(self):
        shell = self.ssh.invoke_shell()
        shell.send('n\n')
        f = open("config_netconf.txt")
        cmd = f.readlines()
        for i in cmd:
            shell.send(i)
            sleep(2)
        dis_this = shell.recv(999999).decode()
        print(dis_this)
        self.ssh.close()
```

在以上代码中，f = open("config_netconf.txt") 表示打开文件 config_netconf.txt；cmd = f.readlines() 表示对 config_netconf.txt 这个文件的内容进行逐行读取。

```
   for i in cmd:
    shell.send(i)
            sleep(2)
```

以上代码定义了一个循环语句，将 config_netconf.txt 的内容输入到设备的命令行，即配置设备的 NETCONF 服务。

（8）定义 netconf_connect(): 方法，用于建立 NETCONF 连接：

```
def netconf_connect(host, port, user, password):
    return manager.connect(host=host,
                           port=port,
                           username=user,
                           password=password,
                           hostkey_verify = False,
                           device_params={'name': "huawei"},
```

```
                        allow_agent = False,
                        look_for_keys = False)
```

定义函数 netconf_connect(host, port, user, password)。函数输入的4个参数为NETCONF主机的IP地址、端口号、NETCONF用户名和密码。函数返回ncclient的manager.connect()方法。manager.connect()方法的作用是建立NETCONF连接。

（9）运行主函数：

```
if __name__ == '__main__':
    joinlabs = datacom(service_ip,ssh_user,ssh_pass)
    joinlabs.ssh_config()
    netconf = netconf_connect(service_ip,port,netconf_user,netconf_pass)
    netconf.edit_config(target='running',config=XMLS)
```

在以上代码中，首先，执行joinlabs.ssh_config()，即调用datacom()类的ssh_config()方法，并且输入SSH登录的IP地址、用户名及密码；然后，执行netconf = netconf_connect(service_ip,port,netconf_user,netconf_pass)，赋值为netconf，输入netconf的参数，建立netconf连接；最后，执行netconf.edit_config(target='running',config=XMLS)，将在变量中定义的XMLS文件通过edit_config()方法发送到设备的running配置文件。